Molecular and Cellular Biology of Insulin-like Growth Factors and Their Receptors

Molecular and Cellular Biology of Insulin-like Growth Factors and Their Receptors

Edited by
Derek LeRoith
National Institutes of Health
Bethesda, Maryland

and

Mohan K. Raizada
University of Florida
Gainesville, Florida

Springer Science+Business Media, LLC

Library of Congress Cataloging in Publication Data

Molecular and cellular biology of insulin-like growth factors and their recep-
tors / edited by Derek LeRoith and Mohan K. Raizada.
      p.   cm.
   Proceedings of a symposium held at the University of Florida, College of Medicine,
Gainesville, Fla., Jan. 23-24, 1989.
   Includes bibliographies and index.

   1. Somatomedin — Physiological effect — Congresses. I. LeRoith, Derek, 1945-
II. Raizada, Mohan K.
   [DNLM: 1. Insulin — congresses. 2. Receptors, Insulin — congresses. 3. Receptors,
Neurohumor — congresses. 4. Somatomedins — congresses. WK 820 M7175 1989]
QP552.S65M65   1989
612′.015756 — dc20
DNLM/DLC   89-16106
for Library of Congress   CIP

ISBN 978-1-4684-5687-5   ISBN 978-1-4684-5685-1 (eBook)
DOI 10.1007/ 978-1-4684-5685-1

Proceedings of a symposium on Molecular and Cellular Aspects of Insulin and
Insulin-like Growth Factors I/II: Implications for the Central Nervous System,
held January 23-24, 1989, in Gainesville, Florida

© 1989 Springer Science+Business Media New York
Originally published by Plenum Press, New York in 1989
Softcover reprint of the hardcover 1st edition 1989

PREFACE

        An essential element in the development and functional integrity of all organisms is
intercellular communication. This is achieved by the secretion of soluble messenger
molecules which subsequently interact with receptor-effector pathways in the responsive
cells. Hormones are traditionally defined as chemical messengers synthesized by endocrine
glands. Unlike hormones produced by endocrine glands, growth factors are hormone-related
substances produced by many tissues and play an important role in controlling growth and
development. While the exact physiological roles of growth factors have yet to be
elucidated, they play important roles in the regulation of cellular proliferation and/or
differentiation during ontogenesis, growth and differentiation.

        During recent years there has been a substantial increase in research related to
peptide growth factors, their receptors, and modes of action. With the discovery and
characterization of numerous growth factors, it became clear that these growth factors had
multiple features in common with classic hormones as well as with oncogenes. Furthermore,
there are distinct families of growth factors based either on structural or functional
similarities.

        One family, the insulin-related growth factors, includes insulin and insulin-like
growth factors (IGF) I and II; its members exhibit similarities both structurally and
functionally, at the level of the ligands as well as their receptors. Given the virtual
explosion of research in this area, the publication of this book is indeed timely. The
articles are written by many of the world's experts in the field and were compiled following
a very successful symposium on these topics held at the University of Florida, College of
Medicine, Gainesville, Florida, in January, 1989. The purpose of this book is to afford the
reader state-of-the-art developments in the molecular and cellular aspects of this interesting
family of growth factors.

        The format of the book is in many respects arbitrary, since many of the chapters
discuss aspects of the IGFs that include topics from more than one section. The first
section covers general aspects of the physiology of the insulin-related growth factors.
Traditionally, insulin and the IGFs were thought to be the unique products of the pancreas
and liver, respectively. Recent evidence suggests that insulin and especially the IGFs are
synthesized by several other tissues and may therefore have autocrine or paracrine modes
of action. Thus, Daughaday's discussion on the original "somatomedin hypothesis" and Van
Wyk's treatment of the potential autocrine/paracrine roles of IGFs is an appropriate
introduction. In addition, there are articles describing IGF-I's interaction with the
hypothalamus and pituitary and its role in growth hormone homeostasis. Examples of
defects in this axis, such as the African pygmy, are included, as is a discussion of the
possible role of the somatomedin inhibitors in disease states.

        The next section highlights the recent advances in elucidating the molecular structures
of the various components of the IGF system. Recombinant DNA technology has not only
helped in the structural analyses of the genes but has also provided investigators with very
sensitive tools for studying the control of gene expression. The initial articles outline the
structure of the human IGF-I and IGF-II genes, as well as the interesting transcriptional
diversity in rat IGF-I gene expression. The following articles clearly demonstrate the

widespread expression of IGF genes in several normal tissues as well as the physiological regulation of IGF gene expression. Since growth factors have obvious implications for both normal and abnormal tissue growth, some examples of their possible roles in cancer are included.

As with all signal messenger molecules, growth factors produce their biological effects by interacting with specific cellular receptors. This interaction is determined by the structure of the ligand and the receptor. Thus we have included articles covering the structural and functional relationships of these receptors. Since the physiological and neuromodulatory role of IGFs in the growth and differentiation of the central nervous system has been steadily gaining momentum, we have also included several articles discussing the expression of IGF receptors in nervous tissue as well as IGF's effect on receptor autophosphorylation and tyrosine kinase activity. This area will be of special interest to readers involved in studies on the role of IGFs in various neurodegenerative diseases.

The final section is devoted to cellular and biological actions of the IGFs. It has become increasingly clear that IGF-specific binding proteins have important regulatory effects on IGF action. Thus, articles describing the recent cloning and sequencing of the binding proteins are featured early. Following are some interesting examples of the biological roles of the IGFs in normal physiology.

Derek LeRoith, M.D., Ph.D.

Mohan K. Raizada, Ph.D.

ACKNOWLEDGMENTS

Financial support for the symposium was obtained from the Interdisciplinary Center for Biotechnology Research, University of Florida; The Upjohn Company; the Kroc Foundation; Squibb-Novo, Inc.; Merck Sharp and Dohme Research Laboratories; as well as the Departments of Physiology, Pharmacology, Neuroscience, Pediatrics, and Pathology at the University of Florida. We greatly appreciate their support.

The secretarial and organizational assistance of Pia Jacobs, Gayle Butters, and Victoria La Placa made this project both enjoyable and very successful. Kevin Fortin's editorial assistance is greatly appreciated.

CONTENTS

SOMATOMEDINS: A NEW LOOK AT OLD QUESTIONS

William H. Daughaday

Washington University School of Medicine
Department of Medicine
St. Louis, MO 63110

Endocrine and Autocrine/Paracrine Actions of IGFs

The somatomedins were originally recognized as GH dependent serum factors which mediated GH action on cartilage in-vitro, while the source of these serum factors was at first unknown, the liver was widely suspected. McConaghey and Sledge[1] in 1970 reported that rat livers released somatomedins bioactivity during perfusion and the release was increased by GH. This finding has been confirmed with more specific assay methods and extended to studies of isolated hepatocytes[2]. Modern molecular biological probes have confirmed GH regulation of IGF-I gene expression in rat and mouse liver. The role of the liver as the major source of serum IGF-II is less well established in man because most of the studies of IGF-II gene expression have been done in rats and mice. In this species, IGF-II mRNA drops to a very low level in all tissues except choroid plexus and meninges[3]. The fact that IGF-II is not elevated in acromegaly and only decreases to about 50% of normal in hypopituitary patients is indicative of reduced GH dependence.

It is now known that many tissues of the young rat, other than liver, contain IGF-I in excess of that attributable to the content of blood[4] and express the IGF-I gene (5). This has led to the widely held view that GH action on extrahepatic tissues is exerted by stimulating local production of IGF-I which acts in an autocrine/paracrine mode. It should be emphasized, however, that the level of IGF-I mRNA in these extrahepatic tissues is less than 1% of that in liver and it is unknown whether the local synthesis of IGF-I is sufficient to sustain normal growth and differentiation. While an answer to this important question is not possible with most tissues, sufficient information exists for skeletal tissues to indicate that IGF-I action in an endocrine mode is essential for full restoration of normal growth parameters in hypophysectmized rats.

First of all, the addition of growth hormone in-vitro to hypophysectomized rat and chick embryo cartilage and to fetal rat produces little or no stimulation of anabolic and mitogenic parameters while these same parameters are readily increased by IGF-I (for review see 6). Some confusion arises because dissociated rat and chicken chondrocytes and dissociated rat osteoblasts are highly responsive to GH added in-vitro. These findings suggest that the matrix in which cells exist in tissues markedly decreases their responsiveness to GH.

The major challenge to the lack of direct GH effects on skeletal tissues has come from Isacksson and his various collaborators. They injected GH directly into

1

one tibial epiphyseal plate and observed increased linear growth as compared to the contralateral tibia injected with saline[7,8]. Later, it was observed by Schleuter et al[9] that the unilateral effects of GH could be blocked by antisera to IGF-I (somatomedin-C). This strongly suggested that the local growth promoting effects of growth hormone are mediated by IGF-I. This and other evidence support a local action of GH on skeletal growth mediated by IGF-I.

These observations do not, however, exclude a role for IGF-I acting as an endocrine agent on skeletal tissues. I have emphasized, elsewhere, that the amount of tibial growth achieved by very large local concentrations of GH in these experiments was only 12 to 22% of the maximal growth which can be obtained by parenteral GH administration[6].

I suggest that both endocrine and autocrine/paracrine modes of IGF-I action contribute to normal skeletal growth.

In the GH deficient child growth velocity is decreased about 50% and this decrement is largely attributed to decreased IGF-I action (Fig. 1). It is unknown how much of this decrement is the result of decreased endocrine and how much is attributable to decreased autocrine/paracrine modes of IGF-I action.

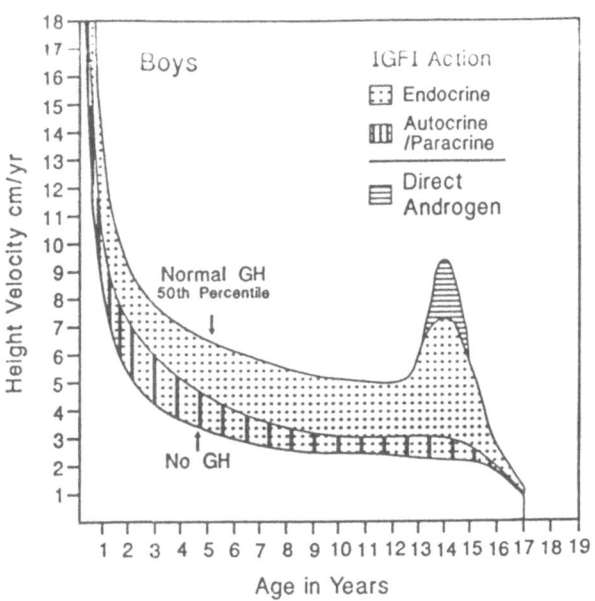

Figure 1. The contribution of GH dependent and direct androgen dependent stimulation of growth velocity is shown. The GH dependent contribution to growth velocity is shown as mediated by IGF-I. This contribution is divided by a hypothetical line into a sector contributed by endocrine action and a sector contributed by the autocrine/paracrine actions of IGF-I on skeletal growth. Also shown is a sector at puberty indicating that androgens contribute to growth velocity directly as well as by increased GH secretion. (Reprinted by permission from Endocrine Reviews 1989, in press).

The fact that serum IGF-I rises progressively in childhood and abruptly at puberty makes it likely that there is a proportionate increase in IGF-I action by the endocrine mode and this is indicated in the figure.

GH Independent Expression of IGFs

Hypophysectomized rats do not grow while the GH deficient child continues to grow at about 50% of normal velocity. This persistent GH independent growth may be related to the continued, albeit reduced, expression of the IGF-II gene and to a lesser extent the IGF-I gene.

In fetal life, at time of rapid growth, serum IGF-I levels are low by adult standards in most species. One factor responsible for low serum IGF-I is the low expression of hepatic GH receptors in the fetus. While it was formerly thought that IGF-I was not important in fetal growth, it is now known that IGF-I and to an even greater extent, IGF-II are expressed in many extrahepatic fetal tissues. Fetal GH deficiency does not greatly affect growth or fetal serum IGF-I consistent with GH independent regulation of gene expression in the fetus.

Other hormones may regulate the expression of the IGF-I genes in special tissues. Insulin and hypernutrition, acting on the liver can maintain serum IGF-I levels and normal growth despite decreased GH secretion in certain patients after craniotomy for craniopharyngioma. Gonadal expression of IGFs is greatly affected by gonadotropins and adrenal expression of IGF-I by corticotropin.

Recent evidence has linked isolated hypertrophy of kidney[10] and muscle[11] to increased expression of an IGF-I gene independent of changes in serum IGF or GH concentrations. A similar activation occurs in repair of vascular damage and wound healing. The mechanism of this GH unrelated activation of IGF gene expression is not known, but other growth factors such as EGF and PDGF may participate in IGF gene activation.

Tumor Associated IGF Synthesis

The possibility that IGFs might be of significance in the growth and systemic effects of neoplasms has a long and controversial history. This was first suggested in 1974 by Megyesi et al[12] who observed that sera of certain patients with tumor associated hypoglycemia had increased concentrations of peptide which competed in an IGF rat liver membrane RRA. It is now known that this RRA measures predominantly IGF-II. Increased serum IGF-II was later confirmed by my group[13] using a different IGF-II RRA. However, Widmer et al[14] were unsuccessful in confirming these findings with several different assay systems.

Hyodo et al[15] in 1977 found that an adrenal carcinoma released an IGF-II-like peptide into conditioned medium. The following year DeLarco and Todero[16] partially purified IGF-II related peptides from conditioned medium of a fibrosarcoma cell line. This subject has attracted much recent attention. Increased amounts of IGF-II mRNA and in some cases IGF-II peptide, have been detected in a great variety of mesenchymal tumors (reviewed by Daughaday and Rotwein[6]). In addition, increased expression of IGF-I has also been reported in certain tumors. The possible significance of IGFs in tumor growth and metabolism remain as yet undefined, but at least some tumor cell lines require IGF-I for growth[17].

The systemic consequences of tumor production of IGF-II have thus far been limited to hypoglycemia. In a particularly instructive case of recurrent hypoglycemia in a woman with a thoracic leiomyosarcoma we found that the tumor contained a greatly increased concentration of IGF-II, most of which was in the "big" form of incompletely processed pro-IGF-II[18]. A similar proportion of "big" IGF-II occurred in the serum. As in earlier reports, there was clear evidence of feedback inhibition of GH secretion with resulting hyposomatotropism which undoubtedly greatly sensitized the patient to IGF-II induced hypoglycemia. The hyposomatotropism also resulted in a loss of the high affinity 150 k IGF binding protein complex; IGF-II transport was nearly entirely of a 40-50 k molecular weight complex which more readily can transverse capillary membranes and deliver their peptide to tissue receptor sites.

REFERENCES

1. P. McConaghey and C. G. Sledge. Production of sulphation factor" by the perfused liver. Nature 225:1249 (1970).
2. E. M. Spencer. Synthesis by cultured hepatocytes of somatomedin and its binding protein. Febs. Lett. 99:157 (1979).
3. M. A. Hynes, P. J. Brooks, J. J. Van Wyk and P.K. Lund. Insulin-like growth factor II messenger ribonucleic acids are synthesized in the choroid plexus of the rat brain. Mol. Endo 2:47 (1988).
4. A. J. D'Ercole, A. D. Stiles, and L. E. Underwood. Tissue concentrations of somatomedin C: Further evidence for multiple sites of synthesis and paracrine or autocrine mechanisms of action. Proc. Natl. Acad. Sci. USA 81:935 (1984).
5. L. J. Murphy, G. I. Bell, and H. G. Friesen. Tissue distribution of insulin-like growth factor I and II messenger ribonucleic acid in the adult rat. Endocrinology 120:1279 (1987).
6. W. H. Daughaday. A personal history of the origin of the somatomedin hypothesis and recent challenges to tis validity. Perspect. Biol. Med. (in press).
7. O. G. P. Isaksson, J.-O. Jansson, I. A. M. Gause. Growth hormone stiumates longitudinal bone growth directly. Science 216:1237 (1982).
8. J. Isgaard, A. Nilsson, A. Lindahl, J. O. Jansson, and G. P. Isaksson. Effects of local administration of GH and IGF-I on longitudinal bone growth in rats. Am. J. Physiol. 250:E367 (1986).
9. N. L. Schlechter, S. M. Russell, E. M. Spencer, and C. S. Nicoll. Evidence suggesting that the direct growth-promoting effect of growth hormone on cartilage in vivo is mediated by local production of somatomedin. Proc. Natl. Acad. Sci. USA 83:7932 (1986).
10. A. D. Stiles, I. R. S. Sosenko, A. J. D'Ercole, and B. T. Smith. Relation of kidney tissue somatomedin C/insulin-like growth factor I to postnephrectomy renal growth in the rat. Endocrinology 117:2397 (1985).
11. E. Jennische and H. A. Hansson. Regenerating skeletal muscle cells express insulin-like growth factor I. Acto Physiol Scand. 130:327 (1987).
12. K. Megyesi, C. R. Kahn, J. Roth, and P. Gorden. Hypoglycemia in association with extrapancreatic tumors: Demonstration of elevated plasma NSILA-s by a new radioreceptor assay. J. Clin. Endocrinol. Metab. 38:931 (1974).
13. W. H. Daughaday, B. Trivedi, and M. Kapadia. Measurement of insulin-like growth factor II by a specific radioreceptor assay in serum of normal individuals, patients with abnormal growth hormone secretion, and patients with tumor-associated hypoglycemia. J. Clin. Endocrinol. Metab. 53:289 (1981).
14. T. Hyodo, K. Megyesi, C. R. Kahn, J. P. McLean, and H. G. Friesen, Adrenocortical carcinoma and hypoglycemia. Evidence for production of nonsuppressible insulin-like activity by the tumor. J. Clin. Endocrinol. Metab. 44:1175 (1977).
15. U. Widmer, C. Schmid, E. R. Froesch. Effects of insulin-like growth factors on chick embryo hepatocytes. Acta Endocrinol. 108:237 (1985).
16. J. E. DeLarco and G. J. Todaro. A human fibrosarcoma cell line producing multiplication stimulating activity (MSA)-related peptides. Nature 272:356 (1978).
17. Y. Nakanishi, J. L. Muishine, P. G. Kasprzyk, R. B. Natale, R. Maneckjee, I. Avis, A. M. Treston, A. F. Gazdar, J. D. Minna, and F. Cuttitta. Insulin-like growth factor-I can mediate autocrine proliferation of human small cell lung cancer cell lines in vitro. J. Clin. Invest. 82:354 (1988).
18. W. H. Daughaday, M. A. Emanuele, M. H. Brooks, A. L. Barbato, M. Kapadia, and P. Rotwein. Insulin-like growth factor II synthesis and secretion by a leiomyosarcoma with associated hypoglycemia. N. Engl. J. Med. 319:1434 (1988).

AUTOCRINE AND PARACRINE EFFECTS OF THE SOMATOMEDINS/INSULIN-LIKE GROWTH FACTORS

Judson J. Van Wyk[*] and P. Kay Lund[#]
Departments of Pediatrics[*] and Physiology[#]
University of North Carolina School of Medicine
Chapel Hill, N.C., 27599

INTRODUCTION

The somatomedins were initially discovered and purified on the basis of their insulinomimetic effects in fat and muscle, their mediation of the action of growth hormone on proteoglycan synthesis in cartilage, and their mitogenic effects in fibroblast cultures. We have subsequently learned that IGF-I and IGF-II act locally by autocrine and paracrine mechanisms, and that they play important regulatory roles in both cell division and in the expression of a vast and rapidly expanding list of differentiated cell functions (1). The great diversity of biological effects attributed to the somatomedins makes it tempting to generalize that these substances must play some important "housekeeping role" that most eukaryotic cells require to function optimally. The lengthening list of IGF mediated events has therefore not served to clarify the biological role of the somatomedins, but rather has made it more difficult to place in perspective the physiologic role of the somatomedins in the hierarchy of other regulatory hormones and peptides.

The holy grail sought by many investigators is some basic action of the somatomedins that is common to all of their effects. Even though a unitary mechanism of somatomedin action is still beyond our grasp, our quest can be well served by reviewing the biologic effects thought to be mediated by the insulin-like growth factors, by indicating some of the mechanisms by which they act, and by examining some of the different roles played by IGF-I and IGF-II.

EFFECTS OF INSULIN-LIKE GROWTH FACTORS ON CELL PROLIFERATION

The somatomedins were first described as growth factors for skeletal tissue, and only gradually was it recognized that they stimulate DNA synthesis and cell proliferation in a wide variety of cell types. Responsive cells are of diverse embryologic origin and derived from species ranging from amphibia to man. Gordon Sato and his colleagues showed that very few of the cell types that he studied would proliferate in serum-free defined media unless micromolar concentrations of insulin were present (2). At such high concentrations, however, insulin can interact with the type I somatomedin receptor and function as a somatomedin surrogate (3); indeed, in most cell types, nanomolar concentrations of IGF promote cell replication just as effectively as micromolar concentrations of insulin.

Some cell types, however, do not appear to require added insulin or somatomedin, and some of these have now been shown to produce somatomedins which then function in an autocrine manner. This was first documented in human fibroblasts. Since these cells appear to grow equally well in somatomedin deficient serum from hypopituitary patients and in normal serum, it was generally believed that they are not dependent on growth hormone or somatomedin. David Clemmons and I, however, were able to demonstrate that when rates of entry into DNA synthesis were carefully measured during a single cell cycle, somatomedin-deficient platelet-poor plasma from a hypopituitary patient promoted a slightly slower rate of entry into DNA synthesis than did plasma from normal individuals (4). These rates of entry were normalized if the plasma from the hypopituitary patient were preincubated either with growth hormone or with IGF-I.

Since we wished to know the effect of growth hormone was a direct effect or secondary to stimulating the production of IGF-I, we then did the following experiment: human fibroblasts were first made competent with platelet-derived growth factor (PDGF) (5), and then placed in media containing hypopituitary plasma fortified with either 10 ng/ml of growth hormone or 5 ng/ml of SM-C. The media were then changed every 2 hours for 76 hours to remove any mitogens produced by the cells. Under these conditions growth hormone no longer was able to normalize the rate of entry into the S phase of the cell cycle, whereas Sm-C was active.

This led us to conclude that growth hormone was acting through the stimulation of somatomedin production. This was confirmed by direct measurements by RIA of Sm-C secreted into the culture medium. This production was increased by growth hormone or by platelet derived growth factor, and the combination of PGDF + growth hormone was additive (6).

Immunoneutralization Studies with Monoclonal Antibodies

The development of monoclonal antibodies capable of neutralizing endogenously produced somatomedins, or, in the case of human cell lines, blocking access to the type I receptor, has made it much easier to document the role of endogenously produced insulin-like growth factors. Our hybridoma cell line secreting the anti-somatomedin antibody designated sm1.20 was cloned from splenocytes obtained from a mouse that had been previously immunized in-vivo with IGF-I and then boosted by in-vitro incubation for 5 days in thymocyte conditioned medium containing pure IGF-I (7). This antibody was of the IgG kappa isotype and crossreacted with IGF-I and to a lesser extent with IGF-II, but not with insulin.

The specificity and effectiveness of sm1.20 were demonstrated by pretreating quiescent BALB/c 3T3 cells with PDGF and then placing them in defined media containing either EGF + insulin, or EGF + IGF-I in the presence of increasing concentrations of sm1.20. Under these circumstances the antibody blocked the effect of EGF + IGF-I on DNA synthesis, but not the effect of EGF + insulin (8). Surprisingly, however, the monoclonal antibody not only inhibited the combination of Sm-C + EGF in BALB/c 3T3 cells, but also inhibited the mitogenic effect of EGF by itself in the absence of exogenous IGF. This suggested that the cells were endogenously producing substances that cross reacted with the antibody. Subsequent analysis of the media in which these cells were grown confirmed that small amounts of IGF-I were produced by BALB cells themselves. Additional immunoneutralization studies such as these advanced the concept that somatomedins are made by many different cell types in culture, and that part of the effect of other growth factors and hormones may be secondary to their ability to stimulate IGF production in their target tissues.

Role of IGFs in the action of TSH on thyroid cells

These concepts have recently been elegantly documented in the case of the thyroid cell line, FRTL-5. This is a cloned rat cell line that responds to thyroid stimulating hormone (TSH) and to thyroid stimulating immunoglobulins from patients with Graves' disease by both cell multiplication and by many of the differentiated functions of thyroid-follicular cells (9). These responses are thought to be almost entirely cAMP mediated and are mimicked by exposure to dibutyryl cAMP, forskolin, and other agents that act through this pathway (10). Recombinant IGF-I also stimulates DNA synthesis and cell replication in these cells, although it does not increase intracellular cAMP levels.

Studies carried out in the Thorndike Laboratories of Drs. Alan Moses and the late Sidney Ingbar have elegantly explored the relationships between TSH and the IGFs on DNA synthesis in the FRTL-5 cell line (11).[1] They showed that the effects of IGF-I and bTSH are at least additive on thymidine incorporation in FRTL-5 cells. They also found that conditioned media from FRTL-5 cells also stimulates thymidine incorporation and potentiates the effect of TSH. They therefore used specific RIAs for hIGF-I and rIGF-II to test the possibility that these cells might be secreting somatomedins into the media and that the mitogenic effect of TSH might be mediated by its stimulatory effect on somatomedin production. They found that the IGF-I content of conditioned media from TSH treated FRTL-5 cells was less than 25 pg/ml, whereas the IGF-II content exceeded 500 ng/ml (11). Furthermore, a mutant of this cell line that grows in the absence of TSH contained 2500 ng/ml of IGF-II by RIA!

Final proof that the mitogenic effects of TSH and cAMP are mediated through IGF production was obtained by immuno-neutralization studies with antibody Sm-1.20b. Although we knew that this antibody cross-reacted with IGF-II, we did not know how well it might neutralize IGF-II in culture. Moses found that 10 ug/ml of sm1.20b fully blocked the mitogenic effects of 50 ng/ml of MSA or 50 ng of recombinant human IGF-II on thymidine incorporation in FRTL-5 cells, but, as suspected, had no effect on the mitogenic effect of 10 ug/ml of insulin. Their crucial finding, however, was that Sm 1.20 b was able to block about 90% of the mitogenic effects produced by 1×10^{-10}M TSH, and about 40% the effect of 1×10^{-9}M TSH (Figure 1A). The antibody also blocked the stimulatory effects of cholera toxin, forskolin, and Graves' IgG (figure 1B). Moses and his coworkers have similarly been able to block the effect of IGFs and TSH on FRTL-5 cells with "pregnancy protein 12", an IGF binding protein isolated from amniotic fluid (12).

----

footnote 1: We are indebted to Dr. Moses and his group for allowing us to use some of their unpublished as well as published data in the FRTL-5 cell line.

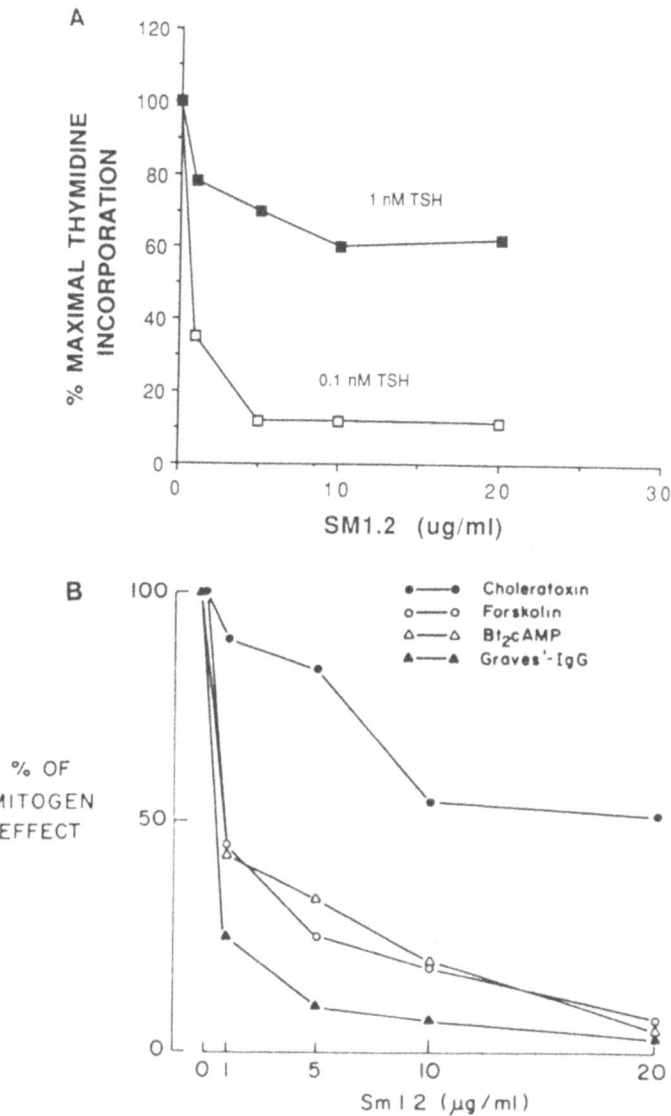

Figure 1A. Effects of increasing concentrations of sm1.20b on the mitogenic response to bTSH; 1B. or to cholera toxin (10 pg/ml), forskolin $(10^{-5}M)$, dibutyryl cAMP $(10^{-4}M)$, or Graves' IgG (2 mg/ml) (11). Responses are expressed as a percent of those obtained with each agent in the absence of the antibody. From J.Clin.Invest. <u>82</u>: 1546, 1988 with permission.

Figure 2 summarizes the relationship between TSH, cAMP, and IGF-II in stimulating cell proliferation in FRTL-5 cells. The foregoing studies provide strong evidence that at least part of the mitogenic effects of TSH and thyroid stimulating immunoglobulins are secondary to their stimulation of IGF-II production through a cyclic AMP mediated pathway. A far less likely explanation, however, is that endogenously produced IGF-II is permissive for a more direct mitogenic effect of TSH.

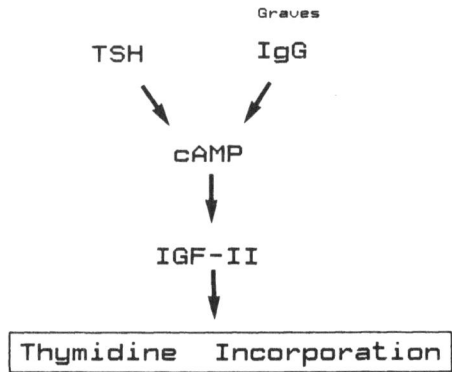

Figure 2. Proposed mechanism by which TSH and thyroid stimulating immunoglobulins stimulate DNA synthesis in FRTL-5 cells. This scheme does not exclude the possibility that TSH has additional actions to potentiate the effect of IGF-II.

The additive effects of TSH and IGFs on DNA synthesis are most economically explained by the summation of effects produced by exogenous and endogenously produced IGF. In some of these studies, however, the effects appear to be more than additive, and additional studies are required to rule out synergism at some other level. In this regard it is perhaps noteworthy that Gershengorn and his colleagues found that various combinations of phorbol esters, TSH, and IGFs stimulated diacyl glycerol in FRTL-5 cells proportionately to their effects on DNA synthesis (13).

EFFECTS OF SOMATOMEDINS ON DIFFERENTIATED CELL FUNCTIONS

Somatomedins have now been shown to have a profound effect on an ever increasing list of differentiated cell functions. In many of these actions their effect by themselves can be demonstrated only in the presence of other growth factors or hormones; indeed, amplification of the effect of other agonists is a common theme of somatomedin action.

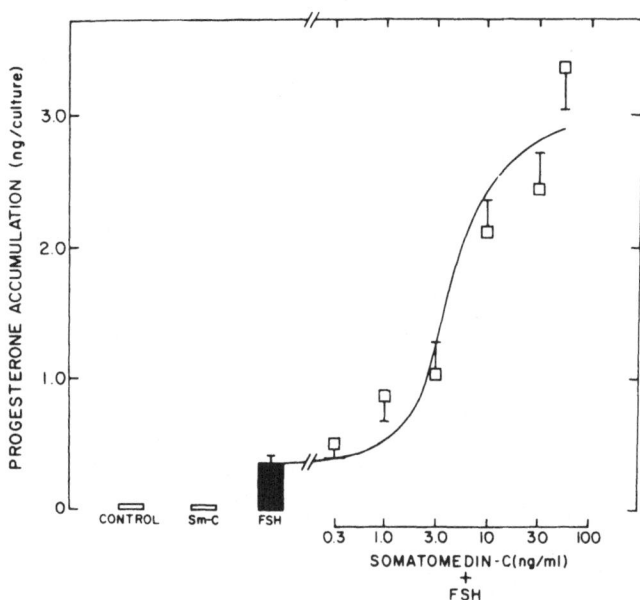

Figure 3. On the ordinate is plotted the accumulation of progesterone over a 72 hour period in cultures of granulosa cells obtained from stilbesterol primed hypophysectomized rats. The production of progesterone is not influenced by IGF-I by itself; however, the effect of 20 ng/ml of FSH is feeble unless the culture is supplemented with IGF-I (17). From Endocrinol. 115, 1227, 1984, with permission.

Effect of IGFs on differentiation of chick lens

Beebe et.al. found that lens epithelial cells from chick embryos undergo rapid elongation and synthesis of the specific lens protein delta crystallin when incubated in the presence of vitreous humor or high concentrations of insulin (14). They later found that IGF-I can substitute for vitreous humor in stimulating lens fiber differentiation, and that the "lentropic" activity of vitreous humor, but not that of insulin, could be neutralized by our anti-somatomedin antibody Sm 1.20 (15). This is a pure example of the effect of IGFs on differentiation since no cellular proliferation occurs under these circumstances.

Effect of IGFs on rat granulosa cells

The effect of IGFs on rat granulosa cells was discovered by Dr. Eli Adashi who found IGF-I to be a powerful potentiator of the actions of FSH. In this cell type FSH stimulates a wide range of responses including the induction of progesterone biosynthesis, induction of LH receptors, induction of aromatase activity, and stimulation of glycosoaminoglycan synthesis (16). In figure 3 is shown the effect of IGF-I in potentiating the effect of FSH on progesterone production. The peak effect of Sm-C on progesterone accumulation is reached at 48 hours. Insulin produces similar potentiation, but at concentrations that are several orders of magnitude higher.

There appear to be multiple mechanisms by which IGFs potentiate the actions of FSH on gonadal function. In a manner similar to the effect of TSH in the thyroid, FSH induces its effects in the ovary through a cyclic nucleotide pathway. Adashi et al showed that IGF-I potentiated the accumulation of cAMP in the media of granulosa cells exposed to cholera toxin to a degree that was roughly proportional to its potentiation of

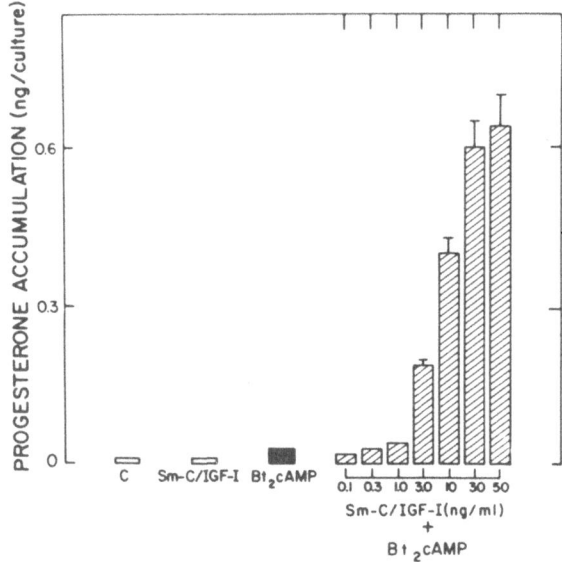

Figure 4. Effect of increasing concentrations of IGF-I on progesterone accumulation in rat granulosa cells cultured for 72 hours with 3×10^{-4}M dibutyryl cAMP. IGF-I at concentrations up to 50 ng/ml did not increase progesterone accumulation above basal values. The maximal stimulatory concentration of dibutyryl cAMP (10^{-3}M) on progesterone production was not further enhanced by 50 ng/ml IGF-I. From Endocrinol. 122, 1583, 1988 with permission.

progesterone accumulation (18). They suggested that the amplifying effect of IGF-I on the actions of FSH in granulosa cells might not be entirely due to the amplification of cAMP production, however, since IGF-I also amplifies the effect of submaximal concentrations of diButyryl cAMP on progesterone production (19) (Figure 4).

Effects on other types of gonadal cells

Rat granulosa cells, as used in the above studies, are incapable of undergoing mitosis, and hence all of these effects are due to stimulation of differentiated cell functions. In cultures of porcine and bovine granulosa cells, however, IGF-I stimulates both proliferation and differentiation (20)(21). Veldhuis et.al. showed that in porcine granulosa cells one of the major effects of IGF-I is to stimulate cholesterol side chain cleavage (22).

IGF-I is produced in rat Sertoli cells where it may act in an autocrine manner to potentiate the effects of gonadotropins (23). IGFs produced by Sertoli cells may also act in a paracrine manner on pachytene spermatocytes to provide somatomedin-like peptides for the early stages of spermatogenesis (24). IGFs also appear to potentiate the effect of LH on androgen production by Leydig cells (25)(26).

WHAT DIFFERENT ROLES ARE PLAYED BY IGF-I and IGF-II?

Further insights on the functions of the insulin-like growth factors in the living animal can be gained by reviewing evidence bearing on the different roles played by IGF-I and IGF-II in regulating growth and differentiation. As a point of departure, we propose that the respective biological roles of IGF-I and IGF-II are determined primarily by where and under what circumstances the genes coding for the 2 peptides and their receptors are expressed. This hypothesis is based on 3 major assumptions: first, virtually all of the biological effects produced by one of these peptides can be replicated by the other (although the dosages required may be dissimilar); second, both IGF-I and IGF-II produce most of their biological effects through the type I somatomedin receptor (although not necessarily through the same binding sites) (27), and third, type I receptors are widely distributed in virtually every tissue.

Until recently, our information about in-vivo production of IGF-I and IGF-II has been based on RIA measurements in serum and tissue extracts. Comparisons of IGF-I and IGF-II concentrations in serum reveal differences in responsivity to hormonal and nutritional stimuli,

TABLE I. SOME DIFFERENCES IN EXPRESSION OF IGF-I AND IGF-II mRNAs

	IGF-I		IGF-II	
	Human	Rat	Human	Rat
Expression in fetus				
Liver	+	+	+++++	+++++
Brain	+	+	+/-	++++
Extra-hepatic	+	+	+++++	+++++
Expression in adult				
Liver	+++++	+++++	+	+
Brain	ND	+	ND	+++
Extra-hepatic	+	+	ND	-/+
GH dependence				
Liver	ND	++++	ND	-
Brain	ND	++	ND	+
Extra-hepatic	ND	+	ND	-/+

ND = not done; - = not detected or no detectable regulation

----

in developmental profiles, and across species. In both human and rodent
serum IGF-I concentrations are much more stringently regulated by growth
hormone and nutritional status than are those of IGF-II. Serum IGF-I
levels are low in the fetus, rise to high levels during the peak of the
adolescent growth spurt, and then progressively decline through adult
life. IGF-II exhibits a different developmental pattern since serum
levels of IGF-II are high during fetal life and do not exhibit the adolescent surge. The relationships between IGF-I and IGF-II also differ
between species. Serum levels of IGF-II in adult humans are 3-4 times
higher than those of IGF-I, while in adult rat serum concentrations of
IGF-I may be as much as 40 times higher than those of IGF-II.

Although RIA studies of serum and tissue IGFs have provided much
information about in vivo regulation of IGF-I and IGF-II, their interpretation has sometimes been suspect because available antibodies have
often lacked complete specificity, and because RIAs carried out in the
presence of interfering binding proteins may give artifactual results that
go unrecognized. Recently, the availability of cDNA probes has made it
possible to obtain a more direct picture of IGF-I and IGF-II synthesis
in different tissues as a function of developmental status and hormonal
influences. Measurements of mRNAs for IGF-I and IGF-II have confirmed

distinct differences in the expression of the genes for IGF-I and IGF-II, thereby providing insights into their respective biological roles. Some species differences in IGF expression during development and some tissue specific differences with respect to regulation by GH are shown in Table I.

Whereas liver and peripheral non-hepatic tissues of human and rat fetus show expression of both IGF-I and IGF-II, and in general greater expression of IGF-II mRNAs, the fetal brain of man and rat differ (Figures 5 and 6). In the rat, Lund et al (28) demonstrated high levels of IGF-II

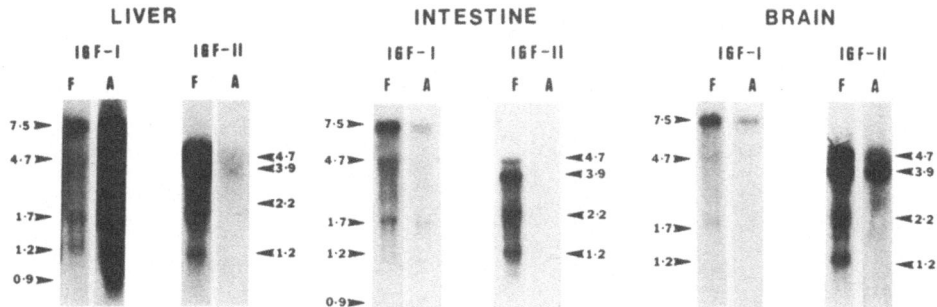

Figure 5. Autoradiograms of Northern blots of poly A+ RNAs from day 18 gestation fetal and adult rat liver, intestine, and brain hybridized with ^{32}P labelled rat IGF-I cDNA (left) and human IGF-II cDNA (right). 20ug of each poly A+ RNA sample was applied to each lane. Numbered arrows show estimated sizes of hybridizing mRNAs in kilobases (KB). Exposure times were 24 h at -100°C with intensifying screens

expression in fetal brain and lower levels of IGF-I. In contrast, in the human fetus Han et al found detectable IGF-I expression in cortex, hypothalamus and brainstem between 16 and 20 weeks of gestation, but no detectable IGF-II expression in cortex and hypothalamus and only trace expression of IGF-II in the brain stem (29) (Figure 6).

In contrast to liver and peripheral tissues of the rat, where expression of IGF-I but not IGF-II is regulated by GH, the rat brain exhibited GH dependence of both IGF-I and IGF-II mRNAs (30) (Figure 7).

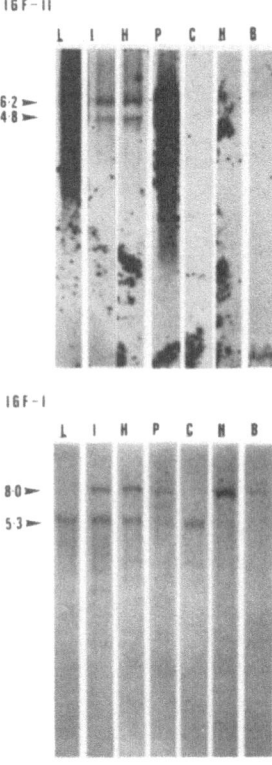

Figure 6. Autoradiograms of Northern blots showing IGF-II and IGF-I mRNAs detected in 10 ug aliquots of poly A+ RNAs from human fetal Liver, Intestine, Heart, Pancreas, Cortex, Hypothalamus, and Brainstem. Numbered arrows indicate estimated sizes of mRNAs in kilobases. Exposure time was 18 h for IGF-II blots. Longer exposures did not reveal hybridizing mRNAs in cortex and hypothalamus although trace amounts of IGF-II mRNA were detected in brainstem after a 108 h exposure. Exposure time for IGF-I blots was 96 h.

IN SITU HYBRIDIZATION HISTOCHEMISTRY

Although Northern blot hybridizations and other quantitative hybridization approaches have recently added much to our understanding of the regulation of IGF-I and II expression, such an approach cannot provide information about cellular localization of IGF-I and II expression in tissues comprised of heterogeneous cell populations. Application of in situ hybridization histochemistry to localize IGF-I and IGF-II mRNAs has provided some insight into cellular sites of IGF expression. Han, et al found that mesenchymal cells and connective tissues were predominant sites of IGF-I and IGF-II mRNA expression in a wide range of human fetal tissues, including muscle, lung, liver, kidney, and other organs (31). The eye was the only region of the human CNS where IGF mRNAs were present

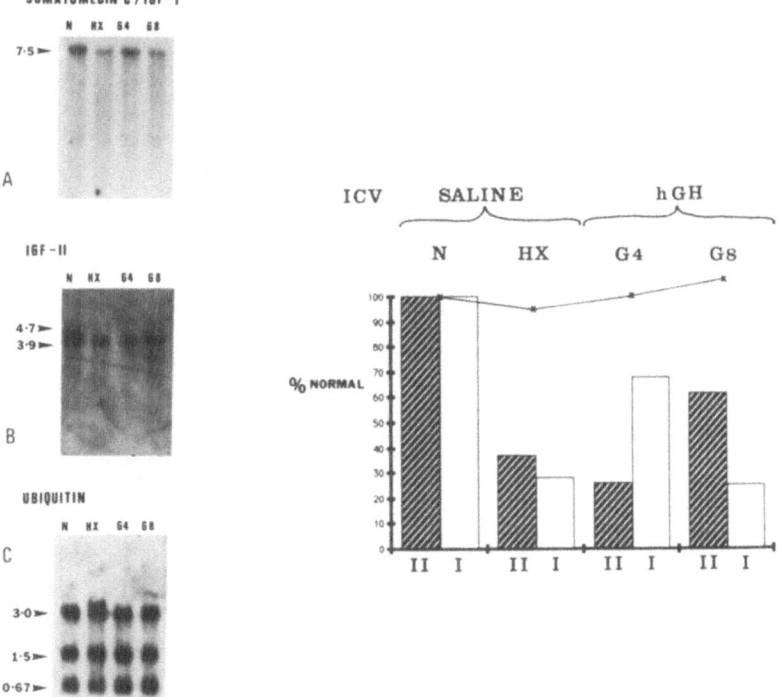

Figure 7. Northern blots of 25 ug poly A+ RNAs extracted from whole brains of normal (N), hypophysectomized (HX), and hypophysectomized rats 4h or 8h after ICV injection of hGH (G4 & G8). At the left are the estimated sizes of hybridizing mRNAs. A: Blots hybridized with a rat IGF-I genomic DNA probe; B: Blots hybridized with a human IGF-II cDNA probe. Blots in both A and B were exposed to x-ray film for 48 h. C: Blots shown in A were stripped of hybridized IGF-II cDNA and rehybridized with a human ubiquitin cDNA probe to control for the amount of RNA applied to each lane. In the right panel are histograms showing the relative abundance of IGF-I and IGF-II mRNAs estimated by densimometric densitometric scanning of blots shown on the left. From Mol. Endocrinol 1, 233, 1987 with permission.

in sufficient concentrations to be detected by in situ hybridization, and both IGF-I and IGF-II mRNAs were again localized to cells of mesenchymal origin, in this instance the sclera (Figure 8).

These findings, and similar findings of Beck et al in the fetal rat (32) provide evidence that IGFs produced in fetal connective tissues might exert paracrine actions on neighbouring cell types. In adult rat brain, a similar paracrine action of IGF-II is suggested by recent findings that the choroid plexus is the primary site of IGF-II synthesis with no detectable mRNA in other brain cell types (Figure 9) (33).

The failure to detect IGF synthesis in many cell types of non-mesenchymal origin by in situ hybridization is difficult to reconcile with

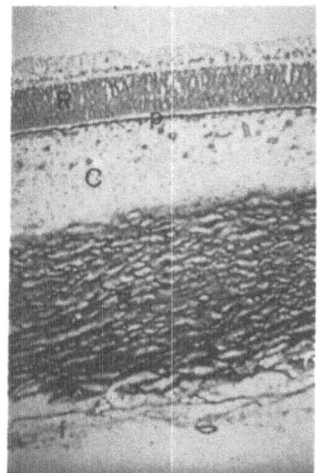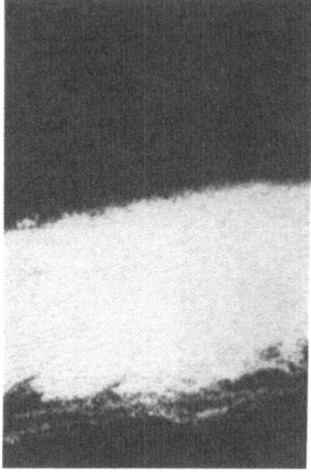

Figure 8. LEFT: Photograph through section of human fetal eye to indicate morphology. The tissues labelled by letters are: sclera (S), retina (R), choroid (C), pigment layer (P), and fovea (Y). RIGHT: Dark field photograph of autoradiogram of section on left after hybridization with a ^{32}P labelled oligomer specific for IGF-II mRNA. Sections were exposed to Kodak NTB3 photoemulsion for 48h to visualize the hybridization signal. Note that the probe hybridizes to sclera and not to other regions.

the demonstrated synthesis of IGFs during in-vitro culture of a wide variety of cell types such as cultured neuronal cells and hepatocytes. It is possible that negative in situ data for non-mesenchymal cell types in the fetus may simply reflect a low abundance of IGF mRNA or lack of accessibility of IGF mRNA for detection in situ. The cellular sites of IGF expression postnatally have not been extensively studied as yet, and differences might emerge compared with the fetus. In addition, it is important to keep in mind that cells in culture are separated from influences of neighbouring cell types which could alter their pattern of IGF synthesis compared with that in vivo. These caveats should be considered when comparing in situ data on tissues collected in vivo with the well documented in vitro expression of IGFs in differing cell types .

SUMMARY AND CONCLUSIONS

Studies such as those reviewed here are rapidly enlarging and focusing our views on the physiological roles of IGF-I and IGF-II. For example, the "somatomedin hypothesis of growth hormone action", which was originally based on classical endocrine models, become securely validated only when it was recognized that growth hormone acts indirectly by stimulating the production of IGF-I in local tissues near the epiphyseal growth plate.

Similarly, we have had to amend the briefly held notion that IGF-II functions primarily as the fetal form of somatomedin. We now find that in some species IGF-II is the predominant somatomedin in the nervous system, osteoblasts of trabecular bone (36)(37), and now the thyroid. Undoubtedly this list will rapidly expand.

IGF-II 31mer

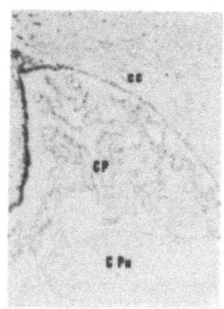 

IGF-II 31mer + RNAse

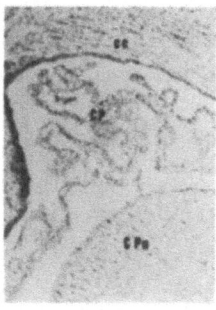 

Figure 9. Hybridization of an IGF-II oligomer to Choroid Plexus within a Coronal Section of the Rat Brain at the Level of the Preoptic Area.
Top: Bright field photograph (left) shows the choroid plexus (CP) within the lateral ventricle, and adjacent corpus callosum (cc) and caudate putamen (CPu). Dark field (right) shows accumulation of silver grains, indicative of hybridized probe, over the choroid plexus, and only randomly distributed silver grains over other brain regions.
Bottom: An adjacent section of the one shown at the top that was pretreated with 20 ug/ml ribonuclease A for 10 min at 37 C before hybridization. Bright field (left) shows the same brain regions as above, and dark field (right) illustrates that RNAse pretreatment abolised hybridization of IGF-II probe to choroid plexus. Exposure of sections to photoemulsion was for 14 days at 4 C. Scale bar = 200 um. From Mol. Endocrinol 2, 471, (1988) with permission.

Also, we have only recently discovered that insulin-like growth factors may mediate the actions of several hormones other than growth hormone. As we have seen, IGF-II appears to mediate the growth promoting role of Thyrotropin in thyroid follicular cells, and IGF-I appears to be a true "estromedin" in the uterus. These fragmentary new observations are tantalizing, and we can anticipate that the next few years will see many revisions and embellishments of these themes.

ACKNOWLEDGMENTS

Supported by R01-DK01022-33 (JJVW). We are indebted to Ms Susan Giles and Mr. Burton J. Balfour for assistance in preparing the manuscript.

REFERENCES

1. Van Wyk JJ (1984) The somatomedins: biological actions and physiologic control mechanisms. In C.H. Li (ed): Hormonal Proteins and Peptides. New York: Academic Press, pp. 81-125.

2. Barnes D, and G Sato (1980) Methods for growth of cultured cells in serum-free medium. Anal. Biochem. 102: 255-279

3. Van Wyk JJ, Graves DR, Casella SJ, and S Jacobs: Evidence from monoclonal antibody studies that insulin stimulates deoxyribonucleic acid synthesis through the type-I somatomedin receptor. J Clin Endocrinol Metab 61:639-643, 1985.

4. Clemmons DR, and JJ VanWyk (1981) Somatomedin-C and platelet derived growth factor stimulate human fibroblast replication. J Cell Physiology 106:361-367.

5. Pledger WJ, Leof EB, Chow BB, Olashaw H, O'Keefe EJ, Van Wyk JJ, and WR Wharton: Initiation of cell cycle traverse by serum derived growth factors. In: Sato GH, Pardee AB, Sirbasku DA (eds), Growth of Cellsin Hormonally Defined Media, Cold Spring Harbor Laboratory, Cold Spring Harbor, pp. 259-273, 1982.

6. Clemmons DR, Van Wyk JJ and LE Underwood (1981) Hormonal control of somatomedin production by human fibroblasts. J Clin Invest 67:10.

7. Gillespie GY, Van Wyk JJ, Underwood LE, and ME Svoboda (1987) Derivation of monoclonal antibodies to human somatomedin C/insulin-like

growth factor I. In D Barnes and D Sirbasku (eds):Methods in Enzymology; Peptide Growth Factors. Orlando,Fl: Academic Press, 146: 207-216.

8. Russell WE, Van Wyk JJ, and WJ Pledger (1984) Inhibition of the mitogenic effects of plasma by a monoclonal antibody to somatomedin-C. Proc Natl Acad Sci USA 81:2389-2392.

9. Ambesi-Impiombato FS, Parks LAM, and HG Coon (1980): Culture of hormone dependent functional cells from rat thyroids. Proc. Nat. Acad. Sci. U.S.A. , 77:3455-3459.

10. Tramantano D, Cushing GW, Moses AC and SH Ingbar (1986): Insulin-like growth factor-I stimulates the growth of rat thyroid cells in culture and synergizes the stimulation of DNA synthesis induced by TSH and Graves'-IgG. Endocrinology, 119(2):940-942.

11. Maciel RMB, Moses AC, Villone G, Tramontano D, and SH Ingbar (1988): Demonstration of the production and physiological role of insulin-like growth factor II in rat thyroid follicular cells in culture. J. Clin. Invest. 82(Nov.):1546-1553.

12. Frauman AG, Tsuzaki S, and AC Moses: The binding characteristics and biological effects in FRTL-5 cells of PP-12, an insulin-like growth factor (IGF) binding protein purified from human amniotic fluid. (in press)

13. Brenner-Gati L, Berg KA, Gershengorn MC (1988) Thyroid-stimulating hormone and insulin-like growth factor-1 synergize to elevate 1,2-Diacylglycerol in rat thyroid cells. J. Clin. Invest. 82(Sept.):1144-1148.

14. Beebe DC, Feagans DE, Jebens HAH (1980) Lentropin: a factor in vitreous humor which promotes lens fiber differentiation. Proc.Natl.Acad.Sci USA 77:490.

15. Beebe DC, Silver MH, Belcher KS, Van Wyk JJ, Svoboda ME, and PS Zelenka (1987) Lentropin, a protein that controls lens fiber formation, is related functionally and immunologically to the insulin-like growth factors. Proc.Natl.Acad.Sci. USA 84:2327-2330.

16. Adashi EY, Resnick CE, D'Ercole AJ, Svoboda ME, and JJ Van Wyk (1985) Insulin-like growth factors as intraovarian regulators of granulosa cell growth and function. Endocrine Reviews 6:400-420.

17. Adashi EY, Resnick CE, Svoboda ME, and JJ VanWyk (1984) A novel role for somatomedin-C in the cytodifferentiation of the ovarian granulosa cell. Endocrinology 115:1227-1229.

18. Adashi EY, Resnick CE, Svoboda ME, and JJ Van Wyk (1986) Somatomedin-C as an amplifier of follicle-stimulating hormone action: enhanced accumulation of adenosine 3'5'- cyclic monophosphate. Endocrinol. 118:149-155.

19. Adashi EY, Resnick CE, Hernandez ER, May JV, Knecht M, Svoboda ME, and JJ VanWyk (1988) Insulin-like growth factor-I as an amplifier of follicle-stimulating hormone action: studies on mechanism(s) and site(s) of action in cultured rat granulosa cells. Endocrinol. 122,1583

20. Savion N, Lui GM, Laherty R, Gospodarowicz W (1981) Factors controlling proliferation and progesterone production by bovine granulosa cells in serum-free medium. (1981)Endocrinol 109:409-420.

21. Hammond JM, English HF (1987). Regulation of deoxyribonucleic acid synthesis in cultured porcine granulosa cells by growth factors and hormones. Endocrinology 120:1039-1046.

22. Veldhuis JD, Demers LM (1985) A role for somatomedin-C as a differentiating hormone and amplifier of hormone action on ovarian cells: studies with synthetically pure somatomedin-C and swine granulosa cells. Biochem Biophy Res Comm 130:234-240.

23. Smith EP, Svoboda ME, Van Wyk JJ, Kierszenbaum AK, and LL Tres (1987) Chemical characteristics of a somatomedin-like peptide and its binding protein secreted by cultured rat Sertoli cells. Endocrinol. 120:186-193.

24. Tres LL, Smith EP, Van Wyk JJ, and AL Kierszenbaum (1986) Immunoreactive Sites and Accumulation of Somatomedin-C in Rat Sertoli Spermatogenic Cell Co-cultures. Expt. Cell Research 162:33-50.

25. Kasson BG, and AJ Hsueh (1987) Insulin-like growth factor-I augments gonadotropin-stimulated androgen biosynthesis by cultured rat testicular cells. Mol.Cell.Endocrinol. Jul. 52(1-2):27-34.

26. Perrard-Sapori MH, Chatelain PC, Rogemond N, and JM Saez (1987) Modulation of Levdig cell functions by culture with Sertoli cells or with Sertoli cell-conditioned medium: effect of insulin, somatomedin-C and FSH. Mol.Cell.Endocrinol. Apr. 50(3):193-201.

27. Casella SJ, Han VK, D'Ercole AJ, Svoboda ME, and JJ VanWyk (1986) Insulin-like growth factor binding to the type I somatomedin receptor:

Evidence for two high affinity binding sites. J.Biol. Chem. 261:9268-9273.

28. Lund PK, Moats-Staats BM, Hynes MA, Simmons JG, Jansen M, D'Ercole AJ, and JJ VanWyk (1986) Somatomedin-C/IGF-I and IGF-II mRNA's in rat fetal and adult tissues. Jour. Biol. Chem. 261:14539-14544.

29. Han VKM, Lund PK, Lee DC, and AJ D'Ercole (1988) Expression of somatomedin/insulin-like growth Factor messenger ribonucleic acids in the human fetus:Identification, characterization, and tissue distribution. J. Clin. Endorinol Metab 6(2):422-429.

30. Hynes MA, Van Wyk JJ, Brooks PJ, D'Ercole AJ, Jansen M, and PK Lund (1987) Growth hormone dependence of somatomedin-C/insulin-like growth factor I and insulin-like growth factor II messenger ribonucleic acids. Mol Endocrinol 1: 233-242.

31. Han VKM, D'Ercole AJ, and PK Lund (1987) Cellular localization of somatomedin (insulin-like growth factor) messenger RNA in the human fetus. Science 236:193.

32. Beck F, Samani NJ, Penschow JD, Thorley B, Tregar GW, and JP Coghlan (1987) Histochemical localization of IGF-I and IGF-II mRNA in the developing rat embryo. Development 101: 175-184.

33. Hynes MA, Brooks PJ, Van Wyk JJ, and PK Lund, (1988) Insulin-like growth factor II messenger ribonucleic acids are synthesized in the choriod plexus of the rat brain. Mol.Endocrinol 2:47-54.

34. Casella SJ, Smith EP, Van Wyk JJ, Joseph DR, Hynes MA, Hoyt EC and PK Lund (1987) Isolation of rat testis cDNAs encoding an insulin-like growth factor-I precursor. DNA 6:325-330.

35. Hynes MA, Brooks PJ, Van Wyk JJ, and PK Lund (1988) Insulin-like growth factor II messenger ribonucleic acids are synthesized in the choroid plexus of the rat brain. Mol.Endocrinol 2:47-54.

36. Mohan S, Jennings JC, Linkhart TA, and DJ Baylink (1988) Primary structure of human skeletal growth factor: homology with human insulin-like growth factor-II. Biochim Biophys Acta 961: 44-55

37. Frolik CA, Ellis EF, and DC Williams (1988) Isolation and characterization of insulin-like growth factor-II from human bone.Biochem Biophys Res Commun 151: 1011-8.

ROLE OF IGFs IN GROWTH HORMONE HOMEOSTASIS

Michael Berelowitz

State University of New York
Endocrine Div; HSC T15, 060
Stony Brook, New York   11794

Neuroendocrine Control of Growth Hormone

The classic postulate by Harris of neuroendocrine control of the anterior pituitary gland predicted that factors would be identified within the hypothalamus that were transported via capillaries of the hypophyseal-portal vessels to regulate pituitary hormone secretion. Two such factors have recently been identified that regulate growth hormone secretion - growth hormone releasing factor (GRF) (1) and the inhibitory peptide, somatostatin (SRIF) (2) (Fig 1). SRIF is widely synthesized and distributed through the hypothalamus and extra-hypothalamic brain (3). Somatostatinergic neurons that originate in perikarya in the periventricular nucleus of the hypothalamus have axons that project via the retrochiasmatic region to terminate in nerve endings in the external layer of the median eminence (4). Manipulation of growth hormone and steroid hormone homeostasis influence preprosomatostatin mRNA (5,6) and SRIF content (7) primarily in this pathway and lesions along the pathway deplete the median eminence of SRIF (8). It thus appears likely that this tract represents the hypophysiotropic somatostatinergic pathway. GRF is localized to the hypothalamus where it is present in perikarya within the arcuate and ventromedial nuclei (9) with axons that pass to the median eminence.

Presynaptic stimulation of median eminence nerve endings results in release of both GRF and SRIF into capillaries of the hypophyseal portal vessels (10) via which they reach the anterior pituitary to influence synthesis and secretion of growth hormone.

GRF receptors have been identified in anterior pituitary membrane preparations (11) and partly characterized by ligand affinity cross-linking (12). Binding of GRF to its receptor results in stimulation of growth hormone synthesis (as determined by increases in mRNA and protein concentration) and secretion (13). This process appears to be mediated intracellularly by signalling pathways dependent on adenylyl cyclase and protein kinase A activation and on phospholipase C

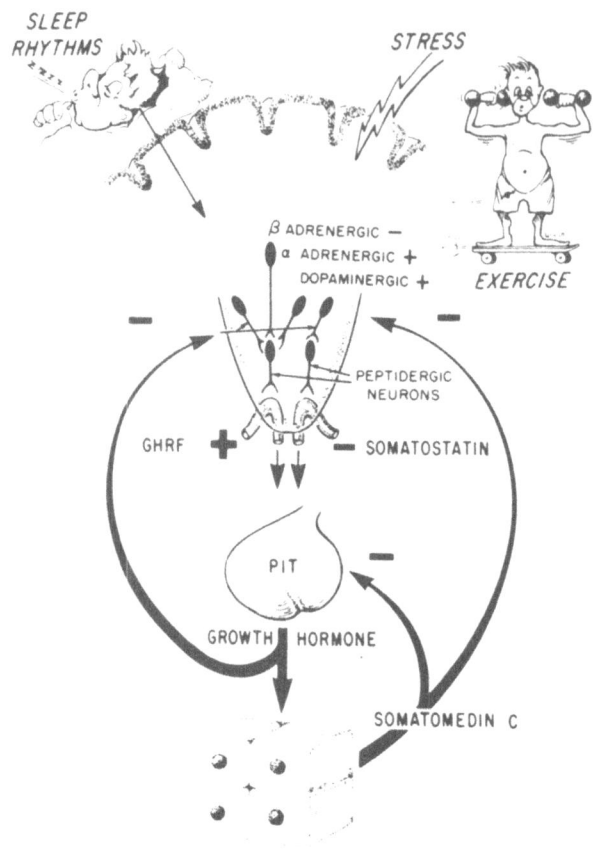

Figure 1

Schematic representation of hypothalamic-pituitary growth hormone - somatomedin-C (IGF-I) axis. Regulation by environmental influences via neurotransmitter/neuropeptide fibres and by long- and short-loop hormonal feedback signals are depicted. (From: Regulation of growth hormone secretion and its disorders. In: Clinical Neuroendocrinology Edition 2. JB Martin and S. Reichlin Eds. (1987), with permission).

mediated increases in membrane phosphoinositol turnover with protein kinase C activation (14). SRIF too exerts its effects on growth hormone after binding to a partially characterized membrane receptor (15). This receptor is linked to the inhibitory guanyl nucleotide binding protein complex, Gi (16). Activation of the SRIF receptor following SRIF binding inhibits growth hormone secretion via two mechanisms: inhibition of adenylyl cyclase production and cAMP-mediated growth hormone secretion and by membrane hyperpolarization, inhibition of calcium influx and inhibition of non-cAMP dependent growth hormone secretion (17). SRIF is thus a powerful inhibitor of basal and stimulated growth hormone secretion. No consistent effects of SRIF have been demonstrated to date on growth hormone synthesis.

In summary then, growth hormone secretion is regulated by opposing actions of two hypothalamic peptides, GRF and SRIF. Physiological patterns of growth hormone secretion depend on the complex regulation and orchestration of GRF-SRIF secretion, interaction and action.

GRF-SRIF Interaction

Growth hormone secretion is characteristically pulsatile in all species studied to date (18). It seems clear that this pattern of intermittent secretion is important for full biologic expression of growth hormone action. Thus pulsatile growth hormone administration results in enhanced insulin-like growth factor 1 (IGF-I) synthesis and secretion and growth rate compared to continuous administration (19). GRF and SRIF secretion must be coordinated to produce intermittent growth hormone secretion. Experimental neutralization of GRF influence on the pituitary (using GRF antiserum) abolishes growth hormone secretory peaks (20). This suggests that GRF secretion is associated with pulses of GH secretion. Evidence for SRIF involvement is more subtle. Injection of GRF during a trough of growth hormone secretion is less effective in stimulating growth hormone than during a pulse (21). This suggests the existence of inhibitory (perhaps SRIF) tone during troughs of growth hormone secretion. Neutralization of SRIF tone (by SRIF antiserum) reduces this inhibitory influence and raises trough growth hormone levels (21). Tannenbaum has thus proposed a model (Fig 2) for the final common pathway of growth hormone secretion that predicts rhythmic secretion of GRF that is 180 degrees out of phase with SRIF secretion with a frequency equal to that of resultant growth hormone pulses (21). This model has been confirmed by direct measurement in the rat of hypophyseal-portal venous immunoreactive GRF and SRIF concentration (10). The complexity of GRF-SRIF interaction in regulation of pulsatile growth hormone demands coordination of their secretion and thus further levels of control - peripheral and central.

Feedback Regulation of GRF and SRIF

Autoregulation of the GRF-SRIF-growth hormone system, like other hypothalamic-pituitary axes, is partly dependent on negative feedback by peripheral effectors. Growth hormone deficiency (induced experimentally by hypophysectomy with hormone replacement or by passive immunoneutralization) results in an increase in hypothalamic GRF synthesis (mRNA and peptide) (22) and a decrease in periventricular nucleus SRIF synthesis and hypothalamic SRIF secretion (23). This can be reversed by growth hormone replacement (albeit incompletely in some studies) (22). Thus growth hormone, or factors dependent on growth hormone (for example IGFs), modulate hypothalamic GRF and SRIF activity. IGF-I (somatomedin-C), a peripheral mediator of many of growth hormone's mitogenic actions, stimulates hypothalamic SRIF secretion over a physiologic dose range (Fig 3) (24). This suggests that the effect of growth hormone on hypothalamic SRIF could be mediated indirectly by IGF-I. Direct effects of GH on hypothalamic SRIF secretion have however been demonstrated (23); thus for this axis it seems that both

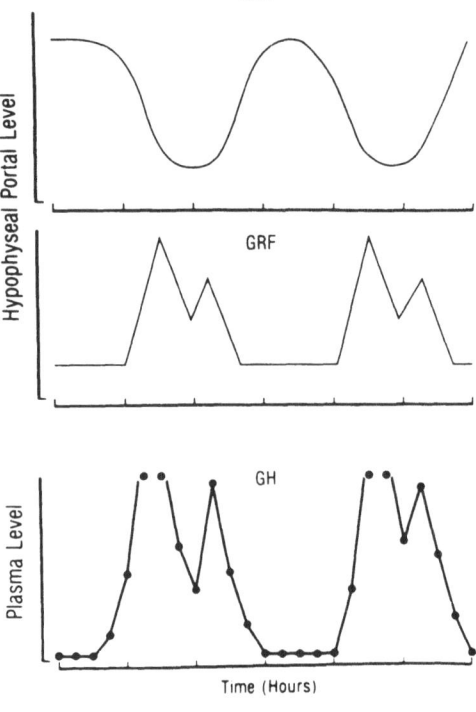

Figure 2

Schematic representation of postulated rhythmic secretion of
SRIF and GRF into hypophyseal portal blood, with the net
result on plasma GH profile. (From Tannenbaum GS and Ling N
(21), with permission. Copyright by The Endocrine Society).

short-loop (growth hormone) and long-loop (IGF-I) feedback
regulate hypothalamic GRF and SRIF. In addition an
ultrashort-loop feedback system (?interneuronal) exists
within the hypothalamus whereby GRF and SRIF modulate each
other's secretion (25). It is thus possible that a
GRF/growth hormone pulse stimulates SRIF/diminishes GRF
thereby inhibiting growth hormone and instituting a new
growth hormone cycle. Central signals must exist that
maintain cyclicity and mediate physiologic growth hormone
responses to environmental signals.

Neurotransmitter Control of GRF and SRIF

The hypothalamus appears to act as a coordination center
in which internal and external environmental influences
modulate endocrine systems. GRF and SRIF, as outlined above,
provide a final common pathway for growth hormone regulation
but both peptides are profoundly regulated by neurotransmitter and neuropeptide inputs to the hypothalamus (Fig 1).

Alpha-2-adrenergic influences stimulate (via GRF) while beta adrenergic influences inhibit (via SRIF) growth hormone secretion (26). Epinephrine rather than norepinephrine is the likely adrenergic transmitter in these pathways.

The cholinergic system is also important physiologically in growth hormone regulation. Basal and sleep-related growth hormone release and growth hormone secretion induced by a wide variety of stimuli (including GRF) are blocked by cholinergic muscarinic M1 receptor antagonists (27). It has been suggested that this occurs through disinhibition of SRIF release from the hypothalamus and that the cholinergic role is therefore primarily one of SRIF inhibition.

Other less well characterized neurotransmitter systems may play a role in modulating hypothalamic GRF/SRIF. Dopamine receptor agonists stimulate growth hormone secretion in man yet dopamine antagonists are without effect on spontaneous secretion. Furthermore dopamine itself suppresses growth hormone release following stimulation by arginine or hypoglycemia suggesting a complex interplay of hypothalamic and pituitary actions for this pathway (28). Indeed, effects of dopamine on both GRF (29)and SRIF (30) secretion have been demonstrated in vitro. Serotonin, GABA and histamine pathways also influence growth hormone secretion but their physiologic roles remain unclear.

A variety of peptides with putative neuromodulatory functions have been shown experimentally to influence growth hormone secretion through effects on GRF or SRIF. These include neurotensin, substance P (31), bombesin (32), glucagon (33) and corticotropin releasing factor (34). Once again a physiologic role has not been demonstrated.

From the data reviewed above it is apparent that a hierarchy exists whereby environmental influences alter neurotransmitter/ neuropeptide input to the hypothalamus thus modulating GRF/SRIF output and growth hormone secretion. Growth hormone exerts peripheral actions directly and indirectly via IGF-I mediation. Both growth hormone and IGF-I autoregulate the system via classical negative feedback.

Role of IGF in Growth Hormone Homeostasis

The somatomedin hypothesis predicted that mitogenic actions of growth hormone are not direct but rather indirect effects mediated by somatomedins (35). The somatomedin family of polypeptides has subsequently been characterized and identified as growth factors IGF-I and IGF-II. While it was originally considered that these factors acted as classical hormones following release into the circulation from the liver it now is clear that IGFs are synthesized in many tissues (36). Since many of these tissues are also sites of IGF action it has been postulated that locally synthesized IGF may be locally active. IGF-I, in particular, is synthesized and stored throughout adult mammalian tissues notably brain (especially hypothalamus), pituitary, liver, kidney, muscle and bone (37). In these tissues IGF-I synthesis (mRNA and peptide content) is growth hormone

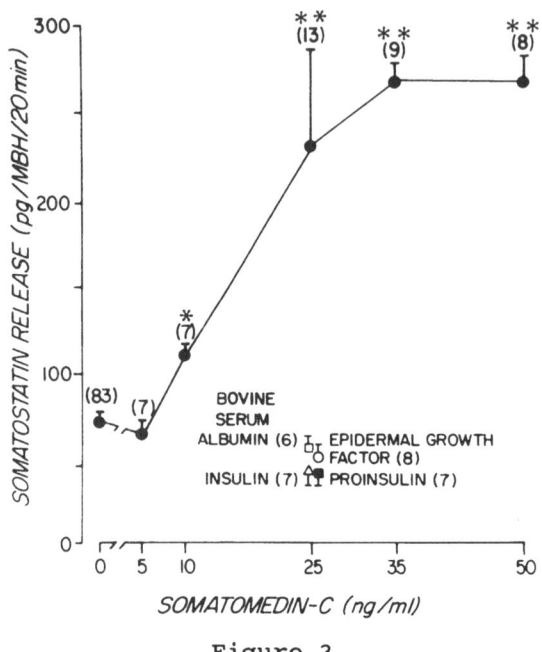

Figure 3

Effect of somatomedin-C (IGF-I) on somatostatin release from
incubated rat medial basal hypothalamus in vitro. (From
Berelowitz M et al (24), copyright 1981 by the AAAS).

dependent and regulated (37). It thus seems that growth
hormone regulates synthesis of IGF-I in many sites of tissue
growth hormone action. From the perspective of negative
feedback of growth hormone it is of interest to evaluate
possible interactions between growth hormone and hypothalamic
or pituitary IGF-I synthesis/secretion.

IGF-I as a hypothalamic regulator of growth hormone

Possible involvement of IGF-I in growth hormone negative
feedback at the hypothalamic level could be conceived in
three ways. First, as a hormone. In this mode circulating
IGF-I would be the likely signal. This mechanism of action
would represent a delayed influence of growth hormone rather
than an acute feedback response since a time-lag of up to 3
hours exists between growth hormone alteration and resultant
IGF-I perturbations (38). It would also represent an insensitive or integrated feedback signal since major growth
hormone alterations must take place to induce changes in
serum IGF-I levels (39). Second, locally synthesized IGF
could act within the hypothalamus to regulate SRIF/GRF output
and thereby mediate growth hormone feedback regulation. This
is a feasible mechanism since all the components for such a
signalling system exist within the hypothalamus. IGF-I is
synthesized and stored within the mammalian hypothalamus(40).

Though discrete hypothalamic nuclear localization of IGF-I or its mRNA has not been described it seems clear that IGF-I is synthesized locally in glia alone or in both glia and neurons (41). In addition, synthesis of IGF-I within the hypothalamus is regulated by growth hormone (22). Hypothalamic IGF-I mRNA is reduced in hypophysectomized rats and restored by growth hormone replacement (22). Finally IGF-I receptors are present in the hypothalamus (42) and IGF-I has been shown to exert physiologic effects - stimulating protein and RNA synthesis (in glia and neurones) and DNA synthesis (in glia) (43,44) that can lead to neurotransmitter release. Thus an IGF-I receptor - effector system exists within the hypothalamus that is responsive to perturbation of growth hormone. Centrally administered IGFs have been shown to inhibit growth hormone secretion (45) an effect that could be SRIF mediated since IGF-I stimulates SRIF secretion in vitro from incubated hypothalamus (24). Such findings provide a circumstantial local role of hypothalamic IGF-I in the regulation of growth hormone. They also raise two interesting possibilities. First, could hypothalamic IGF-I mediate non-feedback influences, for example nutrient signals, on growth hormone secretion? Second, what about IGF-II? An IGF-II receptor - effector system is also in place in the hypothalamus and hypothalamic IGF-II mRNA is growth hormone regulated (22). All these postulates can now be tested.

Finally IGF-I could be released from the hypothalamus in response to feedback stimulation and could act directly on the pituitary as a neuroendocrine regulator of growth hormone.

IGF-I as a Pituitary Regulator of Growth Hormone

IGF-I is a powerful inhibitor of pituitary growth hormone secretion in vitro (24). As outlined above, if IGF-I plays a physiologic role in the regulation of pituitary growth hormone this could occur via circulating peptide (endocrine or neuroendocrine) or locally. In the pituitary, as in other putative sites of IGF action, IGF is synthesized and the peptide is stored (46). It is unclear as to which of the pituitary cells synthesize IGF-I. IGF-I and IGF-II receptors are also present in crude pituitary membrane preparations and these receptors share the biochemical and functional characteristics of peripheral IGF-I and IGF-II receptors (47,48). IGF-I inhibits growth hormone transcription (49), translation (50) and secretion (24,50), apparently following binding and activation of the IGF-I receptor. In keeping with a possible intrapituitary regulatory role of IGF-I on growth hormone synthesis and secretion is the finding that growth hormone, in turn, regulates pituitary IGF-I synthesis (51). This suggests a close autoregulatory interaction (autocrine) between IGF-I and growth hormone in pituitary cells.

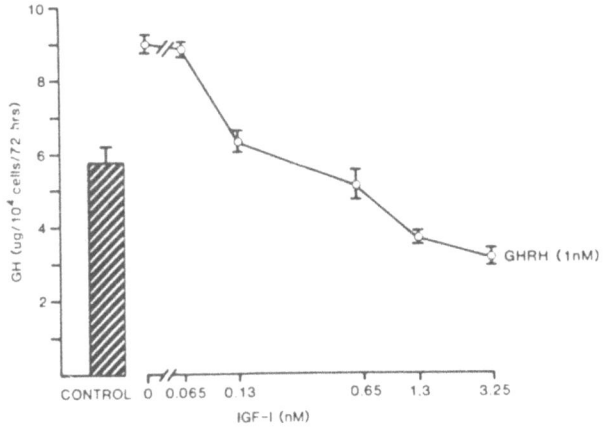

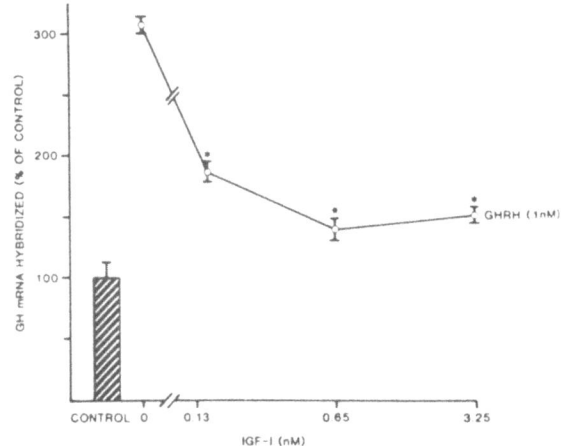

Figure 4

Influence of IGF-I (0-3.25nM) on GRF (GHRH, lnM) - stimulated
growth hormone secretion (above) and mRNA content (below).
(From Yamashita and Melmed (49) with permission. Copyright
by The Endocrine Society).

In this review the regulation of growth hormone has been
outlined. Neuropeptide and neurotransmitter signals modulate
hypothalamic secretion of the regulatory peptides GRF/SRIF to
co-ordinate and regulate growth hormone secretion. As is the
case with other pituitary hormones growth hormone secretion
is pulsatile. This implies complex orchestration of GRF /
SRIF and growth hormone. Many peripheral actions of growth
hormone, including negative feedback autoregulation, are
mediated by IGF-I. As in other tissue sites of action IGF-I
may exert its feedback influence on growth hormone locally -
within the hypothalamus by altering SRIF (and perhaps GRF)
secretion and within the pituitary by directly modifying
growth hormone synthesis and secretion. Much remains to be
determined regarding the physiologic importance and
regulatory diversity of these systems.

32

References

1)   Rivier J, Spiess J, Thorner M, Vale W.
        Characterization of a growth hormone releasing factor
        from a human pancreatic islet tumor, Nature 300:275
        (1982).

2)   Brazeau P, Vale W, Burgus R, Ling N, Butcher M,
        Rivier J, Guillemin R. Hypothalamic polypeptide that
        inhibits the secretion of immunoreactive growth
        hormone, Science 179:77 (1973).

3)   Brownstein M, Arimura A, Sato H, Schally AV, Kizer
        JS. The regional distribution of somatostatin in the
        rat brain, Endocrinology 96:1456 (1975).

4)   Epelbaum J, Tapia Arancibia L, Herman JP, Kordon C,
        Palkovits M. Topography of median eminence
        somatostatinergic innervation, Brain Research 230:412
        (1981).

5)   Rodgers KV, Vician L, Steiner RA. The effect of
        hypophysectomy and growth hormone administration on
        pre-prosomatostatin messenger ribonucleic acid in the
        periventricular nucleus of the rat hypothalamus.
        Endocrinology 122:586 (1988).

6)   Werner H, Koch Y, Baldino F, Gozes I. Steroid
        regulation of somatostatin mRNA in the rat
        hypothalamus, J Biol Chem 263:7666 (1988).

7)   Nakagawa K, Ishizuka T, Obara T, Matsubara M, Akikawa
        K. Dichotomic action of glucocorticoids on growth
        hormone secretion Acta Endocrinol (kbh) 116:165
        (1987).

8)   Critchlow V, Rice RW, Abe K, Vale W. Somatostatin
        content of the median eminence in female rats with
        lesion-induced disruption of the inhibitory control
        of growth hormone secretion, Endocrinology 103:817
        (1978).

9)   Vandepol CJ, Leidy JW, Finger TE, Robbins RJ.
        Immunohistochemical localization of GRF-containing
        neurons in rat, Neuroendocrinology 42:143 (1986).

10)   Plotsky PM, Vale W. Patterns of growth hormone-
        releasing factor and somatostatin secretion into the
        hypophyseal-portal circulation of the rat, Science
        230:461 (1985).

11)   Seifert H, Perrin M, Rivier J, Vale W. Growth
        hormone releasing factor binding sites in rat
        anterior pituitary membrane homogenates: modulation
        by glucocorticoids, Endocrinology 117:424 (1985).

12)   Zysk JR, Cronin MJ, Anderson JM, Thorner MO. Cross-
        linking of a growth hormone releasing factor-binding
        protein in anterior pituitary cells, J Biol Chem
        261:16781 (1986).

13)   Barinaga M, Bilezikijan LN, Vale W, Rosenfeld MG, Evans RM. Independent effects of growth hormone releasing factor on growth hormone release and gene transcription, Nature 314:279 (1985).

14)   Cronin M, Summers S, Sortino M, Hewlett E. Protein kinase C enhances growth hormone releasing factor (1-40)-stimulated cyclic AMP levels in anterior pituitary, J Biol Chem 261:13932 (1986).

15)   Reubi JC, Perrin M, Rivier J, Vale W. High affinity binding sites for somatostatin to rat pituitary, Biochemical and Biophysical Research Communications 105:1538 (1982).

16)   Koch BD, Dorflinger LJ, Schonbrunn A. Pertussis toxin blocks both cyclic AMP-mediated and cyclic AMP-independent actions of somatostatin, J Biol Chem 260:13138 (1985).

17)   Koch BD, Schonbrunn A. Characterization of the cyclic AMP-independent actions of somatostatin in GH cells, J Biol Chem 263:226 (1988).

18)   Tannenbaum GS, Martin JB. Evidence for an endogenous ultradian rhythm governing growth hormone secretion in the rat, Endocrinology 98:562 (1976).

19)   Brook CGD, Hindmarsh PC, Stanhope R. Growth and growth hormone secretion, J Endocr 119:179 (1988).

20)   Wehrenberg WB, Brazeau P, Luben R, Bohlen P, Guillemin R. Inhibition of pulsatile secretion of growth hormone by monoclonal antibodies to the hypothalamic growth hormone releasing factor (GRF), Endocrinology 111:2147 (1982).

21)   Tannenbaum GS, Ling N. The interrelationship of growth hormone (GH)-releasing factor and somatostatin in generation of the ultradian rhythm of GH secretion, Endocrinology 115:1952 (1984).

22)   Wood TL, Berelowitz M, McKelvy JF. Regulation of IGF-I and IGF-II mRNAs in the male rat hypothalamus after hypophysectomy, Society for Neuroscience Abstracts Vol 14, part 1:72 (1988).

23)   Berelowitz M, Firestone SL, Frohman LA. Effects of growth hormone excess and deficiency on hypothalamic somatostatin content and release and on tissue somatostatin distribution, Endocrinology 109:714 (1981).

24)   Berelowitz M, Szabo M, Frohman LA, Firestone SL, Chu L, Hintz RL. Somatomedin-C mediates growth hormone negative feedback by effects on both the hypothalamus and the pituitary, Science 212:1279 (1981).

25)   Aguila MC, McCann SM. The influence of hGRF, CRF, TRH and LHRH on SRIF release from median eminence fragments, Brain Research 348:180 (1985).

34

26)   Martin JB. Neural regulation of growth hormone
        secretion, New England Journal of Medicine 228:1384
        (1973).

27)   Casanueva FF, Villanueva L, Cabranes JA, Cabezas-
        Cerrato J, Fernandez-Cruz A. Cholinergic mediation
        of growth hormone secretion induced by arginine,
        clonidine and physical exercise in man, J Clin
        Endocrinol Metab 59:526 (1984).

28)   Tallo D, Malarkey WB. Adrenergic and dopaminergic
        modulation of growth hormone and prolactin secretion
        in normal and tumor bearing human pituitaries in
        monolayer culture, J Clin Endocrinol Metab 53:1278
        (1981).

29)   Wakabayashi I, Miyazawa Y, Kanda M, Miki N, Demura R,
        Demura H, Shizume K. Stimulation of immunoreactive
        somatostatin release from hypothalamic synaptosomes
        by high K^+ and dopamine, Endocrinol Jpn 24:601
        (1977).

30)   Katajima N, Chihara K, Abe H, Okimura Y, Fujii Y,Sato
        M, Shakutsui S, Watanabe M, Fujita T. Effects of
        dopamine on immumoreactive growth hormone - releasing
        factor and somatostatin secretion from rat
        hypothalamic slices perifused in vitro, Endocrinology
        124:69 (1989).

31)   Aronin N, Coslovsky R, Leeman SE. Substance P and
        neurotensin: Their roles in the regulation of
        anterior pituitary function, Annual Review of
        Physiology 48:537 (1986).

32)   Murphy WA, Lance VA, Heiman ML, Hocart SJ, Coy DH.
        Prolonged inhibition of growth hormone secretion by
        peripheral injection of bombesin is mediated by
        somatostatin in the rat, Endocrinology 117:1179
        (1985).

33)   Katakami H, Kato Y, Matsushita N, Shimatsu A, Imura
        H. Involvement of hypothalamic somatostatin in
        glucagon-induced suppression of growth hormone
        secretion in conscious rats, Peptides 4:849 (1983).

34)   Rivier C, Vale W. Corticotropin-releasing factor
        acts centrally to inhibit growth hormone secretion in
        rat, Endocrinology 114:2409 (1982).

35)   Daughaday WH, Herington AC, Phillips LS. The
        regulation of growth by endocrines, Annual Review of
        Physiology 37:211 (1975).

36)   Lund P, Moats-Staats B, Hynes M, Simmons J, Jansen M,
        D'Ercole A, Van Wyk J. Somatomedin-C/Insulin-like
        growth factor-I and insulin-like growth factor-II
        mRNAs in rat fetal and adult tissues, J Biol Chem
        261:14539 (1986).

37)   Roberts CT, Lasky SR, Lowe WL, Seaman WT, LeRoith D.
        Molecular cloning of rat insulin-like growth factor I

complementary deoxyribounucleic acids: Differential
messenger ribonucleic acid processing and regulation
by growth hormone in extrahepatic tissues, Molecular
Endocrinology 1:243 (1987).

38) Hall K. Effect of intravenous administration of
human growth hromone on sulphation factor activity in
serum of hypopituitary subjects, Acta Endocrinol
(kbh) 66:491 (1971).

39) Phillips LS, Vassilopoulou-Sellin R. Somatomedins,
New England Journal of Medicine 302:371 and 438
(1980).

40) Rotwein P, Burgess SK, Milbrandt JD, Krause JE.
Differential expression of insulin-like growth factor
genes in rat central nervous system, Proc Natl Acad
Sci USA 85:265 (1988).

41) Adamo M, Werner H, Farnsworth W, Roberts CT, Raizada
M, LeRoith D. Dexamethasone reduces steady state
insulin-like growth factor I messenger ribonucleic
acid levels in rat neuronal and glial cells in
primary culture, Endocrinology 123:2565 (1988).

42) Baskin DG, Wilcox BJ, Figlewicz DP, Dorsa DM.
Insulin and insulin-like growth factors in the CNS,
Trends in Neurosciences 11:107 (1988).

43) Shemer J, Raizada MK, Masters BA, Ota A, Le Roith D.
Insulin-like growth factor-I receptors in neuronal
and glial cells, J Biol Chem 262:7693 (1987).

44) Adamo M, Werner H, Farnsworth W, Roberts CT, Raizada
MK, LeRoith D. Dexamethasone reduces steady state
insulin-like growth factor-I messenger ribonucleic
acid levels in rat neuronal and glial cells in
primary culture, Endocrinology 123:2565 (1988).

45) Tannenbaum GS, Guyda HJ, Posner BI. Insulin-like
growth factors: A role in growth hormone negative
feedback and body weight regulation via brain,
Science 220:77 (1983).

46) Binoux M, Hossenlopp P, Lassarre C, Hardouin N.
Production of Insulin-like growth factors and their
carrier by rat pituitary gland and brain explants in
culture, FEBS Letters 124:178 (1981).

47) Goodyer CG, deStephano L, Lai WH, Guyda HJ, Posner BI.
Characterization of insulin-like growth factor
receptors in rat anterior pituitary, hypothalamus, and
brain, Endocrinology 114:1187 (1984).

48) Rosenfeld RG, Ceda G, Cutler CW, Dollar LA, Hoffman AR.
Insulin and Insulin-like growth factor (Somatomedin)
receptors on cloned rat pituitary tumor cells,
Endocrinology 117:2008 (1985).

49)   Yamashita S, Melmed S. Insulin-like growth factor I
        action on rat anterior pituitary cells: Suppression of
        growth hormone secretion and messenger ribonucleic acid
        levels, Endocrinology 118:176 (1986).

50)   Yamashita S, Melmed S. Insulin-like growth factor I
        regulation of growth hormone gene transcription in
        primary rat pituitary cells, Journal of Clinical
        Investigation 79:449 (1987).

51)   Fagin J, Yamashita S, Melmed S. Ultrashort Feedback
        Regulation of Pituitary insulin-like growth factor-I
        mRNA levels by growth hormone, Clinical Research
        34:544A (1986).

INTERACTION OF IGF WITH THE HYPOTHALAMUS AND PITUITARY

Ron G. Rosenfeld, Ian Ocrant, Karen L.
Valentino, and Andrew R. Hoffman

Departments of Pediatrics and Medicine
Stanford University School of Medicine
Stanford, CA; and
Neurex Corporation, Menlo Park, CA

INTRODUCTION

Postnatal growth, regulated by pituitary growth
hormone, and mediated by IGF-I, is controlled by the
central nervous system. The mechanisms underlying the
regulation of this system are under intense study in many
laboratories. By analogy with other pituitary hormones,
GH secretion may be controlled by negative feedback,
either by GH itself, or by the IGFs produced in diverse
tissues in response to GH.[1] This feedback signal may be
received by the hypothalamus, the pituitary gland, or
possibly by higher brain centers. Responses to these
regulatory signals would necessarily require mediation by
specific receptors for IGFs. Such receptors have been
localized and characterized within both the hypothalamus
and pituitary gland; the results of these studies are
reviewed herein.

IGF RECEPTORS IN THE HYPOTHALAMUS

IGF receptors were first demonstrated in hypothalamic
microsomal membrane preparations by competitive binding
with $[^{125}I]$IGF-I and -II.[2-4] These studies revealed the
existence of both type 1 and type 2 receptors in the
hypothalamus. The presence of these receptors in the
hypothalamus[5] and median eminence[6] was confirmed using
autoradiography and digital image analysis of radioligand
bound to tissue sections. The median eminence consists of
the most ventral portion of the arcuate nucleus and
contains axons from hypothalamic neurons which transmit
pituitary releasing factors to the hypophyseal portal
system and neurohypophysis. IGF receptors in these

locations may influence a variety of hormonal systems, including those controlling growth.

Brain parenchyma is heterogeneous, containing both vascular tissue, supporting glial cells of multiple types, and neurons. In order for IGFs to mediate feedback regulation of GH secretion, it is necessary for neurons to be capable of responding to this signal. However, evidence for IGF receptors on hypothalamic neurons has been lacking until recently, when primary cultures of adult rat hypothalamic neurons were affinity labeled with [^{125}I]IGF-I and -II, demonstrating both type 1 and 2 IGF receptors on neuronal plasma membranes.[7] Using immunohistochemical techniques, we have further determined that only a small proportion of hypothalamic neurons in primary culture express type 2 receptors. This suggests that only a minority of such cells is equipped to respond to IGF-II. Recent work in our laboratory suggests that hypothalamic neurons express more type 2 receptor immunoreactivity in the fetus than in the adult.[8] Based on immunostaining for type 2 receptors in fetal rat brain sections, these neurons appear to be located in the region of the arcuate nucleus and ventromedial hypothalamic nucleus. Since these regions contain pathways for both GHRH and SRIF secretion, hypothalamic IGF receptors are strategically located to participate in the regulation of GH secretion.

ORIGIN OF IGFs IN THE HYPOTHALAMUS

The brain parenchyma is isolated and protected from the influences of circulating substances. This protection results from tight junctions formed between brain capillary endothelial cells in conjunction with other structures, such as glial end foot processes, thus constituting the blood brain barrier. Generally, this barrier prevents passive diffusion of large molecules, such as proteins, and smaller hydrophilic molecules. The question thus arises as to how IGFs, as large hydrophilic molecules, gain access to hypothalamic receptors, if they originate from peripheral sources. Possible solutions to this problem are:

1) Transportation of IGFs across the blood brain barrier

2) IGF access via regions in which the blood brain barrier is incomplete or nonexistent, as in the circumventricular organs[9]

3) Local synthesis of IGFs in response to some other feedback signal

Several reports address the first possibility. Brain microvessels have been documented to have both types of IGF receptor, which bind and internalize [^{125}I]IGF-I and -II *in vitro*.[10,11] However, transport of intact peptides across the antiluminal surface of the brain microvasculature has not been adequately proven.

The second possibility has been explored, albeit more thoroughly with insulin than with the IGFs. Specifically, insulin has been shown to gain direct access from plasma to the hypothalamic circumventricular organs.[9] Projections from these regions extend to several other nuclei in the hypothalamus and anterior pituitary gland. By analogy, circulating IGFs may, therefore, gain hypothalamic access and influence GH secretion in this manner. The observation that nude rats harboring IGF-II secreting tumors had decreased IGF-I levels is consistent with this hypothesis.[12,13] GRF-stimulated GH secretion remained normal in these rats, suggesting that IGF-I suppression may be mediated by inhibition of pulsatile GRF secretion, and not by direct suppression of the pituitary gland; direct peripheral inhibition of IGF-I secretion by IGF-II is an alternative explanation. Additionally, intracerebroventricular injection of IGF into freely moving rats has been shown to inhibit GH secretion and decrease food intake, while control injections were without effect.[14,15] These studies suggest that IGFs, from sources outside brain parenchyma, can influence brain activity.

As for the third possible signalling mechanism, that of local IGF synthesis in response to signals from the periphery, several studies have documented the ability of brain tissue to synthesize IGF-I and -II mRNA and peptide.[16-31] The choroid plexus, in particular, is a major site for the synthesis of IGF-II. From there, IGF-II may communicate with regions of the brain in contact with CSF, but lacking an effective CSF-brain barrier; such regions are found in the paraventricular regions of the hypothalamus surrounding the third ventricle.[9] The existence and nature of peripheral signals resulting in IGF synthesis by CNS tissues remains largely unexplored, although a few reports suggest that GH itself may mediate this signal. Additional studies will be required before the details of the GH feedback pathway, as well as the regulation and additional actions of IGFs in the CNS, are fully understood.

IGF RECEPTORS IN THE PITUITARY GLAND

IGF-I and -II receptors have been demonstrated in a variety of human and murine pituitary systems and models, including: 1) rat pituitary membrane preparations,[3] 2) primary dispersed cultures of rat pituitary cells,[32] 3)

1) Generally, most IGF-II/M6P receptors are found intracellularly,[36] consistent with lysosomal transport functions. However, certain tissues, like intermediate lobe parenchyma, appear to have most of their

type 2 receptors located on the cell surface, a finding difficult to reconcile with an exclusive function in lysosomal targeting.

2) IGF-II receptors from different tissues migrate at different apparent M_r, as shown by SDS-PAGE, suggesting the possibility of tissue-specific functions.[7,41] The type 2 receptor may, like other receptors (such as epinephrine receptors), have multiple roles (depending on location), multiple ligands (resembling GABA receptors in this regard), and multiple intracellular signalling pathways (as with insulin receptors).

3) Additionally, recent reports describe cellular activities, such as calcium transport, which appear to be modulated by the binding of IGF-II to its specific receptor.[42]

4) Many cell lines and tissues elaborate large quantities of IGF binding proteins (see below), which are so efficient in sequestering IGFs, that little, if any, free peptide has ever been found in extracellular fluids. For this reason, it is difficult to understand why a cell would undertake inefficient and energetically costly removal of IGF-II, using type 2 receptors.

The association of type 2 receptors with intermediate lobe cell surfaces suggests additional possible functions.[35] The primary function of the intermediate lobe of the pituitary is the secretion of alpha-melanotropin, and beta-endorphin.[43] The presence of large quantities of type 2 binding activity and immunoreactivity in the intermediate lobe of the pituitary, the non-somatotrophic parenchymal cells of the anterior pituitary, and hypothalamic neurons suggests that IGF-II may be involved in the regulation of the synthesis or secretion of pro-opiomelanocortin (POMC) gene derivatives, such as MSH and endorphin in the intermediate lobe of the pituitary, ACTH in the anterior lobe, and endorphins and other POMC products in the hypothalamus.

CHARACTERIZATION OF HYPOTHALAMIC AND PITUITARY IGF RECEPTORS

Insulin and IGF-I and -II receptors from brain parenchyma exhibit higher electrophoretic mobility (lower apparent M_r) when compared with receptors from other tissues.[7,41,44-46] Additional variant forms are found in fetal tissues (see table). Studies with deglycosylation enzymes suggest, but do not prove, that variant receptors have the same primary structure, but are differently glycosylated. What these differences mean in functional terms has not yet been determined. However, alternative forms suggest alternative, tissue-specific functions. Recent studies indicate that both neurointermediate lobe and anterior pituitary IGF-I and -II receptors have higher apparent M_r than those from brain tissue, and have migration pattern identical to that of receptors from peripheral tissue sources.[35] Using primary cultures of adult rat hypothalamic neurons, we have reported high M_r (peripheral tissue type) IGF-I and -II receptors.[7] We cannot, however, exclude modification of receptor phenotypic expression by the tissue culture conditions employed. Affinity labeling studies on hypothalamic receptors isolated from microsomal membrane preparations have not been reported.

Table 1. Apparent M_r of Rat IGF Receptors

Tissue	Type 1[a]	Type 2[a]
liver (adult)	140K	250K
liver (fetal)	140K	250K
brain (adult)	127K	240K
brain (fetal)	130K, 120K	250K
astrocytes (neonatal)	140K	250K
neurons (adult)	140K	250K
ant. pituitary (adult)	140K	250K
NI lobe pituitary (adult)	140K	250K

[a]M_r estimates from SDS-PAGE vary among individual experiments and published reports. Since the magnitude of these differences is small, adjacent lane comparisons in gels are necessary for validation. The apparent M_rs reported above are based on averages from multiple experiments employing standard adult rat liver and brain microsomal membrane preparations. These standards were used in adjacent lane comparisons to determine the relative apparent M_r for receptors from other tissues and cells.

A word of caution is required concerning the use of Scatchard analysis to estimate receptor affinity and quantity. Virtually all tisues contain insulin receptors, as well as type 1 and 2 IGF receptors. Furthermore, many tissues, including brain and pituitary (see below), produce IGF binding proteins that compete for radioligand. Additionally, contamination of tissue extracts with membrane-associated binding protein can occur.[47] As a result of the combined contribution of these different binding moieties, curvilinear Scatchard plots often result. Interpretation of these curves is difficult, because without further separation or manipulation of the different binding activities, one cannot determine what part of the curve represents the receptor of interest. The same problem applies to quantitative densitometric autoradiography of tissue sections. Since biologically active receptors do not have to be abundant to exert significant biological effects, and since the composition of the buffer system used in displacement studies has a major influence on receptor affinity, Scatchard analysis should be reserved for binding affinity estimation in highly purified preparations, and for determination of the influence of different experimental conditions on receptor concentrations. However, if receptor quantitation is desirable, and if highly purified receptor preparations are unavailable, other methods, such as radioimmunoassay, ELISA, or immunoblot assay should be used in preference to Scatchard analysis, in order to eliminate the confounding influences of binding proteins and homologous receptors.

IGF BINDING PROTEINS PRODUCED BY THE PITUITARY AND HYPOTHALAMUS

Both IGF-I and IGF-II normally circulate in plasma complexed to large (apparent M_r = 150,000) and small (apparent M_r = 30,000) binding proteins (BPs).[48,49] These BPs are capable of binding IGF with both high specificity and affinity, and appear to modulate access of IGF to specific receptors. In 1981, Binoux et al[16], employing explants of rat anterior and neurointermediate lobes, first demonstrated production of BPs by pituitary tissue. By gel filtration chromatography of explant conditioned media, they estimated an apparent M_r of 40,000 for these BPs, but noted that the chromatographic pattern suggested structural heterogeneity. More recently, Shiu and Paterson[30] reported that metabolically labeled rat pituitary explants and cultures which were immunoprecipitated with an anti-IGF-II antibody, demonstrated a protein of apparent M_r= 33,000. If one subtracts the 8700 molecular weight of the IGF-II species identified in rat pituitary in this study, the resulting putative IGF BP has an apparent M_r = 24-25,000.

We have recently confirmed the presence of IGF BPs in conditioned media (CM) from primary cultures of both rat anterior pituitary (AP) and neurintermediate lobe (NI), and have characterized these BPs by affinity labeling, deglycosylation studies and Western ligand blots.[50] When either [^{125}I]IGF-I or -II are cross-linked to CM from rat AP cultures, multiple small BPs are identifiable, ranging

in apparent M_r from 21,000 to 35,000. Similar sized BPs
are found in CM from NI cell cultures, although the
relative distribution of these BPs is different. In NI CM,
the predominant IGF BP has an apparent M_r of 27,000 by
cross-linking and 26,000 by Western ligand blots. This BP
appears to be a minor component in AP CM, although it is
possible that the larger AP BPs represent glycosylated
variants of this 27K BP. Whether the AP and NI BPs are
post-translational modifications of the same protein(s)
remains to be determined. However, it is of interest that
the 27K BP, the major BP in NI CM, appears identical in
size, in its lack of glycosylation, and in its relatively
high affinity for IGF-II to the BP produced by BRL-3A and
BRL-3A2 rat hepatoma cells.

It is of further note, that a similar binding protein
is produced by cultured hypothalamic neurons and by
astrocytes.[51] Additionally, the most prominent BP in human
CSF migrates at an apparent $M_r = 34,000$, is nonglycosylated, and also has an affinity for IGF-II 10-20
fold greater than its affinity for IGF-I.[52,53] These
preliminary findings strongly suggest that the CNS
contains a third class of IGF BPs, which is produced by
anterior pituitary cells, neurointermediate cells,
hypothalamic neurons and glia, and which can be found in
high concentrations in CSF. This binding protein appears
to be structurally related, if not identical, to the BP
produced by BRL-3A cells. However, this BP appears to be
structurally and functionally unrelated to the type 2 IGF
receptor, which is present in high concentrations in the
CNS, pituitary, choroid plexus, and ependymal lining of
the ventricles.[8,36] Antibodies directed against the rat
type 2 IGF receptor can neither immunoprecipitate the
cross-linked BP, nor block binding of [^{125}I]IGF to the
BP.[36,50,54]

The role of IGF BPs in normal pituitary physiology
and throughout the CNS remains uncertain, since the
biological actions of the IGFs, themselves, in the
pituitary and CNS remain to be established. It would seem
reasonable that locally produced BPs would be capable of
binding IGFs synthesized locally or transported to the
pituitary/CNS from the systemic circulation. The BPs could
thereby modulate access of both IGF-I and IGF-II to their
specific receptors on neurons, glia and pituitary cells.

IGF AND INSULIN REGULATION OF PITUITARY FUNCTION

Numerous studies have demonstrated that the GH-
dependent somatomedins, IGF-I and -II, are able to feed
back to inhibit GH secretion.[1,15,34,55-64] These effects
appear to be mediated both by direct action at the
somatotroph and by indirect action at the hypothalamus,
where IGF-I has been shown to inhibit GHRH release[65] and

to stimulate somatostatin secretion.[66] While both IGF-I and IGF-II have been shown to inhibit GH release from cultured rat and human somatotrophs, the ED_{50} for IGF-I appears to be significantly lower,[56,61] suggesting that IGF-II directed-inhibition of GH is probably mediated through the type 1 IGF receptor. Feedback inhibition by IGF-I at the level of the somatotroph can be observed even in the fetus.[63,67,68] The mechanism and degree of IGF-I directed inhibition of pituitary GH secretion differs from that of the other major GH-release inhibitory factor, somatostatin. Not only does somatostatin appear to be a more potent inhibitor of GH release, but it also appears to be more effective, completely abolishing GHRH-induced GH secretion at sub-nanomolar concentrations.[69] Using a reverse hemolytic plaque assay, Hoeffler et al[70] demonstrated that while somatostatin inhibited GH release from most, if not all, cultured rat somatotrophs, IGF-I prevented GH release from only a relatively small subpopulation of GH-producing cells. In experiments in which new protein synthesis was prevented by the addition of cycloheximide, Sheppard and Bala[69] reported that IGF-I was no longer able to block GH release, while somatostatin was still an effective inhibitor. These results indicate that IGF-I preferentially inhibits the release of newly synthesized GH.

Although the second messenger mediating IGF-I's inhibitory action at the somatotroph is unknown, IGF-I was able to inhibit both cyclic AMP-directed and protein kinase C-mediated GH synthesis and secretion, suggesting that IGF-I acted on the GH gene by a mechanism that cannot be overridden by these two intracellular messengers.[71] Yamashita and Melmed have recently demonstrated that IGF-I could inhibit both rat[72] and human[73] GH gene transcription. These investigators transfected a 2.6 kilobase human GH gene into human choriocarcinoma cells and showed that IGF-I could inhibit GH gene transcription. They concluded that there is a cis-acting, non-tissue specific regulatory element on the human GH gene which is responsive to IGF-I, and that this element resides in the 0.5 kb 5'-flanking region or in an intron.

While both in vitro and acute in vivo exposure to IGF-I reliably results in inhibition of GH release, the effect of chronically elevated levels of IGF on GH physiology is still unclear. To address this issue, Wilson et al[12,13] implanted nude rodents with IGF-II secreting 18-54,SF cells. These animals had persistently elevated levels of IGF-II, generally >100 ng/ml, but neither pituitary GH mRNA nor GHRH-elicited GH release was different from non-implanted controls. Of interest, however, was the observation that serum IGF-I levels were significantly decreased in this model, suggesting that IGF-II can directly inhibit IGF-I secretion. (An alternative possibility is that increased serum IGF-II levels resulted in increased displacement of IGF-I from plasma binding proteins, resulting in enhanced degradation of IGF-I). Furthermore, since it appears that IGF-I is more potent than IGF-II as an inhibitor of GH release in vitro, it is possible that the reciprocal decrease in serum IGF-I levels seen in transplanted rats was responsible for the observed lack of GH suppression. In a different animal model, Mathews et al[74] created a line of transgenic mice which carry the human IGF-I gene and which overexpress IGF-I peptide. These animals demonstrated selective organomegaly, with the brain growing 50% heavier than in control mice. GH mRNA accumulation was reduced to approximately 40% of control, but circulating GH levels were below the level of detection, indicating that long-term exposure to high levels of IGF-I inhibits GH synthesis and release.

The role of the IGFs in the regulation of GH secretion has also been investigated in humans. While Keligman and Frohman[75] and Loche et al[76] have found an inverse correlation between IGF-I levels and the peak GH response to GHRH, others have not been able to confirm this finding.[77] When a single injection of IGF-I was given to normal volunteers, hypoglycemia promptly appeared.[78] When compared to a similar level of insulin-induced hypoglycemia, the rise in serum GH was not as great following IGF-I administration (although this difference did not reach statistical significance). This suggests that elevated serum IGF-I potentially inhibited the somatotroph response to hypoglycemia. Longer-term administration of IGF-I appears to be capable of reducing serum GH levels to below the level of detection (J. Zapf, personal communication). It is of note that recent reports of patients with severe hypoglycemia associated with the production and secretion of IGF-II by neoplasms have indicated an impaired GH counterregulatory response to hypoglycemia.[79,80] After removal of the IGF-II producing tumor, the GH response to hypoglycemia was restored, thereby providing further evidence that high levels of IGF-II can inhibit GH release.[81]

While insulin, like IGF-I, has been shown to inhibit GH gene transcription, synthesis and secretion from cultured rat pituitary cells,[82-85] it appears to be less potent than IGF-I, as indicated by perfusion experiments with rat pituitary cells.[61] However, in some cases, insulin has been shown to be capable of inhibiting GH release from cultured human somatotroph adenoma cells.[34] Isaacs et al[86] have studied the regulation of rat and human GH gene expression by insulin. Using GH_3 rat pituitary tumor cells, they showed that insulin could both upregulate and downregulate rat GH gene expression, depending upon the metabolic state of the cell (i.e., depending upon whether media and hormones were replaced daily). They subsequently showed that when the human GH gene was transfected into a similar rat tumor cell line (GC), its regulation by insulin was also determined by the metabolic status of the cells.[87] In cultures of dispersed human GH-secreting adenoma cells, insulin alone had no effect on GH gene expression, but did blunt the stimulatory effects of glucocorticoids. Prager and Melmed[88] reported that insulin inhibited the transcription of the human GH gene which had been transfected into either GC or HeLa cells. In experiments employing a truncated GH gene, they determined that cis-acting regulatory sequences on the 497 base pair 5'-flanking region were necessary for the human GH gene to respond to insulin.

Preliminary data also suggest that insulin and the IGFs may play a role in the regulation of prolactin secretion. In GH$_3$ cells, physiologic concentrations of insulin increased prolactin mRNA levels and peptide production. Insulin also partially antagonized the suppressive effects of glucocorticoids on prolactin gene expression.[89] Significantly higher concentrations of IGF-I were required to see a similar effect in these cells, and even at very high concentrations, IGF-I did not alter prolactin gene transcription in normal rat pituitary cells.[72] IGF-I has, however, been shown to stimulate the synthesis and secretion of prolactin from cultured human decidual cells.[90]

CONCLUSIONS

The role(s) of both IGF-I and IGF-II in the development and maintenance of the central nervous system remains largely unresolved at this time. While mRNA for both IGF-I and -II has been localized in the brain, and its ontogeny evaluated, and while receptors for IGF-I and -II have been similarly localized and characterized, the specific actions of the IGFs in the CNS are still speculative. These issues are further complicated by the identification of specific IGF binding proteins in both the rat and human brain.

Despite these caveats, a good case can be made for an important biological role for IGF-I (and, perhaps, IGF-II) in the regulation of pituitary function. Specific receptors for both IGF-I and -II have been identified in both the anterior pituitary and the neurointermediate lobe, and both in vitro and in vivo studies have demonstrated that IGF-I can inhibit GH gene transcription, as well as GH synthesis and release. Even here, however, several critical questions remain unanswered:

1) How physiologically relevant is the ability of IGF-I (and IGF-II) to inhibit GH secretion?

2) Are the feedback actions of IGF-I, IGF-II and insulin all mediated through the type 1 IGF receptor?

3) Do the IGFs also regulate GH synthesis and secretion at an hypothalamic level?

4) What other pituitary/hypothalamic functions are regulated by the IGFs?

5) What is the role of the type 2 IGF receptor in pituitary/hypothalamic development and function?

6) What role do IGF binding proteins play in IGF regulation of pituitary function?

These questions clearly warrant further investigation into the nature of IGF interactions throughout the central nervous system.

ACKNOWLEDGEMENTS

Supported by NIH grants DK 36054 and DK 28229. RGR is a recipient of a Research Career Development Award (DK 01275) from the NIH. IO is a recipient of a National Research Service Award (DK 08075) from the NIH.

REFERENCES

1. R. G. Rosenfeld and A. R. Hoffman, Insulin-like growth factors and their receptors in the pituitary and hypothalamus, in: Insulin, Insulin-like Growth Factors, and Their Receptors in the Central Nervous System," M. K. Raizada, M. I. Phillips, and D. LeRoith, eds., Plenum Press, New York (1987).
2. V. R. Sara, K. Hall, H. Von Holtz, R. Humbel, B. Sjogren, and L. Wetterberg, Evidence for the presence of specific receptors for insulin-like growth factors 1 (IGF-1) and 2 (IGF-2) and insulin throughout the adult human brain, Neurosci. Lett. 34:39 (1982).
3. C. G. Goodyer, L. De Stephano, W. H. Lai, H. J. Guyda, and B. I. Posner, Characterization of insulin-like growth factor receptors in rat anterior pituitary, hypothalamus, and brain, Endocrinology 114:1187 (1984).
4. S. Gammeltoft, G. K. Haselbacher, R. E. Humbel, M. Fehlmann, and E. Van Obberghen, Two types of receptors for insulin-like growth factors in mammalian brain, EMBO J. 4:3407 (1985).
5. M. A. Lesniak, J. M. Hill, W. Kiess, M. Rojeski, C. B. Pert, and J. Roth, Receptors for insulin-like growth factors I and II: Autoradiographic localization in rat brain and comparison to receptors for insulin, Endocrinology 123:2089 (1988).

6. N. J. Bohannon, D. P. Figlewicz, E. S. Corp, B. J. Wilcox, D. Porte, Jr., and D. G. Baskin, Identification of binding sites for an insulin-like growth factor (IGF-I) in the median eminence of the rat brain by quantitative autoradiography, Endocrinology 119:943 (1986).

7. I. Ocrant, K. L. Valentino, L. F. Eng, R. L. Hintz, D. M. Wilson, and R. G. Rosenfeld, Structural and immunohistochemical characterization of insulin-like growth factor I and II receptors in the murine central nervous system, Endocrinology 123:1023 (1988).

8. K. L. Valentino, I. Ocrant, and R. G. Rosenfeld, Developmental expression of insulin-like growth factor II receptor-immunoreactivity in the rat CNS (submitted for publication).

9. M. Van Houten, B. I. Posner, B. M. Kopriwa, and J. R. Brawer, Insulin binding sites in the rat brain: In vivo localization to the circumventricular organs by quantitative radioautography, Endocrinology 105:666 (1979).

10. K. R. Duffy, W. M. Pardridge, and R. G. Rosenfeld, Human blood-brain barrier insulin-like growth factor receptor, Metabolism, 37:136 (1988).

11. R. G. Rosenfeld, H. Pham, B. T. Keller, R. T. Borchardt, and W. M. Pardridge, Demonstration and structural comparison of receptors for insulin-like growth factor-I and -II (IGF-I and -II) in brain and blood-brain barrier, Biochem. Biophys. Res. Commun. 149:159 (1987).

12. D. M. Wilson, J. A. Thomas, T. E. Hamm, Jr., J. Wyche, R. L. Hintz, and R. G. Rosenfeld, Transplantation of insulin-like growth factor-II-secreting tumors into nude rodents, Endocrinology 120:1896 (1987).

13. D. M. Wilson, S. N. Perkins, J. A. Thomas, S. Seelig, S. A. Berry, T. E. Hamm, Jr., A. R. Hoffman, R. L. Hintz, and R. G. Rosenfeld, Effects of elevated serum insulinlike growth factor-II on growth hormone and insulinlike growth factor-I mRNA and secretion, Metabolism 38:57 (1989).

14. T. J. Lauterio, L. Marson, W. H. Daughaday, and C. A. Baile, Evidence for the role of insulin-like growth factor II (IGF-II) in the control of food intake, Physiol. Behav. 40:755 (1987).

15. G. S. Tannenbaum, H. J. Guyda, and B. I. Posner, Insulin-like growth factors: a role in growth hormone negative feedback and body weight regulation via brain, Science 220:77 (1983).

16. M. Binoux, P. Hossenlopp, C. Lassare, and N. Hardouin, Production of insulin-like growth factors and their carrier by rat pituitary gland and brain explants in culture, FEBS Lett. 124:178 (1981).

17. G. K. Haselbacher, N. E. Schwab, O. Pasi, and R. E. Humbel, Insulin-like growth factor II (IGF-II) in human brain: regional distribution of IGF-II of higher molecular mass form, Proc. Natl. Acad. Sci. USA 82:2153 (1985).

18. E. C. Hoyt, J. J. Van Wyk, and P. K. Lund, Tissue and

development specific regulation of a complex family of rat insulin-like growth factor I messenger ribonucleic acids, Mol. Endocrinol. 2:1077 (1988).

19. M. B. Soares, D. N. Ishii, and A. Efstratiadis, Developmental and tissue-specific expression of a family of transcripts related to rat insulin-like growth II mRNA, Nucl. Acids Res. 13:1119 (1985).

20. M. B. Soares, A. Turken, D. Ishii, L. Mills, V. Episkopou, S. Cotter, S. Zeitlin, and A. Efstratiadis, Rat insulin-like growth factor II genes: a single gene with two promoters expressing a multitranscript family, J. Mol. Biol. 192:737 (1986).

21. P. K. Lund, B. M. Moats-Staats, M. A. Hynes, J. G. Simmons, M. Jansen, A. J. D'Ercole, and J. J. Van Wyk, Somatomedin-C/insulin-like growth factor-I and insulin-like growth factor-II mRNAs in rat fetal and adult tissues, J. Biol. Chem. 261:14539 (1986).

22. J. Scott, J. Cowell, M. E. Robertson, L. M. Priestly, R. Wadley, B. Hopkins, J. Pritchard, G. I. Bell, L. B. Rall, C. F. Graham, and T. J. Knott, Insulin-like growth factor-II gene expression in Wilms' tumour and embyonic tissues. Nature 317:260 (1985).

23. J. A. Romanus, Y. W-H. Yang, S. O. Adams, A. N. Sofair, L. Y-H. Tseng, S. P. Nissley, and M. M. Rechler, Synthesis of insulin-like growth factor II (IGF-II) in fetal rat tissues: Translation of IGF-II ribonucleic acid and processing of pre-pro-IGF-II, Endocrinology 122:709 (1988).

24. L. K. Andersson, D. Edwall, G. Norstedt, B. Rozell, A. Skottner, and H-A. Hansson, Differing expression of insulin-like growth factor I in the developing and in the adult rat cerebellum, Acta Physiol. Scand. 132:167 (1988).

25. T. Ueno, K. Takahashi, T. Matsuguchi, H. Endo, and M. Yamamoto, Transcriptional deviation of the rat insulin-like growth factor II gene initiated at three alternative leader-exons between neonatal tissues and hepatomas, Biochim. Biophys. Acta 950:411 (1988).

26. V. K. M. Han, P. K. Lund, D. C. Lee, and A. J. D'Ercole, Expression of somatomedin/insulin-like growth factor ribonucleic acids in the human fetus: identification, characterization, and tissue distribution, J. Clin. Endocrinol. Metab. 66:422 (1988).

27. P. Rotwein, K. M. Pollock, M. Watson, and J. D. Milbrandt, Insulin-like growth factor gene expression during rat embryonic development, Endocrinology 121:2141 (1988).

28. J-C. Irminger, K. M. Rosen, R. E. Humbel, and L. Villa-Komaroff, Tissue-specific expression of insulin-like growth factor II mRNAs with distinct 5' untranslated regions. Proc. Natl. Acad. Sci. USA 84:6330 (1987).

29. A-C. Sandberg, C. Engberg, M. Lake, H. Von Holst, and V. R. Sara, The expression of insulin-like growth factor I and insulin-like growth factor II genes in the human fetal and adult brain and in glioma, Neurosci. Lett. 93:114 (1988).

30. R. P. C. Shiu and J. A. Paterson, Characterization of

insulin-like growth factor II peptides secreted by explants of neonatal brain and of adult pituitary from rat, Endocrinology 123:1456 (1988).

31. F. Beck, N. J. Samani, J. D. Penschow, B. Thorley, G. W. Tregear, and J. R. Coghlan, Histochemical localization of IGF-I and -II mRNA in the developing rat embryo, Development 101:175 (1987).

32. R. G. Rosenfeld, G. Ceda, D. M. Wilson, L. A. Dollar, and A. R. Hoffman, Characterization of high affinity receptors for insulin-like growth factors I and II on rat anterior pituitary cells, Endocrinology 114:1571 (1984).

33. R. G. Rosenfeld, G. P. Ceda, C. W. Cutler, L. A. Dollar, and A. R. Hoffman, Insulin and insulin-like growth factor (somatomedin) receptors on cloned rat pituitary tumor lines, Endocrinology 117:2008 (1985).

34. G. P. Ceda, A. Hoffman, G. Silverberg, D. M. Wilson, and R. G. Rosenfeld, Regulation of growth hormone release from cultured human pituitary adenomas by somatomedins and insulin, J. Clin. Endocrinol. Metab. 60:1204 (1985).

35. I. Ocrant, K. L. Valentino, A. R. Hoffman, R. L. Hintz, D. M. Wilson, and R. G. Rosenfeld, Structural characterization and immunohistochemical localization of receptors for insulin-like growth factor II in the rat pituitary gland, Neuroendocrinol. (in press).

36. K. L. Valentino, H. Pham, I. Ocrant, and R. G. Rosenfeld, Distribution of insulin-like growth factor II receptor immunoreactivity in rat tissues, Endocrinology 122:2753 (1988).

37. F. Stylianopoulou, A. Efstratiadis, J. Herbert, and J. Pintar, Pattern of the insulin-like growth factor II gene expression during rat embryogenesis, Development 103:497 (1988).

38. D. O. Morgan, J. C. Edman, D. N. Standring, V. A. Fried, M. C. Smith, R. A. Roth, and W. J. Rutter, Insulin-like growth factor II receptor as a multifunctional binding protein, Nature 329:301 (1987).

39. R. A. Roth, Structure of the receptor for insulin-like growth factor II: The puzzle amplified. Science 239:1269 (1988).

40. W. Kiess, J. F. Haskell, L. Lee, L. A. Greenstein, B. E. Miller, A. L. Aarons, M. M. Rechler, and S. P. Nissley, An antibody that blocks insulin-like growth factor (IGF) binding to the type II IGF receptor is neither an agonist nor an inhibitor of IGF-stimulated biologic responses in L6 myoblasts, J. Biol. Chem. 262:12745 (1987).

41. A. McElduff. P. Poronnik, and R. C. Baxter, The insulin-like growth factor-II (IGF-II) receptor from rat brain is of lower apparent M_r than the IGF-II receptor from rat liver, Endocrinology 121:1306 (1987).

42. I. Kojima, I. Nishimoto, E. Ogata, and R. Rosenfeld, Evidence that type II insulin-like growth factor receptor is coupled to calcium gating system, Biochem. Biophys. Res. Commun. 154:9 (1988).

43. B. A. Eipper and R. E. Mains, Biosynthesis of pro-adrenocorticotropin/endorphin and related peptides, Endocr. Rev. 1:1 (1980).

44. K. A. Heidenreich and D. Brandenburg, Oligosaccharide heterogeneity of insulin receptors. Comparison of N-linked glycosylation of insulin receptors in adipocytes and brain, Endocrinology 118:1835 (1986).

45. J. Shemer, M. K. Raizada, B. A. Masters, A. Ota, and D. LeRoith, Insulin-like growth factor I receptors in neuronal and glial cells. Characterization and biological effects in primary culture. J. Biol. Chem. 262:7693 (1987).

46. S. K. Burgess, S. Jacobs, P. Cuatrecasas, and N. Sahyoun, Characterization of a neuronal subtype of insulin-like growth factor I receptor, J. Biol. Chem. 262:1618 (1987).

47. M. A. Sturm, C. A. Conover, H. Pham, and R. G. Rosenfeld, Insulin-like growth factor receptors and binding proteins in rat neuroblastoma cells, Endocrinology 124:388 (1989).

48. J. L. Martin and R. C. Baxter, Insulin-like growth factor-binding protein from human plasma. Purification and characterization. J. Biol. Chem. 261:8754 (1986).

49. S. Hardouin, P. Hossenlopp, B. Segovia, D. Seurin, G. Portolan, C. Lassarre, and M. Binoux, Heterogeneity of insulin-like growth factor binding proteins and relationships between structure and affinity. I. Circulating forms in man, Eur. J. Biochem. 170:121 (1987).

50. R. G. Rosenfeld, H. Pham, Y. Oh, and I. Ocrant, Characterization of insulin-like growth factor binding proteins in cultured rat pituitary cells, Endocrinology (in press).

51. I. Ocrant, H. Pham, Y. Oh, and R. G. Rosenfeld, Insulin-like growth factor binding proteins produced by cultured rat neurons and astrocytes (submitted for publication).

52. P. Hossenlopp, D. Seurin, B. Segovia-Quinson, and M. Binoux, Identification of an insulin-like growth factor-binding protein in human cerebrospinal fluid with a selective affinity for IGF-II, FEBS Lett. 208:439 (1986).

53. R. G. Rosenfeld, H. Pham, C. A. Conover, R. L. Hintz, and R. C. Baxter, Structural and immunological comparison of insulin-like growth factor binding proteins of cerebrospinal and amniotic fluids, J. Clin. Endocrinol. Metab. (in press).

54. R. G. Rosenfeld, D. Hodges, H. Pham, P. D. K. Lee, and D. R. Powell, Purification of the insulin-like growth factor II (IGF-II) receptor from an IGF-II-producing cell line, and generation of an antibody which immunoprecipitates and blocks the type 2 IGF receptor, Biochem. Biophys. Res. Commun. 138:304 (1986).

55. H. Abe, M. E. Molitch, J. J. Van Wyk, and L. E. Underwood, Human growth hormone and somatomedin C suppress the spontaneous release of growth hormone

in unanesthetized rats, _Endocrinology_ 113:1319 (1983).

56. P. Brazeau, R. Guillemin, N. Ling, J. Van Wyk, and R. Humbel, Inhibition par le somatomedines de la secretion de l'hormone de croissance stimuleepar le facteur hypothalamique somatocrinine (GRF) ou le peptide de synthese hpGRF, _C.R. Acad. Sci. [D] (Paris)_ T295:651 (1982).

57. C. G. Goodyer, L. De Stephano, H. J. Guyda, and B. I. Posner, Effects of insulin-like growth factors on adult male rat pituitary function in tissue culture, _Endocrinology_ 115:1568 (1984).

58. M. S. Sheppard and R. M. Bala, Insulin-like growth factor inhibition of growth hormone secretion, _Can. J. Physiol. Pharmacol._ 64:525 (1986).

59. S. Yamashita and S. Melmed, Insulin-like growth factor I action on rat anterior pituitary cells: Suppression of growth hormone secretion and messenger ribonucleic acid levels, _Endocrinology_ 118:176 (1986).

60. S. Melmed and S. Yamashita, Insulin-like growth factor I action on hypothyroid rat pituitary cells: Suppression of triiodothyronine-induced growth hormone secretion and messenger ribonucleic acid levels, _Endocrinology_ 118:1483 (1986).

61. G. P. Ceda, R. G. Davis, R. G. Rosenfeld, and A. R. Hoffman, The growth hormone (GH) releasing hormone (GHRH)-GH-somatomedin axis: Evidence for rapid inhibition of GHRH-elicited GH release by insulin-like growth factors I and II, _Endocrinology_ 120:1658 (1987).

62. G. P. Ceda, B. Narog, R. G. Rosenfeld, and A. R. Hoffman, The role of insulin-like growth factors and insulin in the regulation of growth hormone secretion, _in_ "Endocrinology '85", G. M. Molinatti and L. Martini, eds., Elsevier Science Publishers. Amsterdam (1986).

63. C. G. Goodyer, S. Marcovitz, J. Hardy, Y. Lefebre, H. Guyda, and B. I. Posner, Effect of insulin-like growth factors on human fetal, adult normal and tumor pituitary function in tissue culture, _Acta Endocrinol. (Kbh)_ 112:49 (1986).

64. S. Yamashita, M. Weiss, and S. Melmed, Insulin-like growth factor I regulates growth hormone secretion and messenger ribonucleic acid levels in human pituitary tumor cells, _J. Clin. Endocrinol. Metab._ 63:730 (1986).

65. T. Shibasaki, N. Yamauchi, M. Hotta, A. Masuda, T. Imaki, H. Demura, N. Ling, and K. Shizume, In vitro release of growth hormone-releasing factor from rat hypothalamus: Effect of insulin-like growth factor I, _Regulatory Peptides_ 15:47 (1986).

66. M. Berelowitz, M. Szabo, L. A. Frohman, S. Firestone, L. Chu, and R. L. Hintz, Somatomedin-C mediates growth hormone negative feedback by effects on both the hypothalamus and the pituitary, _Science_ 212:1279 (1981).

67. M. M. Blanchard, C. G. Goodyer, J. Charrier, and B. Barenton, In vitro regulation of growth hormone (GH) release from ovine pituitary cells during fetal and neonatal development: Effects of GH-releasing factor, somatostatin, and insulin-like growth factor I, Endocrinology 122:2114 (1988).

68. F. De Zegher, M. Bettendorf, S. L. Kaplan, and M. M. Grumbach, Hormone ontogeny in the ovine fetus: XXI. The effect of insulin-like growth factor I on plasma fetal growth hormone, insulin and glucose concentrations, Endocrinology 123:658 (1988).

69. M. S. Sheppard and R. M. Bala, Cycloheximide blocks insulin-like growth factor I but not somatostatin inhibition of growth hormone secretion, Can. J. Physiol. Pharmacol. 65:515 (1987).

70. J. P. Hoeffler, S. A. Hicks, and L. S. Frawley, Existence of somatotrope subpopulations which are differentially responsive to insulin-like growth factor I and somatostatin, Endocrinology 120:1936 (1987).

71. S. Morita, S. Yamashita, and S. Melmed, Insulin-like growth factor I action on rat anterior pituitary cells: Effects of intracellular messengers on growth hormone secretion and messenger ribonucleic acid levels, Endocrinology 121:2000 (1987).

72. S. Yamashita and S. Melmed, Insulinlike growth factor I regulation of growth hormone gene transcription in primary rat pituitary cells, J. Clin. Invest. 79:449 (1987).

73. S. Yamashita, J. Ong, and S. Melmed, Regulation of human growth hormone gene expression by insulin-like growth factor I in transfected cells, J. Biol. Chem. 262:13254 (1987).

74. L. S. Mathews, R. E. Hammer, R. R. Behringer, A. J. DiErcole, G. I. Bell, R. L. Brinster, and R. D. Palmiter, Growth enhancement of transgenic mice expressing human insulin-like growth factor I, Endocrinology 123:2827 (1988).

75. M. Kelijman and L. A. Frohman, Enhanced growth hormone (GH) responsiveness to GH-releasing hormone after dietary manipulation in obese and nonobese subjects, J. Clin. Endocrinol. Metab. 66:489 (1988).

76. S. Loche, M. Cappa, P. Borrelli, A. Faedda, A. Crino, S. G. Cella, R. Corda, E. E. Muller, and C. Pintor, Reduced growth hormone response to growth hormone-releasing hormone in children with simple obesity: Evidence for somatomedin-C mediated inhibition, Clin. Endocrinol. 27:145 (1987).

77. G. Plewe, C. Schneider, V. Kurtz, G. Nolken, U. Krause, and J. Beyer, Variability of GH response to GRF shows no relationship with peripheral somatostatin and somatomedin C levels, Horm. Metabol. Res. 17:481 (1985).

78. H-P. Guler, J. Zapf, and E. R. Froesch, Short-term metabolic effects of recombinant human insulin-like growth factor I in healthy adults, N. Engl. J. Med. 317:137 (1987).

79. W. H. Daughaday, M. A. Emanuele, M. H. Brooks, A. L. Barbato, M. Kapadia, and P. Rotwein, Synthesis and secretion of insulin-like growth factor II by a leiomyosarcoma with associated hypoglycemia, N. Engl. J. Med. 319:1434 (1988).
80. L. Axelrod and D. Ron, Insulin-like growth factor II and the riddle of tumor-induced hypoglycemia, N. Engl. J. Med. 318:1477 (1988).
81. D. Ron, A. C. Powers, M. R. Pandian, J. E. Godine, and L. Axelrod, Increased insulin-like growth factor II (IGF-II) production and consequent suppression of growth hormone secretion: A dual mechanism for tumor induced hypoglycemia, J. Clin. Endocrinol. Metab. (in press).
82. S. Melmed, Insulin suppresses growth hormone secretion by rat pituitary cells, J. Clin. Invest. 73:1425 (1984).
83. S. Melmed and S. M. Slanina, Insulin suppresses triiodothyronine-induced growth hormone secretion by GH_3 rat pituitary cells, Endocrinology 117:532 (1985).
84. S. Melmed, L. Nielson, and S. Slanina, Insulin suppresses rat growth hormone messenger ribonucleic acid levels in rat pituitary tumor cells, Diabetes 34:409 (1985).
85. S. Yamashita and S. Melmed, Insulin regulation of rat growth hormone gene transcription, J. Clin. Invest. 78:1008 (1986).
86. R. E. Isaacs, D. G. Gardner, and J. D. Baxter, Insulin regulation of rat growth hormone gene expression, Endocrinology 120:2022 (1987).
87. R. E. Isaacs, P. R. Findell, P. Mellon, C. B. Wilson, and J. D. Baxter, Hormonal regulation of expression of the endogenous and transfected human growth hormone gene, Mol. Endocrinol. 1:569 (1987).
88. D. Prager and S. Melmed, Insulin regulates expression of the human growth hormone gene in transfected cells, J. Biol. Chem. 263:16580 (1988).
89. D. Prager, S. Yamashita, and S. Melmed, Insulin regulates prolactin secretion and messenger ribonucleic acid levels in pituitary cells, Endocrinology 122:2946 (1988).
90. K. M. Thraikill, A. Golander, L. E. Underwood, and S. Handwerger, Insulin-like growth factor I stimulates the synthesis and release of prolactin from human decidual cells, Endocrinology 123:2930 (1988).

FEEDBACK REGULATION OF GROWTH HORMONE GENE

EXPRESSION BY INSULIN-LIKE GROWTH FACTOR I

Diane Prager, Shlomo Melmed and James Fagin

Cedars-Sinai Medical Center
Division of Endocrinology and Metabolism
8700 Beverly Blvd.
UCLA School of Medicine
Los Angeles, CA 90048

INTRODUCTION

Growth hormone (GH) is quantitatively the main hormone secreted by the pituitary and is under hormonal control emanating from both central and peripheral sources.

The human GH (hGH) gene cluster consists of 5 genes encoding hGH and human chorionic somatomammotropin (hCS) within 50 kilobases on band q22-24 of chromosome 17. The hGH-N gene comprising 5 exons and 4 introns is expressed in the pituitary where it codes for a 22 kDa protein consisting of 191 amino acids. Approximately 10% of pituitary GH is present as a 20 kDa variant lacking amino acid residues 32-46 (2,3), and probably arising as a result of an alternate splicing mechanism (1). A second hGH gene designated hGH-variant (hGH-V) has been described (1). It differs from the hGH-N gene by 35 scattered base substitutions in the coding region resulting in 15 amino acid changes (4). hGH-V is expressed at very low levels in the placenta (5). This chapter focuses on the feedback regulation of GH gene expression by insulin-like growth factor I (IGF-I).

Regulation of GH expression

GH (as well as other hormones and growth factors) is required for normal linear growth. It also has lipolytic, anabolic and diabetogenic properties (6). The GH gene is under dual stimulatory and inhibitory hypothalamic control. Growth hormone releasing hormone (GHRH) secreted by the hypothalamus acts on the somatotroph cell to stimulate GH gene transcription (7,8) and secretion (9-12).

Somatostatin (SRIF) is synthesized in the medial preoptic area of the hypothalamus and inhibits the secretion of pituitary GH (13). Galanin, a 29 amino acid peptide (14) found in the central and peripheral neurons of several species including humans (15), elicits an increase in plasma GH levels in conscious male rats (16), and in ovariectomized female rats (17) after intracerebroventricular injection. Synthethic galanin given intravenously has also been reported to raise plasma GH levels in humans (18). Galanin fails to stimulate GH release from perifused rat anterior pituitary cells (17), suggesting an indirect mechanism of action on the somatotroph. Triiodothyronine, hydrocortisone and insulin act directly on the pituitary somatotroph to influence rat GH gene transcription and secretion (19-21). In addition, hydrocortisone enhances GHRH stimulation of the somatotroph (22).

GH binds to somatogenic receptors in the liver and stimulates the hepatic production of IGF-I (23). The interaction of GH and IGF-I appears to be necessary to fully promote longitudinal growth. There is evidence suggesting that GH may preferentially promote chondrocyte differentiation whereas IGF-I induces cell proliferation in the growth plate (24). Since IGF-I is the major target hormone for GH, there has been considerable interest in defining a possible negative feedback effect of IGF-I upon GH secretion.

IGF-I is a 70 amino acid polypeptide with a molecular weight of 7650 and a structure similar to that of proinsulin and IGF-II (25,26). The IGF-I gene is present as a single copy gene in the human genome (27,28). IGF-I is secreted predominantly by the liver but the gene is also expressed in multiple tissues (29-32). IGF-I is found in the circulation as a peptide-carrier protein complex which is well-described elsewhere in this volume. IGF-BP's have been reported to possess both stimulatory (33) and inhibitory (34) effects upon IGF-I action at the cellular level (35).

Effects of IGF-I upon the hypothalamus

Both GH and IGF-I act rapidly on the hypothalamus to stimulate SRIF release which in turn inhibits pituitary GH secretion via a negative short-feedback loop (36). IGF-I may also deplete hypothalamic GHRH content (37), further reducing positive stimulation of GH secretion.

Effects of IGF-I upon the pituitary

IGF-I also acts directly upon the somatotroph to inhibit both basal and stimulated pituitary GH gene expression. Both pituitary tumor cells, as well as normal pituitary cells possess abundant binding sites for IGF-I, IGF-II and insulin (38-40). This observation lends further credence to a physiologic role for the insulin-like growth factors in regulating pituitary function.

Rat GH mRNA positive cells are first detected in the fetal pituitary on day 19 of gestation (41). These immunoreactive GH cells are found scattered throughout the anterior lobe (41). Plasma corticosterone levels are transiently elevated during this period of somatotroph differentiation (42-42), possibly also contributing to induction of GH expression. Pretreatment of fetal and neonatal ovine pituitary cells with 100 nM IGF-I for 3 days reduces GHRH-stimulated GH secretion (44). IGF-I also suppresses GH secretion in the ovine fetus in vivo (45) and in mid-gestation human fetal cultures (46). IGF-I secretion is detected in fetal and neonatal ovine pituitary cultures (44), whilst IGF-I mRNA has been detected in neonatal ovine pituitary cells (44).

In vivo effects of IGF-I on the GH axis

Autoregulation of GH secretion has been observed in rats (47-52), rhesus monkeys (53) and man (54,55). GH may regulate its own secretion by acting on the hypothalamus to increase SRIF release, or by a direct effect on pituitary cells (56-61). Peripherally, pituitary GH stimulates IGF-I generation and this peptide participates in the feedback regulation of GH both at the hypothalamic and pituitary levels (36, 62-66). Rats with GH-secreting transplantable tumors have suppressed GH pituitary mRNA compared to control animals (67). This effect is selective, as pituitary   actin mRNA, a cytoskeletal mRNA, is not altered. The observed autoregulation of GH mRNA thus may be occuring at the level of the hypothalamus or pituitary. SRIF however, does not directly suppress GH mRNA levels in vitro. Hypothalamic GRH depletion in GH-secreting tumor-bearing rats has also been reported (37). GHRH has been shown to increase GH transcription and mRNA levels (7). GH has not been consistently shown to directly inhibit in vitro pituitary GH secretion. Intraventricular injections of partially purified IGF-I were shown to suppress the in vivo episodic secretion of rat GH (68,69). The action of IGF-I may therefore occur at the level of the hypothalamus, by stimulating SRIF or decreasing GHRH secretion, as well as by direct action on the pituitary.

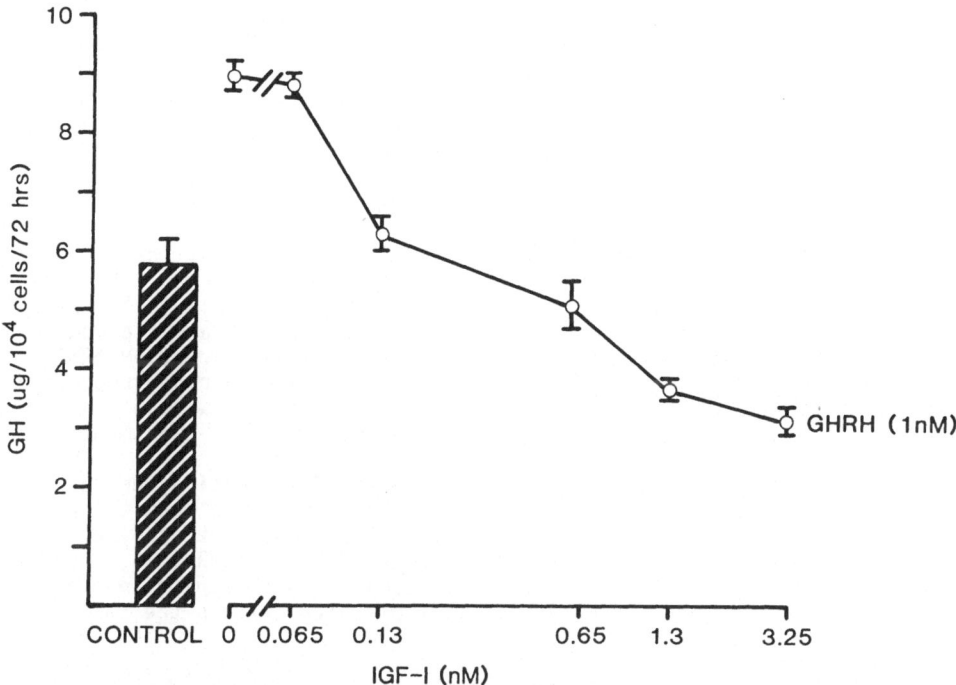

Fig. 1. Effects of varying doses of IGF-I on GHRH-induced GH stimulation. Pituitary cells were grown for 72 h in the presence of a constant dose of GHRH (1 nM) with the indicated doses of IGF-I. Control wells received no added GHRH or IGF-I. Each point depicts the mean ± SEM of triplicate wells. Adapted from Yamashita S, Melmed S. Endocrinology 118:176, 1986.

In vitro effects of IGF-I on the rat somatotroph cell

The effects of IGF-I have been examined on basal and stimulated pituitary GH secretion and on GH mRNA levels. To directly test the effects of IGF-I on somatotroph function, primary rat pituitary cells were cultured in vitro and treated with IGF-I (64). IGF-I inhibited the accumulation of GH in the medium of rat anterior pituitary monolayer cultures for up to 72 hrs (64). Maximal suppression was achieved with 3.25 nM IGF-I. IGF-I (3.25 nM) suppressed both basal and GHRH-stimulated GH secretion in a time dependent fashion. IGF-I suppressed basal GH levels by 50% after 24 h and by 70% by 48 h. GHRH stimulation of GH was abolished by IGF-I and after 48 h, GH secretion from GHRH treated cells was lower than basal secretion by control cells not exposed to GHRH (Figure 1). This observed inhibitory effect of IGF-I is specific, as other growth factors including epidemal growth factor (EGF) and fibroblast growth factor (FGF), had no effect on GH secretion. The fractional release of GH (medium GH/total GH) was decreased in the presence of IGF-I. Prolactin secretion was not altered in the same cells, indicating that the effects of IGF-I on GH are selective.

When GH mRNA levels were measured in cells treated with IGF-I, the peptide suppressed GH mRNA levels by 65%. Significant suppression of GHRH-stimulated GH mRNA levels was achieved with 0.13 nM IGF-I (Figure 2).

T3, another inducer of GH gene expression, was also used to examine the effects of IGF-I on stimulated GH expression (65). T3 (0.25 nM) stimulated GH secretion 7-fold during 72 h in hypothyroid rat pituitary cultures. IGF-I suppressed T3-induced GH secretion, with an ED50 of 0.8 nM at 72 h. IGF-I also attenuated T3-induced GH mRNA levels without altering total pituitary RNA content.

Effect of IGF-I on rat GH gene transcription

To determine whether the effects of IGF-I occurred at the level of GH gene transcription, nuclear run-off assays were performed. Primary pituitary monolayer cultures were treated with IGF-I, cell nuclei were harvested and incubated with [^{32}P] GTP in an in vitro transcription reaction (70). IGF-I (3.25 nM) maximally attenuated nascent GH mRNA synthesis by 70% at 4 h with GH gene transcription remaining suppressed by about 50% at 24 hr. Maximal inhibition was produced with 6.5 nM IGF-I. Prolactin mRNA synthesis was unaffected by IGF-I treatment.

These observations indicate the presence of a negative feedback loop for IGF-I acting directly on the somatotroph in the regulation of GH gene expression. This action occurs at the transcriptional level and also involves direct inhibition of GH secretion. GHRH, T3, hydrocortisone and IGF-I therefore all appear to participate in the modulation of GH gene expression at multiple sites within the rat somatotroph.

Effects of IGF-I on human GH mRNA levels

Dispersed human pituitary adenoma cells exposed to IGF-I showed 65% inhibition of GH secretion (71). IGF-I also blocked the stimulation of GH secretion produced by 1 nM GHRH by 50%. In addition, IGF-I attenuated both basal and GHRH stimulated GH mRNA levels in adenoma cells treated for 48 h. The monoclonal IGF-I receptor antibody IR3, blocked the suppressive effect of IGF-I on GH secretion and mRNA levels. IR3 did not alter basal GH secretion.

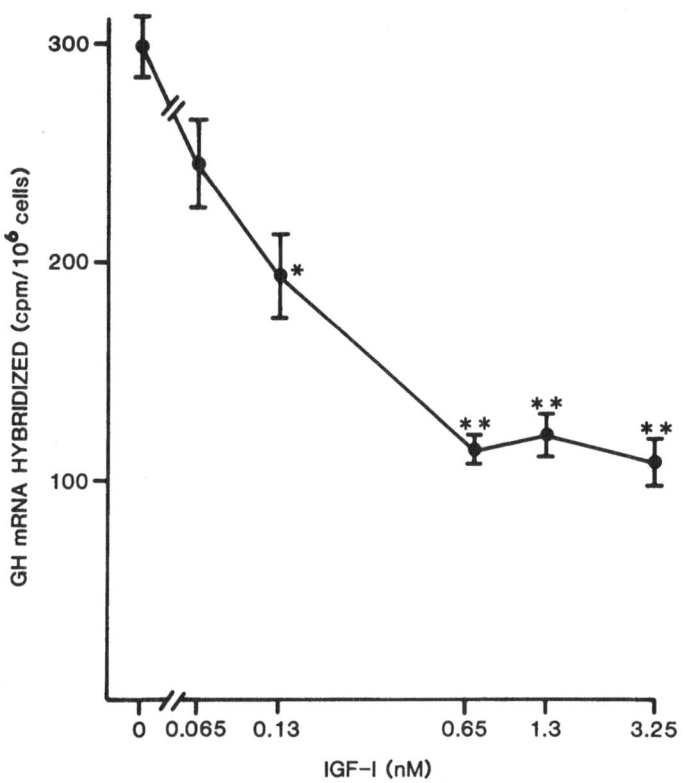

Fig. 2. Dose response of IGF-I on GH mRNA levels. After 72 h of treatment with the indicated doses of IGF-I, cells were taken up, and total RNA was extracted. RNA was immobilized on nitrocellulose and hybridized with rat GH [^{32}P] cDNA. Hybridization signal was quantified by counting radioactivity in each spot. Background counts from non-RNA-containing nitrocellulose spots were less than 60 cpm and were subtracted from the indicated counts per min. Hybridization by RNAse-treated cell extracts was negligible. Each dot blot received 5 ug denatured RNA dissolved in 30 ul sodium-citrate sodium chloride (15x) and formaldehyde (20 ul). Each point represents the mean $\pm$ SEM of triplicate blots. *, P > 0.05;**, P > 0.001. Adapted from Yamashita S, Melmed S. Endocrinology 118:176, 1986.

Regulation of hGH gene expression by IGF-I in transfected cells

To further delineate the molecular mechanism of IGF-I action on hGH gene expression, a human GH gene was cloned (72) and transfected into human choriocarcinoma cells (JEG3) or human HeLa cells. Both these cell lines contain abundant IGF-I receptors (72,73). Unfortunately, cloned rat pituitary tumor cells are relatively insensitive to IGF-I which precluded their use as recipients for the transfection studies. The 2.6 kb hGH gene containing 497 bp of 5' flanking DNA was transfected by CaPO$_4$ co-precipitation. 8-Bromo-cAMP and hydrocortisone stimulated basal GH expression in transfected JEG3 cells by 340%. IGF, 6.5 nM, suppressed stimulated but not basal GH secretion by 50% (72).

Immunoprecipitation of de-novo [35S] methionine labeled GH in
transfected cells yielded a 22 kDa protein band, similar in size to
authentic human GH (72). The specificity of the hGH band was confirmed
in several ways. Non-transfected cells did not incorporate [35S]
methionine into a 22-kDa protein. Excess cold hGH added to the
immunoprecipitation reaction prevented the appearance of the 22 kDa
band. The 125I-labeled hGH standard (NIDDK) co-migrated with the 22
kDa band. Non-immune rabbit serum and anti-bovine albumin serum failed
to precipitate the 22 kDa protein. Treatment of transfected cells with
8-bromo-cAMP and HCT for 72 h stimulated new GH synthesis. IGF-I (6.5
nM) partially suppressed this de-novo GH synthesis (72). The
transfected cells expressed both a 2.2-kb hGH mRNA precursor and the
1.0 kb hGH mature species. IGF-I treatment attenuated both these mRNA
bands (72).

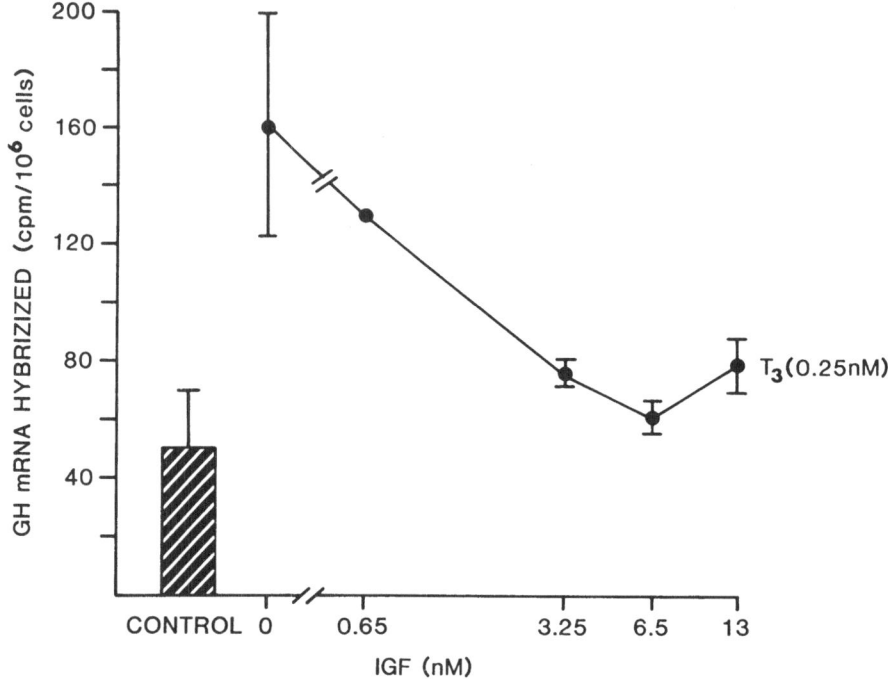

Fig. 3. Effects of T3 and IGF-I on GH mRNA sequences. After
hybridization with [32P] cDNA for rGH, nitrocellulose spots of
immobilized RNA were counted. Each IGF-I treatment point represents
triplicate blots obtained from 5 x 10^5 cells in three to six replicate
wells. Data from three experiments were pooled and expressed as a
percentage of the control value (no added IGF-I or T3)*, P < 0.05; **,
P < 0.001. Adapted from Melmed S, Yamashita S. Endocrinology
118:1483, 1986.

In view of the effects of IGF-I upon GH gene transcription, we
postulate that the 2.6 kb hGH gene contains regulatory elements
responsive to cAMP, hydrocortisone as well as IGF-I in addition to the
recently characterized cis-active tissue specific DNA sequences (74).

To further study the mechanism of IGF-I action we performed 5'
deletions using the hGH promoter region on a human GH reporter gene in
order to identify putative IGF-I responsive DNA sequences. Loss of a
dominant intronic GRE, (75,76) prevented us from utilizing a
heterologous reporter to further study the IGF-I effect. Sequences
extending to -292 or -182 bp of 5' hGH maintained IGF-I responsiveness
during 72 h of incubation (73). Addition of IR3 to the culture
medium 1 h prior to IGF-I treatment blocked the suppressive action of
IGF-I on hGH gene expression. Deletion and reconstitution of the
-292/-270 promoter region with polylinkers did not alter IGF-I
responsiveness. However, further 5' deletion to -132 bp caused a loss
of response to IGF-I. Therefore sequences within 182 bp from the
transcription start site appear to be important for the IGF-I response
of the hGH gene. The hGH transcripts in this transient expression
system were correctly initiated (77). pSV2CAT was co-transfected and
served as an internal control for this transfection study. The
presence of putative suppressor sequences which function in an
enhancer-like fashion was sought utilizing -80/-497 bp of 5' hGH linked
in the opposite orientation to SV40CAT. These sequences may function
as a silencer element in cis but do not appear to be enhancer-like.
Thus the '50 mer sequence (-182/-132) appears to contain at least one
IGF-I responsive region. Studies are in progress to further
characterize the hGH promoter region for IGF-I responsiveness. It is
presently unclear whether the IGF-I/receptor complex, internalized IGF-
I or an IGF-I dependent intracellular transduced signal is responsible
for the observed nuclear effects.

Pituitary IGF-I Gene Expression

As mentioned above, IGF-I circulates in plasma, yet is also
produced in many tissues and by a variety of cell types (29-31). The
relative roles of endocrine versus paracrine or autocrine derived IGF-I
to the bioactivity of the peptide are presently unclear. Adding to the
complexity of this question is the fact that IGF-I binding proteins,
which may interfere with IGF action, are present in the circulation,
and are also produced locally in various tissues. There is good in-
vitro evidence supporting the autocrine mode of action of IGF-I upon
cell proliferation. A monoclonal antibody which binds IGF-I and blocks
its action inhibits DNA synthesis in porcine aortic smooth muscle cells
and fibroblasts (78). The negative feedback of IGF-I upon GH gene
expression could theoretically be exerted, therefore, via blood-borne
delivery of the peptide (endocrine) and/or through the action of IGF-I
produced in the pituitary itself (paracrine/autocrine). The pituitary
was first reported by Binoux to be a source of insulin-like growth
factors in tissue culture explants (79). GH$_3$ cells also express the
IGF-I gene, as demonstrated by the presence of IGF-I mRNA transcripts,
immunohistochemical staining and accumulation of IGF-I immunoreactivity
in vitro (Figure 5). Normal rat pituitary tissue also contains IGF-I
mRNA transcripts (80). However, in order to postulate a role for
locally derived IGF-I in the feedback regulation of GH gene expression,
pituitary IGF-I should be in turn subject to regulation by GH. Several
lines of evidence do in fact support this concept. GH$_3$ cells grown in
thyroid hormone-depleted medium to decrease GH gene expression also
show inhibition of IGF-I mRNA content. After treatment of hypothyroid
pituitary cells with either T3 and/or GH, IGF-I mRNA content and IGF-I
immunoreactivity in the medium is induced in a time-and dose-dependent
fashion (81). These data indicate that IGF-I gene expression in rat
pituitary tumor cells is regulated by GH itself, and that the

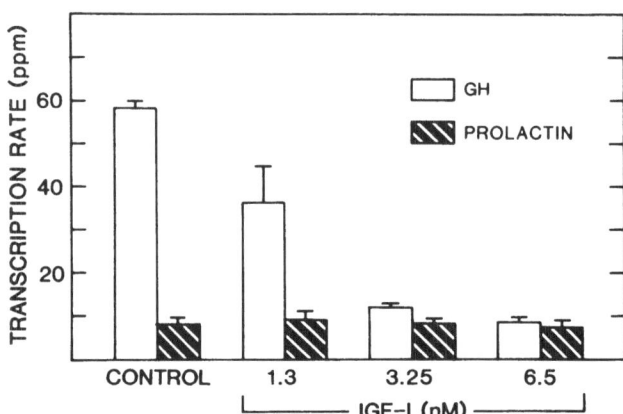

Fig. 4. Dose response of IGF-I effect on GH and prolactin gene transcription. Primary cultures of rat anterior pituitary cells were incubated in Dulbecco's modified Eagle's medium containing fetal calf serum (2.5%). After 3 d, cells were washed and medium replaced with serum-free defined medium with or without added IGF-I for 1-24 h. After the IGF-I treatment, nuclei were isolated and incubated for 45 min in an in vitro transcriptional runoff assay with ^{32}P-labeled GTP as described. New GH mRNA and prolactin mRNA were measured by hybridization against immobilized GH cDNA and prolactin cDNA (2 ug/dot), respectively. Each hybridization reaction also included a companion immobilized pBR322 DNA blot. GH- and prolactin-specific hybridization are present as parts per million of total input [^{32}P]RNA (~ 5 x 10^6 cppm) and the corrections were made for the hybridization efficiency measured by hybridization of ^{3}H-labeled cRNA. Bars represent mean and range of duplicate dots from a representative experiment. Adapted from Yamashita S, Melmed S. J Clin Invest 79:449, 1987.

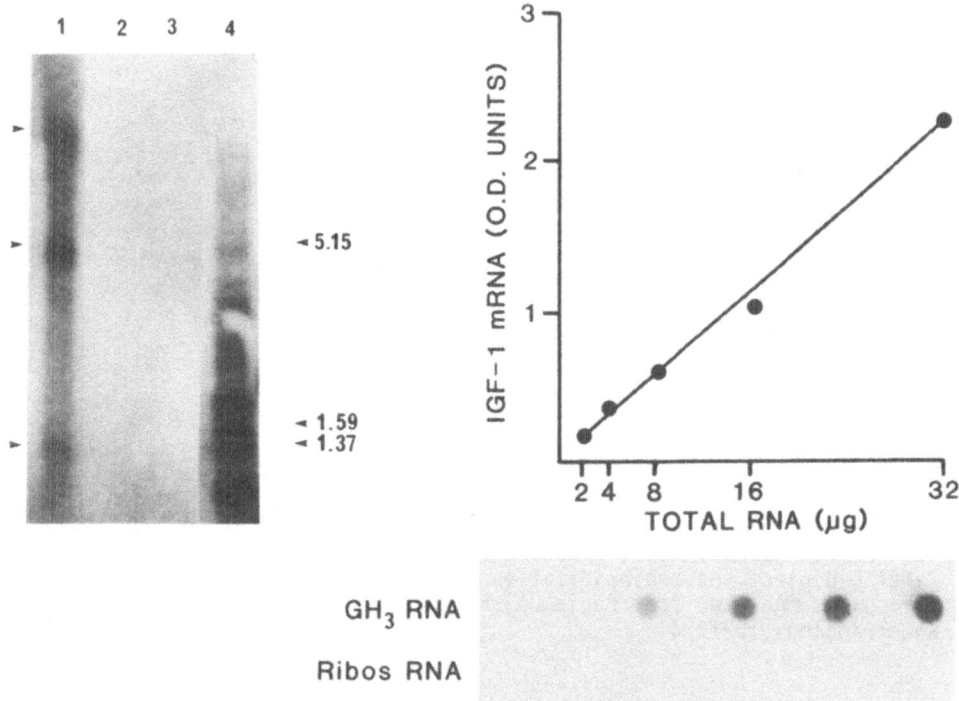

Fig. 5. Total RNA dot blots of GH₃ cells showed a linear hybridization
signal, as quantified by optical densitometry of autoradiographic dots
with [^{32}P]IGF-I cDNA. Ribosomal RNA did not hybridize with the IGF
cDNA probe (right panel). Northern analysis of poly(A) RNA of GH₃
(lane 1), 3T3 fibroblasts (lane 2), and JEG-3 cells (lane 3) is shown
in the left panel. Fifteen micrograms of poly (A) RNA extracts of each
cell type were electrophoresed in a 1% agarose gel, transferred to
nitrocellulose paper, and hybridized with [^{32}P]IGF-I cDNA (5 x 10^6
cpm). ^{32}P end-labeled Hind III and Hind III-EcoRI   DNA digests (lane
4) were used as markers. Arrows indicate mRNA transcript sizes of
approximately 7.7, 5.3, and 1.3 kb. Adapted from Fagin JA et al, 1987,
Endocrinology 120:2037.

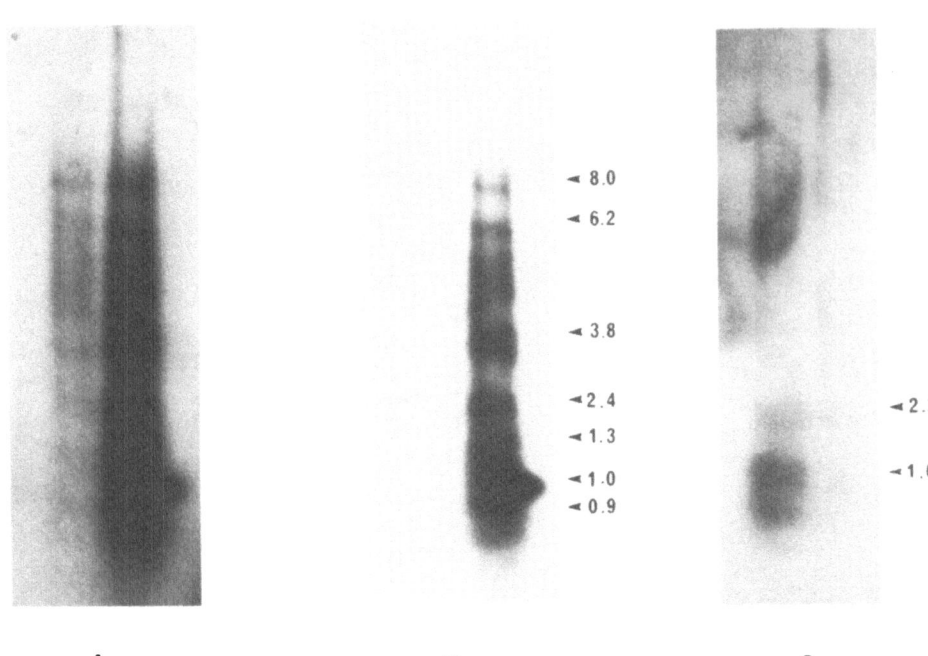

Fig. 6. Results of Northern gel electrophoresis of 35 ug poly (A) RNA derived from pituitaries pooled from control (C) and tumor-bearing animals (T). Panel A shows hybridization with the IGF-I cDNA probe. Panel B represent a shorter autoradiographic exposure of the same blot. Panel C depicts the rehybridization of the blot to rat GH cDNA after stripping. Adapted from Fagin JA, Brown A, Melmed S, 1988, Endocrinology 122:2204.

expression of these two genes may be mutually interacting within the same cell or neighboring cells. Pituitary tissue IGF-I mRNA content was also induced in vivo in thyroidectomized rats by treatment for 48 h with T3 or GH (Fagin JA, Melmed S, unpublished data). Further evidence supporting GH-dependency of pituitary IGF-I gene expression was shown in rats harboring ectopic somatotrophic tumors (80). Pituitary IGF-I mRNA transcripts were induced in a time-dependent fashion in response to increasing serum GH levels in rats with ectopic GH-producing tumors. Interestingly, eutopic pituitary GH mRNA levels decreased in the tumor-bearing rats (Fig 6). These in vivo data suggest that pituitary IGF-I gene expression may be regulated by circulating GH.

Since both pituitary and serum IGF-I content are regulated coordinately, it is not possible in vivo to discriminate the relative contributions of local vs systemically available IGF-I to the negative feedback action upon GH biosynthesis. We have attempted to address this question in primary cultures of rat pituitary cells. If pituitary cell IGF-I production is involved in negative feedback regulation of GH, factors capable of blocking IGF-I action in vitro would lead to increased GH secretion. Addition of monoclonal antibody SmC 1.2 (78) to serum free medium was associated with marginal changes in GH secretion into the conditioned medium. Conversely, a purified human IGF binding protein which inhibits IGF bioactivity in other systems (82), did induce GH secretion in primary pituitary cells in a time and dose-dependent fashion (Fagin JA, Melmed S, unpublished observations). The reason for these divergent findings is not clear. The concentrations of immunoreactive IGF-I in the culture medium of primary pituitary cultures is low (0.5 to 2.5 ng/ml), and would be expected to only minimally alter GH gene expression. This may be significant in that blocking the bioactivity of these low concentrations of IGF-I is unlikely to yield clear-cut changes in GH secretion. Therefore there is presently no good in vitro or in vivo model to fully document and conclusively prove the existence of a paracrine/autocrine feedback loop between IGF-I and GH within the pituitary. However, the fact that pituitary IGF-I expression is GH-dependent both in vivo and in vitro suggests that paracrine interactions may indeed be important in the mutual feedback regulation of these two peptides.

REFERENCES

1. Miller WL, Eberhardt NL 1983 Structure and evolution of the growth hormone gene family. Endocrinol Rev 4:97.
2. Lewis VJ, Dunn JT, Bonewald LF, Seavey BK, Vanderlaan WP 1978 A naturally occuring structural variant of human growth hormone. J Biol Chem 253:2687.
3. Lewis VJ, Bonewald LF, Lewis VJ 1980 The 20,000-dalton variant of human growth hormone location of the amino acid deletions. Biochem Biophys Res Commun 92:511.
4. Seeburg PH 1982 The human growth hormone gene family. Nucleotide sequences show recent divergence and predict a new polypeptide hormone DNA 1:234.
5. Frankenne F, Rentier-Delrve F, Scippo ML, Martial J, Hennen G 1987 Expression of the growth hormone variant gene in human placenta. J Clin Endocrinol Metab 64:635.
6. Thorner MO, Vance ML 1988 Growth hormone. J Clin Invest 82:745.
7. Barinaga M, Yamamoto G, Rivier C et al 1983 Transcriptional regulation of growth hormone gene expression by growth hormone releasing factor. Nature 306:84.
8. Barinaga M, Bilezikjian LM, Vale MW, Rosenfeld MG, Evans RM 1985 Independent effects of growth hormone releasing factor on growth hormone release and gene transcription. Nature 314:279.
9. Thorner MO, Peryman RL, Cronin MJ, et al 1982 Somatotroph hyperplasia: Successful treatment of acromegaly by removal of a pancreatic islet tumor secreting a growth hormone-releasing factor. J Clin Invest 70:965.
10. Guilllemin RR, Brazeau P, Bohlen P et al 1982 Growth hormone-releasing factor from a human pancreatic tumor that caused acromegaly. Science 218:585.
11. Ling N, Zeytin F, Bohlen P et al 1985 Growth hormone releasing factors. Annu Rev Biochem 54:403.
12. Frohman LA, Jansson JO 1986 Growth hormone-releasing hormone. Endocr Rev 7:223.

13. Brazeau P, Vale W, Burgus R et al 1973 Hypothalamic polypeptide that inhibits the secretion of immunoreactive pituitary growth hormone. Science 179:77-79.
14. Takemoto K, Rokaeus A, Joruvav H, Mc Donald TJ, Mutt V 1983 Galanin-a novel biologically active peptide from porcine intestine. FEBS Lett 164:124.
15. Ch'ng JLC, Christofides ND, Anand P, Gibson SJ, Allen YS, Su HC, Takemoto K 1985 Distribution of galanin immunoreactivity in the central nervous system and the responses of galanin-containing pathways to injury. Neuroscience 16:343.
16. Murakami Y, Kto Y, Koshiyama H, Inove T, Yanaihara N, Imura H 1987 Galanin stimulates growth hormone (GH) secretion via GH-releasing factor (GRF) in conscious rats. Evr J Pharmacol 136:415.
17. Ottlecz A, Samson WK, McCann SM 1986 Galanin: evidence for a hypothalamic site of action to release growth hormone. Peptides 7:51.
18. Baver FE, Ginsburg L, Venetikov M, MacKay DJ, Burrin JM, Bloom SR 1986 Growth hormone release in man induced by galanin, a new hypothalamic peptide. Lancet 2:192.
19. Yaffe BM, Samuels HH 1984 Hormonal regulation of the growth hormone gene: Relationship of the rate of transcription to the level of nuclear thyroid hormone-receptor complexes. J Biol Chem 259:6284.
20. Evans RM, Birnberg NC, Rosenfeld MG 1982 Glucocorticoid and thyroid hormones transcriptionally regulate growth hormone gene expression. Proc Natl Acad Sci USA 79:7659.
21. Yamashita S, Melmed S 1986 Insulin regulation of rat GH gene transcription. J Clin Invest 78:1008.
22. Wehrenberg WB, Baird A, Ling N 1983 Potent interaction between glucocorticoids and growth hormone-releasing factor in vivo. Science 221:556.
23. Clemmons DR, Van Wyk JJ 1984 Factors controlling blood concentrations of somatomedin-C. Clin Endocrinol Metab 13:113.
24. Isgaard J, Moller C, Isaksson OGP, Nilsson F, Mathews LS, Norstedt G 1988 Regulation of insulin-like growth factor messenger ribonucleic acid in rat growth plate by growth hormone. Endocrinology 122:1515.
25. Rinderknecht E, Humber RE 1978 The amino acid sequence of human insulin-like growth factor I and its structural homology with proinsulin. J Biol Chem 253:2769.
26. Klapper DG, Svoboda ME, Van Wyk J 1983 Sequence analysis of somatomedin-C: Confirmation of identity with insulin-like growth factor I. Endocrinology 112:2215.
27. Ullrich A, Berman CH, Dulet J, Gray A, Lee JM 1984 Isolation of the human insulin-like growth factor I gene using a single synthetic DNA probe. EMBOJ 3:361.
28. Tricoli JV, Rall RB, Scott J, Bell GI, Shows TB 1984 Localization of insulin-like growth factor genes to human chromosomes 11 and 12. Nature 310:784.
29. D'Ercole AJ, Stiles AD, Underwood LE 1984 Tissue concentrations of somatomedin-C: Further evidence for multiple sites of synthesis and paracrine/autocrine mechanisms of actin. Proc Natl Acad Sci USA 81:935
30. Lund PK, Moats-Staats BM, Hynes MA, D'Ercole AJ, Jansen M, Van Wyk JJ 1986 Somatomedin-C/IGF-I and IGF-II mRNAs in rat fetal and adult tissues. J Biol Chem 261:14539.
31. Murphy LJ, Bell GI, Friesen HG 1987 Tissue distribution of insulin-like growth factor I and II messenger ribonucleic acid in the adult rat. Endocrinology 120:1279.
32. Fagin JA, Pixley S, Slanina S et al 1987 Insulin-like growth factor I gene expression in GH$_3$ rat pituitary cells: mRNA content, immunocytochemistry and secretion. Endocrinology 120:2037.
33. Elgin RG, Busby WH, Clemmon DR 1987 An insulin-like growth factor

(IGF) binding protein enhances the biologic response to IGF-I. Proc Natl Acad Sci USA 84:3254.

34. Flier JS, Usher P, Moses AC 1986 Monoclonal antibody to the Type I insulin-like growth factor (IGF-I) receptor blocks IGF-I receptor-mediated DNA synthesis: Clarification of the mitogenic mechanism of IGF-I and insulin in human skin fibroblasts. Proc Natl Acad Sci USA 83:664.

35. Conover CA, Liu F, Powell D, Rosenfeld RG, Hintz RL 1989 Insulin-like growth factor binding proteins from cultured human fibroblasts: Characterization and hormonal regulation. J Clin Invest 83:852.

36. Berelowitz M, Szabo M, Frohman LA et al 1981 Somatomedin-C mediates growth hormone negative feedback by effects on both the hypothalamus and the pituitary. Science 212:1279.

37. Ono M, Miki N, Miyoshi H 1985 Decreased content and release of hypothalamic GH-releasing factor in rats bearing GH producing GH_3 tumors. 67th Ann Mtg of the Endo Soc, Baltimore, MD p 126 (Abstract).

38. Rosenfeld RG, Ceda G, Wilson DM et al 1984 Characterization of high affinity receptors for insulin-like growth factors I and II on rat anterior pituitary cells. Endocrinology 114:1571.

39. Goodyer CG, Stephano LD, Lai WH et al 1984 Characterization of insulin-like growth factor receptors in rat anterior pituitary hypothalamus and brain. Endocrinology 114:1187.

40. Rosenfeld RG, Ceda G, Cutler CW et al 1985 Insulin and insulin-like growth factor receptors on cloned rat pituitary tumor cells. Endocrinology 117:2008.

41. Nagomi H, Suzuki K, Enomoto H, Ishikawa H 1989 Studies on the development of growth hormone and prolactin cells in the rat pituitary gland by in situ hybridization. Cell Tissue Res 255:23.

42. Dupouy JP, Coffingny LL, Marge S 1975 Maternal and fetal corticosterone levels during late pregnancy in rats. J Endocrinol 65:347.

43. Hemming FJ, Begeot M, Dubois MP, Dubois PM 1984 Fetal rat somatotrophs in vitro: Effects of insulin, cortisol and growth hormone-releasing factor on their differentiation: A light and electron microscopic study. Endocrinology 114:2107.

44. Silverman Bl, Bettendorf M, Kaplan SL, Grumbach MM, Miller WL 1989 Regulation of growth hormone (GH) secretion by GH-releasing factor, somatostatin and insulin-like growth factor I in ovine fetal and neonatal pituitary cells in vitro. Endocrinology 124:84.

45. de Zegher F, Bettendorf M, Kaplan SL, Grumbach MM 1988 Hormone ontogeny in the ovine fetus. XXI The effect of insulin-like growth factor I on fetal plasma growth hormone, insulin and glucose concentrations. Endocrinology 122:658.

46. Goodyer CG, Marcovitz S, Hardy J, Lefevre Y, Guyda HJ, Posner BI 1986 Effect of insulin-like growth factors on humal fetal, adult normal and tumor pituitary function in tissue culture. Acta Endocrinol (Copenh) 112:49.

47. Macleod RM, De Witt GW, Smith MC 1966 Hormonal properties of transplanted pituitary tumors and their relation to the pituitary gland. Endocrinology 79:1149.

48. Peake GT, Mariz IK, Daughaday WH 1968 Radioimmunoassay of growth hormone in rats bearing somatotropin producing tumors. Endocrinology 83:714.

49. Sawano S, Arimura A, Bowers CY et al 1967 Effect of CNS depressants, dexamethasone and growth hormone on the response of growth hormone-releasing factor. Endocrinology 81:1410.

50. Katz SH, Molitch ME, McCann SM 1969 Effects of hypothalamic implants of GH on anterior pituitary weight and GH concentration. Endocrinology 85:725.

51. Tannenbaum GS 1980 Evidence for autoregulation of GH secretion via the central nervous system. Endocrinology 107:2117.
52. Voogt JL, Clemens JA, Negro-Villar et al 1971 Pituitary GH and hypothalamic GRF after median eminence implantation of ovine or human GH. Endocrinology 88:1363.
53. Sakuma M, Knobil E 1970 Inhibition of endogenous growth hormone secretion by exogenous growth hormone infusion in the rhesus monkey. Endocrinology 86:8904.
54. Abrams RL, Grumbach MM, Kaplan SL 1971 The effect of human groth hormone on plasma growth hormone, cortisol, glucose and free fatty acid response to insulin: Evidence for growth hormone autoregulation in man. J Clin Invest 50:940.
55. Mendelson WB, Jacobs LS, Gillin JC 1983 Negative feedback suppression of sleep-related growth hormone secretion. J Clin Endocrinol Metab 56:486.
56. Sheppard MC, Kronheim S, Pimstone BL 1978 Stimulation by growth hormone of somatostatin release from the rat hypothalamus in vitro. Clin Endocrinol (Oxf) 9:583.
57. Hoffman DL, Baker BL 1977 Effects of treatment with growth horone on somatostatin in the median eminence of hypophysectomized rats. Proc Soc Exp Biol Med 156:265.
58. Patel YC 1979 Growth hormone stimulates hypothalamic somatostatin. Life Sci 24:1589.
59. Molitch ME, Hlivyak LE 1980 Growth hormone short-loop feedback: Anatomic specificity of growth hormone stimulation of hypothalamic somatostatin concentration. Horm Metab Res 12:559.
60. Berelowitz M, Firestone SI, Frohman LA 1981 Effects of growth hormone excess and deficiency on hypothalamic somatostatin content and release and on tissue somatostatin distribution. Endocrinology 109:714.
61. Chihara K, Minamitani N, Kaji H et al 1981 Intraventricularly injected growth hormone stimulates somatostatin release into rat hypophysial portal blood. Endocrinology 109:2279.
62. Brazeau P, Guillemin R, Ling N et al 1982 Inhibition by somatomedins of growth hormone secretion stimulated by hypothalamic growth hormone releasing factor (somatocrinin, GRF) on the synthetic peptide hpGRF. C R Acad Sci [D] (Paris) 295:651.
63. Goodyer CC, De Stephano L, Guyda HJ et al 1984 Effects of insulin-like growth factor on adult male rat pituitary function in tissue culture. Endocrinology 115:1568.
64. Yamashita S, Melmed S 1986 Insulin-like growth factor I action on rat anterior pituitary cells: Suppression of growth hormone secretion and messenger ribonucleic acid levels. Endocrinology 118:176.
65. Melmed S, Yamashita S 1986 Insulin-like growth factor I action on hypothyroid rat anterior pituitary cells: Suppression of 3, 5, 3'-triiodothyronine-induced growth hormone secretion and messenger ribonucleic acid levels. Endocrinology 118:1483.
66. Goodyer CG, Marcovitz S, Hardy J et al 1986 Effect of insulin-like growth factors on human foetal, adult normal and tumour pituitary function in tissue culture. Acta Endocrinolo 112:49.
67. Yamashita S, Slanina S, Kado H et al 1986 Autoregulation of pituitary growth hormone messenger RNA levels in rats bearing transplantable mammosomatotrophic pituitary tumors. Endocrinology 118:915.
68. Abe H, Molitch Me, Van Wykk JJ et al 1983 Human growth hormone and somatomedin-C suppresses the spontaneous release of growth hormone in unanesthetized rats. Endocrinology 113:1319.
69. Tannenbaum GS, Guyda HJ, Posner BI 1983 Insulin-like growth factors: A role in growth hormone negative feedback and body weight regulatin via brain. Science 220:77.
70. Yamashita S, Melmed S 1987 Insulin-like growth factor I regulation of

growth hormone gene transcription in primary rat pituitary cells. J Clin Invest 79:449.

71. Yamashita S, Weiss M, Melmed S 1986 Insulin-like growth factor I regulates growth hormone secretion and messenger ribonucleic acid levels in human pituitary tumor cells. JCEM 63(3)730.

72. Yamashita S, Ong J, Melmed S 1987 Regulation of human growth hormone gene expression by insulin-like growth factor I in transfected cells. J Biol Chem 262:13254.

73. Prager D, Gebremedhin S, Melmed S 1989 Insulin-like growth factor I responsive DNA elements on the human growth hormone gene promoter. (submitted)

74. Lefevre C, Imagawa M, Dana S, Grindlay J, Bodner M, Karin M 1987 Tissue-specific expression of the human growth hormone gene is conferred in part by binding of a specific trans-acting factor. EMBO Journal 6:971.

75. Moore DD, Marks AR, Buckley DI, Kapler G, Payvar F, Goodman HM 1985 The first intron of the human growth hormone gene contains a binding site for glucocorticoid receptor. Proc Natl Acad Sci USA 82:699.

76. Slater EP, Rabenau O, Karin M, Baxter JJ, Beatro M 1985 Glucocorticoid receptor binding and activation of a heterologous promoter by dexamethasone by the first intron of the human growth hormone gene. Mol Cell Biol 5:2984.

77. Prager D, Melmed S 1988 Insulin regulates expression of the human growth hormone gene in transfected cells. J Biol Chem 263(32):16580.

78. Clemmons DR, Van Wyk JJ 1985 Evidence for a functional role of endogenously produced somatomedin-like peptides in the regulation of DNA synthesis in cultured human fibroblasts and porcine aortic smooth muscle cells. J Clin Invest 75:1914.

79. Binoux M, Hossenlopp P, Lassarre C, Hardovin N 1981 Production of insulin like growth factors and their carrier by rat pituitary gland and brain explants in culture. FEBS Letts 124:178.

80. Fagin JA, Brown A, Melmed S 1988 Regulation of pituitary insulin like growth factor I messenger ribonucleic acid levels in rats harboring somatommamotropic tumors: Implications for growth hormone autoregulation. Endocrinology 122:2204.

81. Fagin JA, Fernandez Mejia C, Melmed S 1989 Pituitary insulin like growth factor I gene expression: regulation by triiodothyronine and growth hormone. Endocrinology (in press).

82. Herington AC, Kuffer AD 1981 Identification of a specific inhibitor of non-suppressible insulin-like activity in a partially purified human serum fraction. Endocrinology 109:1634.

GROWTH HORMONE AND IGF ABNORMALITIES OF THE AFRICAN PYGMY

Thomas J. Merimee *

Department of Medicine
University of Florida
Gainesville, FL

INTRODUCTION

Since the elucidation of the structures of insulin-like growth factors I and II (IGF I and IGF II) by Humbel[1,2], and the development of specific radioimmunoassays for each by Zapf and Froesch[3], a number of clinical conditions have been identified that are associated with defects in either the secretion or action of these agents[4,5]. After 25 years of investigating stature of the African pygmy, data have now been obtained indicating a unique genetic variation in the GH-IGF I axis in these subjects. These data will be presented in this article.

Pygmies themselves are unique: living in the central part of Africa south of the Sahara, they now are one of the few surviving groups that live much as they did in Paleolithic times. In Zaire live the Mbuti of the Ituri forest, considered the smallest of them all, and more to the south, the Kivu Twa. The Binga form a western group. The existence of these groups in Africa can be dated to the earliest known historical records. Monuments in early Egypt depict pygmies, and Greek writers including Homer (900 B.C.) and Strabo (24 B.C.) described accurately their geographical location[6].

The cause of short stature in the pygmy was long the subject of speculation prior to our work[7-10], but none of these earlier studies provided definitive hormonal and genetic investigations.

EARLY HORMONAL STUDIES

In our earliest studies of short stature in the African Pygmy, we established that African pygmies normally secrete growth hormone (GH) after provocative stimuli but do not respond to the administration of GH with normal changes in nitrogen, calcium, phosphorus and carbohydrate metabolism[11-14]. When hypopituitary subjects were treated with GH, we demonstrated hypercalsuria, a decrease in serum urea nitrogen and enhanced insulin secretion after glucose[13]. These responses were never seen in pygmy subjects. The change of plasma free fatty acids in pygmies after infusing growth hormone was particularly informative[13]. The mean fasting concentration of plasma free fatty acids (FFA) in Pygmies (0.722 ± 0.10 mM/liter) did not differ significantly from that of controls (0.719 ± 0.10 mM/liter). After the intravenous infusion of 4 mg hGH over a 20-minute period, the mean plasma FFA concentration of the controls rose to 150%

* Send reprint requests to: Thomas J. Merimee, M.D., Professor of Medicine, University of Florida, Department of Medicine, Division of Endocrinology and Metabolism, Box J-226, JHMHC, Gainesville, Florida 32610-0226.

and that of hGH-deficient dwarfs to 178% of basal values after 120 minutes. The mean plasma FFA concentration of the Pygmies, however, rose to only 111% of the basal value after GH, a change that was not significantly different from that observed after saline infusion. This reduced lipolytic effect, unlike the absence of GH's other effects, could only represent direct resistance to growth hormone, per se, ñot a deficiency of IGF I. Growth hormone *in vitro* and *in vivo* increases FFA release from adipose tissue, whereas IGF I has the opposite effect[17,18]. The importance of this observation was only appreciated at a later date.

In 1982 we were able to establish that serum concentrations of IGF I were decreased in adult African pygmies, whereas their IGF II concentrations in serum were normal[15]. We were able to show also that the IGF I concentration in serum did not increase after growth hormone treatment, while such increases were readily demonstrated in other non-pygmy subjects[15,16]. In 1987, we reported data that added substantially to earlier findings, demonstrating that the major

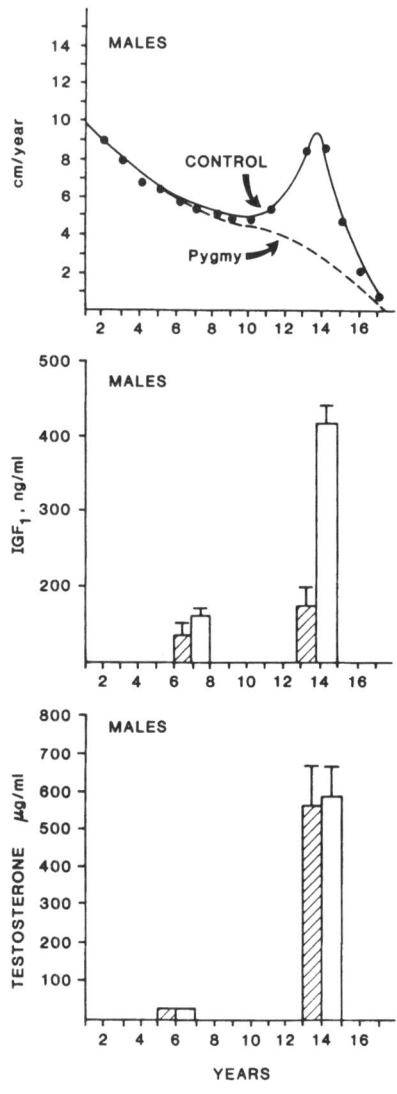

Figure 1.   Growth in cm is shown for controls and pygmies in the upper panel. For the serum concentrations of IGF I (2nd panel) and testosterone (3rd panel) the open bars indicate controls; hatched bars, pygmies. SEMs are given.

74

defect of IGF I secretion in the pygmy occurred primarily during puberty[19]. Our study that conclusively proved this point was catalyzed by two reports which deserve special recognition:

1) Several pygmy children were raised and cared for in Italy well over a century ago in the household of Count Minischalchi-Erizzo. Giglioli, an Italian scientist, summarized the frequent linear measurements made of these pygmies in 1880, and commented specifically that they failed to show a growth spurt during puberty[20]. Cavalli-Sforza later collated these findings, plotted growth curves derived from the initial data and reached precisely the same conclusion[21].

2) In a large series comparing heights of children and adults in 38 ethnic groups, it was found that the shortest children at the age of 3 years were not pygmies, but Mayans (81.1 cm). Pygmy children (89.5 cm) ranked only 18th, yet pygmy adults ranked last in linear height. No comparable change in height status from childhood to adulthood characterizes any other known group[22,23].

The logical inference from the reports cited was that short stature of the pygmy resulted from a failure to accelerate growth during puberty, and that the increase of IGF I which normally occurs in puberty[24] (often to levels similar to those of acromegalics) would not occur. Both of these inferences proved to be true.

For subsequently studying pubertal growth and IGF I secretion, we used data from 600 pygmy subjects to calculate growth curves. This included 290 examined by our group and others previously examined by van de Koppel[25,26]. Subjects of each sex were ordered by age, and groups of 10-14 successive subjects were formed so that the groups would be as homogenous for age as possible. Median age and median stature were calculated for each group and the growth rates were computed from pairs of successive groups. The results in pygmies were compared with those obtained in a similar manner by Tanner, in a large combined study of the British population[27] and by van de Koppel and Hewlett in studies of native Africans (not pygmies)[26]. Although the growth rates reported in the Tanner study vary in different populations,

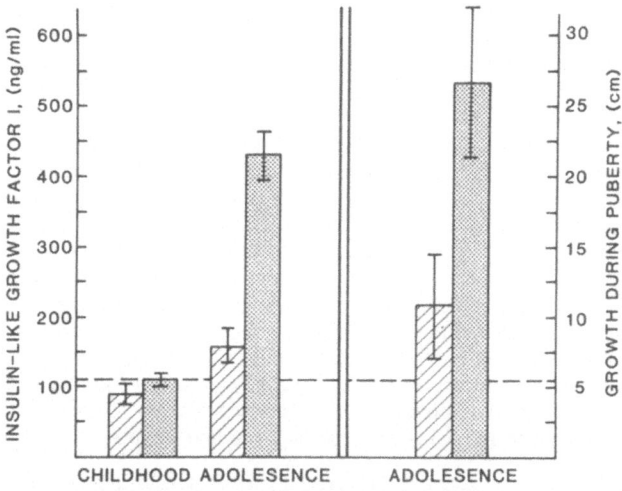

Figure 2.   *The mean serum concentration of IGF I is shown for pygmies and controls (hatched and stippled bars, respectively). The broken line indicates the lower limit of the normal IGF I range (left panel). Growth during puberty is calculated from three consecutive years, and the calculated growth increment with no increase of IGF I is indicated by the broken line (right panel).*

75

Table 1. *Increments of Linear Height and Mean Serum Concentration of IGF I in Normal and Pygmy Males.*

Measurement	Children (ages 3-6)			Adolescents (ages 12.5-15.5)		
	Control	Pygmy	Percent of control	Control	Pygmy	Percent of control
Total Increment* in Linear Ht.	36 ± 4.6	35.2 ± 5.0	97%	26.7 ± 5.6	11.0 ± 3.8	41%
		p = n.s.			$p < .01$	
Mean IGF I	108 ± 11.7	89 ± 14.7	82%	435 ± 37	154 ± 22.4	35%

* Total increment in linear ht. is the sum of 4 consecutive values.

the basic pattern of growth is highly consistent for a variety of ethnic groups. The acceleration of linear height in an American control group which we established closely approximated that reported by van de Koppel and by Tanner. The data showing the key hormonal findings in pygmies before and during puberty are illustrated in Figures 1 and 2. As shown in these figures, adolescent male pygmies did not exhibit a normal increase of IGF I during puberty and these pygmies also failed to exhibit a growth spurt. Control subjects had a substantial increase of IGF I at puberty associated with a period of accelerated growth. The serum IGF I concentration in normal subjects at puberty, as expected, exceeded significantly the mean concentration in normal adults. Pygmy girls (not illustrated) exhibited a similar pattern of growth and IGF I secretion. Testosterone concentrations in serum were similar in prepubertal male pygmies and controls and similarly increased in pubertal pygmies and controls; the same was true of estradiol levels in girls. It should be noted that we were able to show conclusively in earlier studies that neither diet nor environment could explain low serum concentrations of IGF I[19]. The significance of the absent pubertal growth spurt on determining final height is clearly indicated in Table 1.

The decreased IGF I secretion in pygmies, particularly as seen at puberty, could result from a defective promoter gene for the secretion of IGF I or from defective GH receptors. The IGF I gene has not been fully characterized, but this postulate seemed reasonable in view of what had been reported for the IGF II gene: Using two cDNAs, Sussenbach[28] showed that the human IGF II gene contains at least 7 exons. Two different IGF II promoters were identified 19 kilobases apart which were active in a development-specific manner. One promoter was active only during the fetal stage; a second promoter was active in the adult and was located 1.4 kilobases downstream from the insulin gene. Considering the strong homology of IGF I and IGF II structurally, and the similarities in their biological actions, one might predict that promoters exist for the IGF I gene, with one promoter active at puberty. Although studies of the IGF I gene are not complete, Cavalli-Sforza has completed preliminary structural studies of the IGF I gene and finds no difference between the IGF I gene of pygmies and controls.

A decrease of GH receptors or a defect in the number of functioning growth hormone receptors could explain poor secretion of IGF I as well as a decreased free fatty acid response of pygmies to intravenous growth hormone. However, such receptor defects could not be total: pygmies do not exhibit the extreme short stature, the consistent elevation of basal GH concentrations, and the extremely depressed serum concentrations of IGF I (usually to <40 ng/ml) that characterize Laron dwarfism, a condition in which all GH receptors are absent[27,28]. However, since a partial defect might exist, our recent work was directed to investigating this problem.

Table 2.   Effect of GH on IGF I Secretion In Vitro by Transformed B-Lymphocytes.

Subjects	Week 1 IGF I ng/10⁴ cells		Week 3 IGF I ng/10⁴ cells	
Control				
Unstimulated	22.7 ± 4.5		0.59 ± 0.29	
		p < .05		p < .005
GH stimulated	40.4 ± 5.5		*3.36 ± 0.52	
Pygmies				
Unstimulated	31.0 ± 5.8		0.35 ± 0.13	
		n.s.		n.s.
GH stimulated	29.5 ± 5.8		*0.52 ± 0.28	

* Indicates values differ from each other with a p < .005.

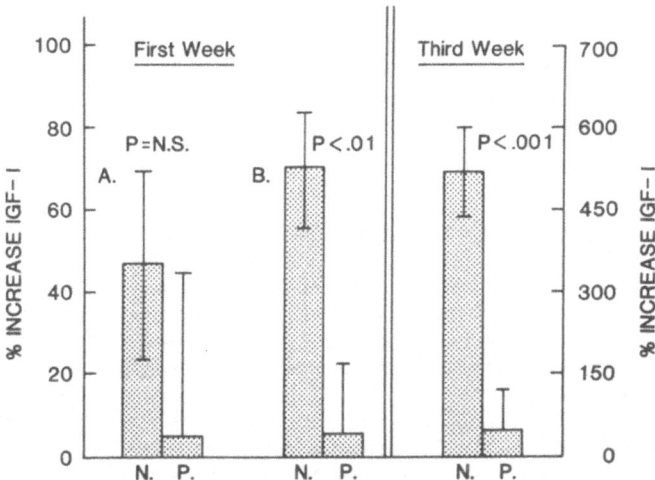

Figure 3.   The effect of growth hormone (150 ng/ml) on the percent increase of IGF I by B-lymphocytes in culture media is shown. N = normal statured controls, P = pygmies. At one week of culturing, A = comparison of all cultures, B = comparison of cultures with similar cell counts.

GH RECEPTORS, ISOLATED CELL RESPONSES TO GROWTH HORMONE

To prove or disprove a clinically significant receptor defect in the pygmy, two sets of studies were needed – the obvious one, i.e. to determine if pygmy cells had a decreased number of growth hormone receptors or an alteration in receptor affinity, and studies to determine if growth hormone stimulation of these cells resulted in less than normal IGF I secretion. Actual studies of the GH receptor or of the GH receptor gene, of course, would be even more definitive. Ideally, initial studies of the receptor on cells should have been performed with tissues known to secrete IGF I and react to GH[29,30], but it was not possible to obtain biopsies from pygmy subjects.

During the course of our studies we had obtained B-lymphocytes from pygmy subjects and immortalized them with EB virus; however, these lymphocytes did not bind a measurable amount of growth hormone, a finding similar to other investigators' experiences with B-lymphocytes from other sources[31]. A failure to demonstrate binding of GH to these cells did not necessarily mean an absence of receptors, but might simply reflect the inadequacy of the methodology. Since this investigative path was blocked, we established a collaborative study with Dr. G. Baumann to measure the amount of growth hormone in pygmy serum that was bound to a newly described growth hormone-carrier protein. This "carrier" for growth hormone is actually the GH-receptor which dissociates from the cell after growth hormone binds to it on the membrane and then circulates as a unit with GH[32-34]. In Laron dwarfs who totally lack GH receptors, no "bound" GH is found in serum or plasma, suggesting the usefulness of this assay in characterizing GH-receptor states[35].

In a blinded study design, plasma samples from 19 pygmies and 11 African controls were incubated with ^{125}I GH and free GH was separated from bound GH by Sephadex G100 chromatography[32]. GH associated with the binding protein, i.e. the circulating GH receptor, was $10.4 \pm 0.8\%$ (mean + standard error of the mean) in pygmy subjects compared to $18.9 \pm 1.6\%$ in the controls. Thus there was nearly a 50% decrease in circulating membrane derived growth hormone receptors in pygmies, consistent with a decrease of cellular GH receptors[35]. Studies of the secretion of IGF I in response to GH by cells isolated from pygmies were consistent with these findings of a decrease of the GH carrier protein and a probable decrease of GH cellular receptors.

To determine responsiveness of pygmy cells to growth hormone, 13 transformed B-lymphocyte lines (6 from pygmies, and 7 from controls) were seeded into separate 75-ml flasks containing 15 ml of RPM-I-1640 media per flask. Each flask received 6×10^3 cells at the time of seeding. Two control and two growth hormone stimulated flasks were utilized for each cell line; at one and three weeks cells were counted, one cell pallet per duplicate frozen at -20°C and the media from each flask concentrated to 3 ml using an amicron-NM5 filter. Assays for IGF I and IGF II in the media concentrates were run at 2 dilutions and in duplicate at each dilution as described in previous publications[15,16].

Table 2 and Figure 3 present the essential data of this study. Transformed B-lymphocytes initially isolated from controls produced significantly more IGF I, both 1 week and 3 weeks after starting the cultures, when stimulated with growth hormone than when not stimulated with growth hormone. Growth hormone did not augment IGF I secretion by pygmy cells. The data demonstrating this difference were unequivocal. At one week of culturing, mean IGF I concentration per flask in GH stimulated cultures was 40.4 ± 5.5 ng/10^4 cells in control subjects of normal stature and in pygmies 22.7 ± 4.5 ng/10^4 cells (p < 0.05). At three weeks, the figures were respectively 3.36 ± 0.52 ng/10^4 cells vs. 0.52 ± 0.28 ng/10^4 cells (p = 0.005). Finally, in addition to obtaining evidence that GH receptors might be decreased and that pygmy cells respond poorly to GH, it has now been possible to characterize the gene for the GH receptor.

Bowcock and Cavalli-Sforza have examined the growth hormone receptor gene (GHR)[36]. Restriction fragment length polymorphisms (RFLPs) in the vicinity of the GHR gene were studied utilizing the probe ghr.501, which contains a large portion of the cDNA of GHR. They found a variant BstNI fragment of 1.55 kilobases (kb) in pygmies instead of the usual band of 0.75 kb. The variant allele was rare or absent in all populations except the pygmies from Zaire. The frequencies of this 1.55 kb gene allele in pygmies and 4 other world populations are given in Table 3.

The differences between allele frequencies shown in Table 3 of the pygmy population compared with other non-pygmy populations were all significant with a p < 0.001. The RFLP was in Hardy Weinberg equilibrium in all tested populations. These last data thus further support a defect in the GH receptor in the pygmy, and strongly indicate that the primary site of this defect is the GHR gene.

On the basis of the data summarized in the studies we have performed, plus data from collaborators, we believe now that the African pygmy clearly has a deficiency in the number of growth hormone cellular receptors; this may be due to variant GH receptors in the pygmy which

Table 3.   Allele Frequency for the 1.55 Gene Allele.

Population	N (Genes)	Allele frequency
Zaire Pygmy	98	.33
Non Pygmy-Black African	42	.02
Caucasoid	74	.00
Melanese	31	.03
Oriental (Chinese & Japanese)	160	.06

differ in their binding characteristics for GH.   This growth hormone receptor defect probably results in defective secretion of IGF I, particularly at puberty, and reasonably accounts for the final short stature of the pygmy and the metabolic aberrations that have been reported.

ACKNOWLEDGEMENTS

Supported in part by a grant (AM 18130) from the National Institute of Arthritis, Metabolism, and Digestive Diseases, National Institutes of Health; by General Clinical Research Center Program of the Division of Research Resources, National Institutes of Health (RR 00082). Also supported by a contract with the Department of Health and Rehabilitative Services of the State of Florida for the University of Florida Diabetes Research Education and Treatment Center.

REFERENCES

1.   E. Rinderknecht and R.E. Humbel, The amino acid sequence of human insulin-like growth factor I and its structural homology to proinsulin, *J. Biol. Chem.* 253:2769-76 (1978).

2.   E. Rinderknecht and R.E. Humbel, Primary structure of human insulin-like growth factor II, *FEBS Lett.* 89:283-6 (1978).

3.   J. Zapf, H. Walter, and E.R. Froesch, Radioimmunological determination of insulin-like growth factors I and II in normal subjects and in patients with growth disorders and extrapancreatic tumor hypoglycemia, *J. Clin. Invest.* 68:1321-30 (1981).

4.   K. Megyesi, C.R. Kahn, J.V. Roth, and P. Gorden, Hypoglycemia associated with extrapancreatic tumors: demonstration of elevated plasma NS TLA-S by a new radioreceptor assay, *J. Clin. Endocrinol. and Metab.* 38:931-4 (1974).

5.   R.W. Furlanetto, L. Underwood, J.J. Van Wyk, and A.J. D'Ercole, Estimation of somatomedin-C in normals and patients with pituitary disease by radioimmunoassay, *J. Clin. Invest.* 60:648 (1977).

6.   D.L. Rimoin, T.J. Merimee, D. Rabinowitz, and V.A. McKusick, Genetic aspects of clinical endocrinology, Proc. 1967 Laurentian Hormone Conf., *Rec. Prog. Hormone Res.* 24:365 (1968).

7.   H.S. Kupperman, "Human Endocrinology," Philadelphia, PA (1963), p. 31.

8.   R. Gates, *Acta Genet. Med. Gemmell.* 7:159 (1958).

9.   P. Schumacher, "Die Kivu Pygmaen.   Memoires de l'Institut Royal Colonial Belge, Science, Morales et Politiques," collection in 4 volumes (1951).

10.   J.L. Ghesquiere and M.J. Karvonen, Some anthropometric and functional dimensions of the pygmy (Kivo Twa), *Annals of Human Biol.* 8:119-34 (1981).

11.   T.J. Merimee, D.L. Rimoin, D. Rabinowitz, L.L. Cavalli-Sforza, and V.A. McKusick, Metabolic studies in the african pygmy, *Trans. Assoc. Am. Physicians* 81:221 (1968).

12.   T.J. Merimee, D.L. Rimoin, J.D. Hall, and V.A. McKusick, A metabolic and hormonal basis for classifying ateliotic dwarfs, *Lancet*, 2:963 (1965).

13.   T.J. Merimee, D.L. Rimoin, E. Pennett, and L.L. Cavalli-Sforza, Metabolic studies in the african pygmy, *J. Clin. Invest.* 51:395-401 (1972).

14.   T.J. Merimee, J. Zapf, and E.R. Froesch, Insulin-like growth factors in pygmies and subjects with the pygmy tract: Characterization of the metabolic actions of IGF I and IGF II in man, *J. Clin. Endocrinol. and Metab.* 55:1081-7 (1982).

15. T.J. Merimee, J. Zapf, and E.R. Froesch, Dwarfism in the pygmy: An isolated deficiency of insulin-like growth factor I, *New Eng. J. Med.* 305:965-8 (1981).

16. T.J. Merimee, J. Zapf, and E.R. Froesch, Insulin-like growth factors in pygmies and subjects with the pygmy trait: Characterization of the metabolic actions of IGF I and IGF II in man, *J. Clin. Endocrinol. and Metab.* 55:1081-7 (1982).

17. J. Zapf, E. Schenle, and E.R. Froesch, Insulin-like growth factors I and II: Some biological actions and receptor binding characteristics, *Eur. J. Biochem.* 87:285 (1978).

18. P.H. Henneman, A.P. Forbes, M. Moldawen, F. Dempsey, and E.L. Carroll, Effects of human growth hormone in man, *J. Clin. Invest.* 60:1223 (1960).

19. T.J. Merimee, J. Zapf, B. Hewlett, and L.L. Cavalli-Sforza, Insulin-like growth factors in pygmies: The role of puberty in determining final stature, *New Eng. J. Med.* 316: 906-11 (1987).

20. E.H. Giglioli, Ulteriori notizie enterno ai negriti, *Boll. Sci. Nat. Soc. Adriat.* 5:404-11 (1880).

21. L.L. Cavalli-Sforza, ed., "The African Pygmy," Academic Press, New York (1986), p. 310-410.

22. H.V. Meredith, Research between 1960 and 1970 on the standing height of young children in different parts of the world, *Adv. Child Dev. Behav.*, 12:1-59 (1978).

23. H.V. Meredith, Research between 1950 and 1980 on unborn-rural differences in body size and growth rate of children and youths, *Adv. Child Dev. Behav.* 17:83-138 (1982).

24. R.M. Bala, J. Lopatka, A. Leung, E. McCoy, and R.G. McArthur, Serum immunoreactive somatomedin levels in normal adults, pregnant women, children at various ages, and children with constitutionally delayed growth, *J. Clin. Endocrinol. and Metab.* 52:508-12 (1981).

25. J. van de Koppel, Preliminary report on central african project, *in:* "Basic Problems in Cross Cultural Psychology," Poorping, ed., Swets and Zeitlinger, Amsterdam (1970), p. 282-8.

26. J. van de Koppel and B. Hewlett, Growth of Aka pygmies and Bagandus of the Central African Republic, *in:* "African Pygmies," L.L. Cavalli-Sforza, ed., Academic Press, New York (1985), p. 95-102.

27. J.M. Tanner, The uses and abuses of growth standards, *in:* "Human Growth: A Comprehensive Treatise," F. Faulkner and J.M. Tanner, eds., Plenum Press, New York (1985) p. 280-5.

28. P. de Pagter-Holthuizen, M. Jansen, F.M. van Schaik, R. van de Kammon, C. Oosterwisk, J.L. Van de Brande, and J.S. Sussenbach, The human insulin-like growth factor II gene contains two development-specific promoters, *FEBS Lett.* 214:259-64 (1987); L.L. Cavalli-Sforza, Personal Communication, April, 1987.

29. A.J. D'Ercole, A.D. Stiles, and L.E. Underwood, Tissue concentrations of Somatomedin-C: Further evidence for multiple sites of synthesis and paracrine or autocrine mechanisms of action, *Proc. Nat'l. Acad. Sci. USA* 81:935-9 (1984).

30. D.R. Clemmons, L.E. Underwood, and J.J. Van Wyk, Hormonal control of immunoreactive somatomedin production by cultured human fibroblasts, *J. Clin. Invest.* 67:10-9 (1981).

31. R.L. Hintz, A.V. Thorsson, G. Enberg, and K. Hall, IGF II binding on human lymphoid cells, *Biochem. Biophys. Res. Comm.* 118:774 (1984).

32. G. Baumann, M.W. Stolar, K. Amburn, C.P. Barsano, and B.C. DeVries, A specific growth hormone binding protein in human plasma: Initial characterization, *J. Clin. Endocrinol. and Metab.* 62:134 (1986).

33. A.C. Herington, S. Ymen, and J. Stevenson, Identification and characterization of specific binding proteins for growth hormone in normal sera, *J. Clin. Invest.* 77:1817 (1986).

34. A.C. Herington, S. Ymen, and J. Stevenson, Affinity purification and structural characterization of a specific binding protein for human growth hormone in human serum, *Biochem. Biophys. Res. Comm.* 139:150 (1986).

35. G. Baumann, M. Shaw, and R.J. Winter, Absence of the plasma growth hormone-binding protein in Laron type dwarfism, *J. Clin. Endocrinol. and Metab.* 65:814-16 (1987).

36. A. Bowcock, L.L. Cavalli-Sforza, W. Wood, T. Jenkins, B. Hewlett, and T.J. Merimee, The growth hormone receptor gene in the african pygmy, *Fed. Proc.* in press (1989).

SOMATOMEDIN INHIBITORS

L.S. Phillips, S. Goldstein, and J.D. Klein

Department of Medicine
Emory University School of Medicine
Atlanta, GA

IMPORTANCE OF THE IGF (SOMATOMEDIN) INHIBITORS

Direct measurement of growth hormone (GH) in problems of disordered growth has revealed conditions in which growth is poor despite the presence of GH levels which are normal to elevated (1). Studies of the potential involvement of the insulin-like growth factors/somatomedins (IGFs) in these GH-resistant states have indicated (a) that there is regulation of IGF generation by factors other than GH, and (b) that there is regulation of IGF action by other circulating factors, the IGF inhibitors. In malnutrition, diabetes, glucocorticoid excess, and renal failure, GH-resistant growth failure is associated with low circulating levels of IGF activity as measured by cartilage bioassay (Table 1).

Table 1. Decreased IGF activity in conditions of "poor growth despite GH"

Conditions	Control IGF activity	Condition IGF Activity
malnutrition	.97 ± .06 U/ml	.36 ± .06
diabetes	.87 ± .09	.02 ± .01
steroid excess	1.02 ± .09	.35 ± .07
renal failure	.80 ± .12	.21 ± .03

IGF activity was measured by porcine cartilage bioassay in sera from rats (malnutrition, diabetes) and humans (glucocorticoid excess, renal failure). Adapted from references (2-5). Mean ± SEM.

In these conditions, poor growth and low IGF activity could be due either to low circulating levels of IGFs, or--if IGF levels were normal--to IGF resistance, due either to circulating antagonists or to unresponsiveness of target tissues. Since early studies (2,3) showed that cartilage from starved or diabetic animals was responsive to IGFs in normal serum, attention became focused on IGF levels and the potential contributions of circulating IGF-antagonistic factors.

In order to determine whether GH-resistant growth failure and low IGF activity was due to a decrease in IGFs or to an increase in IGF inhibitors, it was necessary to develop techniques to evaluate the separate contributions of these factors. Present knowledge is derived largely from chromatographic separations of IGFs and IGF inhibitors based on differences in apparent size, followed by bioassay of inhibitor activity and bioassay or immunoassay of IGFs. With these approaches, examinations of human disease and animal models indicate that all four disorders are associated with increases in levels of circulating IGF inhibitors, and variable decreases in circulating IGFs (Table 2).

Table 2. Levels of IGFs and IGF inhibitors in disorders of "poor growth despite GH".

Disorder	IGF	hiMW-inhibitor	loMW-inhibitor
Diabetes	low	high	high
Malnutrition	low	high	?
Glucocorticoid excess	normal	high	?
Renal failure	normal	normal	high

Sera from humans (glucocorticoid excess, renal failure) or rats (diabetes, malnutrition) were subjected to size exclusion chromatography at neutral and/or acid pH, and levels of IGFs and IGF inhibitors determined by a combination of bioassay and/or immunoassay in active fractions. Adapted from references (4-7).

Thus, although diabetes and malnutrition are associated with both reduced levels of IGFs and increased inhibitors, poor growth in conditions of glucocorticoid excess or renal failure appears to be due largely to an increase in circulating IGF inhibitors. It should be noted that present methodology permits estimation of inhibitor levels only by the ability of serum fractions to antagonize tissue stimulation by IGFs in vitro. While such approaches have revealed the presence of both high- and low-MW IGF inhibitors, the methodology does not provide information as to either the number of inhibitor molecules present or the locus of antagonism of IGF action.

RECOGNITION OF IGF INHIBITORS

The presence of inhibitory factors in serum samples was hypothesized after observation that sera from humans or animals with severe diabetes or malnutrition exhibited aberrant behavior in IGF bioassays. As shown in Figure 1, increasing concentrations of normal rat serum provide progressive stimulation of sulfate uptake by porcine costal cartilage in vitro. Serum from fasted rats produces cartilage stimulation which is qualitatively similar, although cartilage sulfate uptake is proportionately lower at each concentration of serum. In contrast, the response to diabetic rat serum is qualitatively different; increasing concentrations of diabetic rat serum produce a progressive decrease, rather than an increase, in cartilage sulfate uptake, reducing cartilage sulfate uptake to levels below that of cartilage exposed to buffer alone (8). In combination experiments, addition of samples of whole diabetic rat serum to whole normal rat serum also reduced the ability of the normal rat serum to stimulate cartilage, lowering calculated potency of contained IGF activity (3); similar observations have been reported by Van den Brande et al with samples of plasma from malnourished children (9).

Such findings first indicated (a) that some serum samples contain a factor or factors which can antagonize tissue stimulation by IGFs in normal serum, and (b) that such inhibitory factor(s) may have direct effects on target tissue independent of the IGFs. Since studies have demonstrated that inhibitory activity can be detected more readily on the basis of antagonism

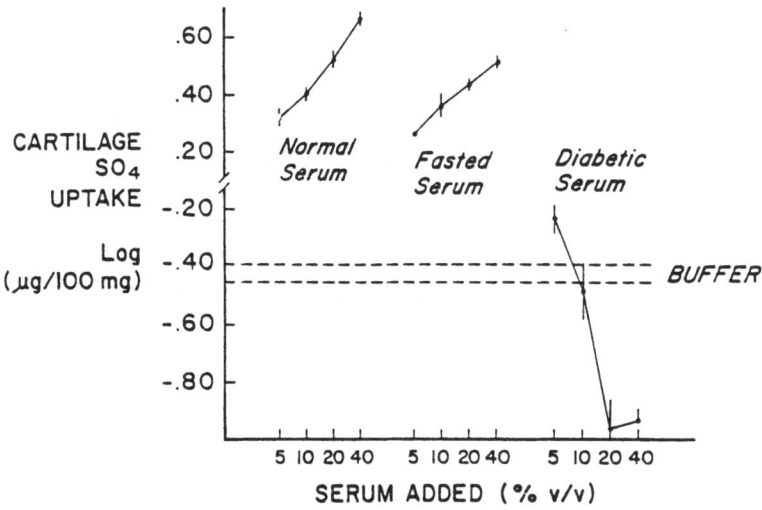

Figure 1. Effect of increasing concentrations of rat serum on sulfate uptake by porcine cartilage. From reference (8), with permission. Mean ± SEM.

of IGF-stimulated tissue responses rather than by blunting of basal tissue responses in the absence of added IGFs (10), our laboratory has maintained the operational designation as "IGF inhibitors". However, the ability of inhibitor-enriched sera to blunt basal tissue activity probably indicates that the "IGF inhibitors" either (a) antagonize IGF action at autocrine/paracrine levels, or (b) have the broader properties of "growth" or "tissue" inhibitors and are not specific antagonists of IGF action.

Since attempts at purification have not progressed sufficiently to permit biological studies with highly purified factors, the IGF inhibitors have been examined largely as partially purified preparations; studies have usuallly involved active fractions from chromatographic separations of serum samples from patients or animals with conditions of "poor growth despite GH". In the process of separating active from inactive serum constituents, it has become evident (a) that inhibitors are present in the circulation of normal humans and animals as well as in conditions of "poor growth despite GH" (7), and (b) that some IGF inhibitors in serum are not metabolically regulated (4,6). For example, fractionation of normal human serum on Sephadex G-50 at pH 2.4 reveals both IGFs and inhibitors (Figure 2 (5)). Fractionation of normal rat serum on Sephadex G-200 at pH 2.4 reveals IGF inhibitory activity corresponding to MW ~30,000, reduced in apparent quantity compared to IGF inhibitory activity of similar size in diabetic serum (7). However, closer examination of fractions containing IGF inhibitory activity indicates that elution of "metabolically regulated" IGF inhibitor activity is preceded by IGF inhibitor activity (of higher MW) which is not metabolically regulated. Thus, normal humans treated with prednisone exhibit a ~100% increase in "metabolically regulated" IGF activity, but no change in "unregulated" inhibitor activity (4). Similarly, starvation or streptozotocin-induced diabetes in rats produces 3- to 20-fold increases in "metabolically regulated" IGF inhibitor activity, but does not change "unregulated" inhibitor activity (6).

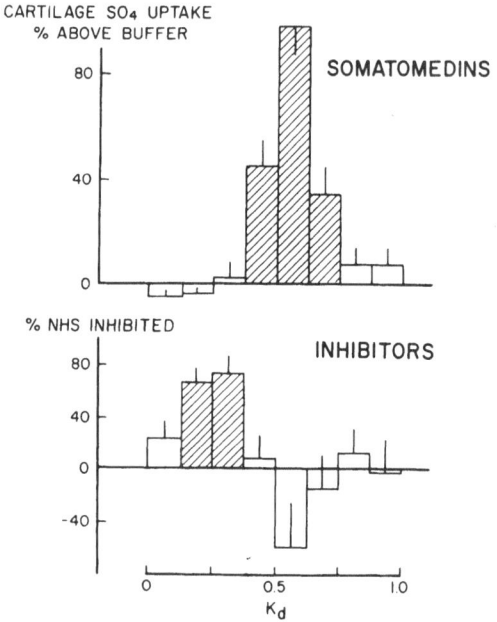

SEPHADEX G-50 pH 2.4

Figure 2. IGFs and IGF inhibitors in sera of eight normal adults. IGFs were measured by the ability to stimulate sulfate uptake by rat costal cartilage in vitro, expressed as increase above buffer levels, and inhibitors by the ability to antagonize stimulation in incubations containing added normal human serum (NHS). From reference (5), with permission. Mean ± SEM.

The relationship between "unregulated" and "regulated" IGF inhibitor activity remains to be elucidated. Vassilopoulou-Sellin et al (11) have described the presence in liver extracts of IGF inhibitory activity which has apparent MW greater than 30,000, and which does not appear to be metabolically regulated; unregulated IGF inhibitor activity in the the liver may contribute to unregulated inhibitor activity in the circulation, since such high-MW inhibitory activity is released by the liver during perfusion studies (12). Presumably, potential relationships between "unregulated" and "regulated" IGF inhibitors will be delineated once the respective factors have been purified and characterized, and can be detected by sensitive and specific antisera or by finding specific biologic properties; our preliminary studies suggest antigenic overlap between inhibitors of ~30,000 MW and those of >30,000 MW in serum samples. The discussion which follows will be focused entirely on the biological properties of metabolically regulated IGF inhibitors, as it seems likely that these factors are probably of greater importance to endocrine/metabolic function than are the unregulated IGF inhibitors.

BIOLOGIC PROPERTIES

The properties of the high-MW IGF inhibitor have been examined most extensively with fractions of serum from glucocorticoid-treated humans (4) and streptozotocin-diabetic animals (13). IGF antagonism by such fractions parallels antagonism by whole serum (7). As shown in Figure 3, both diabetic rat serum and an inhibitor-enriched serum fraction can antagonize cartilage stimulation by whole normal rat serum, insulin, and an IGF-enriched serum fraction. Similar antagonism has been shown for stimulation of glucose oxidation to CO_2 in segments of epididymal fat pads (Figure 4), and for glucose incorporation into glycogen in rat hemidiaphragm in vitro. Whole diabetic rat serum has also been shown to antagonize insulin stimulation of both adipose tissue and muscle in vivo (Figure 5).

Such blunting of tissue responses can be reversed completely after 4 hr exposure to IGF inhibitors, but reversal is less extensive after 24 hr exposure (14). Lineweaver-Burk analysis indicates that inhibitor interactions with both IGFs and insulin are noncompetitive (13), consistent with the hypothesis that the inhibitors are not specific IGF antagonists.

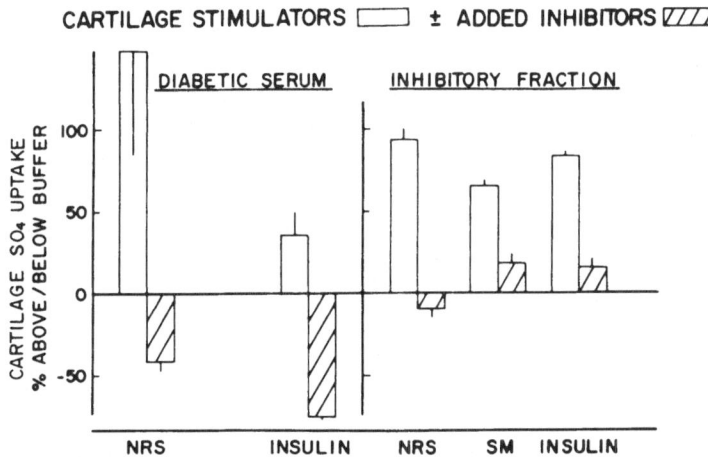

Figure 3. Effect of combined cartilage stimulators and inhibitors on sulfate uptake by hypophysectomized rat cartilage in vitro. From reference (13), with permission. Mean ± SEM.

Much less is known about the low-MW IGF inhibitor which is present at increased concentrations in the circulation of streptozotocin-diabetic rats and humans with end-stage renal disease (7,10). Serum fractions with low-MW IGF-inhibitor activity decrease IGF stimulation of sulfate, uridine, and thymidine uptake by cartilage and also decrease insulin stimulation of glucose oxidation in isolated adipocytes, glucose and amino acid transport in cultured muscle cells, and glucose incorporation into lipid in hepatocytes in primary culture. Since the low-MW IGF inhibitor is of sufficiently small size to permit transplacental passage, there has been interest in the potential of this factor to produce abnormalities in fetal growth and development. As shown in Figure 6, small quantities of serum fractions enriched in low-MW inhibitory activity produce both decreased growth and developmental abnormalities in a mouse embryo culture system, suggesting a potential role in the etiology of congenital malformations found in offspring of women with poorly controlled diabetes during pregnancy (15).

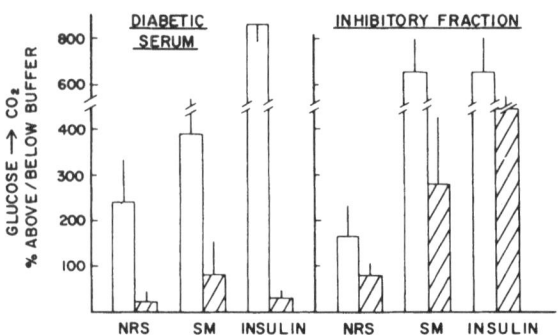

Figure 4. Effect of tissue stimulators with or without added inhibitors on glucose oxidation to CO_2 in segments of rat epididymal fat pads. From reference (13), with permission. Mean $\pm$ SEM.

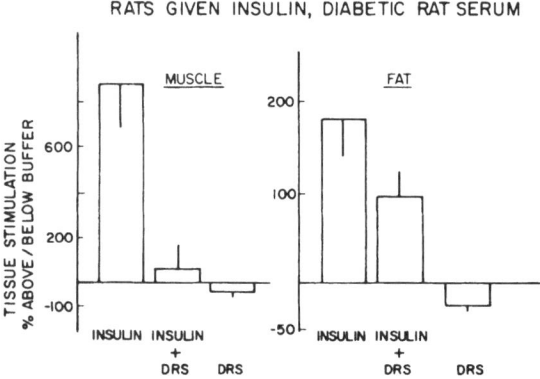

Figure 5. Antagonism of insulin action in vivo by inhibitors in whole diabetic rat serum. Rats were given $1\text{-}^{14}C$-glucose and insulin and/or whole diabetic rat serum (DRS) ip, and sacrificed 1 hr later, with measurement of glucose incorporation into glycogen in diaphragm and into lipid in epididymal fat pad. Results are expressed as percent of tissue activity from animals injected with glucose but without insulin or DRS. From reference (13), with permission. Mean $\pm$ SEM.

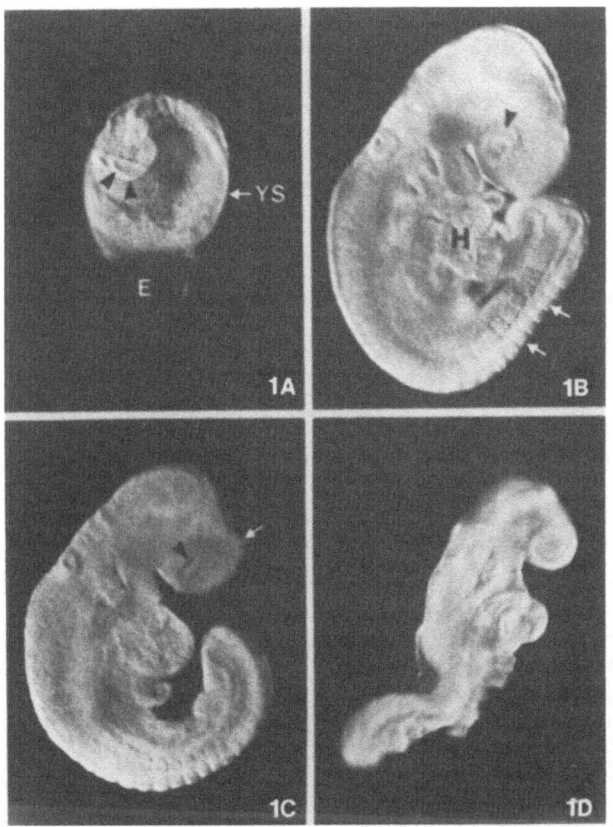

Figure 6. "1A": Pre-culture early somite(3-5) stage mouse embryo. Headfold
(arrowheads); visceral yolk sac (YS); ectoplacental cone(E). "1B": Embryo
cultured from the early somite state for 24 h in control medium showing
normal growth and development during this peroid in vitro. 18-19 somites
(arrows); heart (H); optic vesicle (arrowheads). "1C": Early somite embryo
cultured for 24 h in 3 ul/ml of low-MW IGF inhibitor from gel filtration of
diabetic rat serum on Sephadex G-25 at pH 7. Growth is reduced and the
embryo is malformed. Hemorrhage (arrow); optic vesicle (arrowhead). "1D":
Early somite embryo cultured for 24 h in 5 ul/ml of IGF inhibitor. The
embryo is more adversely affected than the one in Figure "1C" and proper
formation of body curvature has failed to occur. From reference (15), with
permission.

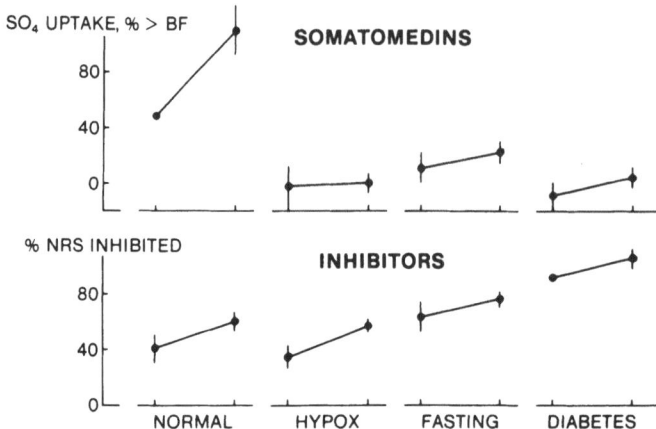

Figure 7. IGFs and high-MW IGF inhibitors in normal, hypophysectomized (hypox), fasted, and diabetic rats. Lines connect points denoting bioassay responses to two concentrations of IGFs (upper) and inhibitors (lower); in each case, the lower concentration is on the left. Serum was fractionated at pH 3.0 to separate IGFs (MW ~7500) from inhibitors (MW ~30,000). IGF activity was measured by stimulation of cartilage SO_4 uptake [above buffer level (BF)] at 0.4% and 2% sample, and inhibitor activity was measured by the ability of samples (at 2% and 8%) to decrease stimulation by 1% added normal rat serum (NRS). From reference (6), with permission. Mean ± SEM.

REGULATION

The regulation of the high-MW IGF inhibitor has been explored most extensively in models of malnutrition and diabetes mellitus in rats. As shown in Figure 7, rats subjected to three days of starvation or two days of streptozotocin-induced diabetes exhibit a fall in IGF activity comparable to that found in hypophysectomized rats, but also a rise in IGF inhibitor activity which is not found in hypophysectomized rats with similar IGF levels (6). In general, animals with severe diabetes exhibit higher levels of IGF inhibitor activity than animals with comparable weight loss due to dietary deprivation, suggesting that the abnormal hormonal milieu associated with diabetes may contribute to this differential. Further study of similar animal models over time and with progressive severity of metabolic decompensation indicates that the IGF inhibitors may be more sensitive than the IGFs to alterations in metabolic status. Thus, fasted rats exhibit both a fall in IGFs and a rise in IGF inhibitors, but refeeding produces earlier normalization of IGF inhibitor than IGFs (16). Similarly, rats with graded severity of streptozotocin-induced diabetes exhibit a rise in levels of IGF inhibitor before levels of IGFs fall with progressive increases in glucose and ß-hydroxybutyrate. Such findings suggest that IGFs and IGF inhibitors have separate regulatory mechanisms, consistent with the observation that IGF inhibitors rise without accompanying alterations in levels of IGFs in situations of glucocorticoid excess (9) and renal failure (10).

To examine more directly the hypothesis that IGFs and IGF inhibitors are controlled by different mechanisms, Hofert et al (17) determined levels of both factors in streptozotocin-diabetic rats with and without adrenalectomy and glucocorticoid replacement. Adrenalectomized diabetic animals exhibited hyperglycemia comparable to intact diabetic animals, but had reduced weight loss and levels of ß-hydroxybutyrate which were comparable to values found in normal control animals; administration of glucocorticoids increased weight

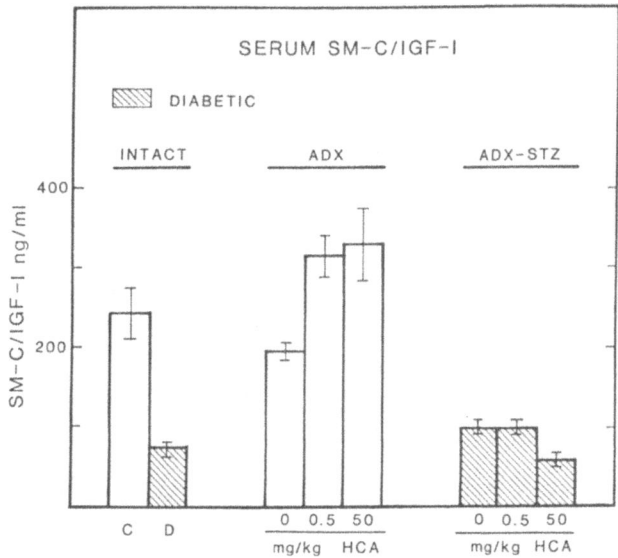

Figure 8. Serum IGF-1 in diabetic (STZ) and nondiabetic intact (C) and adrenalectomized (ADX) rats, with or without administration of hydrocortisone acetate (HCA) 0.5 or 50 mg/kg sc, measured by RIA after separation from carrier proteins by gel permeation HPLC under acidic conditions. Mean ± SEM. From reference (17), with permission.

loss and levels of ß-hydroxybutyrate to values found in intact diabetic animals. As shown in Figure 8, adrenalectomy with or without glucocorticoid replacement did not affect the depression in IGF-1 associated with streptozotocin-induced diabetes. In contrast, the diabetes-associated rise in IGF inhibitor was completely absent in adrenalectomized-diabetic animals, but restored by adminstration of glucocorticoids (Figure 9). These findings (a) indicate that glucocorticoids play at least a permissive role in the diabetes-associated rise in IGF inhibitor, (b) identify circumstances in which fluctuations in levels of inhibitor appear to be independent of changes in IGF-1, and (c) suggest that IGF-1 and inhibitor are regulated by separate mechanisms.

Considerably less is known about the regulation of the low-MW IGF inhibitor found in diabetes and kidney failure. Our preliminary studies suggest that levels of the low-MW inhibitor fall with a single hemodialysis treatment in patients with end-stage renal disease, consistent with a comparable decrease on dialysis of serum samples in vitro (18).

ORIGIN

Although the origin of the low-MW IGF inhibitor is unknown, it appears that the high-MW IGF inhibitor originates in the liver. Early studies by Vassilopoulou-Sellin et al (19) revealed that while perfused livers from normal rats released net cartilage-stimulating (IGF) activity into the perfusion medium, livers from fasted rats released net cartilage-inhibiting

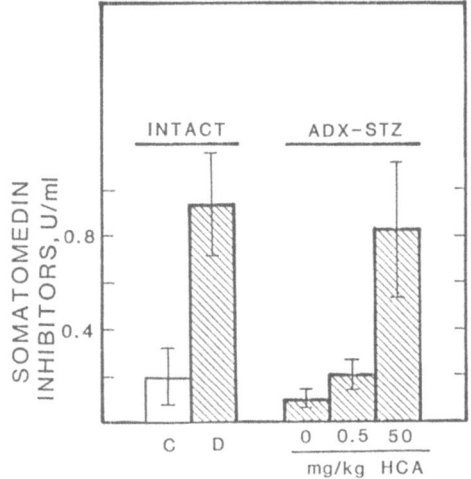

Figure 9. Requirement of glucocorticoids for the diabetes-associated rise in serum IGF inhibitor. Groups were treated as in Figure 8. After fractionation of serum by HPLC under acidic conditions, IGF inhibitor activity was measured by parallel-line bioassay according to the ability of samples to antagonize cartilage stimulation by IGFs in normal serum, and potency was expressed relative to that activity in the serum of intact STZ-diabetic rats. Mean ± SEM. From reference (17), with permission.

activity instead, suggesting the presence of IGF inhibitors. The time-course of the hepatic release of IGF and IGF inhibitor activity in response to fasting and refeeding has recently been studied by Goldstein et al (12). Rats were subjected to three days of fasting, followed by 24 hours of refeeding. Animals sacrificed at various time points were subjected to in situ 2-hour recirculating liver perfusion, IGF and IGF inhibitor separated by size exclusion HPLC at pH 3.0, and levels of both IGFs and IGF inhibitor measured by bioassay. Fasting was associated with a decrease in hepatic release of IGFs, together with an increase in hepatic release of IGF inhibitor. However, with refeeding, there was a prompt decrease in release of IGF inhibitor, followed by a more delayed restoration of IGF release. Since progressive normalization of IGF inhibitor release was more rapid than the rise in release of IGFs, these results suggest regulation by separate mechanisms. And since restoration of both IGFs and IGF inhibitor preceded comparable changes in the circulation of similar animals (Figure 10), the liver perfusion studies are also consistent with the hypothesis that the liver contributes both IGFs and IGF inhibitors to the circulation.

After Vassilopoulou-Sellin et al (14) initially suggested that insulin might regulate hepatic release of IGF inhibitor, the contributions of both insulin and cortisol were examined more closely by Binoux et al (20-22), using a liver explant culture system. Binoux et al found that release of IGF inhibitor activity was decreased slightly by GH, decreased substantially by insulin, and increased significantly by cortisol; the increased release of inhibitor in incubations with added cortisol is consistent with the ability of glucocorticoids to increase inhibitor levels in vivo (9) and to facilitate release of IGF inhibitor activity in perfused liver systems. Thus, Hofert et al (17) found that the release of IGF inhibitor activity by perfused livers from diabetic-adrenalectomized rats was quite low, but increased substantially when comparable animals were treated with glucocorticoids. These findings suggest that glucocorticoids may act in the liver to promote the diabetes-associated rise in IGF inhibitor levels seen in vivo.

90

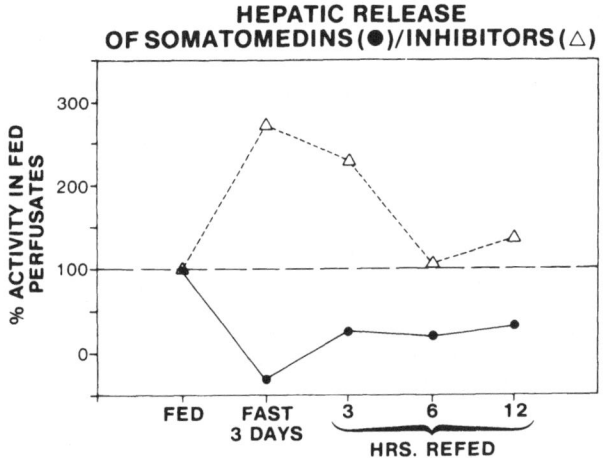

HEPATIC RELEASE
OF SOMATOMEDINS (●)/INHIBITORS (△)

CIRCULATING SOMATOMEDINS (●)/INHIBITORS (△)

Figure 10. (Above) Release of IGF activity and IGF inhibitor activity by perfused livers of normal, fasted, and fasted/refed rats. (Below) IGF activity and IGF inhibitor activity in serum of comparable animals. Liver perfusates and serum from experimental animals were chromatographed on TSK-2000 at pH 3.0, with collection of IGF inhibitor activity at K_{av} 0.42-0.67, and IGF activity K_{av} 0.67-0.92. IGF activity was measured according to stimulation of sulfate uptake in vitro by cartilage of hypophysectomized rats, and inhibitor activity by the ability of samples to antagonize cartilage stimulation in incubations containing added normal rat serum. From reference (12), with permission.

While the precise mechanisms which underlie hepatic production of IGF inhibitor remain to be established, a relationship to ketogenesis is suggested by correlations between circulating levels of IGF inhibitor and ß-hydroxybutyrate. Experiments utilizing both animals with graded severity of streptozotocin-induced diabetes (23) as well as streptozotocin-diabetic animals with or without concomitant adrenalectomy and steroid replacement (17) have both revealed stronger correlations of levels of IGF inhibitor with ß-hydroxybutyrate than with glucose, an alternate metabolic fuel. To explore further the potential relationship to ketogenesis, Gagliardi and Goldstein, working in our laboratory, examined IGF inhibitor levels in streptozotocin-diabetic animals given metabolic inhibitors to block various stages of the production and utilization of metabolic fuels (24). Inhibition of gluconeogenesis decreased glucose levels to normal but did not affect levels of IGF inhibitor. Blockade of lipolysis lowered levels of free fatty acids to normal, did not affect plasma glucose, and decreased but did not normalize ß-hydroxybutyrate, but levels of IGF inhibitor did not fall significantly. In contrast, blockade of fatty acid oxidation raised levels of free fatty acids but lowered levels of ß-hydroxybutyrate to normal, associated with a significant fall in levels of IGF inhibitor to values comparable to those found in normal control animals. In combination, these findings are consistent with the hypothesis that hepatic production of IGF inhibitor is related to ketogenesis.

ISOLATION OF IGF INHIBITORS

Relatively little is known about the nature of the low-MW IGF inhibitor. To date, the most extensive examinations of this factor have involved fractionation of serum from patients with end-stage renal disease, pre-hemodialysis. Active fractions from Sephadex G-25 chromatography (MW ~1,000) are devoid of detectable sodium, potassium, glucose, or phosphorus, and have broad antianabolic activities: antagonism of serum-stimulated cartilage uptake of sulfate, thymidine, and uridine (10), together with blunting of insulin-stimulated glucose oxidation in adipocytes, transport of amino acids and glucose in myocytes, and formation of lipid in hepatocytes. Based on these observations, it seems unlikely that IGF inhibitor activity cannot be attributed to the presence of glucose, amino acids, sulfate, or a comparable small molecule. Since exposure to proteolytic enzymes decreases IGF inhibitor activity (10), it seems likely that the factor is a peptide. This is also supported by ninhydrin reactivity of active fractions. However, further characterization has been slow because low-MW IGF inhibitor preparations are dialyzable and are bound poorly by the chemical groups involved in conventional chromatographic separations. Such properties suggest that the low-MW IGF inhibitor is different from factors such as nonesterified fatty acids and lysophospholipids, which are found in human plasma and inhibit NaK-ATPase (25). Low-MW IGF inhibitory activity (of size similar to that of the circulating low-MW inhibitor) is also found in normal urine, indicating that the inhibitor may be a protein breakdown product and accumulate in the circulation when renal function is defective.

There is better understanding of the high-MW IGF inhibitor found at increased concentrations in conditions of malnutrition, diabetes, or glucocorticoid excess. Although free fatty acids and glucocorticoids may have direct effects on the target tissues used in IGF inhibitor bioassays, elevated levels of inhibitor are found in conditions where levels of free fatty acids are normal (nicotinic acid treatment), and contributions from glucocorticoids have been ruled out by Unterman et al (4), who showed that radioactive tracer glucocorticoids added to serum samples containing IGF inhibitor activity elute from Sephadex columns much later than the active fractions containing high-MW IGF inhibitor activity. Combined studies from our laboratory and the laboratories of Salmon et al (26,27), and Binoux et al (20-22) suggest that the inhibitor is a protein, based on susceptibility to inactivation by heat and by exposure to proteolytic enzymes, with apparent molecular weight approximately 30,000 daltons. More recently, we have used a combination of size exclusion and ion exchange chromatography, hydrophobic interaction chromatography, and reversed-phase HPLC to isolate a peptide which appears to be the high-MW IGF inhibitor. The most highly-purified preparations exhibit a doublet pattern on SDS-PAGE with apparent MW 32,000, consist of a single peak on analytical C_8 and C_{18} reversed-phase HPLC, and have N-terminal and internal amino acid sequence which is different from the reported sequences for the peptides encoded by the cDNAs for the high-MW and low-MW IGF binding proteins (28-31). Consistent with such lack of similarity to the sequences for the IGF binding proteins, our highly purified preparations of high-MW IGF inhibitor do not react with binding protein antisera in Western blots, and do not bind ^{125}I-IGF-1 in ligand blots. We are now in the process of isolating the cDNA for the high-MW IGF inhibitor, and developing monoclonal and polyclonal antibodies to characterize further the nature and regulation of the high-MW IGF inhibitor.

SUMMARY

1. Two circulating IGF inhibitors have been identified and partially characterized: a ~30,000 MW factor(s) and a ~1,000 MW factor(s). The IGF inhibitors appear to contribute to diminished growth and anabolic processes in disorders of "poor growth despite GH": diabetes, malnutrition, glucocorticoid excess, and renal failure.

2. The IGF inhibitors are recognized operationally on the basis of inhibition of IGF action on target tissues, but the inhibitors can also blunt tissue processes in the absence of added IGFs.

3. In the intact animal, high-MW IGF inhibitor(s) increases in conditions of diabetes, nutritional deprivation, or glucocorticoid excess; inhibitor levels appear to be more sensitive than IGF levels to changes in metabolic status. Circulating levels of the low-MW IGF inhibitor are increased in patients with end-stage renal disease and in animals with severe diabetes mellitus.

4. Both the high- and low-MW IGF inhibitors have broad abilities to antagonize both IGF and insulin action on cartilage (synthesis of proteoglycan, RNA and DNA), as well as anabolic actions of insulin on muscle and adipose tissue. In mouse embryo culture systems, both inhibitors decrease embryonic growth and increase developmental anomalies.

5. The high-MW IGF inhibitor appears to originate in the liver, where its release is controlled according to metabolic status, perhaps with a relation to ketogenesis. Although the origin of the low-MW IGF inhibitor is unknown, this factor may be a protein breakdown product.

6. Biochemical characterization studies indicate that the high-MW IGF inhibitor is a 32,000 MW protein with relative high pI, with amino acid sequence different from that found in the IGF binding proteins. Relatively little is known about the biochemical nature of the low-MW IGF inhibitor, but it appears to be different from other low-MW inhibitory factors isolated from human biological fluids.

Further studies of the IGF inhibitors should shed light not only on the pathophysiology of human disease, but also on fundamental underlying mechanisms related to anabolic/catabolic balance.

ACKNOWLEDGEMENTS

This work is supported in part by awards DK-33475 and DK-34785 from NIH. We thank Mss. A. Salim and M.L. Mojonnier for help in manuscript preparation.

REFERENCES

1. Phillips, L.S. and Unterman, T.G. Somatomedin activity in disorders of nutrition and metabolism. Clinics Endocrinol Metab 13:145-188, 1984
2. Phillips, L.S. and Young, H.S. Nutrition and somatomedin. I. Effect of fasting and refeeding on serum somatomedin activity and cartilage growth activity in rats. Endocrinology 99:304-314-438-446, 1976
3. Phillips, L.S. and Young, H.S. Nutrition and somatomedin. II. Serum somatomedin activity and cartilage growth activity in streptozotocin-diabetic rats. Diabetes 25:516-527, 1976
4. Unterman, T.G. and Phillips, L.S. Glucocorticoid effects on somatomedins and somatomedin inhibitors. J Clin Endocrinol Metab 61:618-626, 1985
5. Phillips, L.S., Fusco, A.C., Unterman, T.G. and del Greco, F. Somatomedin inhibitor in uremia. J Clin Endocrinol and Metab 59:764-772, 1984
6. Goldstein, S., Stivaletta, L.A. and Phillips, L.S. Separation of somatomedins and somatomedin inhibitors by size exclusion high-performance liquid chromatography. J Chromatography 339:388-393, 1985
7. Phillips, L.S., Belosky, D.C., Young, H.S. and Reichard, L.A. Nutrition and somatomedin. VI. Somatomedin activity and somatomedin inhibitory activity in serum from normal and diabetic rats. Endocrinology 104:1519-1524, 1979
8. Phillips, L.S. "Somatomedin Inhibitors" (Proceedings of the International Workshop on Advances in Research on Human Growth, Airlie, VA. (In press, 1989)
9. Van den Brande, J.L. and Du Caju, M.V.L. Plasma somatomedin activity in children with growth disturbances. In: Advances in Human Growth Hormone Research, edited by Raiti, S. Washington: DHEW , 1974, p. 98-126.
10. Phillips, L.S., Belosky, D.C. and Reichard, L.A. Nutrition and somatomedin. V. Action and measurement of somatomedin inhibitor(s) in serum from diabetic rats. Endocrinology 104:1513-1518, 1979
11. Vassilopoulou-Sellin, R., Foster, P.L., Oyedeji, C.O. and Samaan, N.A. Cartilage sulfation inhibitor from rat liver partial characterization of properties and biologic action. Metabolism 37:38-45, 1988
12. Goldstein, S. and Phillips, L.S. Nutrition and Somatomedin XVIII. Nutritionally-regulated release of somatomedins and somatomedin inhibitors from perfused livers in rats. Metabolism (in press, 1989)
13. Phillips, L.S. and Scholz, T.D. Nutrition and somatomedin. IX. Blunting of insulin-like activity by inhibitor in diabetic rat serum. Diabetes 31:97-104, 1982
14. Phillips, L.S., Vassilopoulou-Sellin, R. and Reichard, L.A. Nutrition and somatomedin. VIII. The "somatomedin inhibitor" in diabetic rat serum is a general growth cartilage inhibitor. Diabetes 28:919-924, 1979
15. Sadler, T.W., Phillips, L.S. and Goldstein, S. Inhibition of mouse embryonic growth and development by a somatomedin inhibitor(s). Diabetes 35:861-865, 1986
16. Phillips, L.S., Goldstein, S. and Gavin, III, J.R. Nutrition and somatomedin XVI. Somatomedins and somatomedin inhibitors in fasted and refed rats. Metabolism 37:209-216, 1988
17. Hofert, J.F., Goldstein, S. and Phillips, L.S. Glucocorticoids are required for elevation of somatomedin inhibitors in diabetic rats. Metabolism (in press, 1989)
18. Phillips, L.S. and Kopple, J.D. Circulating somatomedin activity and sulfate levels in adults with normal and impaired kidney function. Metabolism 30:1091-1095, 1981
19. Vassilopoulou-Sellin R., Phillips, L.S. and Reichard, L.A. Nutrition and somatomedin VII. Regulation of somatomedin activity by the perfused rat liver. Endocrinology 101:260-267, 1980
20. Binoux, M., Lassarre, C. and Hardouin, N. Somatomedin production by rat liver in organ culture. I. Studies on the release of insulin-like growth

94

factor and its carrier protein measured by radioligand assays. Effects of growth hormone, insulin and cortisol. Acta Endocrinol 99:422-430, 1982

21. Binoux, M., Lassarre, C. and Seurin, A. Somatomedin production by rat liver in organ culture. II. Studies of cartilage sulphation inhibitors released by the liver and their separation from somatomedins. Acta Endocrinol 93:83-90, 1980

22. Binoux, M., Hossenlopp, P., Lassarre, C. and Seurin, D. Somatomedin production by rat liver in organ culture. III. Validity of the technique. Influence of the released material on cartilage sulphation. Effects of growth hormone and insulin. Acta Endocrinol 93:73-82, 1980

23. Phillips, L.S., Fusco, A.C. and Unterman, T.G. Nutrition and somatomedin. XIV. Altered levels of somatomedins and somatomedin inhibitors in rats with streptozotocin-induced diabetes. Metabolism 34:765-770, 1985

24. Gagliardi, A.R.T., Goldstein, S. and Phillips, L.S. Somatomedin inhibitor in diabetic rats: relation to ketogenesis and gluconeogenesis. Clin Res 36:23A, 1988 (Abstract)

25. Kelly, R.A., O'Hara, D.S., Mitch, W.E. and Smith, T.W. Identification of NaK-ATPase inhibitors in human plasma as nonesterified fatty acids and lysophospholipids. J Biol Chem 261:11704-11711, 1986

26. Salmon, W.D., Jr. Interaction of somatomedin and a peptide inhibitor in serum of hypophysectomized and starved, pituitary-intact rats. Adv Metab Disord 8:183-199, 1975

27. Salmon, Jr., W.D., Holliday, L.A. and Burkhalter, V.J. Partial characterization of somatomedin inhibitor in starved rat serum. Endocrinology 112:360-370, 1983

28. Brewer, M.T., Stetler, G.L., Squires, C.H., Thompson, R.C., Busby, W.H. and Clemmons, D.R. Cloning, characterization, and expression of a human insulin-like growth factor binding protein. Biochem Biophy Res Commun 152:1289-1297, 1988

29. Brinkman, A., Groffen, C., Kortleve, D.J., van Kessel, A.G. and Drop, S.L.S. Isolation and characterization of a cDNA encoding the low olecular weight insulin-like growth fac3.nding protein (IBP-1). EMBO 7:2417-2423, 1988

30. Lee, Y-L., Hintz, R.L., James, P.M., Lee, P.D.K., Shively, J.E. and Powell, D.R. Insulin-like growth factor (IGF) binding protein complementary deoxyribonucleic acid from human HEP G2 hepatoma cells: predicted protein sequence suggests an IGF binding domain different from those of the IGF-1 and IGF-11 receptors. Molec Endocrinol 2:404-411, 1988

31. Wood, W.I., Cachianes, G., Henzel, W.J., Winslow, G.A., Spencer, S.A., Hellmiss, R., Martin, J.L., Baxter, R.C. Cloning and expression of the growth hormone-dependent insulin-like growth factor-binding protein. Molec Endocrinol 2:1176-1185, 1988

THE STRUCTURE AND EXPRESSION OF THE HUMAN INSULIN-LIKE

GROWTH FACTOR GENES

Pieternella Holthuizen

Laboratory for Physiological Chemistry
State University of Utrecht
Vondellaan 24^a, 3521 GG Utrecht
The Netherlands

INTRODUCTION

The insulin-like growth factors (IGF) are a group of peptides involved in mammalian growth and development (37). The complete structures of two major species were first elucidated by Rinderknecht and Humbel (27,28). IGF-I is a basic peptide of 70 amino acids which mediates the growth promoting action of growth hormone and plays an important role in postnatal growth. The biological action of IGF-II, a 67 amino acids long peptide, is less obvious. It is thought to stimulate a variety of growth promoting processes in the fetus but several studies also indicate a function for IGF-II in the postnatal stage (4).

The IGFs are synthesized in many tissues, implicating an intricate mechanism of regulation of expression. In addition to the circulating levels of peptide, which exert an endocrine mode of action, local synthesis has been described suggesting autocrine and paracrine actions for the IGFs as well (11).

This review will summarize the data on the structure and expression of the human IGF genes, with the main emphasis on the IGF-II gene.

HUMAN IGF-I

In 1983 the first cDNA for IGF-I was isolated by M. Jansen from a human adult liver cDNA library (19). Since then aditional IGF-I cDNAs have been reported by several other groups (24,29). Nucleotide sequence analysis of the cDNAs indicates that IGF-I is synthesized as a larger precursor molecule which undergoes processing at both the amino-terminus (signal peptide) and the carboxyl-terminus (E domain) to produce the 70 amino acids mature IGF-I protein.

To date the initiation site of translation has not been determined unambiguously. There are three methionine codons (AUG) present at positions -48, -25 and -22 relative to the first amino acid of mature IGF-I. Comparison of the putative IGF-I amino terminal peptide with signal peptides of related proteins such as insulin, IGF-II and relaxin, favors Met -25 or Met -22 as the actual initiation

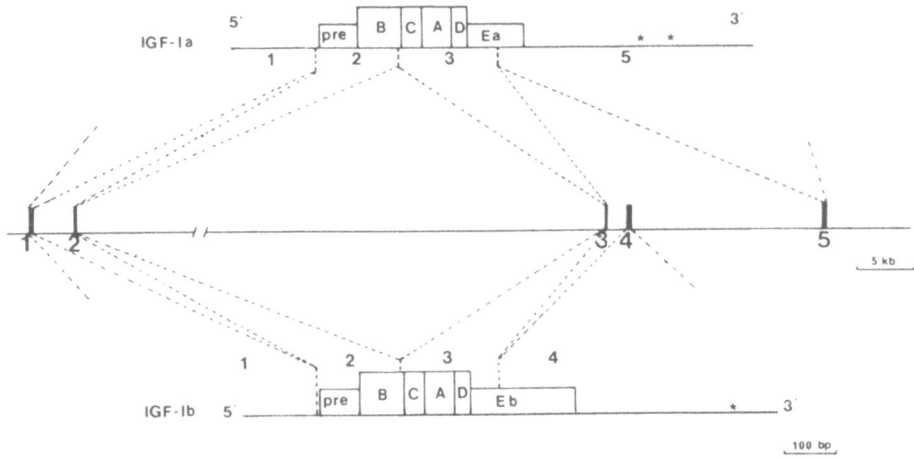

Figure 1. Schematic representation of the human IGF-I gene and the two identified types of transcripts, IGF-Ia and IGF-Ib. The coding regions of the IGF-I precursors are indicated in blocks, showing Met -25 as the initiation of translation codon. The exons are numbered 1 - 5, asterisks in the 3' non-translated regions mark polyadenylation signals.

site. However, as demonstrated by Rotwein et al. using an in vitro transcription/translation system, initiation of translation from Met - 48 seems to be the preferred site (31).

In addition two cDNAs encoding partially different carboxyl-terminal extensions (E domains) of the IGF-I precursor have been identified (Figure 1). The cDNA for IGF-Ia encodes an E domain of 35 amino acids and the cDNA for IGF-Ib encodes an E domain of 77 amino acids long (19,29).

The human gene for IGF-I has been mapped on chromosome 12 (3,15,35). It consists of at least five exons spanning a region of more than 95 kb of chromosomal DNA (6,30,36). The structure of the human IGF-I gene is shown in Figure 1 as well as the two characterized cDNAs. IGF-Ia cDNA consists of exons 1, 2, 3 and 5, while IGF-Ib cDNA is derived from exons 1, 2, 3 and 4 (6,30).
From these results it is clear that two different types of IGF-I precursors arise by alternative splicing of the primary transcript. Interestingly, the differential splicing for the IGF-I gene influences the structure of the precursor protein. It is tempting to speculate that these different precursor proteins may serve specific biological functions.

Expression studies, mainly by Northern blot analysis have been performed by several groups (2,11,14,17,29,32). Multiple mRNA species are present which range from 7.6 kb to 1 kb. These IGF-I mRNAs can be detected in a variety of adult tissues as well as in several tumor tissues but only weakly in fetal tissues. This supports the notion that IGF-I is mainly functional in postnatal growth.

Finally alternative polyadenylation signals have been detected in the 3' untranslated regions of the gene which may explain in part the wide array of IGF-I mRNA species.

In summary, human IGF-I has many interesting features. Its gene is exceptionally large and regulation of expression can be directed on various levels by alternative splicing, differential polyadenylation and extensive precursor processing. Furthermore these processes of regulation of expression can act in a tissue-specific and development-specific manner.

HUMAN IGF-II

Initially with the employment of IGF-I cDNAs, followed by more specific probes, a variety of IGF-II cDNAs have been isolated (1,7,8,20,25,33). Unlike IGF-I, where different precursor proteins can be formed, all IGF-II cDNAs encode an 180 amino acids long precursor protein. The 67 amino acids mature IGF-II is flanked by an amino terminus of 24 amino acids encoding a signal peptide and a carboxyl-terminus encoding a relatively long E domain peptide of 89 amino acids. Analysis of the various IGF-II cDNAS revealed that the 5' non-coding regions diverge extensively indicating differential usage of alternative 5' untranslated exons (7,8,34).

The human gene is located on the tip of the short arm of chromosome 11 (3,5,26,35). It is situated very closely to the insulin gene and spans a region of 30 kb of chromosomal DNA. The complete genomic structure is shown in Figure 2. The gene consists of 8 exons of which exons 5, 6 and 7 contain the coding sequences and exons 1, 2, 3, 4 and 4B are all 5' untranslated sequences which are differentially expressed. Initiation of transcription can take place in a tissue- as well as development-specific way at three different promoters, named P1, P2 and P3 which precede exons 1, 4 and 4B respectively (2,5,6,7,8,9,10,13,18,21,32,34).

Analysis of the mRNAs for IGF-II has shown a complicated pattern of regulation of expression. In Figure 2 a composite map of the IGF-II gene and the various mRNAs transcribed from it are indicated. The 5.3 kb mRNA is initiated at exon 1, which is preceded by promoter region P1 that is active in adult human liver tissue only. Initiation of transcription at the second promoter region P2, preceding exon 4, generates a 6.0 kb mRNA transcript which can be detected in many human fetal tissues as well as in several adult non-liver tissues and a variety of liver cell lines. A second, much smaller mRNA species initiating at P2 can also be detected. This 2.2 kb mRNA is generated by alternative polyadenylation at an internal site in exon 7 (8). The amount of 2.2 kb mRNA species can vary extensively among tissues.

Finally a third promoter region P3 folowed by a small exon 4B, located between exons 4 and 5 was identified. Initiation of transcription at P3 gives rise to a 4.8 kb mRNA species which can be detected at low levels in most human fetal tissues and a number of adult non-liver tissues 7,8,10,11,12,16,32,33). An example of Northern blot hybridization with an IGF-II specific probe can be seen in Figure 3.

The length of the various 5' untranslated regions for IGF-II mRNAs range from only 100 bp (4.8 kb species) to almost 1200 bp (6.0 kb species). This may have some physiological implications such as influencing their translatability. Furthermore, these results clearly demonstrate the tissue-specific and developmentally regulated activation of the three promoters of the human IGF-II gene. Specific trans-acting regulatory factors may be involved in the triggering or deactivation of these three promoters.

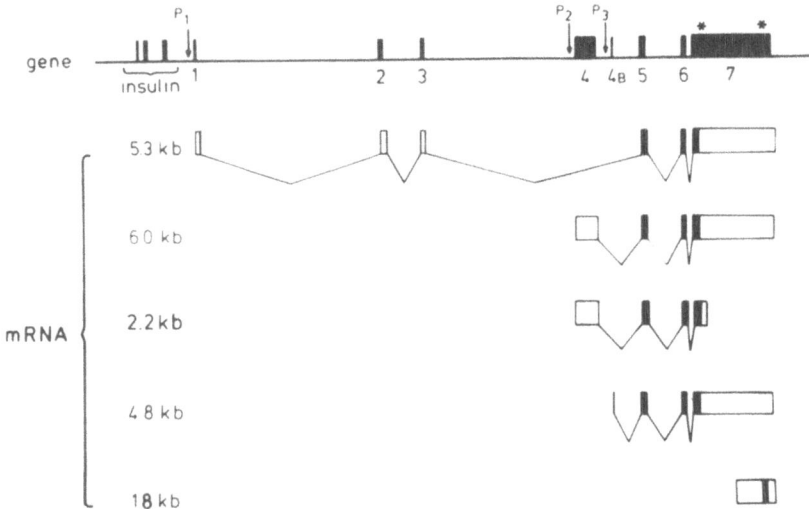

Figure 2. Schematic representation of the human IGF-II gene and the various mRNA transcripts. The human IGF-II gene is located very close to the insulin gene on chromosome 11. The gene consists of 8 exons, named 1 - 4, 4B and 5 - 7. The coding regions are indicated as black boxes in the mRNAs. Polyadenylation sites are marked by asterisks.

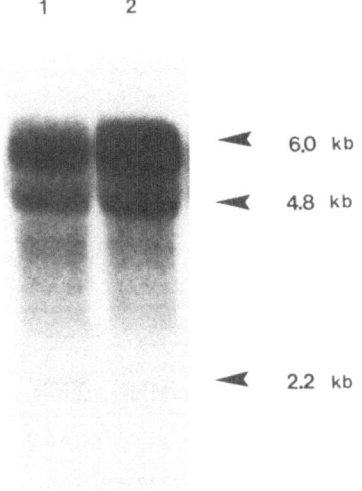

Figure 3. Northern blot of total RNA isolated from two human hepatoma cell lines (22). Lane 1, HepG2 ; lane 2 Hep3B. Each lane contains 10 µg of RNA hybridized with an IGF-II specific cDNA probe (20).

Not only does the IGF-II gene have a complicated structure at the
5' end, the 3' end structure also reveals some interesting features.
Recently the complete nucleotide sequence of the 3' end of the human
IGF-II gene was determined (8). The entire 3' region is contained
within exon 7 and is almost 4 kb long. The longer IGF-II mRNAs of 6.0,
5.3 and 4.8 kb all contain the entire exon 7 sequence, 3879 nucleotides
of which are untranslated followed by a poly(A) tail. The shorter 2.2
kb IGF-II mRNA species terminate at the first polyadenylation signal at
positions 714-719 in exon 7. Why the first polyadenylation signal is
not used more frequently is not clear, but it may be shielded through
the existence of secondary structures in the mRNA.

                  3559
    GGGGCTGGCATGACCCCGGGGGTCGTCCATGCCAGTCCGCCTCAGTCGCAGAGGGTCCCT
         MetThrProGlyValValHisAlaSerProProGlnSerGlnArgValPro

                  3619
    CGGCAAGCGCCCTGTGAGTGGGCCATTCGGAACATTGGACAGAAGCCCAAAGAGCCAAAT
    ArgGlnAlaProCysGluTrpAlaIleArgAsnIleGlyGlnLysProLysGluProAsn

                  3679
    TGTCACAATTGTGGAACCCACATTGGCCTGAGATCCAAAACGCTTCGAGGCACCCCAAAT
    CysHisAsnCysGlyThrHisIleGlyLeuArgSerLysThrLeuArgGlyThrProAsn

                  3739
    TACCTGCCCATTCGTCAGGACACCCACCCACCCAGTGTTATATTCTGCCTCGCCGGAGTG
    TyrLeuProIleArgGlnAspThrHisProProSerValIlePheCysLeuAlaGlyVal

                  3799
    GGTGTTCCCGGGGGCACTTGCCGACCAGCCCCTTGCGTCCCCAGGTTTGCAGCTCTCCCC
    GlyValProGlyGlyThrCysArgProAlaProCysValProArgPheAlaAlaLeuPro

                  3859
    TGGGCCACTAACCATCCTGGCCCGGGCTGCCTGTCTGACCTCCGTGCCTAGTCGTGGCTC
    TrpAlaThrAsnHisProGlyProGlyCysLeuSerAspLeuArgAla

Figure 4. Nucleotide sequence and derived amino acid sequence of the
longest open reading frame in the 1.8 kb mRNA transcript. Note that a
correction on the nucleotide sequence of exon 7 as presented by de
Pagter-Holthuizen et al. (8) is included. An insertion of a 74 base
pair fragment was established by Dr. L. Helman (Bethesda,MD, personal
communication) and in the authors' laboratory and has been included in
Figure 4 from positions 3749 to 3822. As a consequence of the
insertion, the hypothetical 84 amino acid protein encoded by the 1.8 kb
mRNA increases to 113 residues.

    Most surprisingly a novel 1.8 kb mRNA species was detected on
Northern blots of both fetal and adult liver mRNA which consists of an
internal region of exon 7 (8).
This 1.8 kb mRNA was shown to be colinear with exon 7 and its 3'
terminus is identical to that of IGF-II mRNAs. The abundance even seems
to parallel that of IGF-II mRNAs.

Although it does not code for IGF-II, three other open reading frames are present within this mRNA, one is located between positions 2864 and 3040, and a second one between positions 2910 and 3065. They may encode peptides of 59 and 52 amino acid residues, respectively. However, as judged from amino acid composition of these predicted peptides, it is questionable that these two open reading frames are actually translated. The longest reading frame is located between positions 3519 to 3847 and may encode a 113 amino acid peptide (Figure 4). In vitro translation of this mRNA using a rabbit reticulocyte lysate system indeed directs the synthesis of a translation product of approximately 8.3 kDa. This creates a controversy since such a molecular weight would be too small for a 113 amino acid protein. Therefore it remains to be proven that the latter peptide actually is synthesized in vivo and we can only speculate on the possible physiological function of such a protein.

Expression of the human IGF-II gene is very complicated and regulated at various levels.

First, the existence of multiple promoters can direct activation of the gene in a very accurate manner by utilizing different transcription factors which may be development and/or tissue specific. Furthermore the various 5' untranslated leader sequences may influence the efficiency of translation in a considerable way.

Second, the presence of at least two functional polyadenylation signals leads to the synthesis of mRNAs which only differ in the size of their 3' untranslated region, affecting most likely the stability of the mRNAs.

Thirdly and most speculatively, the possibility of coexpression of a second protein, which open reading frame is located within the 3' untranslated region of the IGF-II mRNA, gives many possibilities for a sophisticated regulatory mechanism.

The complex way by which the expression of the IGF genes is regulated is consistent with the view that the IGFs play a wide role in various biological processes which influence growth and development. Production of the IGFs is regulated at different levels, from transcription of the gene to extensive processing of the precursor molecules. With the elucidation of the genomic structures for IGF-I and IGF-II it is now feasible to study the regulatory mechanisms involved in IGF expression.

Acknowledgements

The author thanks Dr. J.S. Sussenbach for critical reading of the manuscript. This work was supported in part by the Netherlands Organization for the Advancement of Pure Research (NWO).

REFERENCES

1. Bell GI, Merryweather JP, Sanchez-Pescador R, Stempien MM, Priestley L, Scott J, Rall LB (1984) Sequence of a cDNA clone encoding human preproinsulin-like growth factor II. Nature 310:775-777

2. Bell GI, Gerhard DS, Fong NM, Sanchez-Pescador R, Rall LB (1985) Isolation of the insulin-like growth factor genes: Insulin-like growth factor II and insulin genes are contiguous. Proc Natl Acad Sci USA 82:6450-6454

3. Brissenden JE, Ullrich A, Francke U (1984) Human chromosomal mapping of genes for insulin-like growth factors I and II and epidermal growth factor. Nature 310:781-784

4. D'Ercole AJ, Stiles AD, Underwood LE (1984) Tissue concentrations of somatomedin C: further evidence for multiple sites of synthesis and paracrine or autocrine mechanisms of action. Proc Natl Acad Sci USA 81:935-939

5. De Pagter-Holthuizen P, Höppener JWM, Jansen M, Geurts van Kessel AHM, Van Ommen GJB, Sussenbach JS (1985) Chromosomal localization and preliminary characterization of the human gene encoding insulin-like growth factor II. Hum Genet 69:170-173

6. De Pagter-Holthuizen P, Van Schaik FMA, Verduijn GM, Van Ommen GJB, Bouma BN, Jansen M, Sussenbach JS (1986) Organization of the human genes for insulin-like growth factors I and II. FEBS Lett 195:179-184

7. De Pagter-Holthuizen P, Jansen M, Van Schaik FMA, van der Kammen RA, Oosterwijk C, Van den Brande JL, Sussenbach JS (1987) The human insulin-like growth factor II contains two development-specific promoters. FEBS Lett 214:259-264

8. De Pagter-Holthuizen P, Jansen M, van der Kammen RA, Van Schaik FMA, Sussenbach JS (1988) Differential expression of the human insulin-like growth factor II gene. Characterization of the IGF-II mRNAs and an mRNA encoding a putative IGF-II-associated protein. Biochim Biophys Acta 950:282-295

9. Dull TJ, Gray A, Hayflick JS, Ullrich A (1984) Insulin-like growth factor II precursor gene organization in relation to the insulin gene family. Nature 310:777-781

10. Han VKM, D'Ercole AJ, Lund PK (1987) Cellular localization of somatomedin (insulin-like growth factor) messenger RNA in the human fetus. Science 236:193-197

11. Han VKM, Hill DJ, Strain AJ, Towle AC, Lauder JM, Underwood LE, D'Ercole AJ (1987) Identification of somatomedin/insulin-like growth factor immunoreactive cells in the human fetus. Pediatric Res 22:245-249

12. Han VKM, Lund PK, Lee DC, D'Ercole AJ (1988) Expression of somatomedin/insulin-like growth factor messenger ribonucleic acids in the human fetus: Identification, characterization and tissue distribution. J Clin Endocrinol Metabol 66:422-429

13. Haselbacher GK, Irminger JC, Zapf J, Ziegler WH, Humbel RE (1987) Insulin-like growth factor II in human adrenal pheochromocytomas and Wilms tumors: expression at the mRNA and protein level. Proc Natl Acad Sci USA 84:1104-1106

14. Huff KK, Kaufman D, Gabbay KH, Spencer EM, Lippman ME, Dickson RB (1986) Secretion of an insulin-like growth factor-I-related protein by human breast cancer cells. Cancer Res 46:4631-4619

15. Höppener JWM, De Pagter-Holthuizen P, Geurts van Kessel AHM, Jansen M, Kittur SD, Antonorakis SE, Lips CJM, Sussenbach JS (1985) The human gene encoding insulin-like growth factor I is located on chromosome 12. Hum Genet 69:157-160

16. Höppener JWM, Steenbergh PH, Slebos RJC, De Pagter-Holthuizen P, Jansen M, Roos BA, Van den Brande JL, Sussenbach JS, Jansz HS, Lips CJM (1987) Expression of insulin-like growth factor -I and -II genes in rat medullary thyroid carcinoma. FEBS Lett 215:122-126

17. Höppener JWM, Mosselman S, Roholl PJM, Lambrechts, Slebos RJC, De Pagter-Holthuizen P, Lips CJM, Jansz HS, Sussenbach JS (1988) Expression of insulin-like growth factor-I and -II genes in human smooth muscle tumours. EMBO J 7:1379-1385

18. Irminger JC, Rosen KM, Humbel RE, Vila-Komaroff L (1987) Tissue-specific expression of insulin-like growth factor II mRNAs with distinct 5' untranslated regions. Biochemistry 84:6330-6334

19. Jansen M, Van Schaik FMA, Ricker AT, Bullock B, Woods KH, Gabbay DE, Nussbaum AL, Sussenbach JS, Van den Brande JL (1983) Sequence of cDNA encoding human insulin-like growth factor I precursor. Nature 306:609-611

20. Jansen M, Van Schaik FMA, Van Tol H, Van den Brande JL, , Sussenbach JS (1985) Nucleotide sequences of cDNAs encoding precursors of human insulin-like growth factor II (IGF-II) and an IGF-II variant. FEBS Lett 179:243-246

21. Jansen M, De Pagter-Holthuizen P, Höppener JWM, Sussenbach JS, Van den Brande JL (1987) Somatomedin gene structure and expression. In: Highlights on endocrinology. Eds. C. Christiansen, BJ Riis, Kopenhagen.

22. Knowles BB, Howe CC, Aden DP (1980) Human hepatocellular carcinoma cell lines secrete the major plasma proteins and hepatitis B surface antigen. Science 209:497-499

23. Kozak M (1987) An analysis of 5'noncoding sequences from 699 vertebrate messenger RNAs. Nucleic Acids Res 15:8125-8132

24. LeBouc Y, Dreyer D, Jaeger F, Binoux M, Sondermeyer P (1986) Complete characterization of the human IGF-I nucleotide sequence isolated from a newly constructed adult liver cDNA library. FEBS Lett 196:108-112

25. LeBouc Y, Noguiez P, Sondermeyer P, Dreyer D, Girard F, Binoux M (1987) A new 5'-non-coding region for human placental insulin-like growth factor II mRNA expression. FEBS Lett 222:181-185

26. Mannens M, Slater RM, Heyting C, Bliek J, De Kraker J, Coad N, De Pagter-Holthuizen P, Pearson PL (1988) Molecular nature of genetic

changes resulting in loss of heterozygosity of chromosome 11 in
Wilms' tumours. Hum Genet 81:41-48

27. Rinderknecht E, Humbel RE (1978a) The amino-acid sequence of human
    insulin-like growth factor I and its structural homology with
    proinsulin. J Biol Chem 253:2769-2775

28. Rinderknecht E, Humbel RE (1978b) Primary structure of human
    insulin-like growth factor II. FEBS Lett 89:283-286

29. Rotwein P (1986) Two insulin-like growth factor I messenger RNAs
    are expressed in human liver. Proc Natl Acad Sci USA 83:77-81

30. Rotwein P, Folz RJ, Gordon JI (1987) Biosynthesis of human
    insulin-like growth factor I. The primary translation product of
    IGF-I mRNA contains an unusual 48 amino acid signal peptide. J Biol
    Chem 262:11807-11812

31. Rotwein P, Pollock KM, Didier DK, Krivi GG (1986) Organization and
    sequence of the human insulin-like growth factor I gene. J Biol
    Chem 261:4828-4832

32. Schofield PN, Tate VE (1989) Development regulation of IGF-II
    transcription. In press.

33. Shen SJ, Daimon M, Wang CY, Jansen M, Ilan J (1986) Isolation of
    an insulin-like growth factor II cDNA with a unique 5' untranslated
    region from human placenta. Proc Natl Acad Sci USA 85:1947-1951

34. Sussenbach JS (1989) The gene structure of the insulin-like growth
    factor family. Progress Growth Factor Res 1:33-48

35. Tricoli JV, Rall LB, Scott J, Bell GI, Shows TB (1984) Localization
    of insulin-like growth factor genes to human chromosomes 11 and 12.
    Nature 310:784-786

36. Ullrich A, Berman CH, Dull TJ, Gray A, Lee JM (1984) Isolation of
    the human insulin-like growth factor I gene using a synthetic DNA
    probe. EMBO J 3:361-364

37. Zapf J, Schmid C, Froesch ER (1984) Biological and immunological
    properties of insulin-like growth factors I and II. Clin Endocrinol
    Metab 13:3-30

TRANSCRIPTIONAL DIVERSITY IN RAT INSULIN-LIKE GROWTH FACTOR I GENE EXPRESSION

Charles T. Roberts, Jr., William L. Lowe, Jr., and
Derek LeRoith

Section of Molecular and Cellular Physiology
Diabetes Branch, Building 10, Room 8S-243
National Institutes of Health, Bethesda, MD 20892

Insulin-like growth factor I (IGF-I), or Somatomedin C, is a member
of a family of insulin-like peptides which also includes insulin itself,
insulin-like growth factor II (IGF-II) or multiplication-stimulating
activity (MSA), and, in some schemes, relaxin (1,2,3). The IGFs are
similar to proinsulin in that they contain B and A domains separated by a
C peptide sequence, but differ in that the mature IGF-I proteins retain
their C peptide moeities and therefore consist of single polypeptide
chains. Additionally, the mature forms of the IGFs contain a carboxyterminal D domain not found in insulin (2,4,5). Finally, as discussed in
more detail below with respect to IGF-I, cDNA sequences corresponding to
IGF-I and IGF-II mRNAs suggest the presence of an additional carboxyterminal E peptide predicted to be a component of the IGF-I and IGF-II
pro-hormones (6,7,8,9).

IGF-I is known to have short-term metabolic effects such as
enhanced glucose uptake and glycogen synthesis, as well as long-term
growth-promoting effects (as assayed by increased DNA synthesis). As
such, it is thought to mediate many, but not all, of the effects of
growth hormone (GH) during development. Althought the liver accounts for
the majority of circulating IGF-I (10), it is clear that many extrahepatic tissues contain significant amounts of hormone which is probably
the result of local biosynthesis (11). Thus, IGF-I undoubtedly has
important autocrine or paracrine functions in addition to any classical
endocrine roles.

As part of our study of rat IGF-I gene expression, we have isolated
a number of IGF-I cDNAs from an adult rat liver cDNA library using a
human IGF-I cDNA as a probe (12,13). The sequences of these clones and
the results of hybridization analyses of RNA using these clones as probes
have revealed three aspects of the transcriptional diversity of the rat
IGF-I gene. Firstly, IGF-I mRNAs contain multiple 5'-untranslated regions
(UTRs) which may encode alternate forms of pre-pro-IGF-I. Secondly,
differential splicing in the E peptide-coding region produces RNAs
encoding alternate forms of pro-IGF-I. Finally, the rat IGF-I gene is
transcribed into several size classes of mRNA, all of which are larger
than necessary to encode the IGF-I precursor. In this chapter we will
discuss these types of transcriptional diversity, their relationship to

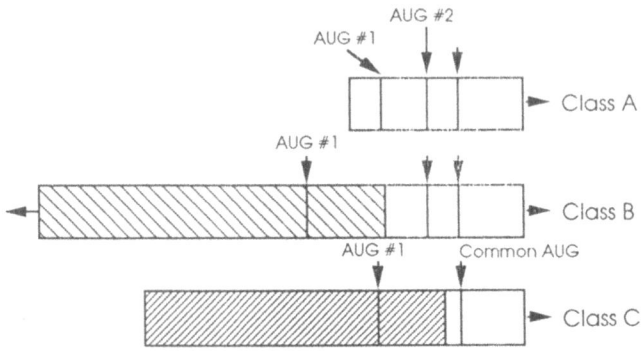

Figure 1. Schematic respresentation of the alternative 5'-UTRs
present in rat IGF-I cDNA clones. The common AUG is at MET-22
with respect to the amino acid sequence of the mature peptide.

one another and the regulation of this diversity by the major regulator
of IGF-I biosynthesis, GH.

    Rat IGF-I cDNAs isolated from an adult rat liver cDNA library were
found to contain three divergent UTRs (termed class A, B, or C) preceed-
ing the region encoding the mature peptide (13). As depicted schematically in figure 1, all the cDNAs were identical through the 15 bases
upstream of the codon specifying the Met-22 of the IGF-I signal peptide
amino acid sequence. At this point, the class A and B sequences diverged
from the class C sequence. The class A and B sequences diverged from
each other 57 bases further upstream of the divergence point of these
sequences from the class C sequence. Recently, a variant of the class C
sequence has been described which has a 186-base deletion in the region
upstream of the divergence of this sequence from the class A and B
sequences (14). Since IGF-I mRNAs are transcribed from a single-copy
gene in the rat (15), this diversity must arise from differential
splicing of the primary transcript, possibly in conjunction with multiple
promoters as is the case in the rat IGF-II gene (16). In this context,
it should be noted that an intron does occur in the rat gene at the
position at which the class A and B 5'-UTR sequences diverge from the
class C sequence (15). A clear understanding of the genetic basis for
the diversity in 5'-UTR sequences will require more extensive characterization of rat genomic sequences upstream of those exons which encode the
mature peptide.

    Two aspects of the various 5'-UTR sequences merit further discus-
sion. As illustrated in Figure 1, all three sequences contain upstream
AUG codons which are in frame with the sequence encoding the mature IGF-I
peptide, and there are no intervening termination codons. Thus, initiation at these upstream AUG codons in vivo could give rise to a collection
of pre-pro-IGF molecules which could conceivably be differentially processed or secreted. Indeed, in vitro translation of a human IGF-I mRNA
sequence (extremely homologous to the rat class C sequence) was found to
initiate at an upstream AUG, giving rise to an IGF-I precursor containing
a 48-amino acid signal sequence (17). It must be borne in mind, then,
that certain portions of these putative UTRs may be translated in vivo.

    The sequence of the class A 5'-UTR is particularly intriguing in that the 31 bases unique to the class A sequence are part of a 40-base sequence which is complementary to a 41-base segment of the 3'-UTR sequence found in all rat IGF-I mRNAs. This sequence arrangement would allow annealling of the 5' and 3'-UTR sequences, forming an RNA duplex which would include the most upstream AUG in the class A 5'-UTR sequence. This secondary structure could potentially inhibit translation of this IGF-I mRNA species in vivo, unless the RNA duplex was dissociated by some type of RNA helicase activity. Such activities have been characterized in Xenopus (18,19) and in mammalian cells (20). Thus, the divergent 5'-UTRs found in rat IGF-I mRNAs suggest the possibility of heterogeneity in IGF-I precursors, as well as possible translational control of one species of IGF-I mRNA (13).

In order to obtain further evidence for the possible physiological significance of this transcriptional diversity in 5'-UTR sequences, we evaluated the tisses distribution and GH sensitivity of IGF-I mRNAs with the various 5'-UTRs. This was done by simultaneously determining the steady-state levels of IGF-I mRNAs containing the class A, B, and C 5'-UTRs with sensitive solution hybridization/RNase protection assays (21). These experiments used an antisense RNA probe complementary to 224 bases of coding region and 98 bases of the class A 5'-UTR (Figure 2).

Hybridization of this probe to aliquots of total RNA, followed by RNase digestion and denaturing polyacrylamide gel electrophoresis, would be expected to yield protected probe fragments 322, 297, and 241 bases in length, corresponding to IGF-I mRNAs with the class A, B, or C 5'-UTR. Additionally, the intensity of the bands (after correction for band length) reflects the relative levels of the three transcripts in each sample. The rationale for this approach is illustrated in Figure 2, which also shows the results obtained with normal liver RNA.

Using this technique, we have evaluated the relative abundance of IGF-I mRNAs with the various 5'-UTRs in several hepatic and extra-hepatic tissues from hypophysectomized rats treated with rat GH or with control vehicle alone. The results of this analysis are shown in Figure 3. With respect to the tissue distribution of the 5'-UTR variants, it can be seen that liver contains class A, B, and C IGF-I mRNAs, while lung and kidney contain only the class A and C variants. Only the class C variant was detectable in heart. In terms of relative abundance, the class C variant was the most prevalent in all four tissues, the class A variant was present in significant amounts in liver and less so in lung and kidney, and the class B variant was restricted to liver, where it was relatively rare. The quantitation of the effect of GH on the levels of the various 5'-UTR-containing IGF-I mRNAs is shown in Table 1. The steady-state levels of the class C transcripts were increased 1.7- to 3-fold in each tissue by GH, whereas the levels of the class A and B transcripts were elevated 6-8-fold in liver. Class A transcripts in extra-hepatic tissues were, however, unresponsive to GH. Thus, GH differentially regulates alternate splicing in the 5'-UTR region of the rat IGF-I gene in a tissue-specific manner.

The possibility of alternate splicing in the IGF-I E peptide coding region was first suggested by the characterization of human IGF-I cDNAs which contained alternative C-terminal E peptide coding sequences and 3'-untranslated regions (6,22). The first reported sequence subsequently was termed IGF-Ia, and the variant, IGF-Ib (22). The derived E-peptide

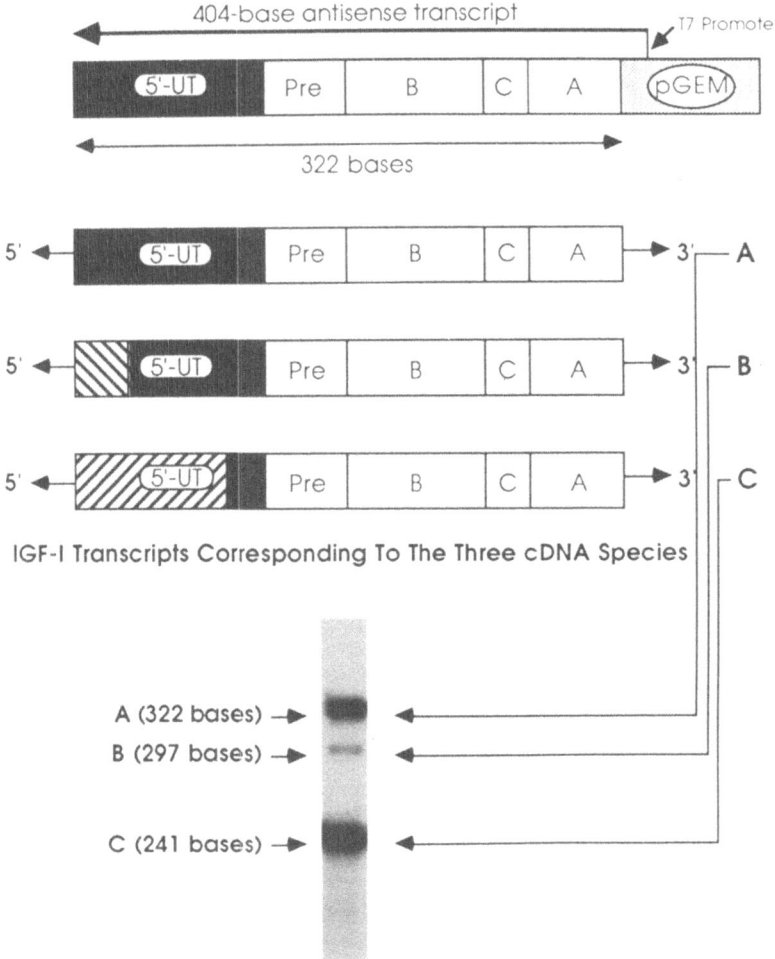

SCHEMATIC REPRESENTATION OF SOLUTION HYBRIDIZATION QUANTITATION OF THE THREE 5'-UNTRANSLATED REGIONS IN IGF-I TRANSCRIPTS

404-base antisense transcript

T7 Promoter

5'-UT | Pre | B | C | A | pGEM

322 bases

5' ◄ 5'-UT | Pre | B | C | A ► 3' ─ A

5' ◄ 5'-UT | Pre | B | C | A ► 3' ─ B

5' ◄ 5'-UT | Pre | B | C | A ► 3' ─ C

IGF-I Transcripts Corresponding To The Three cDNA Species

A (322 bases) →
B (297 bases) →
C (241 bases) →

Autoradiogram of bands protected by the IGF-I antisense transcript

Figure 2. Rationale for solution hybridization/RNase protection assays described in the text. The results shown at the bottom of the figure are typical of those obtained with normal liver RNA from a 50-day-old rat.

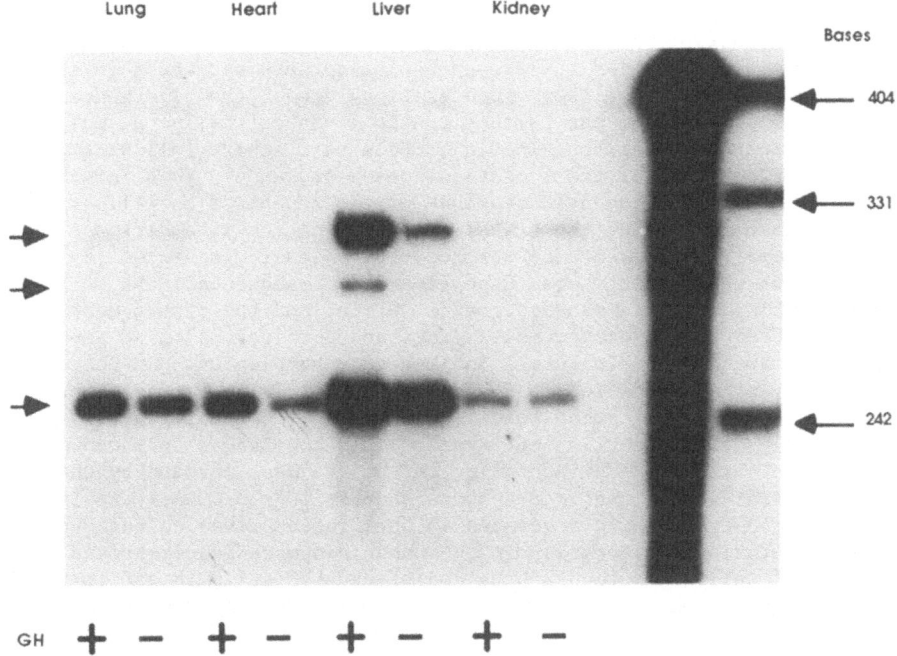

Figure 3. Solution hybridization/RNase protection analysis of aliquots of total RNA from various tissues of hypophysectomized and GH-treated rats. The three arrows at the left illustrate the positions of protected probe bands corresponding to IGF-I mRNAs with the class A (top), class B (middle), or class C (bottom) 5'-UTRs.

Table 1. Effect of GH on IGF-I 5'-UTR variants

IGF-I transcript	Liver	Kidney	Lung	Heart
Class A				
(322 bases)	7.7 ± 2.5	1.0 ± 0.3	1.1 ± 0.1	—
Class B				
(297 bases)	6.4 ± 3.1	—	—	—
Class C				
(241 bases)	2.2 ± 0.9	1.7 ± 0.3	2.0 ± 0.3	2.9 ± 0.4

Values represent the relative increase in the steady-state levels of class A, B, and C IGF-I mRNAs due to GH treatment.

amino acid sequences were identical for the first 16 residues, at which
point the Ia and Ib E peptide sequences diverged. The subsequent
elucidation of the human IGF-I gene sequence (23) revealed that the
alternate Ia and Ib mRNAs arose from the mutually exclusive splicing of
either exon 4 (IGF-Ib) or exon 5 (IGF-Ia) to exons 2 and 3 (which encode
the remainder of the pro-hormone.

Subsequently, characterization of multiple mouse (24) and rat (12)
IGF-I cDNAs also revealed the presence of alternate sequences in the E
peptide-coding region, although the exact nature of the divergence
appeared to differ from the human situation. Specifically, a small
proportion of the cDNAs contained a 52-base pair insert following the
codon for the Lys 16 of the E peptide coding sequence. This insert
alters the derived amino acid sequence of the E peptide as well as the
reading frame, such that the translation termination site differs
depending upon the presence or absence of the insert sequence. By
analogy with the variant human E peptide-coding sequences, the
insert-lacking sequence was designated IGF-Ia, and the insert-containing
sequence IGF-Ib. The human, mouse, rat, and porcine (25) Ia E peptides
are all 35 amino acids in length and are very homologous. On the other
hand, the rat and mouse IGF-Ib E peptides are 41 amino acids long as
compared with their 77-amino-acid-long human IGF-Ib counterpart;
additionally, the rodent Ib peptides are only homologous to the human
sequence in the region corresponding to the residues encoded by the
52-base pair insert found in the mouse and rat IGF-Ib cDNAs. Analysis of
the human and rat genomic sequences in this region revealed that an
apparent mutational alteration in the human sequence had destroyed the
original 3' splice junction at the end of exon 4 (still intact in the rat
genomic sequence) so that the human exon 4 is an extended version of the
original 52-base pair exon 4, and includes sequences which encode the
unique human IGF-Ib E peptide as well as an alternative 3'-UTR (26).
This situation is illustrated in Figure 4.

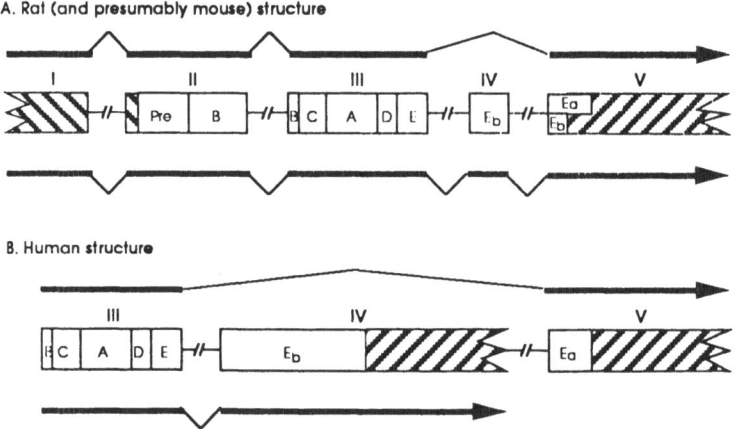

Figure 4. Schematic representation of alternate splicing in
the E-peptide coding region of rodent and human IGF-I genes.
The splicing patterns above the gene structure represents
IGF-Ia transcripts and those below, IGF-Ib transcripts.

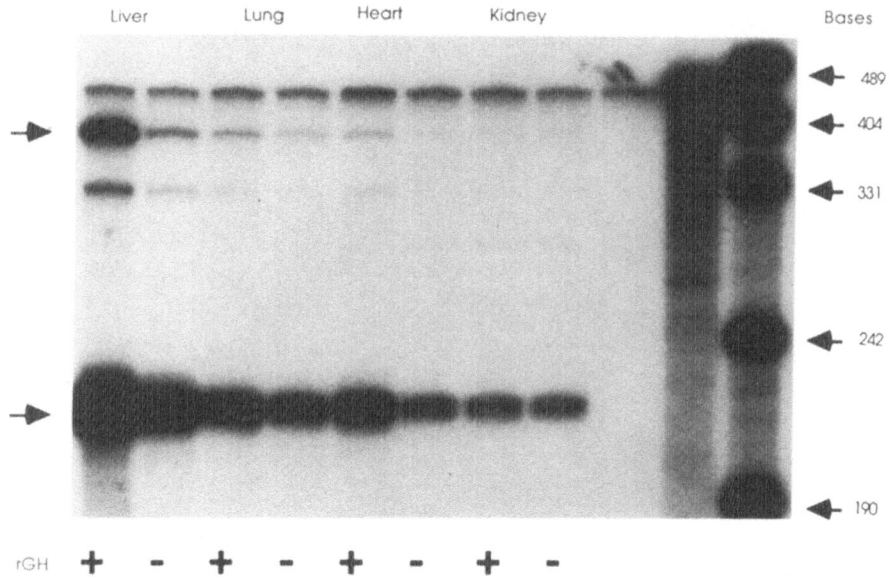

Figure 5. Solution hybridization/RNase protection analysis of aliquots of
total RNA from various tissues of hypophysectomized and GH treated
rats. The arrow at the left represent the position of protected probe
bands corresponding to IGF-Ib (upper) and IGF-Ia (lower) mRNAs.

Table 2. Effect of GH on steady-state levels of IGF-Ib and Ia mRNAs

IGF-I Transcript (Protected Band Length)	Liver	Heart	Kidney	Lung
IGF-Ib (386 bases)	8.1 ± 1.7	3.2 ± 0.4	1.3 ± 0.3	1.8 ± 0.1
IGF-Ia (224 bases)	2.4 ± 0.7	3.2 ± 0.1	1.3 ± 0.4	2.6 ± 0.6

In order to analyze the transcriptional regulation of alternate
splicing in the E-peptide coding region of the rat IGF-I gene, we
employed a solution hybridization/RNase protection assay similar in
concept to the procedure used to analyze the differential splicing of
alternate 5'-UTRs described in a previous section. In this case, the
probe consisted of an antisense RNA complementary to mRNA sequences
encoding a portion of the A peptide, the D domain, the IGF-Ib E domain
and 172 bases of the 3'-UTR. Hybridization of this probe to total RNA
would be expected to result in the appearance of protected bands 376
bases (corresponding to IGF-Ib mRNAs) or 224 and 100 bases (corresponding
to IGF-1a mRNAs) in length. Figure 5 shows the results of hybridization
of this probe to aliquots of RNA from various tissues of hypophysectomized rats treated with rat GH or with control vehicle alone. It is clear
that both IGF-Ia and IGF-Ib mRNAs are present in every tissue, and that
IGF-Ia mRNA is the predominant species (>90%) in each case. The
quantitation of the effect of GH on the steady-state levels of IGF-Ia and
IGF-Ib mRNAs is shown in Table 2. Both mRNA species were increased
slightly in every case by GH, although IGF-Ib mRNA levels in liver were
significantly more stimulated by GH treatment.

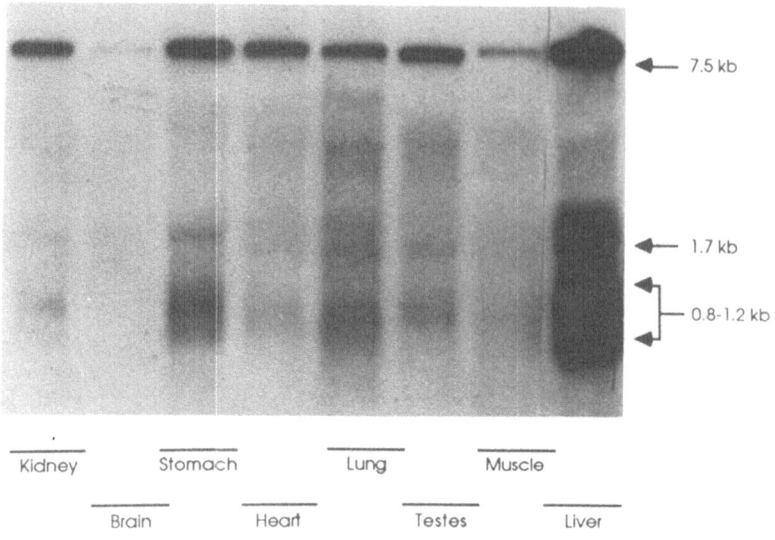

Figure 6. Northern analysis of IGF-I mRNA size classes
in RNA from various tissues of 50-day-old rats.

A comparison of the tissue specificity and control by GH of splicing
in the 3'-coding region as compared to splicing in the 5'-UTR/pre-peptide
coding region reveals that the former is much more coordinate than the
latter. Thus, alternative splicing in these two separate regions of the
rat IGF-I gene can be differentially regulated by tissue-specific and
hormonal factors.

The third aspect of the transcriptional diversity in rat IGF-I gene
expression is the generation of multiple sizes of IGF-I mRNA, presumably
by differential processing of the primary transcript. The possible
contribution of multiple promoters to this heterogeneity is impossible to
assess at the present time. As shown in Figure 6, multiple size classes
of IGF-I mRNA occur in every tissue examined. The major species is ~7.5
kilobases in length, two species 0.8-1.2 and ~1.7 kilobases in length are
the next most abundant, and a small amount of a 4.7-kilobase species is
detectable in some tissues. All of these mRNAs are longer than necessary
to encode any forms of pre-pro-IGF-I described to date, and therefore
must contain (at least in the case of the larger mRNA species) extensive
5'-or 3'-UTR sequences. All of the size classes of IGF-I are
polyadenylated (W.L.L., Jr., unpub. observations), although poly(A) tails
probably do not contribute significantly to the overall size of the
various size classes of IGF-I mRNA.

Finally, hybridization of Northern blots of liver RNA with probes
specific for various 5'-UTR sequences or for IGF-I mRNAs encoding
alternate E-peptides gave results similar to those obtained with a probe
complementary to sequences encoding the mature IGF-I peptide (21,26).
Thus, those mechanisms which give rise to the multiple size classes of
IGF-I mRNA must involve segments of the primary transcript which lie
outside of these sequences in the immediate vicinity of the pre-pro-IGF-I
coding region. A clearer understanding of the overall organization and
splicing of the IGF-I gene will, therefore, require significantly more
structural information than has been described to date.

References

1. T.L. Blundell, S. Bedarker, and R.E. Humbel, Tertiary structures, receptor binding, and antigenicity of insulin-like growth factors. Fed. Proc. 42:2592 (1983).
2. E.Rinderknecht and R.E. Humbel, The amino acid sequence of human insulin-like growth factor I and its structural homology with proinsulin. J. Biol. Chem. 253:2769 (1978a).
3. M.O. Dayhoff, Atlas of Protein Sequence and Structure, p. 150, (National Biomedical Research Foundation, Washington, D.C.) (1978).
4. D.C. Klapper, M.E. Svoboda, and J.J. Van Wyk, Sequence analysis of somatomedin-C: confirmation of identity with insulin-like growth factor II. FEBS Letters 89:283 (1983).
5. E. Rinderknecht and R.E. Humbel, Primary structure of insulin-like growth factor II. FEBS Letters 89:283 (1978b).
6. M. Jansen, F.M.A. van Schaik, A.T. Ricker, B. Bullock, D.E. Woods, K.H. Gabbay, A.L. Nussbaum, J.S. Sussenbach, and J.L. van den Brande, Sequence of cDNA encoding human insulin-like growth factor I precursor. Nature 306:609 (1983).
7. A. Ullrich, C.H. Berman, T.J. Dull, A. Gray, and J.M. Lee, Isolation of the human insulin-like growth factor I gene using a single synthetic DNA probe. EMBO Journal 3:361 (1984).
8. G.I. Bell, J.P. Merryweather, R. Sanchez-Pescador, M.M. Stempien, L. Preistly, S. Scott, and L. B. Rall, A cDNA clone encoding human preproinsulin-like growth factor II. Nature 310:775 (1984).
9. H.R. Whitfield, C.B. Bruno, R. Frunzio, J.R. Terrell, S.P. Nissley, and M.M. Rechler, Isolation of a cDNA clone encoding rat insulin-like growth factor-II precursor. Nature 312:277 (1984).
10. E.R. Froesch, C. Schmid, J. Schwander, and J. Zapf, Actions of insulin-like growth factors. Ann. Rev. Physiol. 47:443 (1985).
11. A.J. D'Ercole, A.D. Stiles, and L.E. Underwood, Tissue concentrations of somatomedin C: Further evidence for multiple sites of synthesis and paracrine or autocrine mechanisms of action. Proc. Nat. Acad. Sci. (USA) 81:935 (1984).
12. C.T. Roberts, Jr., S.R. Lasky, W.L. Lowe, Jr., W.T. Seaman, and D. LeRoith, Molecular cloning of rat insulin-like growth factor I complementary deoxyribonucleic acids: differential messenger ribonucleic acid processing and regulation by growth hormone in extrahepatic tissues. Mol. Endocrinol. 1:243 (1987).
13. C.T. Roberts, Jr., S.R. Lasky, W.L. Lowe, Jr., and D. LeRoith, Rat IGF-I cDNAs contain multiple 5'-untranslated regions. Biochem. Biophys. Res. Comm. 146:1154 (1987).
14. A. Shimatsu and P. Rotwein, Sequence of two rat insulin-like growth factor I mRNAs differing within the 5'-untranslated region. Nuc. Acids Res. 15:7196 (1988).
15. A. Shimatsu and P. Rotwein Mosaic evolution of the insulin-like growth factors. J. Biol. Chem. 262:7894 (1987).
16. L. Chiariotti, A.L. Brown, R. Frunzio, D.R. Clemmons, M.M. Rechler, and C.B. Bruni, Structure of the rat insulin-like growth factor II transcriptional unit. Mol. Endocrinol. 2:1115 (1988).
17. P. Rotwein, R.J. Folz, and J.J. Gordon, Biosynthesis of human insulin-like growth factor I (IGF-I). J. Biol. Chem 262:11807 (1987).
18. M.R. Rebagliati, and D.A. Melton, Antisense RNA injections in fertilized frog eggs reveal an RNA duplex unwinding activity. Cell 48:599 (1987).
19. B.L. Bass and H. Weintraub, A developmentally regulated activity that unwinds RNA duplexes. Cell 48:607 (1987).

20. R.W. Wagner and K. Nishikawa, Cell cycle expression of RNA duplex unwindase activity in mammalian cells. Mol. Cell. Biol. 8:770 (1988).

21. W.L. Lowe, Jr., C.T. Roberts, Jr., S.R. Lasky, and D. LeRoith, Differential expression of alternative 5'-untranslated regions in mRNAs encoding rat insulin-like growth factor I. Proc. Natl. Acad. Sci. (USA) 84:8946 (1987).

22. P. Rotwein, Two insulin-like growth factor I messenger RNAs are expressed in human liver. Proc. Natl. Acad. Sci. (USA) 83:77 (1986).

23. P. Rotwein, K.M. Pollock, D.K. Didier, and G.G. Krivi, Organization and sequence of the human insulin-like growth factor I gene. J. Biol. Chem. 261:4828 (1986).

24. G.I. Bell, M.M. Stempien, N.M. Fong, and L.B. Rall, Sequence of liver cDNAs encoding two different mouse insulin-like growth factor I precursors. Nuc. Acids Res. 14:7873 (1986).

25. A. Tavakkol, F.A. Simmen, and R.C.M. Simmen, Porcine insulin-like growth factor I (pIGF-I): complementary deoxyribonucleic acid cloning and uterine expression of messenger ribonucleic acid encoding evolutionarily conserved IGF-I peptides. Mol. Endocrinol. 2:674 (1988).

26. W.L. Lowe, Jr., S.R. Lasky, D. LeRoith, and C.T. Roberts, Jr., Distribution and regulation of rat insulin-like growth factor I messenger ribonucleic acids encoding alternative carboxyterminal E peptides: evidence for differential processing and regulation in liver. Mol. Endocrinol. 2:528 (1988).

PHYSIOLOGICAL REGULATION OF INSULIN-LIKE GROWTH FACTOR EXPRESSION

P. Rotwein, D. DeVol, P. Lajara, P. Bechtel and M. Hammerman

Washington University School of Medicine
Departments of Medicine and Genetics
St. Louis, Mo. 63110

I. Introduction

The insulin-like growth factors (IGF) I and II comprise a structurally homologous pair of circulating polypeptides of fundamental importance in mammalian growth processes (1). Although IGF-I is a major mediator of growth hormone (GH) action and is predominantly expressed during postnatal life, while IGF-II is synthesized primarily during fetal development (1,2), the precise roles of either peptide in promoting tissue growth remain at best incompletely understood. In addition the relative contributions of circulating versus locally-produced IGFs toward organ or tissue growth also are unknown. With the advent of cloned probes for mammalian IGF genes (3-8) it has become possible to more accurately assess the biosynthesis of IGFs I and II in various tissues and thus potentially to determine the roles of these growth factors as paracrine, autocrine or endocrine agents. Summarized in this chapter are results of several experiments designed to analyze the mechanisms involved in the control of IGF synthesis in vivo. Our data indicate that during growth of skeletal muscle both systemic and local factors can regulate the production of mRNAs for IGFs I and II. In the kidney peripheral GH administration stimulates IGF-I mRNA in collecting duct cells, with a consequent increase in IGF-I.   During renal hypertrophy following unilateral nephrectomy IGF-I production also is enhanced in collecting duct, but in the absence of any changes in steady-state levels of IGF-I mRNA. Thus both transcriptional and translational mechanisms appear to modify IGF-I biosynthesis during renal growth, while during skeletal muscle hypertrophy the regulation of IGF-I and IGF-II expression occurs primarily at the levels of gene transcription and mRNA abundance.

II. IGF Gene Expression During Skeletal Muscle Hypertrophy

In initial experiments we have used a model of chronic GH excess to study the potential roles of the IGFs in muscle growth. Adult female Wistar rats harboring transplanted GH_3 cells grow while their saline injected littermates do not. The weight of many organs and tissues, including muscle, heart, and liver, increased in the GH-stimulated animals. In skeletal muscle (gastrocnemius) IGF-I mRNA increased 8-fold and IGF-II mRNA 6-fold in the treated rats as measured by a specific solution-hybridization assay (9). No change was seen in mRNA abundance in the control animals. A comparable increase was found in cardiac muscle. In liver, although GH induced the accumulation of IGF-I mRNA, IGF-II mRNA remained nearly undetectable (10). These observations support other studies suggesting a role for both IGF-I and IGF-II as locally-derived factors involved in muscle growth processes (11-13), and also demonstrate that in muscle as well as in the brain IGF-II gene expression is not limited to fetal life (9).

The experiments described in the preceding paragraph examined the effects of GH on IGF gene expression over a period of several weeks. We next focused on more acute regulation, that which occurs over days, in two experimental models: muscle hyperpertrophy after GH treatment to hypophysectomized (hypox) rats, and work-induced muscle hypertrophy.

Systemic GH Treatment Induces IGF mRNAs: As anticipated GH treatment (100 ug/day) to hypox rats was effective in restoring growth. Rats injected with GH gained 27 g in body weight over the 9 day study period, while the weights of vehicle-injected control animals remained constant. Steady-state levels of IGF-I mRNA in soleus and plantaris muscles, as measured by a nuclease protection assay (9), increased 2.5-3.0-fold relative to the controls, which were unchanged. These results are similar to the observations of others (5,14-16). In soleus but not in plantaris IGF-II mRNA also increased 3.0-fold after GH treatment. The reason for this difference in responsiveness between the two muscles is unknown, but may reflect differences in fiber type, in abundance of satellite cells or in other cell types within muscle (17).

Local Regulation of IGF Gene Expression During Muscle Growth: The experiments outlined above establish that a systemic agent - GH - regulates the abundance of IGF mRNAs in skeletal muscle, and suggest that GH-induced muscle hypertrophy is at least partly a consequence of locally-produced IGFs. Our second series of studies was designed to determine if local regulation of muscle growth is also associated with induction of IGF expression. As first described by Goldberg (18), hypertrophy of soleus and plantaris muscle occurs within days of excision of the gastrocnemius tendon, as the animals use the accessory muscles in locomotion. In our experiments weights of both muscles on the hypertrophied side increased relative to the control, contralateral, limb (soleus: 125 ± 5 vs. 99 ± 3 mg; p < .01, day 8; plantaris: 289 ± 6 vs 237 ± 5 mg; p < .01, day 8).

In both the hypertrophied soleus and plantaris muscles IGF-I mRNA increased 2.5-3.0-fold relative to contralateral controls by day 2 after surgery and remained elevated for the 8 days of the experiment. In plantaris IGF-II mRNA progressively increased during hypertrophy: 2.5-fold on day 2, 4.5-fold on day 4 and 8.0-fold on day 8 relative to control day 2. There was also an increase in IGF-II mRNA in the contralateral leg at day 8. In soleus muscle IGF-II mRNA also rose during hypertrophy, with 7-fold inductions on days 4 and 8. There was no change in IGF-II mRNA in the contralateral soleus muscle.

Although the results of these experiments support a mechanism for regulation of IGF-I (and IGF-II) gene expression independent of GH in skeletal muscle, it is conceivable that GH or other pituitary hormones play a permissive role. To further define the local control of IGF gene

expression in muscle we performed an additional series of studies in
hypox rats. The gastrocnemius tendon was excised as described above,
and animals were analyzed 2, 4 and 8 days after surgery. Again the
contralateral leg served as the control. In agreement with the original
studies by Goldberg (18), compensatory muscle hypertrophy also occurred
in the hypox rats after unilateral excision of the gastrocnemius tendon.
Muscle weight increased in both soleus and plantaris (soleus: $90.3 \pm$
4.0 vs 60.8 ± 2.2, $p < .001$; day 8; plantaris: 188.9 ± 3.9 vs $142.1 \pm$
3.6, $p < .001$, day 8). As in hormonally replete animals, in hypox rats
IGF-I mRNA increased during muscle hypertrophy. A progressive increment
was seen in both muscles: day 2; 4-5-fold; day 4, 8-fold, day 8,
8-12-fold. In the contralateral limb no change in IGF-I mRNA was
evident and levels remained 25-30% of values found in intact control rat
soleus and plantaris muscles. The pattern of IGF-II gene expression
also was similar for both muscles. Unlike hormonally replete animals in
which IGF-II mRNA increased in the contralateral plantaris muscle 8 days
after gastronemius surgery, no changes in IGF-II mRNA were evident in
the nongrowing muscles. In soleus IGF-II mRNA rose 7-fold 4 days after
surgery and remained elevated on day 8. In plantaris the increase was
more modest: 3-fold by day 4 and 2-fold on day 8.

Summary: The studies outlined above demonstrate both local and
systemic regulation of IGF-I and IGF-II gene expression during muscle
growth. The induction of IGF mRNAs in muscle is independent of GH and
of other pituitary hormones since it occurs in hypox rats. Whether IGF-
I and IGF-II mRNA are synthesized by muscle cells, by their satellite
cell precursors, or by fibroblasts or other cell types within muscle,
remains a matter of conjecture. Others have demonstrated that fetal
myoblasts in culture secrete peptides with properties indistinguishable
from those of IGF-I and IGF-II (12). During muscle regeneration
satellite cells appear to synthesize IGF-I, as detected by
immunocytochemistry (19). In addition, several cell types within or
near muscle synthesize IGF-I and IGF-II mRNA (20). Whatever the site of
synthesis of IGFs in muscle, our studies demonstrate a link between
local stimulation of skeletal muscle growth and IGF gene expression, and
suggest that signals produced during muscle hypertrophy act on the genes
encoding IGF-I and IGF-II or their mRNAs to increase IGF mRNA abundance.

III. Expression of IGF-I During Renal Hypertrophy

In 1984 D'Ercole et al (21) showed that hypophysectomy lowers and
GH raises the amount of IGF-I in kidney, suggesting that the kidney is a
site of IGF-I biosynthesis. Stiles et al (22) subsequently demonstrated
that the quantity of IGF-I increased in the remaining kidney following
unilateral nephrectomy compared with the kidney from sham-operated
controls. Taken together these observations support the concept that
both GH-dependent and GH-independent mechanisms may regulate IGF-I
expression in kidney. We have shown recently by a combination of
studies employing RNA hybridization and immunocytochemistry that renal
IGF-I biosynthesis is limited to the principal cells of the collecting
duct (23). Using the same techniques we now have focused on the
regulation of IGF-I expression during kidney growth using the two models
noted above: GH-induced renal hypertrophy and compensatory renal
hypertrophy following unilateral nephrectomy.

Systemic GH Treatment Induces IGF-I mRNA: We treated hypox rats
with GH (200 ug/day) or with vehicle for 9 days, and then isolated
kidneys or collecting ducts for immunohistochemical measurement of IGF-I
and for RNA analysis. After GH administration kidney weight increased

by 40% (460 + 10 vs 330 + 10 mg, n=12 rats, p < .01), and immunostainable IGF-I in collecting duct was enhanced. The abundance of IGF-I mRNA in collecting duct, as measured by solution-hybridization using a probe derived from exon 3 of the rat IGF-I gene, increased 2-fold after GH compared with vehicle-treated animals, but was only 40% of the level found in control, hormonally replete rats. The incomplete restoration of IGF-I mRNA after 9 days of GH injections may reflect the non-physiological mode of administration, the absence of other pituitary hormones, or other factors.

Since the rat IGF-I gene is transcribed and processed into a complex array of mRNAs (5,24,25), we next asked whether the abundance of distinct IGF-I mRNA species is altered by GH treatment. IGF-I mRNAs with at least three different 5' ends and two distinct 3' ends are generated by variable RNA splicing (4,5,24,25). In order to examine the potential effect of GH on IGF-I mRNA processing, we analyzed collecting duct RNA obtained from control, vehicle-treated and GH-treated hypox rats using specific 5' and 3' probes. All three IGF-I transcripts with distinct 5' ends and the two with unique 3' ends are reduced following hypophysectomy and increase following subsequent GH treatment. The magnitude of induction by GH differs slightly for each mRNA species and ranges from 1.5 to 4.0-fold as determined by densitometric scanning (means of 3-4 separate experiments). The three most abundant mRNA species increase to a similar degree (2.5-4.0-fold). We conclude from these findings that GH treatment alters IGF-I gene transcription and/or mRNA stability, but does not appear to exert a major differential effect on specific RNA processing. Regardless of the exact mechanism, it is likely that the renal hypertrophy that accompanies GH administration to hypophysectomized rats is a consequence of increased IGF-I mRNA in collecting duct with subsequent enhanced local production of IGF-I.

Regulation of IGF-I Expression During Compensatory Renal Hypertrophy: As a means to characterize IGF-I expression in a second model of renal hypertrophy, we obtained left kidneys from rats 1, 2, 5 and 14 days following unilateral right nephrectomy or "sham" surgery. In these studies body weights measured post-surgery were not significantly different between nephrectomized and sham-operated rats; but kidney weights were increased at 2, 5 and 14 days post-nephrectomy (p < 0.01, Student's t test (42% increase at 14 days)). These findings are characteristic of the compensatory hypertrophy demonstrated previously by others following unilateral nephrectomy in the rat (22,26). Levels of GH and IGF-I in plasma obtained at the time of sacrifice did not differ significantly between nephrectomized and sham-operated rats.

In order to characterize the distribution of IGF-I in hypertrophied kidneys, we immunostained fixed paraffin-embedded sections of left kidneys obtained following unilateral right nephrectomy or sham surgery. Immunostaining for IGF-I was observed only when serum containing anti-IGF-I antibody was used and was limited to principal cells within cortical and medullary collecting ducts (23). Immunostainable IGF-I was clearly increased only in medullary collecting ducts of kidneys originating from nephrectomized rats compared to sham-operated rats at 5 and 14 days, but not 1 or 2 days following surgery. The intensity of immunostaining was also greater than that in kidneys from control rats. No differences among control, sham or nephrectomized rats were observed in cortical collecting ducts.

To determine whether the mechanism of the post-nephrectomy increase in collecting duct immunostainable IGF-I is a consequence of an increase in IGF-I mRNA, we measured IGF-I mRNA in whole kidneys and isolated medullary collecting ducts obtained at various times post-surgery. In initial experiments we utilized a ^{32}P-labeled probe derived from rat exon 3. The compensatory renal hypertrophy and the increase in immunostainable IGF-I detectable at 5 and 14 days post-nephrectomy were not accompanied by comparable changes in IGF-I mRNA. (In both groups of studies, total RNA isolated from kidney or collecting duct changed by less than 25% in post-nephrectomy vs sham-operated animals). Thus, although we were able to detect a 2-fold increase in IGF-I mRNA from collecting duct under circumstances of a 40% increase in kidney weight after GH administration to hypox rats, we could find no evidence for a similar change under circumstances of a 24% increase in kidney weight 5 days following uninephrectomy or a 42% increase 14 days post-operatively.

As noted above, heterogeneity has been demonstrated at the 5' and 3' ends of rat IGF-I mRNAs. It is possible that during post-nephrectomy renal growth, specific IGF-I mRNA transcripts are increased while others are reciprocally diminished. Relative changes in the proportions of these mRNAs would not be detected using the rat exon 3 probe. Increased immunostainable IGF-I in collecting ducts of hypertrophied kidneys following unilateral nephrectomy might result from a greater abundance of one or more of these IGF-I mRNA species with a subsequent increase in translation into protein. To determine whether this is the case, we analyzed total RNA from collecting ducts of kidneys from sham-operated and uninephrectomized rats using ^{32}P-labeled probes directed against the 5' and 3' ends of IGF-I mRNAs. There were no changes in the abundance of any IGF-I mRNAs after unilateral nephrectomy or after sham surgery. Analysis of data obtained from four independent experiments showed that the abundance of IGF-I mRNA extracted from hypertrophied kidneys ranged from 90-130% of that extracted from kidneys of sham operated controls, and that no consistent differences were evident. Thus, the increased IGF-I detected by immunohistochemistry in collecting duct 5 and 14 days after unilateral nephrectomy cannot be a consequence of a parallel increase in IGF-I mRNA.

Although it has been suggested that immunostainable IGF-I can result from cell surface binding of circulating peptide with subsequent receptor-mediated internalization, we were unable to detect IGF-I receptors in membranes prepared from papillary collecting ducts. In parallel experiments IGF-II receptors were readily measured in these cells. Thus it is unlikely that receptor-mediated uptake of circulating IGF-I accounts for the increase in IGF-I in principal cells of medullary collecting duct during post-nephrectomy compensatory hypertrophy. Rather we favor the hypothesis that translational regulation of IGF-I biosynthesis occurs during renal hypertrophy.

Summary: The experiments described above illustrate two modes of regulation of IGF-I biosynthesis during kidney growth. Both mechanisms are limited to the principal cells of the rat renal collecting duct, the only cell type in kidney in which we detect IGF-I (23). Systemic GH treatment for 9 days to hypox rats leads to a 40% increase in renal weight, a 2-fold induction of IGF-I mRNA, and an increase in immunostainable IGF-I in collecting duct. Following unilateral nephrectomy renal growth is comparable to that induced by GH (24% by 5 days and 42% by 14 days). An increase in immunostainable IGF-I is evident by 5 days, with further enhancement 14 days after uninephrectomy. There is no change in IGF-I mRNA. The increment in immunostainable IGF-I is not a result of receptor-mediated uptake of circulating peptide, since no IGF-I receptors are present in membranes derived from these cells. Thus it appears that principal cells of renal collecting duct synthesize IGF-I in response both to systemic hormonal signals, as reflected in a GH-induced increase in IGF-I mRNA, and to local signals, which probably enhance the translation of one or more species of IGF-I mRNA into protein.

IV. Summary and Perspective

The studies described in this chapter demonstrate some of the complexities involved in the regulation of IGF gene expression during physiological processes. It is clear from the examples presented that despite potential differences in the mechanisms of regulation in both skeletal muscle and kidney the IGFs are intimately associated with tissue growth. The challenge for the future will be to determine the pathways of signal transduction leading from the growth stimulus to the transcriptional and translational responses of IGF-I and IGF-II, and to define the subsequent events at molecular, cellular, and physiological levels.

Acknowledgement: We appreciate the excellent assistance of Janet Seavitte in preparation of this manuscript. These studies were supported by NIH Grants DK37449 and HD20805.

V. References

1. Froesch, E.R., Schmid, C., Schwander, J., and Zapf, J. Action of insulin-like growth factors. Ann. Rev. Physiol. 47: 443-467, 1985.

2. Brown, A.L., Graham, D.E., Nissley, S.P., Hill, D.J., Strain, A.J., and Rechler, M.M. Developmental regulation of insulin-like growth factor II mRNA in different rat tissues. J. Biol. Chem. 261: 13144-13150, 1986.

3. Rotwein, P., Pollock, K., Didier, D., Krivi, G.G. Organization and sequence of the human insulin-like growth factor I gene. J. Biol. Chem. 261: 4828-4832, 1986.

4. Shimatsu, A., and Rotwein, P. Mosaic evolution of the insulin-like growth factors. J. Biol. Chem. 262: 7894-7900, 1987.

5. Roberts, C.T., Lasky, S.R., Lowe, W.L., Seaman, W.T., and LeRoith, D. Molecular cloning of rat insulin-like growth factor I complementary deoxyribonucleic acids: differential messenger ribonucleic acid processing and regulation by growth hormone in extrahepatic tissues. Mol. Endo. 1: 243-248, 1987.

6. Murphy, L.J., Bell, G.I., Duckworth, M.L., and Friesen, H.G. Identification, characterization, and regulation of a rat complementary deoxyribonucleic acid which encodes insulin-like growth factor-I. Endocrinology 121: 684-691, 1987.

7. Soares, M.B., Ishii, D.H., and Efstratiadis, A. Development and tissue-specific expression of a family of transcripts related to rat insulin-like growth factor II mRNA. Nucleic Acids Res. 13: 1119-1134, 1985.

8. Whitfield, H.J., Bruni, C.B., Frunzio, R., Terrell, J.E., Nissley, S.P., and Rechler, M.M. Isolation of a cDNA clone encoding rat insulin-like growth factor II precursor. Nature 312: 277-280, 1984.

9. Rotwein, P., Burgess, S.K., Milbrandt, J.D., and Krause, J.E. Differential expression of insulin-like growth factors in the rat central nervous system. Proc. Nat. Acad. Sci. USA 85: 265-269, 1988.

10. Turner, J., Rotwein, P., Novakofski, J., and Bechtel, P. Introduction of mRNA encoding insulin-like growth factors I and II during growth hormone stimulated skeletal muscle hypertrophy. Am. J. Physiol. 255: E513-517, 1988.

11. Dodson, M.V., Allen. R.E., and Hossner, K.L. Ovine somatomedin, multiplication-stimulating activity, and insulin promote skeletal muscle satellite cell proliferation in vitro. Endocrinology 117: 2357-2363, 1985.

12. Hill, D.J., Crace, C.J., Nissley, S.P., Morrell, D., Holder, A.T., and Milner, R.D.G. Fetal rat myoblasts release both rat somatomedin-C (SM-C)/insulin-like growth factor I (IGF-I) and multiplication-stimulating activity in vitro: partial characterization and biological activity of myoblast-derived Sm-C/IGF-I. Encrinology 117-2061-2072, 1985.

13. Stracke, H., Schulz, A., Moeller, D., Rossol, S., and Schatz, H. Effect of growth hormone on osteoblasts and demonstration of somatomedin-C/IGF-I in bone organ culture. Acta Endocrinologica 107: 16-24, 1984.

14. Murphy, L.J., Bell, G.I., and Friesen, H.G. Tissue distribution of insulin-like growth factor I and II messenger ribonucleic acid in the adult rat. Endocrinology 120: 1279-1282, 1987.

15. Hynes, M.A., Van Wyk, J.J., Brooks, P.J., D'Ercole, A.J., Jansen, M., and Lund, P.K. Growth hormone dependence of somatomedin-C/insulin-like growth factor-I and insulin-like growth factor-II messenger ribonucleic acids. Mol. Endo. 1: 233-242, 1987.

16. Lund, P.K., Moats-Staats, B.M., Hynes, M.A., Simmons, J.G., Jansen, M., D'Ercole, A.J., and VanWyk, J.J. Somatomedin-C/insulin-like growth factor-I and insulin-like growth factor-II mRNAs in rat fetal and adult tissues. J. Biol. Chem. 261: 14539-14544, 1986.

17. Kelly, A.M. Satellite cells and myofiber growth in the rat soleus and extensor digitorum longus muscles. Devel. Biol. 65: 1-10, 1978.

18. Goldberg, A.L. Work-induced growth of skeletal muscle in normal and hypophysectomized rats. Am. J. Physiol. 213: 1193-1198, 1967.

19. Jennische, E., Skottner, A., and Hansson, H.-A. Satellite cells express the trophic factor IGF-I in regenerating skeletal muscle. Acta Physiol. Scand. 129: 9-15, 1987.

20. Han, V.K.M., D'Ercole, A.J., and Lund, P.K. Cellular localization of somatomedin (insulin-like growth factor) messenger RNA in the human fetus. Science 236: 193-197, 1987.

21. D'Ercole, A.J., Stiles, A.D., and Underwood, L.E. Tissue concentrations of somatomedin C: Further evidence for multiple sites of synthesis and paracrine or autocrine mechanisms of action. Proc. Natl. Acad. Sci. USA 81: 935-939, 1984.

22. Stiles, A.D., Sosenko, I.R.S., D'Ercole, A.J., and Smith, B.T. Relation of kidney tissue somatomedin-C/insulin-like growth factor I to postnephrectomy renal growth in the rat. Encdorinology 117: 2397-2401, 1985.

23. Bortz, J.D., Rotwein, P., DeVol, D., Bechtel, P.J., Hansen, V.A., and Hammerman, M.R. Focal expression of insulin-like growth factor I in rat kidney collecting duct. J. Cell Biol. 107: 811-819, 1988.

24. Shimatsu, A., and Rotwein, P. Sequence of two rat insulin-like growth facctor I mRNAs differing within the 5' untranslated region. Nucl. Acids Res. 15: 7196, 1987.

25. Lowe, Jr., W.L., Roberts, Jr., C.T., Lasky, S.R., and LeRoith, D. Differential expression of alternative 5' untranslated regions in mRNAs encoding rat insulin-like growth factor I. Proc. Natl. Acad. Sci. USA 84: 8946-8950, 1987.

26. Fagin, J.A., and Melmed, S. Relative increase in insulin-like growth factor I messenger ribonucleic acid levels in compensatory renal hypertrophy. Endocrinology 120: 718-723, 1987.

DIETARY AND HORMONAL FACTORS AFFECTING THE mRNA LEVEL OF IGF-I IN RAT LIVER

IN VIVO AND IN PRIMARY CULTURES OF RAT HEPATOCYTES

Hisanori Kato, Yutaka Miura, Asako Okoshi, Tsutomu Umezawa,
Shin-Ichiro Takahashi and Tadashi Noguchi

Department of Agricultural Chemistry, Faculty of Agriculture
The University of Tokyo, Bunkyo-ku, Tokyo 113, Japan

INTRODUCTION

Insulin-like growth factor I (IGF-I, or somatomedin C) is a hormone, which has been shown to be important in the anabolism of dietary proteins into body proteins (Spencer, 1983). Plasma IGF-I level is affected by many nutritional factors including energy intake (Isley et al., 1983), dietary protein level (Prewitt et al., 1982) and nutritional quality of dietary proteins (Bolze et al., 1985; Takahashi et al., 1988). IGF-I probably works in coordination with insulin but in a different manner in the anabolism of dietary proteins, because, different from insulin, plasma immunoreactive IGF-I level does not increase significantly after a meal and is primarily affected by the dietary composition of the nutrients. Another evidence of the difference in the response of insulin and IGF-I is the manner of response of these two hormones to streptozotocin administration. Immunoreactive insulin level decreases very quickly after streptozotocin administration whereas IGF-I level is affected gradually, presumably as a secondary effect of insulin deficiency (Takahashi et al., 1988).

In order to elucidate the mechanism of regulation of plasma IGF-I level, we investigated the mRNA level of IGF-I in rat liver in vivo and in primary cultures of rat hepatocytes.

Several authors reported the changes in liver IGF-I mRNA level in rats under various nutritional or hormonal conditions. The results show that energy intake (Emler et al., 1987), growth hormone (Mathews et al., 1986; Roberts et al., 1986; Heynes et al., 1987), insulin (Goldstein et al., 1988), and prolactin (Murphy et al., 1988) affect the mRNA level significantly. However, there are few papers which showed the relationship between the status of protein nutrition and the liver IGF-I mRNA level in rats.

Besides, it has been elucidated that there are not less than 3 molecular species of mRNA in rat liver, which are derived by different splicing of the precursor of IGF-I mRNA. Mathews et al.(1986) reported that growth hormone increases 0.8 kb mRNA of IGF-I in lit/lit mice liver.

Goldstein et al.(1988) showed that the rate of decrease of liver IGF-I mRNA in response to various doses of streptozotocin was different among the mRNA species. However, there are no detailed studies which elucidated the physiological role of these mRNA species. In this context, we studied the nutritional or hormonal factors which affect the mRNA level in liver in vivo or in primary cultures of rat hepatocytes.

EFFECT OF DIETARY PROTEINS ON THE IGF-I mRNA LEVEL IN RAT LIVER IN VIVO

Rats were fed 12% casein diet (C), 12% gluten diet (marginally deficient in lysine and threonine, G), or protein-free diet (PF) for 1 week. Immunoreactive plasma IGF-I level measured directly employing a human IGF-I assay kit (Nichols Institute) was C > G > PF. The level in C-fed rats was 3 times as high as that in PF-fed rats. These values correlated well with the growth rate of these rats.

Total RNA was extracted from the liver of these rats and was analysed by Northern bolt analysis employing rat IGF-I cDNA as a probe. IGF-I mRNA level decreased both in the rats fed G and PF as compared with C-fed rats. Particularly, the decrease in the mRNA classes of 3.6, 4.0 and 7.4 kb was prominent. Thus, the decrease in immunoreactive IGF-I in plasma of G- or PF-fed rats may be due to decreased mRNA level. However, results of nuclear run-off assay showed no significant difference of transcription rate of IGF-I gene in the liver of these rats. This suggests that the decrease in IGF-I mRNA in the liver of G- or PF-fed rats is due to the enhanced degradation of IGF-I mRNA.

EFFECTS OF HORMONES ON THE PRODUCTION OF IGF-I AND ON ITS mRNA LEVEL IN PRIMARY CULTURES OF RAT HEPATOCYTES.

Primary cultures of rat hepatocytes secrete immunoreactive IGF-I into the medium (Scott et al., 1985; Richman et al., 1985). The secreted IGF-I concentration in the medium increased about 3-fold by addition of dexamethasone. In contrast to this, dexamethasone decreased the IGF-I mRNA level in the cells, which is compatible with the report of Adamo et al.(1988). They showed that dexamethasone lowers IGF-I mRNA in primary cultures of rat neuronal and glial cells. Furthermore, nuclear run-off test of cultured hepatocytes elucidated that the transcription rate of IGF-I gene decreased in the presence of dexamethasone.

These results suggest that the translation rate of IGF-I mRNA increases greatly in the presence of dexamethasone in the medium.

Growth hormone, which is known to stimulate production of IGF-I in vivo (Maes et al., 1988), did not show any prominent effects on IGF-I production or IGF-I mRNA level in cultured hepatocytes. However, growth hormone did accelerate the transcription of IGF-I gene, and, when added with insulin, it increased the mRNA level, although even the combination of these hormones did not increase IGF-I secretion. Taking all these results into account, translation rate of mRNA is considered to be critical to the regulation of the synthesis and secretion of IGF-I.

CONCLUSION

The present paper showed how nutritional or hormonal factors affect the plasma immunoreactive IGF-I level or liver mRNA level of IGF-I in rat liver. Some paradoxical results were obtained. For example, growth hormone stimulates the secretion of IGF-I by liver in vivo if animals are well nourished (Maes et al., 1988). However, this hormone did not affect the secretion rate of IGF-I in cultured hepatocytes although it increased the cellular mRNA level with the aid of insulin. On the contrary, dexamethasone enhanced the secretion of IGF-I but depressed the mRNA level of IGF-I. These results suggest the complicated regulation of IGF-I production.

Besides the results shown in the present paper, it has been proved that the association of IGF-I with the plasma binding proteins is also controlled by nutritional factors (Takahashi et al., 1988). At present, our cultured hepatocytes secrete IGF-I in association with the 40 kD binding protein and not with the 150 kD binding protein. Our previous results suggested that IGF-I associated with the 150 kD binding protein is the physiologically active form. Then, how the IGF-I associated with the 40 kD binding protein is transferred to the 150 kD binding protein. This remains to be elucidated.

In conclusion, nutritional or hormonal factors regulate the content and the activity of plasma IGF-I, through controlling the transcription of IGF-I gene, splicing of the precursor of mRNA, mRNA stability, translation rate of mRNA, and the status of binding of this hormone with the plasma binding proteins.

ACKNOWLEDGMENT

The authors gratefully acknowledge Dr. Hiroshi Naito, Professor of Kyoritsu Women's University, and Dr. Judson J. Van Wyk, Professor of University of North Carolina, for their helpful discussions.

REFERENCES

Adamo, M., Werner, H., Farnsworth, W., Roberts, C. T. Jr., Raizada, M., and LeRoith, D. (1988) Dexamethasone reduces steady state insulin-like growth factor I messenger ribonucleic acid levels in rat neuronal and glial cells in primary culture. Endocrinology 123, 2565-2570.

Bolze, M. S., Reeves, R. D., Lindbeck, F. E., and Elders, M. J. (1985) Influence of selected amino acid deficiencies on somatomedin, growth and glycosaminoglycan metabolism in weanling rats. J. Nutr. 115, 782-787.

Elmer, C. A., and Schalch, D. S. (1987) Nutritionally-induced changes in hepatic insulin-like growth factor I (IGF-I) gene expression in rats. Endocrinology 120, 832-834.

Hynes, M. A., Van Wyk, J. J., Brooks, P. J., D'Ercole, A. J., Jansen, M., and Lund, P. K. (1987) Growth hormone dependence of somatomedin C/insulin-like growth factor-I and insulin-like growth factor-II messenger ribonucleic acids. Mol. Endocrinol. 1, 233-242.

Isley, W. L., Underwood, L. E., and Clemmons, D. R. (1983) Dietary components that regulate serum somatomedin-C concentrations in humans. J. Clin. Invest., 71, 175-182.

Maes, M., Amand, Y., Underwood, L. E., Maiter, D., and Ketelslegers, J.-M. (1988) Decreased serum insulin-like growth factor I response to growth hormone in hypophysectomized rats fed a low protein diet: evidence for a postreceptor defect. Acta Endocrinol.(Copenh.) 117, 320-326.

Mathews, L. S., Norstedt, G., and Palmiter, R. D. (1986) Regulation of insulin-like growth factor I gene expression by growth hormone. Proc. Natl. Acad. Sci., U. S. A. 83, 9343-9347.

Murphy, L. J., Tachibana, K., and Friesen, H. G. (1988) Stimulation of hepatic insulin-like growth factor-I gene expression by ovine prolactin: Evidence for intrinsic somatogenic activity in the rat. Endocrinology 122, 2027-2033.

Prewitt, T. E., D'Ercole, A. J., Switzer, B. R., and Van Wyk, J. J. (1982) Relationship of serum immunoreactive somatomedin-C to dietary protein and energy in growing rats. J. Nutr., 112, 144-150.

Richman, R. A., Benedict, M. R., Florini, J. R., and Toly, B. A. (1985) Hormonal regulation of somatomedin secretion by fetal rat hepatocytes in primary culture. Endocrinology 116, 180-188.

Roberts, C. T. Jr., Brown, A. L., Graham, D. E., Seelig, S., Berry, S., Gabbay, K. H., and Rechler, M. M. (1986) Growth hormone regulates the abundance of insulin-like growth factor I RNA in adult rat liver. J. Biol. Chem. 261, 10025-10028.

Scott, C. D., Martin, J. L., and Baxter, R. C. (1985) Production of insulin-like growth factor I and its binding protein by adult rat hepatocytes in primary culture. Endocrinology 116, 1094-1101.

Spencer, E. M. (ed.) (1983) "Insulin-Like Growth Factors/Somatomedins. Basic Chemistry, Biology, Clinical Importance" Walter de Gruyter, Berlin, New York.

Takahashi, S., Kajikawa, M., Umezawa, T., Takahashi, S.-I., Kato, H., Miura, Y., Nam, T. J., Noguchi, T., and Naito, H. (1988) Effects of dietary proteins on the plasma immunoreactive insulin and insulin-like growth factor I/somatomedin C levels in rats. Manuscript submitted.

EXPRESSION OF IGF-I IN LIVER

Thomas R. Johnson and Joseph Ilan

Department of Anatomy, Case Western Reserve University
Cleveland, Ohio   44106

INTRODUCTION

Insulin-like growth factor I(IGF-I) is a basic, 70 amino acid polypeptide with strong structural homologies to insulin. It has been shown to be critical in promoting skeletal growth during childhood and adolescence(Merimee et al., 1987) but available evidence suggests it is a multifunctional peptide hormone. It influences the differentiation of several cell types and can have insulin-like effects in certain circumstances. It is expressed in many tissues and in all probability exerts its growth-promoting and other effects by endocrine, paracrine, and autocrine mechanisms. In the last decade it has become evident that the liver is the primary source of circulating IGF-I in serum although the relative importance of serum-borne and locally synthesized IGF-I is not yet clear. In this essay, we will discuss the regulation of IGF-I expression in liver as it has been studied in various systems, including results from our own laboratory. A number of different models have been used to study hepatic IGF-I expression. These include:(1) In vivo models, principally hypophysectomized rats or the growth hormone(GH) deficient mouse strain C57BL lit/lit; (2) the isolated, perfused rat liver; (3) studies in organ culture; (4) primary hepatocyte cultures; and recently (5) studies employing transgenic mice. It is well to keep the individual failings of these models in mind, since in some cases different results are obtained depending on which system is used. The literature cited reflects our own biases and we have made no attempt to be exhaustive.

STRUCTURAL GENE ORGANIZATION AND TRANSCRIPT DIVERSITY

Available evidence concerning the structure of the IGF-I gene and its RNA transcripts indicates that this gene has an extraordinary diversity of potential regulatory points. The human gene encompasses at least 45 kilobases(kb) of DNA containing 5 exons, and 4 introns which range in size from 1.5 to over 21 kb(Rotwein et al., 1986). Two of these exons code for distinct carboxyterminal extension peptides which do not appear in the mature protein as it is isolated from serum. Their nucleic acid sequences, however,are represented in cDNA libraries derived from rat liver(Roberts et al., 1987). They are termed IGF-Ia or IGF-Ib depending on the absence

(IGF-Ia) or presence(IGF-Ib) of a 52-base insert in the region coding for the E domain of the mature peptide. In addition, 3 different cDNA sequences associated with the same IGF-I mature protein coding sequence have been isolated from rat liver cDNA libraries(Lowe et al., 1987; Roberts et al., 1987). Thus, the potential exists for at least 6 different mRNA species for IGF-I, presumably generated by alternative splicing mechanisms since the gene is single copy in rats and humans(Brissenden et al., 1984; Rotwein et al., 1986). In addition, at least 5 polyadenylation sites have been identified in genomic clones of the IGF-I gene in rats(Shimatsu and Rotwein, 1987), raising the number of possible processed transcripts to 30 or more.

As visualized by Northern blot analysis, hepatic IGF-I RNA transcripts present a complicated picture consisting of a large(7-8.5 kb; hereafter referred to as 7.5 kb) discrete transcript, a group of transcripts ranging from about 0.8-1.2 kb appearing on autoradiograms as a well-defined region, and an additional discrete transcript of about 1.8 kb. These three regions of hybridization are common to rat, human, and mouse RNAs(e.g., Murphy et al., 1987; Wang et al., 1988; Mathews et al., 1986). The mouse appears to lack the 1.8kb band in liver(Unpublished observations; Mathews et al., 1986;), although it is present in mouse-derived cultured adipose cells(Doglio et al., 1987). Some of the heterogeneity of the 0.8-1.2 kb region may be due to the presence of alternatively spliced and/or polyadenylated transcripts as discussed above(Lowe et al., 1987; Lund et al., 1988). A number of other transcript sizes have been reported for hepatic tissue, including 3.2 and 4.5 kb transcripts(Unpublished observations; Roberts et al., 1986).

It is not known if all of these transcripts are used by the liver's translational machinery. In mouse liver, the 0.8-1.2 kb region is represented on polysomes; however, a large fraction of mouse liver IGF-I RNA sequences are unaffected by puromycin treatment of mouse liver polysomes, suggesting that the liver contains a pool of untranslated IGF-I messenger ribonucleoprotein(Unpublished observations). Although the presence of the 7.5 kb transcript in the polysomal fraction could not be definitively established, IGF-I related sequences were detected in all size classes of polysomes. Circumstantial evidence suggests that the 7.5 kb transcript is at least partially processed(Casella et al., 1987) and thus could be a functional mRNA. The abundance of the 7.5 and 0.8-1.2 kb transcript size classes apparently are regulated somewhat differently, and possibly may not have the same control mechanisms in different organisms. For example, refeeding of rats starved for 24 hr resulted in an 18-fold increase in concentration of the 7.5 kb transcript in liver, as opposed to a much smaller increase in the 0.8-1.2 kb region, over the next 24 hours (Emler and Schalch, 1987). Administration of growth hormone(GH) to hypophysectomized rats increased the concentration of all hybridizing species coordinately (Roberts et al., 1986); however, in the C57BL lit/lit mouse, GH administration does not seem to restore the 7.5 kb transcript although the 0.8-1.2 kb region is dramatically increased(Mathews et al., 1986). However, both these hepatic transcript classes are increased coordinately in C57BL wild-type mice which have been hypophysectomized followed by treatment with GH(Unpublished observations).

FACTORS REGULATING HEPATIC IGF-I EXPRESSION

1. Nutrition

The major factors which regulate serum IGF-I concentration, and which are also known to affect hepatic IGF-I expression, are nutritional status, serum growth hormone concentration, and insulin. Since there are ample data from liver perfusion studies indicating that the liver is the main source of serum IGF-I(e.g., Schwander et al., 1983), studies concerning large changes in serum IGF-I levels may be assumed to reflect changes in liver metabolism and expression. In all three instances, detailed regulatory mechanisms are lacking, and it is clear from in vivo studies that these factors do not operate independently of one another. The effect of starvation/refeeding on rat hepatic IGF-I transcript accumulation has already been discussed(Emler and Schalch, 1987). That study also indicated changes in serum IGF-I concentration which paralleled hepatic transcript levels, principally the 7.5 kb transcript. Isley et al.(1983) showed that human IGF-I serum levels were influenced not only by total caloric intake but by dietary composition in normal subjects. In this study, human volunteers were starved for five days. Food was resumed either in the form of a low protein, isocaloric diet; low protein, low calory diet; or normal diet, and these controlled diets continued for an additional five days. Over the course of the 5-day starvation period IGF-I levels declined from 1.85 U/ml to 0.676 U/ml. Upon refeeding, serum IGF-I levels in subjects receiving a normal diet recovered over 5 days to 1.26 U/ml, compared to a more modest increase to 0.90 U/mL in the low protein, isocaloric group and a continued decline in the low protein, low calory group to 0.31 U/ml. The best statistical correlation of serum IGF-I levels was with overall nitrogen balance. It would be very interesting to compare serum IGF levels as measured by bioassay with levels measured by radioimmunoassay in a study such as that of Isley et al.(1983), since there is evidence for increased serum concentration of IGF inhibitors in situations of nutritional deficiency(Phillips, 1986), particularly in insulin-dependent diabetes. A recent, similar study using starved rats which had been refed alternate diets suggested that the effect of diet was not at the GH receptor level, since liver membrane binding sites for GH were unchanged by starvation even though the response to GH, as measured by IGF-I serum levels, was severely impaired in starved animals(Maes et al., 1988).

2.Growth hormone

Control of hepatic IGF-I expression by growth hormone is currently the best understood mode of regulation. In the GH-deficient mouse strain C57BL lit/lit, hepatic IGF-I RNA transcripts are nearly undetectable in the absence of GH therapy, but rise to control(heterozygote) levels following 7 days of GH infusion via osmotic minipump(Mathews et al., 1986). This same study also showed by in vitro nuclear run-on assays that at least part of this increase was due to an increased rate of transcription relative to GH-insensitive message(in this case albumin). Like other polypeptide hormones, including IGF-I, GH acts through a membrane-bound receptor. Nucleotide sequences of cDNA clones corresponding to the human and rabbit liver GH receptor were recently obtained(Leung et al., 1987). They indicate that the receptor is a transmembrane protein of 620 amino acids(with an additional 18 amino acid signal sequence) with no obvious relationships to other known protein sequences except the receptor for prolactin(Boutin et al., 1988). Although there is one report that the GH receptor is phosphorylated on tyrosine residues(Foster et al., 1987), it lacks sequences related to known tyrosine kinases, and shows no structural elements suggestive of known DNA binding proteins. Thus, at present, few clues exist with regard to the molecular signalling mechanism set in motion by the GH/receptor interaction. Interestingly, the clones analyzed by Leung et al.(1987) contained a number of unique 5'-untranslated regions connected to the same coding sequence, suggesting that, like IGF-I, the growth hormone receptor mRNA is subject to control by alternative splicing. In other studies, hepatic GH receptors have been shown to be strongly downregulated for about 12 hr following GH administration in vivo, recovering to normal levels by 24 hr(Maiter et al., 1988). Longer-term studies indicate that hepatic GH receptors can be upregulated by a 1-week exposure to GH (Baxter and Zaltsmann,1984).

Administration of a single dose of GH in vivo to hypophysectomized rats(Murphy et al., 1987) or lit/+ heterozygous mice(Mathews et al., 1986) results in a rapid rise(detectable by 2 hr) in concentration of IGF-I related RNA sequences in liver. In hypophysectomized rats, the levels peak at about 6 hr and have returned to preinjection levels by 18-24 hr.; the data are similar for the mouse. Our laboratory has examined the effects of introducing GH into the incubation medium of primary rat hepatocyte cultures(Johnson et al., 1989). In that situation, the induced accumulation of IGF-I transcripts often appears biphasic. IGF-I transcripts increase in concentration up to about 4 hr following addition of GH-containing medium. This increase is followed by a decline in transcript concentration until about 10 hr following addition of GH-containing medium; and at 24 hr an increased concentration of IGF-I RNA sequences, relative to cells not exposed to GH, is observed. We showed that the increased concentration of IGF-I sequences at 24 hr was probably due to an increased rate of transcription(Johnson et al., 1989;v. infra), similar to the mouse(Mathews et al., 1986), but the mechanism of the early rise in IGF-I transcripts in cultured hepatocytes, and presumably in vivo, remains undefined at present.

There is evidence that GH has different effects on individual transcript classes in liver, and that GH regulation of IGF-I transcript levels may differ from tissue to tissue. As mentioned earlier, rat liver cDNA libraries contain 3 identified 5'-untranslated sequences associated with the same protein coding sequence(Lowe et al., 1987). Termed A,B, and C, these are normally present in liver in the ratio 27.2: 2.3: 70.5. When hypophysectomized rats were treated with GH, all 3 sequences were elevated relative to controls; however, the A and B classes were elevated 6 to 8-fold in liver as opposed to a 2-fold stimulation for the C sequence. Thus, GH treatment dramatically changed the relative transcript abundances in addition to elevating the total level of transcripts. In other tissues, the A class was unaffected by GH treatment. In another study from the same laboratory, it was shown that hepatic IGF-Ib sequences (containing an insert in the E peptide; v. supra) were about 3 times as responsive to GH as IGF-Ia sequences(Lowe et al., 1988). Liver was unique in this respect among the tissues examined. This same study also indicated that the hepatic IGF-Ib sequence was associated with the GH-responsive Class A 5' untranslated region, but that in other tissues it was associated with the Class C 5' untranslated region. These results suggest that IGF-I is regulated by GH by additional mechanisms than increasing transcription, and that such mechanisms may vary in a tissue-specific manner.

In summary, growth hormone regulates IGF-I transcript abundance in liver by at least two general mechanisms: by increasing transcription(Mathews et al., 1986; Johnson et al., 1989), and by affecting RNA processing(Lowe et al., 1987, 1988). Additional mechanisms, e.g. effects on RNA stability, cannot be ruled out. For example, it would be of interest to know if the alternative 5'- untranslated sequences and carboxyterminal extension peptide sequences are affected by pulsatile secretion of GH, since it has been recently shown that in some tissues, although not liver, intermittent GH administration is markedly more effective than continuous infusion in enhancing IGF-I sequence accumulation(Isgaard et al., 1988).

3.Insulin

Insulin, directly or indirectly, is also a major regulator of hepatic IGF-I expression. Poor growth is a well-known consequence of human insulin-deficient diabetes as well as experimentally induced diabetes in rodents, and in both cases circulating levels of IGF-I are low(Phillips and Vasilopoulou-Sellin, 1980; Goldstein et al., 1988). Studies with diabetic humans or rodents with experimentally induced diabetes have shown that the severity of diabetic symptoms is correlated with observed effects on IGF-I. Goldstein et al.(1988) in a study of streptozotocin-treated rats showed that the metabolic index most highly correlated with serum IGF-I protein and hepatic IGF-I RNA levels was serum betahydroxybutyrate rather than serum glucose. At a streptozotocin dose of 72 mg/kg, the rats exhibited marked hypoglycemia but showed serum levels of IGF-I protein and hepatic IGF-I transcripts which were not significantly different from untreated rats. At higher doses, serum and hepatic IGF-I protein levels, hepatic IGF-I transcript levels, and costal cartilage sulphate uptake were all reduced compared to untreated rats. Although this study did not measure GH levels, and it is known that diabetic rats show impaired GH secretion(Tannenbaum, 1981), other studies have shown that GH administration is ineffective in restoring serum IGF-I levels in diabetic rats(Scott and Baxter, 1986). Similar correlations with progressive severity of diabetic symptoms have been observed in humans(Glaser et al., 1987; Lanes et al., 1985).

While it is clear that the profound metabolic disturbances ensuing as a consequence of insulin deficiency could suppress circulating IGF-I levels by multiple mechanisms, there is evidence that insulin has direct effects on hepatic IGF-I expression. Binoux et al.(1982) showed that insulin was as effective as GH in stimulating somatomedin secretion in liver organ culture, although in that study IGF-I was not differentiated from other somatomedins. Scott et al.(1985) reported that insulin stimulated IGF-I secretion by primary rat hepatocyte cultures. At 3 nM insulin, about 33% more IGF-I was secreted over a 24-hour period than at 30 pM insulin. This effect was not observed in cultures prepared from rats made diabetic with streptozotocin (Scott and Baxter, 1986). GH had no effect on circulating IGF-I levels in diabetic rats unless they had received insulin therapy; similarly, GH had no effect _in vitro_ on IGF-I secretion by primary hepatocyte cultures prepared from diabetic rats(Scott and Baxter, 1986). In contrast to these findings, Griffen et al.(1987) showed that administration of low doses of insulin to streptozotocin-diabetic rats _in vivo_ via the hepatic portal vein resulted in a striking increase in serum IGF-I levels, and in addition restored growth as measured by tail length increase and epiphyseal plate width. The same dose of insulin administered via the jugular vein did not have these effects, suggesting that the liver was the target organ for insulin with regard to effects on growth. Insulin delivered via the hepatic portal vein was also more effective in reducing serum glucose levels than when delivered via the jugular vein, although these treated animals were still markedly hyperglycemic(250% of control). Our findings on cultured hepatocytes suggest that insulin or growth hormone is obligatory for IGF-I secretion by hepatocytes; although GH has a much greater effect on IGF-I transcript accumulation than does insulin, both hormones stimulate about the same level of accumulation of IGF-I protein in the culture medium over a 24 hr period under the conditions employed(Johnson et al., 1989;Table I; _v. infra_).

IGF-I EXPRESSION IN CULTURED HEPATOCYTES

These studies and others provide circumstantial evidence that insulin may have direct effects on hepatic IGF-I expression. In order to avoid the complex interrelationships between the pituitary, pancreatic islets, and liver which exist _in vivo_, our laboratory has studied the effects of insulin and GH _in vitro_ using primary hepatocyte cultures

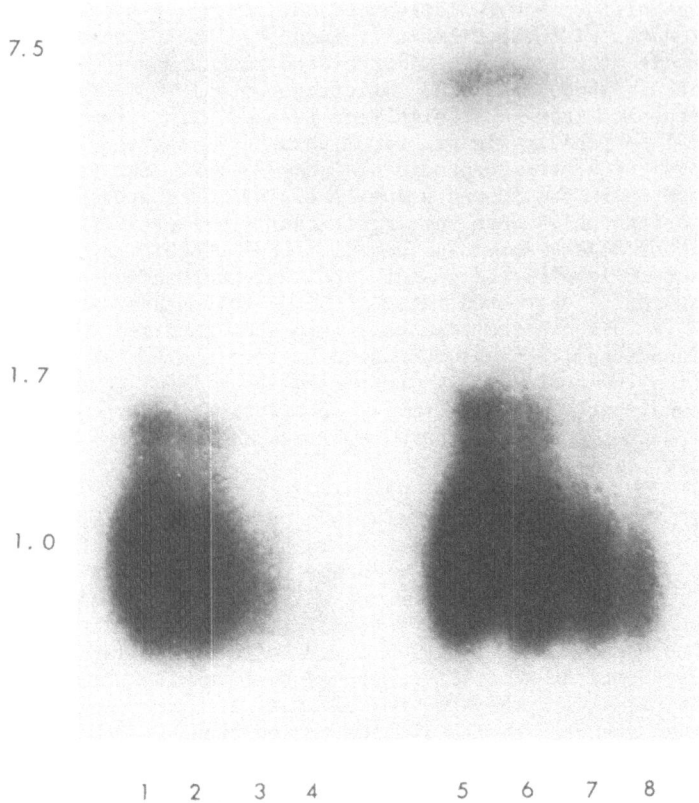

7.5

1.7

1.0

1 2 3 4   5 6 7 8

Figure 1. Effects of insulin and growth hormone on newly synthesized and
longer-term accumulation of IGF-I RNA transcripts in primary rat hepatocyte
cultures. Hepatocytes were plated in serum-free medium containing
insulin(500 ng/ml), triiodothyronine(T3; 1 nM), and hydrocortisone(HC; 10
nM). After overnight incubation cultures were switched to medium containing
insulin, GH, both, or neither in addition to T3 and HC. After 23 hr
additional incubation, cultures were exposed for 1 hr to medium containing
100 uM 4-thiouridine and 5 uC/ml ^{3}H-uridine in addition to the hormonal
supplements. After 1 hr of incubation this medium was aspirated and cells
lysed in guanidine thiocyanate. Newly synthesized, ^{3}H-uridine and
4-thiouridine double-labeled RNA was purified by affinity chromatography
using phenylmercury agarose(Woodford et al., 1988). 80% of the applied
radioactivity bound to the affinity column. Approximately equal amounts of
^{3}H,thiol-labeled RNA was loaded onto the gel(56-79 x 10^3cpm) except for
Lane 4, which received 30 x 10^3 cpm. Lanes 1-4: Thiol-labeled, newly
synthesized RNA isolated from cultures incubated with, in addition to T3
and HC, 500 ng/ml GH and insulin(lane 1); 500 ng/ml GH(lane 2); 500 ng/ml
insulin(lane 3); 1 nM T3 and 10 nM HC only(lane 4). Lanes 5-8:
Unfractionated, total RNA from which thiol-labeled RNA was purified. 10 ug
RNA per lane was loaded. Cultures were incubated with 500 ng/ml insulin and
GH(lane 5); 500 ng/ml GH(lane 6); 500 ng/ml insulin(lane 7); T3 and HC
only(lane 8). Specific activities of the RNA isolated from these 4
conditions was approximately the same(9.9-13 x 10^3 cpm/ug). Following
electrophoresis and transfer to nitrocellulose, the immobilized RNA was
hybridized to a rat IGF-I cDNA probe kindly donated by Dr. Liam
Murphy(Murphy et al., 1987b). Numbers on the left are the sizes(kilobases)
of the major IGF-I transcripts.

prepared from normal rats. One of our interests has been to separate the effects of insulin and GH on IGF-I transcript accumulation and protein secretion, since in the intact rat insulin insufficiency grossly impairs GH secretion(Robinson et al., 1987; Tannenbaum, 1981). In the Northern blot shown in Figure I(from Johnson et al., 1989), the effects of GH and insulin on IGF-I RNA transcript accumulation in cultured rat hepatocytes are shown. Two types of RNA samples were used for this experiment: (1) Total RNA accumulated over 24 hours of exposure to the hormonal supplements(Lanes 5-8); and (2) total RNA synthesized during the last hour of hormonal exposure(Lanes 1-4), which was double-labeled with tritiated uridine and 4-thiouridine and isolated by phenylmercury-agarose affinity chromatography(Woodford et al., 1988). Following hybridization to a probe for IGF-I transcripts and autoradiography, different signal intensities in these latter samples should reflect differences in transcription rates provided that the transcript halflives are significantly longer than the 1-hour 4-thiouridine labeling period. It is apparent from Figure I that insulin does increase IGF-I transcript levels both in newly synthesized(Lane 3) and 24-hour accumulations(Lane 7) of RNA relative to cells not exposed to insulin(Lanes 4 and 8). The same pattern is observed for GH and for the combination of insulin and GH. Since the newly synthesized RNA was labeled with tritium as well as 4-thiouridine, and thus the specific activity of each RNA sample was known, it was possible to estimate the proportion of newly synthesized IGF-I transcripts in the total accumulation by densitometric scanning. Using the approximation that IGF-I transcript levels in cultures treated with insulin represented steady-state levels, this proportion allowed the calculation of 4 hours for the halflife of IGF-I transcripts in the presence of insulin. Thus, while effects of insulin and and GH on transcript stability cannot be ruled out, it appears that in this system insulin and probably GH increase IGF-I gene transcription.

In addition to increasing IGF-I transcript levels in rat hepatocyte primary cultures, insulin and GH also affect secretion of IGF-I protein. Table I(Johnson et al., 1989 and unpublished data) shows the effects of GH and insulin on IGF-I protein accumulation in cells and culture medium over a 24-hr period. In the absence of both hormones, IGF-I is detectable in homogenates of cultured cells but cannot be detected in the culture medium. At this point we do not know if this cellular IGF-I represents protein present before the hepatocytes were prepared for culture, or protein synthesized in culture, or both. In the presence of either GH or insulin, no IGF-I was detected in cellular homogenates but was detected in the culture medium. These data suggest that both GH and insulin facilitate secretion of IGF-I from hepatocytes, and are consistent with the results of Griffen et al.(1987; v. supra) which suggest a significant direct effect of insulin on hepatic IGF-I expression. The data presented in Table I also support those of Scott et al.(1985) concerning the effects of insulin on production of IGF-I protein by rat hepatocyte primary cultures, in that large increases in insulin concentration result in rather modest increases in accumulated IGF-I. In comparing the results presented in Figure I and Table I, it is of interest that approximately the same amount of IGF-I protein was detected in the culture medium when the cells were exposed to GH or insulin, but the relative IGF-I transcript concentration was quite different(GH or the combination of insulin and GH increased IGF-I transcript concentration by a factor of 3 over insulin alone; Johnson et al., 1989). The contribution of protein degradation to these results remains to be established. It is also possible that these hormones have effects on translational efficiency as well as on transcription. As mentioned earlier, adult mouse liver appears to contain a considerable pool of IGF-I mRNA which is not associated with polysomes. The deduced sequence of one rat liver IGF-I mRNA allows, in principle, an extremely stable

TABLE 1

Radioimmunological Detection of IGF-I
in Medium and Cell Homogenates from Hepatocyte Primary Cultures

IGF-I(nmol/liter)

Medium Supplements	Medium	Cells*
T3,HC	Not detected	0.25
T3,HC,In(5 ng/ml)	0.24	ND
T3,HC,In(50 ng/ml)	0.52	ND
T3,HC,In(500 ng/ml)	1.2	Not detected
T3,HC,GH(500 ng/ml)	0.75	Not detected
T3,HC,In,GH(500 ng/ml)	0.71	ND
T3,HC,In (medium only)	Not detected	

Table 1. Accumulation of IGF-I in serum-free medium and cell homogenates from primary hepatocyte cultures. Hepatocytes were isolated by collagenase perfusion and plated in serum-free HCD medium (Isom et al., 1985) containing triiodothyronine(T3; 1 nM), hydrocortisone (HC; 10 nM), and insulin (In; 500 ng/ml). After 24 hr the medium was changed and supplements added as indicated. After 24 hr additional incubation the medium was removed, centrifuged briefly at low speed to remove debris, and analyzed for IGF-I protein. The results are representative of media analyses from three independent cultures(i.e., 3 rats). The results for accumulated IGF-I protein in response to 3 different insulin concentrations are from one culture(duplicate assays of pooled plates).RNA transfer blots prepared from these cultures were hybridized to IGF-I cDNA to verify the activity of GH in promoting IGF-I transcript accumulation(cf.Fig. 1). *Cells (2 x 10[6]) from one of these experiments were scraped from the plates into 1 ml of phosphate-buffered saline and lysed by three cycles of freezing and thawing. Following centrifugation for 1 hr at 100000 x g, the supernatants were assayed for IGF-I. Radioimmunoassays were performed as described(Johnson et al., 1989).ND, no data.

secondary structure formed by basepairing between the 5'- and 3'-untranslated regions(Roberts et al., 1987). A less thermodynamically stable secondary structure introduced into the 5'-untranslated region has been shown to inhibit translation of insulin mRNA in COS cells(Kozak,1986). We are continuing to investigate the regulation of IGF-I expression in primary hepatocyte cultures, which provide an attractive model for studying all the major regulators of serum IGF-I levels.

Acknowledgements. This work was supported by NIH grants AR-20618(to the Northeast Ohio Multipurpose Arthritis Center); CA-43703, and HD-25004; the American Diabetes Association; and the Henry M. and Lillian Stratton Foundation. The superb technical assistance of Ms. Susan D. Rudin was essential to these studies.

LITERATURE CITED

Baxter, R.C. and Z. Zaltsmann. 1984. Induction of hepatic receptor for growth hormone(GH) and prolactin by GH infusion is sex independant. Endocrinology 115: 2009-2014.

Binoux, M., C. Lasarre, and N. Hardouin. 1982. Somatomedin production by rat liver in organ culture. Acta Endocrinol. 99: 422-430.

Boutin, J.M., C. Jolicoeur, H. Okamura, J. Gagnon, M. Edery, M.Shirota, D. Banville, I. Dusanter-Fourt, J. Djiane, and P.A.Kelly. 1988. Cloning and expression of the rat prolactin receptor, a member of the growth hormone/prolactin receptor gene family. Cell 53: 69-77.

Brissenden, J.E., A. Ullrich, and U. Franke. 1984. Human chromosomal mapping of genes for insulin-like growth factors I and II and epidermal growth factor. Nature 310: 781-784.

Casella, S.J., E.P. Smith, J.J. van Wyk, D.R.Joseph, M.A.Hynes, E.C.Hoyt, and P.Kay Lund. 1987. Isolation of rat testis cDNAs encoding an insulin-like growth factor I precursor. DNA 6: 325-330.

Doglio, A., C.Dani, G. Fredrikson, P. Grimaldi, and G. Ailhaud. 1987. Acute regulation of insulin-like growth factor-I gene expression by growth hormone during adipose cell differentiation. EMBO J. 6: 4011-4016.

Emler, C.A. and D.S.Schalch. 1987. Nutritionally-induced changes in hepatic insulin-like growth factor I(IGF-I) gene expression in rats. Endocrinology 120: 832-834.

Foster, C.M., S.Carter-Su , J. Schwartz , and J.A.Shafer. 1987. Proceedings of 16th Annual Meeting of the Endocrine Society.

Glaser, E.W., S. Goldstein, and L. S. Phillips. 1987. Nutrition and somatomedin. Xviii. Circulating somatomedin C during treatment of diabetic ketoacidosis. Diabetes 36: 1152-1160.

Goldstein, S., G.J.Sertich, K.R.Levan, and L.S.Phillips. 1988. Nutrition and somatomedin. XIX. Molecular regulation of insulin-like growth fator-1 in streptozotocin-diabetic rats. Molecular Endocrinology 2: 1093-1100.

Griffen, S.C., S.M. Russel L.S.Katz, and C.S.Nicoll. 1987. Insulin exerts metabolic and growth-promoting effects by a direct action on the liver in vivo: clarification of the functional significance of the portal vascular link between the beta cells of the pancreatic islets and the liver. Proc. Natl. Acad. Sci. USA 84: 7300-7304.

Isgaard, J., L. Carlsson, O.G.P.Isaksson, and J.-O. Jansson. 1988. Pulsatile intravenous growth hormone(GH) infusion to hypophysectomized rats increases insulin-like growth factor I messenger ribonucleic acid in skeletal tissues more effectively than continuous GH infusion. Endocrinology 123: 2605-2610.

Isom, H.C., T. Secott, I Georgoff, C. Woodworth, and J. Mummaw. 1985. Maintenance of differentiated rat hepatocytes in primary culture. Proc. Natl. Acad. Sci. USA 82: 3252-3256.

Johnson, T.R., B.K.Blossey, C.W.Denko, and J. Ilan. 1989. Expression of insulin-like growth factor I in cultured rat hepatocytes: effects of insulin and growth hormone. Molecular Endocrinology 3, in press.

Kozak, M. 1986. Influence of mRNA secondary structure on initiation by eukaryotic ribosomes. Proc. Natl. Acad. Sci. USA 83: 2850–2854.

Lanes, R., B. Recker, P. Fort, and F. Lifshitz. 1985. Impaired somatomedin generation test in children with insulin-dependent diabetes mellitus. Diabetes 34: 156–160.

Leung, D.W., S.A.Spencer, G. Cachianes, R.G.Hammonds C.Collins, W.J.Henzel, R.Barnard, M.J.Waters, and W.I.Wood. 1987. Growth hormone receptor and serum binding protein:purification, cloning, and espression. Nature 330: 537–543.

Lowe, W.L., C.T.Roberts, S.R.Laskey, and D. LeRoith. 1987. Differential expression of alternative 5'-untranslated regions in mRNAs encoding rat insulin-like growth factor I. Proc. Natl. Acad. Sci. USA 84: 8946–8950.

Lowe, W.L., S.R.Laskey, D. LeRoith, and C.T.Roberts. 1988. Distribution and regulation of rat insulin-like growth factor I messenger ribonucleic acids encoding alternative carboxyterminal E-peptides: evidence for differential processing and regulation in liver. Molecular Endocrinology 2: 528–535.

Maes M., Y. Amana, L.E.Underwood, D. Maiter, and J.M. Ketelsleger. 1988. Decreased serum insulin-like growth factor I response to growth hormone in hypophysectomized rats fed a low protein diet: evidence for a post-receptor defect. Acta Endocrinol. 117: 320–326.

Maiter, D., L.E.Underwood, M.Maes, and J.M.Ketelslegers. 1988. Acute down-regulation of the somatogenic receptors in rat liver by a single injedction of growth hormone. Endocrinology 122: 1291–1296.

Mathews, L.S., G. Norstedt, and R.D. Palmiter. 1986. Regulation of insulin-like growth factor I gene expression by growth hormone. Proc. Natl. Acad. Sci. USA 83: 9343–9347.

Merimee, T,J., J. Zapf, B. Hewlett, and L.L. Cavalli-Sforza. 1987. Insulin-like growth factors in pygmies. New England J. Med. 316: 906–911.

Murphy, L.M., G.I.Bell, and H.G.Friesen. 1987a. Tissue distribution of insulin-like growth factor I and II messenger ribonucleic acid in the adult rat. Endocrinology 120: 12798–1282.

Murphy, L.J., G.I.Bell, M.L.Duckworth, and H.G.Friesen. 1987b. Identification, characterization, and regulation of a rat complementary deoxyribonucleic acid which encodes insulin-like growth factor I. Endocrinology 121: 684–691.

Phillips, L.S. and R. Vasillopoulou-Sellin. 1980. Somatomedins. New England J. Med. 301: 371–380; 438–446.

Phillips, L.S. 1986. Nutrition, somatomedins, and the brain. Metabolism 35: 78–87.

Roberts, C.T., A.L. Brown, D.E. Graham. S. Seelig, S. Berry, K.H. Gabbay, and M.M. Rechler. 1986. Growth hormone regulates the abundance of insulin-like growth factor I RNA in adult rat liver. J. Biol. Chem. 261, 10025–10028.

Roberts, C.T., S.R.Laskey, W.L.Lowe, , and D.LeRoith. 1987a. Rat IGF-I cDNA's contain multiple 5'-untranslated regions. Biochem. Biophys. Res. Commun. 146:1154-1159.

Roberts, C.T., Jr., S.R.Laskey, W.L. Lowe, Jr., W.T.Seaman, and D. LeRoith. 1987b. Molecular cloning of rat insulin-like growth factor I complementary deoxyribonucleic acids: differential messenger ribonucleic acid processing and regulation by growth hormone in extrahepatic tissues. Molecular Endocrinology 1: 243-248.

Robinson,I.C.A.F., R.G.Clark, and L.M.S. Carlsson. 1987. Insulin, IGF-I, and growth in diabetic rats. Nature 326: 549.

Rotwein, P., Pollock, K. Didier, D.K., and G. Krivi. 1986. Organization and sequence of the human insulin-like growth factor I gene. J. Biol. Chem. 261, 4828-4832.

Rotwein, P. 1986. Two insulin-like growth factor I messenger RNAs are . expressed in human liver. Proc. Natl. Acad. Sci. USA 83, 77-81.

Rotwein, P., K.M.Pollock, M.Watson, and J.D.Milbrandt. 1987. Insulin-like growth factor gene expression during rat embryonic development. Endocrinology 121: 2141-2144.

Schwander, J.C., C. Hauri, J. Zapf, and E.R.Froesch. 1983. Synthesis and secretion of insulin-like growth factor and its binding protein by the perfused rat liver: dependence on growth hormone status. Endocrinology 113: 297-305.

Scott, C.D.,J.L.Martin, and R.C.Baxter. 1985. Rat hepatocyte insulin-like growth factor I and binding protein: effect of growth hormone in vivo and in vitro. Endocrinology 116: 1102-1107.

Scott, C.D. and R.C.Baxter. 1986. Production of insulin-like growth factor I and its binding protein in rat hepatocyte cultures from diabetic and insulin-treated diabetic rats. Endocrinology 119: 2346-2352.

Shimatsu, A., and P. Rotwein. 1987. Mosaic evolution of the insulin-like growth factors. Organization, sequence, and expression of the rat insulin-like growth factor I gene. J. Biol. Chem. 262: 7894-7900.

Tannenbaum, G.S. 1981. Growth hormone secretory dynamics in streptozotocin diabetes: evidence of a role for endogenous circulating somatostatin. Endocrinology 108: 76-82.

Wang, Chun-Yeh, M. Daimon, S.-J. Shen, G.L. Engleman, and J. Ilan. 1988. Insulin-like growth factor I messenger ribonucleic acid in the developing human placenta and in term placenta of diabetics. Molecular Endocrinology 2: 217-229

Woodford, T.A., R. Schlegel, and A.B.Pardee. 1988. Selective isolation of newly synthesized mammalian mRNA after in vivo labeling with 4-thiouridine or 6-thioguanosine. Anal. Biochem. 171: 166-172.

OVARIAN INSULIN-LIKE GROWTH FACTOR I: BASIC CONCEPTS LEADING TO POTENTIAL CLINICAL OUTLETS

Eli Y. Adashi, Carol E. Resnick, Eleuterio R. Hernandez, Marjorie E. Svoboda, and Judson J. Van Wyk

Division of Reproductive Endocrinology, Department of Obstetrics and Gynecology, University of Maryland, School of Medicine, Baltimore, MD 21201 and Department of Pediatrics, University of North Carolina at Chapel Hill Chapel Hill, North Carolina 27514

INTRODUCTION

Amongst potential novel intraovarian regulators, insulin-like growth factor I (IGF-I) has been the subject of increasingly intense investigation[1]. Taken together, these studies strongly suggest the existence of an intraovarian autocrine control mechanism, wherein IGF-I may serve as the central signal, and the granulosa cell its site of production, reception, and action. In doing so, IGF-I may promote the replication and/or cytodifferentiation of the developing granulosa cell, acting largely (but not exclusively) as an amplifier of gonadotropin action. Furthermore, IGF-I (presumptively of granulosa cell origin) may also provide paracrine input to the adjacent theca-interstitial cell compartment in the interest of coordinated follicular development.

The Ovary as a Site of IGF-I Production

Ovarian production of IGF-I was initially suggested by studies revealing that the immunoreactive (i) IGF-I content of porcine follicular fluid substantially exceeded that encountered in serum[2]. Further evidence consisted of the demonstration of cycloheximide-inhibitable, gonadotropin (and estradiol)-dependent[3] iIGF-I in serum-free media conditioned by cultured porcine granulosa cells[4]. Although the rat ovarian content of iIGF-I appears to be growth hormone-dependent[5], a direct effect remains to be demonstrated. Extending the investigation to the transcriptional level, we have recently shown that that the adult and immature rat ovary (as well as the isolated, immature granulosa cell) is a site of IGF-I gene expression[6] and that it may be subject to gonadotropic regulation. Significantly, of all adult rat organs tested[7], the ovary (O) displays the third highest level of IGF-I gene expression, the uterus (U) and liver (Li) being the most active in this regard. In contrast, human granulosa cells may be a site of IGF-II rather than IGF-I gene expression[8]. These observations and the lack of IGF-II gene expression in the adult rat ovary[6] suggest possible species specificity (Fig. 1). However, profound differences in the experimental conditions may favor IGF-II gene expression as a dedifferentiation (fetal) marker (Fig. 2). More recently, Eden et al. have demonstrated that the concentrations of IGF-I in non-dominant ovarian

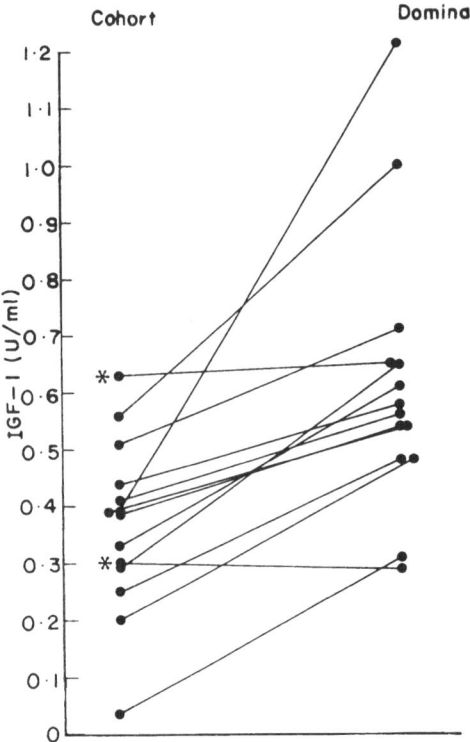

Fig. 1.   IGF-I concentration in dominant and cohort
           follicles. The star refers to 2 patients where
           dominant follicular fluid IGF-I concentrations
           were similar to their cohorts. (Reproduced
           with permission from Eden et al., Clin Endocrinol
           29:327, 1988.)

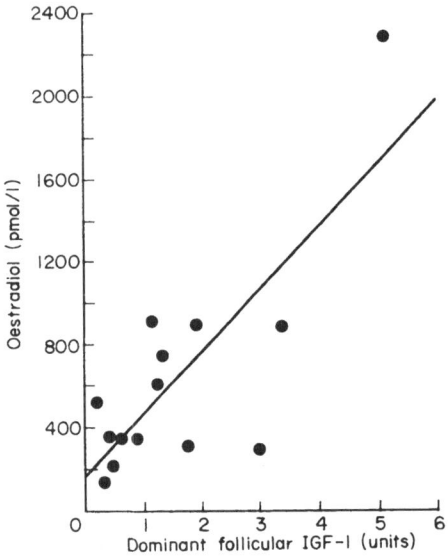

Fig. 2.   Relationship between serum free estradiol 17-β
           and dominant follicular fluid IGF-I (U/follicle).
           The linear regression equation is oestradiol =
           166 + 301 X follicular fluid IGF-I (r = 0.78,
           n = 14, P>0.001). Reproduced with permission
           from Eden et al., Clin Endocrinol 29:327, 1988).

follicles were significantly correlated with those of serum IGF-I and that dominant follicles contained significantly higher concentrations of IGF-I and estradiol then their cohorts[9]. Although detailed studies of ovarian IGF-I gene expression are still lacking, the overall preceding observations are in keeping with the notion that the ovary is a site of IGF-I gene expression and that locally rather than circulatory-derived IGF-I may play a role in ovarian physiology. The cell type(s) responsible for ovarian IGF-I generation remain to be determined.

Ovarian Type I IGF Receptors

Freshly obtained porcine granulosa cells possess high affinity ($Kd = 0.69-2.1X10^{-9}M$), low capacity type IGF binding sites[10,11]. Binding of [^{125}I]IGF-I to FSH-primed rat granulosa cells proved time-, temperature-, and ph-dependent, optimal steady state conditions being achieved following an 8h incubation at 15°C and a pH of 8.0[12]. Although subject to regulation by the cellular density of plating, the binding of [^{125}I]IGF-I to its receptor proved saturable (apparent $Kd = 3.3X10^{-9}M$) as well as reversible, complete or partial tracer displacement being effected by competitive inhibition and dilution, respectively[12]. Scatchard (Fig. 3) and Hill analyses yielded linear plots consistent with

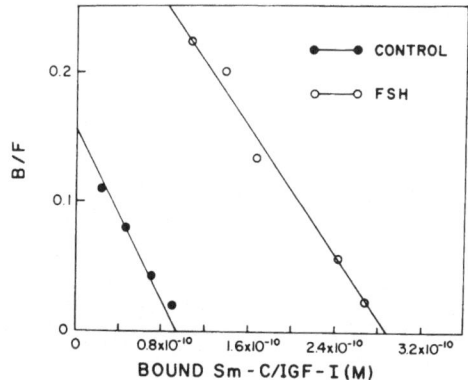

Fig. 3.   Granulosa cell IGF-1 binding capacity and affinity: Effect of FSH. Granulosa cells were initially cultured in the absence or presence of FSH (150 ng/culture). Binding assays were conducted in the presence of a constant amount ($1X10^5$ cpm) of [^{125}I]IGF-I, with or without increasing concentrations (1-100 ng/tube) of its unlabelled counterpart. Scatched plot generated by the binding of IGF-I to its receptor were analyzed using LIGAND software for the determination of receptor binding capacity and affinity. B/F, bound to free ratio. (Reproduced with permission from Adashi et al., Endocrinology 122:194, 1988)

a single class of non-interacting binding sites. Specificity studies revealed the competition for [^{125}I]IGF-I binding to follow a rank order of potency of IGF-I > MSA > insulin, a pattern compatible with a Type I

IGF receptor (Fig. 4). Limited or no displacement was observed for a series of chemically-related and unrelated polypeptides[12].

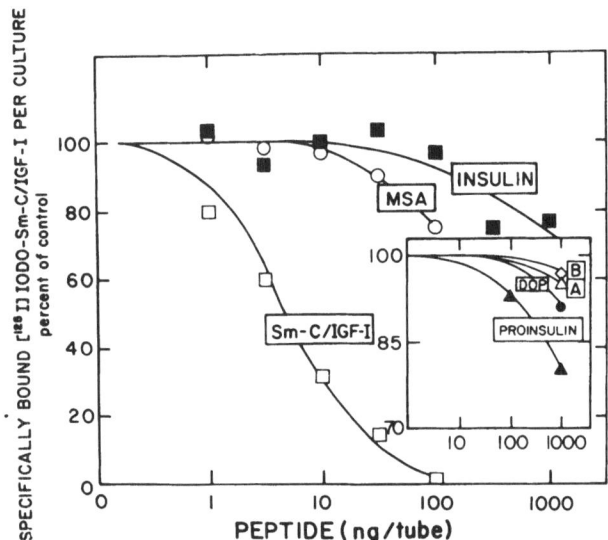

Fig. 4.   Granulosa cell IGF-I binding: specificity.
Granulosa cells initially cultured for 72h in
the presence of FSH (150 ng/culture) were incubated in the presence of a constant amount
(1X10⁵ cpm) of [^{125}I] IGF-I, with or without the
indicated concentrations of IGF-I (1-100 ng/tube),
MSA (multiplication stimulating activity; 1-100
ng/tube), insulin (1-1000 ng/tube), proinsulin
(100 or 1000 ng/tube), DOP (desoctapeptide
insulin; 1000 ng/tube), or insulin chains A
and B (both at 1000 ng/tube). Specific cell-
bound [^{125}I] IGF-I (expressed as a percentage of
the control value) was determined. (Reproduced
with permission from Adashi et al., Endocrinology
122:194, 1988.)

We have recently reported[13] FSH and LH to be capable of upregulating
granulosa cell IGF-I binding, an effect further augmented by growth hormone
but not prolactin. Specifically, we were able to show that FSH is capable
of upregulating IGF-I binding in a time- and dose-dependent (Fig. 5)
fashion and that cAMP, its purported intracellular second messenger, may
play an intermediary role in this regard[13]. Indeed, granulosa cell IGF-I
binding was enhanced following elevation of the intracellular cAMP content
by a series of cAMP-generation agonists, inhibition of cAMP-phosphodiesterase
activity, or the provision of non-degradable cAMP analogs. High dose
forskolin (10^{-5}M), like FSH, proved capable of augmenting IGF-I binding by
itself, while an essentially inert dose (10^{-7}M) synergized with FSH in
this regard. Significantly, heterologous receptor upregulation was not
limited to FSH (Fig. 6), similar increments being observed for luteotropic,
β₂-adrenergic, but not lactogenic granulosa cell agonists[13]. Related
in vivo studies using immature, hypophysectomized, DES-treated rats revealed
that the ability of FSH to upregulate granulosa cell IGF-I binding a) is
not strictly an in vitro phenomenon and that it can be fully reproduced
in vivo; b) is due to enhancement of IGF-I binding capacity rather than
affinity; c) may be subject to diametrically opposed modulation by
somatogenic and GnRH-like granulosa cell agonists (up- and down regulation,

144

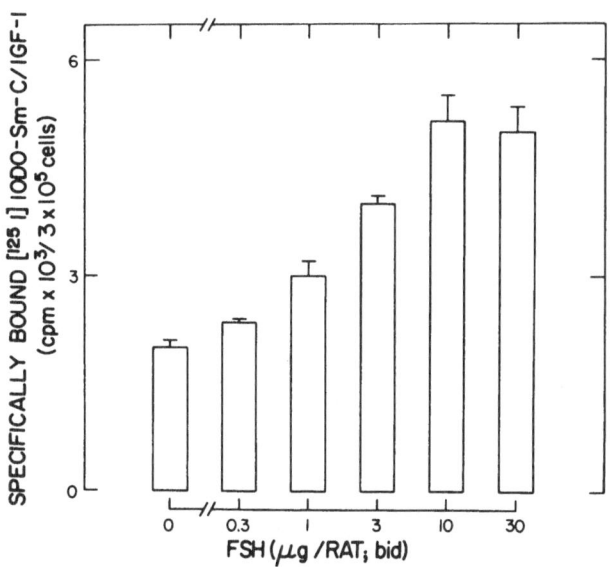

Fig. 5. Effect of in vivo treatment with FSH and granulosa
cell IGF-I binding: dose-dependent. Immature
hypophysectomized, DES-treated rats received
either vehicle or increasing doses of FSH
(0.3-30 μg/rat, twice daily) for 3 consecutive
days beginning the first post operative day.
At the conclusion of this treatment, IGF-I
binding to granulosa cells was evaluated.
Reproduced with permission from Adashi et al.,
Endocrinology 122:1383, 1988).

respectively); and d) is best maintained by gonadotropins but not prolactin.
Inasmuch as gonadotropin-dependence constitutes a unique attribute of the
ovarian granulosa cell, our findings further suggest that the granulosa
cell IGF-I receptor may have thoroughly adopted to its unique environment,
providing the first example of a cell type for which the complement of
IGF-I receptors may be cAMP-dependent.

     Given the pivotal role of FSH in the induction of granulosa cell
receptors for luteotropic and lactogenic ligands, this finding strongly
suggests that the acquisition of IGF-I responsiveness may be part and
parcel of granulosa cell ontogeny. Accordingly, gonadotropins may condition
the cell to respond optimally to IGF-I, thereby conferring selective
advantage upon follicles so endowed.

The Ovary as a Site of IGF-I Action

     Studies carried out in the last several years have clearly established
the granulosa (and more recently the theca-interstitial) cell as sites of
IGF-I action[1]. IGF-I action at the level of the rat, but not porcine
granulosa cell appears largely (but not exclusively) contingent upon its
ability to synergize with pituitary gonadotropins (Fig. 7). These effects
are unaccounted for by enhanced cellular viability, plating efficiency, or
DNA synthesis. Thus, this ability of IGF-I to augment differentiated
phenotypic expression of the developing granulosa cell may be distinct
from its well established growth-promoting property and was thus considered
a novel biologic effect of this polypeptide. In this connection, we have
been able to show that IGF-I is capable of augmenting FSH-supported (but

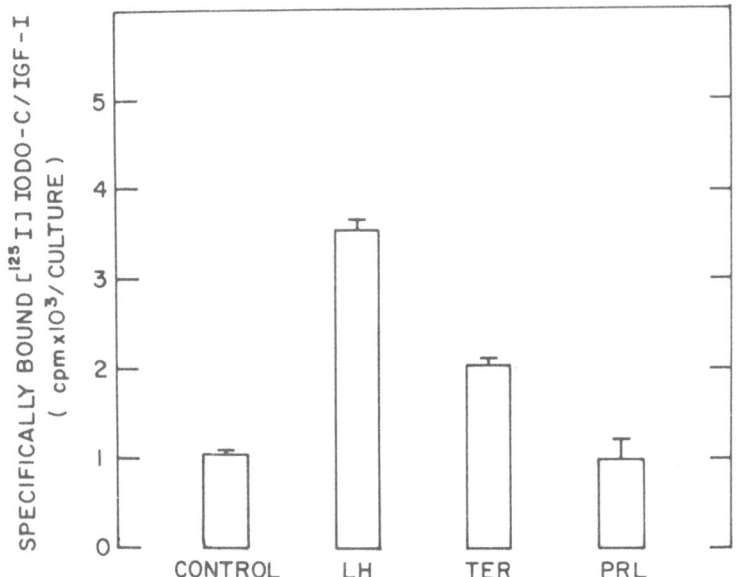

Fig. 6.   Maintenance of granulosa cell IGF-I binding:
effect of luteotropic, lactogenic and β_2-adrenergic
signalling. Granulosa cells were initially
pretreated for 72h in the presence of FSH (150
ng/ml) to induce luteotropic, lactogenic, and
β_2-adrenergic receptors. Thereafter, the media
were thoroughly washed then reincubated for an
additional 72h in the absence or presence of LH
(10 ng/ml), PRL (1 µg), or terbutaline sulfate
(TER; 10^{-6}M). Specific cell-bound [^{125}I] IGF-I
was determined. (Reproduced with permission
from Adashi et al., Endocrinology 122:194, 1988).

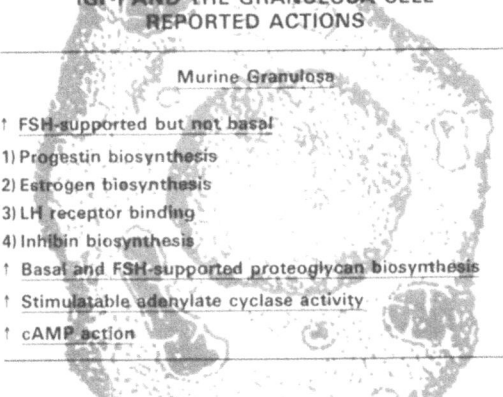

IGF-I AND THE GRANULOSA CELL
REPORTED ACTIONS

Murine Granulosa

↑ FSH-supported but not basal

1) Progestin biosynthesis
2) Estrogen biosynthesis
3) LH receptor binding
4) Inhibin biosynthesis

↑ Basal and FSH-supported proteoglycan biosynthesis

↑ Stimulatable adenylate cyclase activity

↑ cAMP action

Fig. 7.   Reported actions of IGF-I in reference to the rat
granulosa cell.

not basal) progesterone and estrogen biosynthesis as well as the FSH-mediated acquisition of LH receptors. More recently, IGF-I was also found to augment basal as well as FSH-supported proteoglycan biosynthesis[14]. In this respect, IGF-I appeared to exert its classic "sulfation factor" activity at the level of the granulosa cell, the very chondrotropic effect which led to its discovery.

From a mechanistic point of view, IGF-I proved capable of amplifying the FSH transduction sequence at multiple cellular site(s) both proximal and distal to cAMP generation[15]. Although without effect on cAMP break-down (Fig. 8), IGF-I appeared to exert a direct stimulatory effect at the level of the adenylate cyclase complex in the face of unaltered FSH receptor content. However, additional studies will be required to distinguish between direct effect(s) of this peptide on the stimulatory ($G_s = N_s$) or inhibitory ($G_i - N_i$) regulatory membrane proteins, the catalytic (C) cyclase protein, or combinations thereof. In addition,

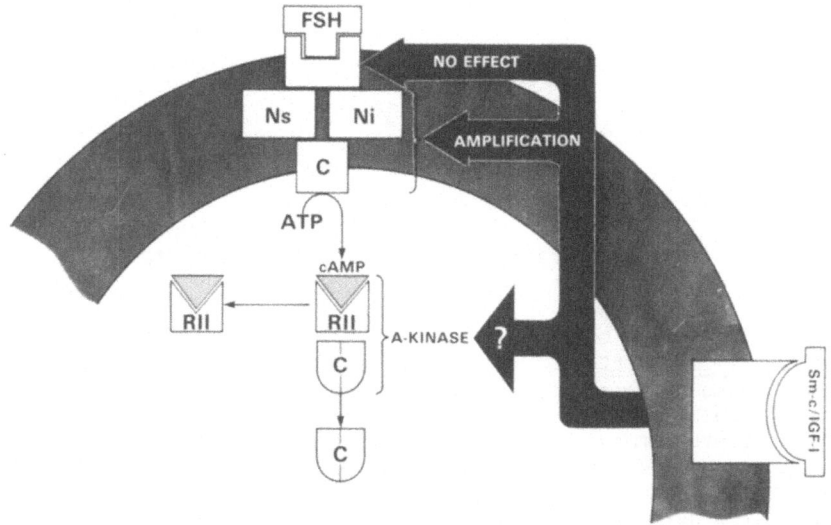

Fig. 8. IGF-I and the granulosa cell: mechanism(s) and site(s) of action.

IGF-I was observed to exert a potent stimulatory effect on cAMP action as reflected in cAMP-supported progesterone biosynthesis. However, additional studies will be required to determine which of the post cAMP events may be involved. The above notwithstanding, there is reason to believe that the ability of IGF-I to enhance FSH hormonal action may not be limited to perturbation of intracellular signalling but may also involve enhanced intercellular communication e.g. granulosa cell clumping and intercellular gap junction formation[16].

We have recently observed that the theca-interstitial, like the granulosa cell, may be a site of IGF-I reception and action and that physiological concentrations of IGF-I may also participate in the regulation of ovarian androgen biosynthesis[17]. As such, these observations are in keeping with the view that IGF-I of granulosa cell origin may not only play an autocrine role but may also serve as one of several signals through which the granulosa cell may communicate in a paracrine fashion with the adjacent theca-interstitial cell compartment. In doing so, the granulosa cell may exert some control over its own destiny by enhancing ovarian androgen provision to suit its aromatizing capabilities and the estrogen requirements of the developing follicle as a whole (Fig. 9). This line of reasoning introduces a level of complexity not previously envisioned implicating IGF-I in the orchestration of the coupling of androgen to estrogen biosynthesis thereby promoting coordinated follicular development.

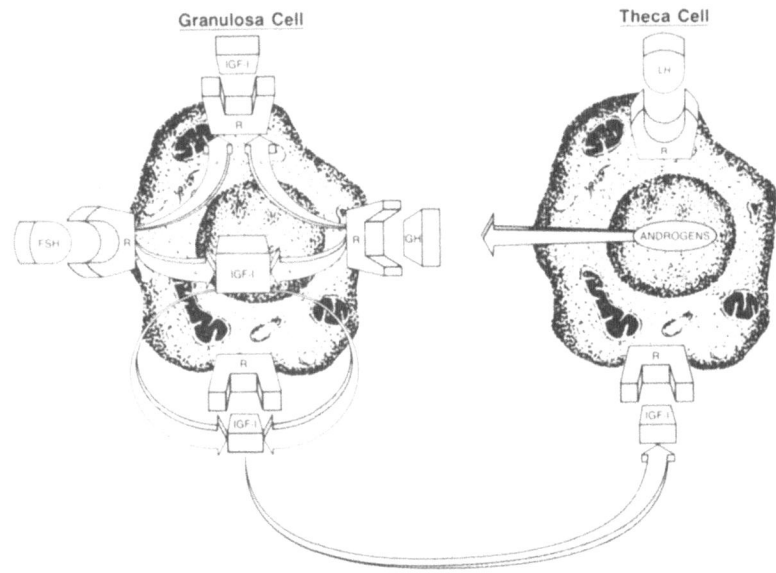

Fig. 9.   The ovarian IGF-I system: intercompartmental
          interaction.

Clinical Implications. The possibility that intragonadal IGF-I may be growth hormone-dependent, and that it might amplify gonadotropin action, could also have a bearing on the otherwise unexplained puberty-promoting effect of growth hormone. This permissive action of growth hormone has been sugested by observations in growth hormone-deficient rats, mice, and humans. In either case, an association appears to exist between isolated growth hormone deficiency and delayed puberty, a process reversed by systemic growth hormone administration to mice and humans. Although a direct effect of growth hormone on the reproductive axis (e.g. enhanced gonadotropin release, or enhanced target tissue response to gonadal steroids) cannot be excluded, it is tempting to speculate that the ability of growth hormone to accelerate pubertal maturation, may be due, at least in part, to the generation of intragonadal IGF and the consequent local amplification of gonadotropin action. In this connection, recent clinical investigation[18] would suggest that the concurrent administration of growth hormone and gonadotropin may in fact result in reduced gonadotropin requirements as might be anticipated from the presumptive intraovarian generation of IGF-I and the consequent amplification of gonadotropin action (Fig. 10). Although additional work will be required, preliminary information would suggest that the addition of growth hormone to ovulation induction regimens may in fact prove useful in terms of overall outcome as measured by overall ovulation and conception rates. In this context, growth hormone may well be applicable in the context of chronic anovulatory conditions as well as in the context of in vitro fertilization procedures.

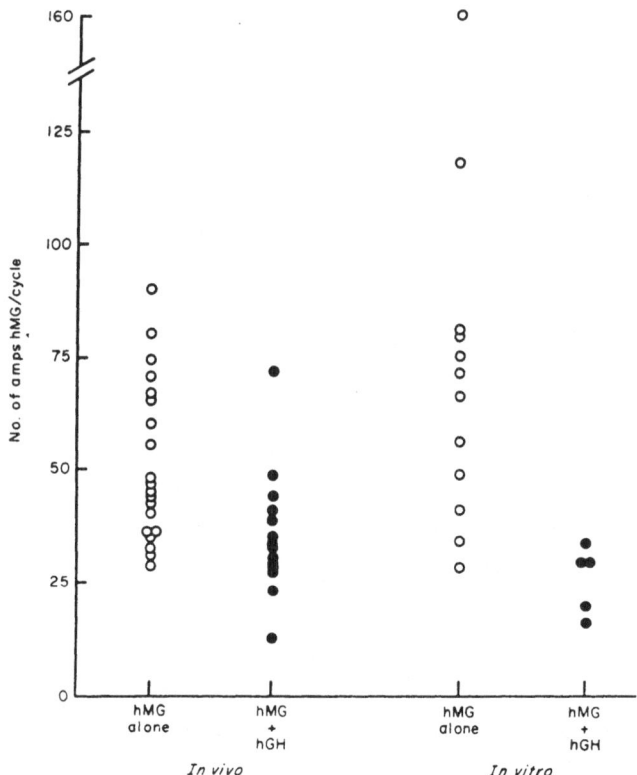

Fig. 10. The number of ampules of hMG (human menopausal
gonadotropins) per cycle needed to induce
follicular growth and maturation in cycles using
hMG alone and the combination of hMG and growth
hormone for in vivo and for in vitro
fertilization. (Reproduced with permission from
Homburg et al., Clin Endocrinol 29:113, 1988).

SUMMARY

     Ovarian IGF-I may potentially subserve at least three central func-
tions: 1) amplification of gonadotropin hormonal action - A key requirement given the exponential nature of follicular development; 2)
integration of follicular development - An essential facet concerned with
the coordination of granulosa-theca cooperation; 3) selection of dominant
follicle(s) - A speculative proposition assuming timely and selective
activation of the IGF-I system in "chosen" follicles.

     Aside from its possible role(s) in the course of established
follicular cycles, IGF-I (and/or IGF-II) may also participate in the early
formation of the follicular apparatus during the late fetal/early neonatal
period. Although the ovary is gonadotropin-independent at that time, we
have previously shown that IGF-I may well interact with VIPergic input now
implicated in the morphodifferentiation of the follicular appartus.
Similarly, IGF-I may be concerned with the promotion of juvenile and
early pubertal follicular development, a time period characterized by
low, ever declining gonadotropin (FSH) levels, but rising titers of serum
growth hormone.

REFERENCES

1. E.Y. Adashi, C.E. Resnick, A.J. D'Ercole, M.E. Svoboda, and J.J. Van Wyk, 1985, Insulin-like growth factors as intraovarian regulators of granulosa cell growth and function, Endocr Rev 6:400-420, (1985).

2. J.M. Hammond, Peptide regulators in the ovarian follicle, Austr J Biol Sci 34:491 (1981).

3. C.J. Hsu and J.M. Hammond, Gonadotropins and estradiol stimulate immunoreactive insulin-like growth factor-I production by porcine granulosa cells in vitro, Endocrinology 120:198 (1987).

4. J.M. Hammond, J.L.S. Baranao, D. Skaleris, A.B. Knight, J.A. Romanus, and M.M. Rechler, Production of insulin-like growth factors by ovarian granulosa cells, Endocrinology 117:2553 (1985).

5. J.B. Davoren, B.G. Kasson, C.H. Li, and A.J.W. Hsueh, Specific insulin-like growth factor (IGF) I- and II-binding sites on rat granulosa cells: relation to IGF action, Endocrinology 119:2155 (1986).

6. E.R. Hernandez, E. Hoyt, J.J. Van Wyk, E.Y. Adashi, The somatomedin-C insulin-like growth factor I (Sm-C/IGF-I) gene is expressed in the rat ovary, Presented at the 69th Annual Meeting of the Endocrine Society, Indianapolis, IN (Abstract #821) (1987).

7. L.H. Murphy, G.I. Bell, and H.G. Friesen, Tissue distribution of insulin-like growth factor I and II messenger ribonucleic acid in the adult rat, Endocrinology 120:1279 (1987).

8. R. Voutilainen, and W.L. Miller, Coordinate tropic hormone regulation of mRNAs for insulin-like growth factor II and the cholesterol side-chain-cleavage enzyme, P450ssc, in human steroidogenic tissues, Pro Natl Acad Sci, USA 84:1590 (1987).

9. R. Homburg, A. Eshel, I Abdalla, and H.S. Jacobs, Growth hormone facilitates ovulation induction by gonadotropins, Clin Endocrinol 29:113-117 (1988).

10. J.L.S. Baranao and J.M. Hammond, Comparative effects of insulin and insulin-like growth factors on DNA synthesis and differentiation of porcine granulosa cells, Biochem Biophys Res Commun 124:484 (1984).

11. J.D. Veldhuis, R.W. Furlanetto, D. Juchter, J. Garmey and P. Veldhuis, Trophic actions of human somatomedin-C/insulin-like growth factor I on ovarian cells: in vitro studies with swine granulosa cells, Endocrinology 116:1235 (1985).

12. E.Y. Adashi, C.E. Resnick, E.R. Hernandez, M.E. Svoboda, and J.J. Van Wyk, Characterization and regulation of a specific cell membrane receptor for somatomedin-C/Insulin-like growth factor I in cultured rat granulosa cells, Endocrinology 122:194-201 (1987).

13. E.Y. Adashi, C.E. Resnick, M.E. Svoboda, and J.J. Van Wyk, Follicle-stimulating hormone enhances somatomedin-C binding to cultured rat granulosa cells: evidence for cAMP-dependence, J of Bio Chem 261:3923-3926 (1986).

14. E.Y. Adashi, C.E. Resnick, M.E. Svoboda, J.J. Van Wyk, V.C. Hascall, M. Yanagishita, Independent and synergistic actions of somatomedin-C in the stimulation of proteoglycan biosynthesis by cultured rat granulosa cells, Endocrinology 118:456 (1986).

15. E.Y. Adashi, C.E. Resnick, J.V. May, M. Knecht, M.E. Svoboda, and J.J. Van Wyk, Somatomedin-C/Insulin like growth factor I as an amplifier of follicle-stimulating hormone action: studies on mechanism(s) and site(s) of action incultured rat granulosa cells, Endocrinology 122:1583-1592 (1987).

16. J.V. May, D.W. Schomberg, S. Gordon, A. Amsterdam, Synergistic effect of insulin and follicle-stimulating hormone on biochemical and morphological differentiation of porcine granulosa cells, Biol Reprod 32:(Suppl) 53 (1985).

17. E.R. Hernandez, C.E. Resnick, M.E. Svoboda, J.J. Van Wyk, D.W. Payne, E.Y. Adashi, Somatomedin-C/insulin-like growth factor I (Sm-C/IGF-I) as an enhancer of androgen biosynthesis by cultured rat ovarian cells, Endocrinology 122:1583-1592 (1987).

18. J.A. Eden, J. Jones, G.D. Carter, and J. Alaghband-Zadeh, A comparison of follicular fluid levels of insulin-like growth factor-1 in normal dominant and cohort follicles, polycystic and multicystic ovaries, Clin Endocrinol 29:327-336 (1988).

REGULATION OF INSULIN-LIKE GROWTH FACTOR-1 GENE EXPRESSION BY ESTROGEN

Liam J. Murphy and Henry G. Friesen

Department of Physiology
Faculty of Medicine
University of Manitoba
Winnipeg, Canada, R3E 0W3

INTRODUCTION

Insulin-like growth factor-I (IGF-I) is a polypeptide of 70 amino acids which has structural similarities to proinsulin and relaxin[1]. In addition to its insulin like actions it is mitogenic for a variety of cultured cells [2]. The growth stimulating actions of growth hormone are thought to be mediated at least in part by insulin-like growth factor-I[2,3]. Expression of IGF-I appears to be GH-dependent in most tissues[4,5,6]. Since IGF-I is able to mimic the effects of GH in the hypophysectomized rat it has been considered to be a relatively specific mediator of GH action[7]. More recently other roles for IGF-I have been proposed. These include local growth in response to various stimuli including partial hepatectomy[8] and partial nephrectomy[9,10]. In addition, IGF-I appears to have some role in differentiation of neural tissue[11], muscle [12] and adipose tissue[12] and appears to synergize with other hormones to enhance responsivity[14,15]. Over the past few years our laboratory has examined the role of IGF-I in estrogen-induced uterine proliferation. Here, we review the evidence that estrogen is the major regulator of uterine IGF-I expression and examine the interaction of estrogen and growth hormone on IGF-I expression in the uterus and liver.

IGF-I EXPRESSION IN THE UTERUS

The uterine response to estrogen is characterized by both cellular hypertrophy and hyperplasia. While this response is easily elicited in vivo it has been difficult to demonstrate an estrogen response in isolated uterine cells in vitro. This paradox has led a number of investigators to suggest that paracrine growth factors may be important in the uterine proliferative response[16]. The observation that estrogen responsiveness of both vaginal and uterine epithelia is dependent upon the presence of appropriate stromal tissue[17] supports the hypothesis that paracrine growth factors elaborated by one cell type may mediate estrogen-induced proliferation in another cell type[18]. A number of growth factors have been isolated from uterine tissue and therefore could potentially act as local estromedins in the uterus[19]. These include, epidermal growth factor [20,21], platelet-derived growth factor [22] and the insulin-like growth factors [23].

In the adult rat, IGF-I mRNA is expressed in most if not all tissues however IGF-I mRNA is more abundant in the uterus than many other tissues[24]. The majority of the IGF-I mRNA is localized to the myometrial layer[23]. This layer of the uterus is also rich in IGF-I immunoreactivity and in IGF-I receptors[25].

Estrogen Induction of Uterine IGF-I mRNA

In the uterus, expression of both IGF-I and its receptor appears to be estrogen responsive [23,25]. Within 1 hour of administration of a single injection of 17 beta-estradiol to pituitary-intact, ovariectomized rats a significant increase in the uterine extractable immunoreactive IGF-I is seen[23]. Maximal concentrations of uterine immunoreactive IGF-I was achieved between 6-12 h after 17 beta-estradiol injection. Acid ethanol extracts of uterine tissue demonstrated parallel displacement curves with authentic recombinant human IGF-I in assays using a series of antisera and monoclonal antibodies raised against serum-derived human IGF-I. Since the serum concentration of IGF-I is considerably higher than the uterine-extractable IGF-I concentration, a major component of the increase in uterine IGF-I almost certainly results from the increase in blood flow to the uterus which follows estrogen administration. However, an increase in uterine IGF-I mRNA abundance was also observed following administration of 17 beta-estradiol [23]. The same groups of IGF-I transcripts seen in RNA extracted from other rat tissues were detected in the uterus and administration of 17 beta-estradiol to ovariectomized rats resulted in a parallel increase in the abundance of each of the three major groups of IGF-I transcripts in the uterus. Maximal accumulation, approximately 14-20 fold compared to untreated rats, was apparent as early as 6 h after 17 beta-estradiol. A similar response is seen in immature rats injected with 17 beta-estradiol. There was no apparent differences in the estrogen-induced IGF-I response in the estrogen "naive" uterus and the uterus of ovariectomized adult rats. In both immature rats and ovariectomized rats the concentrations of IGF-I mRNA in the uterus declines back to basal levels by about 24 hours after estrogen administration. Interestly, in ovariectomized rats treated with daily injections of 17 beta-estradiol, the subsequent injections of 17 beta estradiol tended to elicit a less marked response than the initial injection (Fig. 1).

Localization of IGF-I and its mRNA in the Uterus

Using conventional techniques to prepare separate uterine cell layers, we have detected IGF-I mRNA in both stromal and myometrial fractions however it is not clear from these studies which cell type is responsible for IGF-I expression[23,26]. We were unable to detect IGF-I mRNA in uterine epithelial cells. In the myometrium immunohistochemical techniques localizes IGF-I to the smooth muscle cells. Using in situ hybridizations IGF-I mRNA can be localized to smooth muscle cells however expression is somewhat patchy with only a few cells demonstrating high levels of expression. Other evidence suggests that the smooth muscle cell is likely to be the source of IGF-I mRNA in the myometrial layer. Both IGF-I and IGF-II mRNA are quite abundant in RNA extracted from leiomyomata [27]. It is less clear which cell type is responsible for IGF-I expression in the stroma. However, it is unlike that invading peripheral blood cells are the cause of the increased IGF-I mRNA since IGF-I mRNA is not detectable in mRNA prepared from human or rat peripheral blood mononuclear cells. Furthermore, we can not detect IGF-I mRNA in eosinophils or macrophages prepared by peritoneal lavage even when estrogen is administered to ovariectomized rats prior to peritoneal lavage.

The mechanism whereby estrogen induces an increase in uterine IGF-I mRNA abundance appears to be pituitary-independent since a response of similar or greater magnitude is seen in the ovariectomized, hypophysectomized rat. Although an increase in uterine IGF-I mRNA abundance is seen in ovariectomized, hypophysectomized rats injected with growth hormone this response is less marked and more delayed compared to the estrogen response. Indeed, uterine IGF-I response to estrogen of ovariectomized, pituitary-intact rats is significantly less marked than the response seen in hypophysectomized, ovariectomized rats (Fig. 2). Furthermore, pretreatment of hypophysectomized, ovariectomized rats with growth hormone attentuates the estrogen-induced response. These observations suggest that circulating IGF-I may be able to down-regulate IGF-I expression in response to estrogen. Since

UTERINE IGF-I EXPRESSION

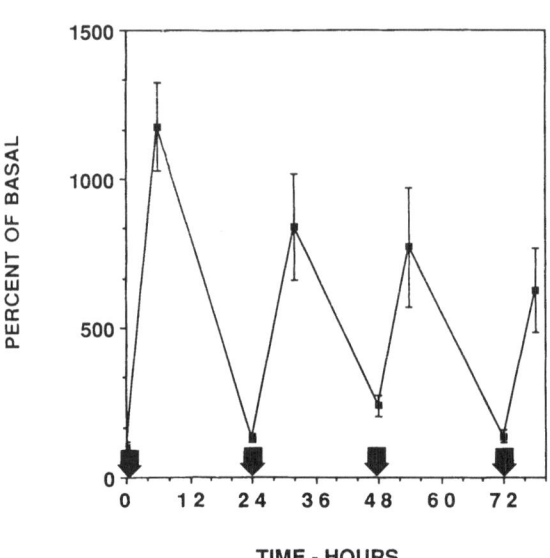

TIME - HOURS

Figure 1. The effect of repeated injections of 17-beta estradiol on uterine IGF-I mRNA accumulation in immature rats. Groups of 12 rats were killed at 6 or 24 hours after a single or repeated intraperitoneal injections of 17-beta estradiol (1 ug/100 g body weight). IGF-I mRNA was quantitated in uterine RNA by slot-blot hybidization. The data represent the mean ± SEM for four separate pools of RNA per time point. Data have been expressed as a percent of the basal levels of IGF-I mRNA in untreated immature rat uteri. Arrows indicate the time of administration of 17-beta estradiol.

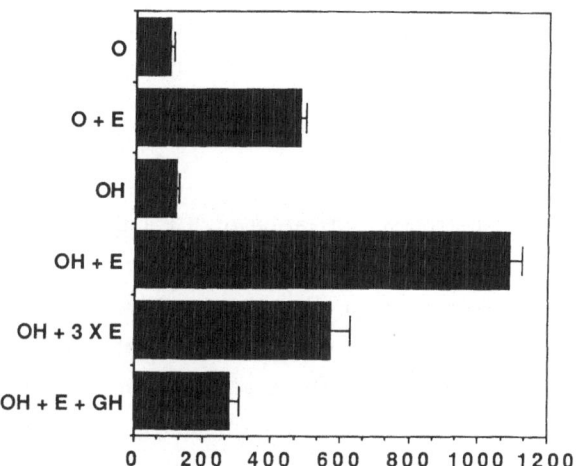

Figure 2. The effect of estrogen on uterine IGF-I mRNA accumulation in pituitary-intact and hypophysectomized, ovariectomized rats. Groups of rats were killed at 6 after a single or repeated intraperitoneal injections of 17-beta estradiol (1 ug/100 g body weight). IGF-I mRNA was quantitated in uterine RNA by slot-blot hybidization. The data represent the mean ± SEM for five or six separate pools of RNA per group. Data have been expressed as a percent of the basal levels of IGF-I mRNA in untreated ovariectomized rats. The abbreviations used are; O, ovariectomized rats, O+E, estrogen-treated ovariectomized rats, OH, ovariectomized, hypophysectomized rats, OH+E, estrogen-treated ovariectomized, hypophysectomized rats, OH+3XE, ovariectomized, hypophysectomized rats which received 3 injections of estrogen at 24 hour intervals, OH+E+GH, ovariectomized, hypophysectomized rats which received human growth hormone (100 ug/rat) together with 17-beta estradiol.

we have not determined whether there are qualitative differences in uterine cell populations which respond to growth hormone and estrogen it is possible that local uterine IGF-I expression in response to growth hormone, in a particular compartment or cell type may modulate the estrogen-induced response in another cell type. Early reports suggested that hypophysectomy results in a delayed uterine mitogenic response to estrogen however this has subsequently been shown to be due to hypothyroidism rather than growth hormone deficiency [28]. Even in hypothyroid rats the early components of the uterine proliferative response to estrogen are normal[28]. Uterine proliferation in response to estrogen is certainly observed in hypophysectomized rats [53] and in growth hormone deficient women.

Mechanisms Involved in Estrogen Activation of IGF-I Expression

The effects of estrogen on uterine IGF-I mRNA accumulation are direct and do not appear to require new protein synthesis. Pretreatment of ovariectomized or immature rats with cycloheximide at doses which reduce uterine protein synthesis to less than 5% of controls does not attenuate the estrogen-induced increase in uterine IGF-I mRNA accumulation[29]. This is in contrast to growth hormone-induction of hepatic IGF-I mRNA accumulation which is sensitive to protein synthesis inhibition[29]. Since estrogen appears to directly activate uterine IGF-I expression we believe that there is likely to be an estrogen responsive element somewhere in the IGF-I gene to which the activated estrogen receptor binds. A search of the published sequences for the human and rat IGF-I genes [30,31] has failed to identify any region with significant similarity to the consensus estrogen responsive element [32]. However, it is clear that there is considerable variability in the sequences of estrogen responsive elements [32,33,34]. Furthermore, the sequences for both human and rat IGF-I genes are incomplete and since the promoter(s) for this gene has not yet been identified it is likely that there are as yet unidentified 5 prime exons.

Recent reports from several laboratories have demonstrated that there is considerable heterogeneity in IGF-I cDNAs [35,36,37] isolated from different tissues. Multiple IGF-I mRNA which vary in the 5' untranslated regions have also been reported [37]. Lowe et al., have demonstrated that there is differential tissue and GH-responsive expression of IGF-I mRNAs with alternative 5' untranslated regions [38]. We have recently isolated from a uterine cDNA library, prepared from immature rats treated with estrogen, an IGF-I cDNA with a unique 5 prime untranslated region. The first 200 bps of this cDNA differ from any of the published IGF-I cDNAs. Isolation of genomic clones containing this sequence may be helpful in localizing the area of the IGF-I gene which confers estrogen responsiveness. Clearly the complexity of the IGF-I gene allows for many levels of regulation, by many different mechanisms.

The Effect of Other Steroid Hormones

The effects of other sex hormones on uterine IGF-I expression have not been extensively investigated. In certain protocols, progesterone appears to inhibit the estrogen effect however a more detailed study is required to determine the interaction between these two hormones in the regulation of uterine IGF-I gene expression. Glucocorticoids which inhibit the accumulation of eosinophils in the uterus [39] do not inhibit the estrogen induced uterine IGF-I response [40]. This, once again, is in contrast to growth hormone-induction of hepatic IGF-I mRNA accumulation which is inhibited by dexamethasone [40].

The Functional Role of IGF-I in the Uterus

IGF-I is a mitogen for a number of different cell lines in culture including fibroblasts and smooth muscle cells [3]. Isolated rat uterine fibroblastoid and smooth muscle cells are also responsive to IGF-I. IGF-I when added to arrested serum-derived cells stimulates DNA synthesis. However, the relevance of this type of experiment to the in vivo role of IGF-I in uterine proliferation is not immediately clear, particularly since cell-cell interaction appears to be important for estrogen-induced uterine proliferation. In organ culture, IGF-I had no significant effect on DNA synthesis (measured as ^{3}H-thymidine incorporation) in uteri from untreated ovariectomized rats whereas a significant enhancement of uterine DNA synthesis by IGF-I was seen in E2 pretreated rats [41]. These observations are analogous to those reported by Zezulak & Green [13]. They found that growth hormone is able to stimulate a clonal expansion of preadipocytes which are more sensitive to IGF-I. A similar situation may exist in the uterus where estrogen may stimulate not only IGF-I expression but a clonal expansion of IGF-I sensitive cells in this tissue. Further studies are necessary to address this question and to determine whether antibodies to IGF-I are able to inhibit estrogen-induced uterine proliferation.

THE EFFECTS OF ESTROGEN ON HEPATIC IGF-I EXPRESSION

A pubertal increase in serum IGF-I is seen in both primates and rodents[42,43] . In man circulating IGF-I concentrations are higher during puberty than at any other time[44]. The observation that serum IGF-I concentration rises in hypogonadal individuals treated with estrogen replacement[45,46] suggest that estrogen may be responsible for the pubertal rise in circulating IGF-I concentration in women. Although the mechanism remains unclear this effect has been attributed by many investigators to an estrogen-induced increase in GH secretion[47,48]. However, a similar increase in serum IGF-I concentration is seen in GH-deficient adolescents[49]. In contrast, pharmacological doses of estrogens have been used to treat tall stature and acromegaly and in this situation estrogen reduces circulating somatomedin levels[50]. In the rat, estrogen appears to retard growth since ovariectomy is associated with increased weight gain and accelerated longitudinal growth[51]. Furthermore, estrogen treatment of ovariectomized rats reduces longitudinal growth[52].

The Acute Effects of Estrogen on Hepatic IGF-I mRNA

In pituitary intact, immature female rats, administration of estrogen has little effect on steady state hepatic IGF-I mRNA levels. Serum IGF-I concentrations and hepatic IGF-I mRNA abundance are similar in rats killed 6 hours after injection of saline or 1 ug of 17-beta estradiol. Interestingly a small, but significant increase in serum IGF-I was seen in ovariectomized, hypophysectomized animals which received identical estrogen treatment[23]. In ovariectomized, hypophysectomized rats, estrogen when administered 3 hours after GH treatment, significantly enhanced the effect of growth hormone on hepatic IGF-I expression[23]. A similar, although less marked effect was apparent when both E2 and GH were administered simultaneously[23]. These studies suggest that estrogen may have some effect acutely on growth hormone signal transduction. Further studies are necessary to determine whether the synergy between estrogen and growth hormone is a result of a specific enhancement of the coupling between GH binding and action, in this case, hepatic IGF-I gene expression or rather the result of non-specific mechanisms such as enhancement of transcription or stabilization of hepatic mRNAs.

The Chronic Effects of Estrogen on Hepatic IGF-I mRNA

In contrast to the acute effects of estrogen, chronic estrogen administration reduces hepatic IGF-I mRNA levels and lowers serum IGF-I concentration. In addition even very small doses of 17-beta estradiol are able to inhibit weight gain and longitudinal growth in ovariectomized rats[53]. In ovariectomized, hypophysectomized rats very small amounts of 17-beta estradiol (100 ng/day) administered daily for 10 days (total dose, 1 ug) inhibits the action of growth hormone in stimulating body weight gain, serum IGF-I and hepatic IGF-I expression[53]. These more recent observations support the original findings of Freudenberger & Hashimoto that ovariectomy accelerates longitudinal growth in the rat[51]. These chronic effects of estrogen contrast with the acute effects described above and with those reported by Shulman et al.[54]. They have recently reported that in pituitary intact, ovarectomized rats which received 0.05 ug/day estradiol benzoate for 28 days, circulating IGF-I levels were significantly higher than sham-operated animals despite similar basal and GRF-stimulated rGH levels[54]. Clearly, further studies are required to address the interaction of estrogen and growth hormone in the regulation of IGF-I expression and skeletal growth.

THE EFFECTS OF ESTROGEN ON IGF-I EXPRESSION IN OTHER TISSUES

The effect of estrogen on IGF-I expression in other tissues has not been extensively investigated. Lippman et al., initially reported that human breast cancer cells synthesize IGF-I in culture[55]. This observation was in contrast to the earlier reports that the apparent IGF-I immunoreactivity in conditioned medium from breast cancer cells was the result of an assay artifact related to an IGF-binding protein[56]. Lippman's group also suggested that expression of IGF-I in human breast cancer cells was estrogen-dependent. They proposed that IGF-I may be one of a number of autocrine mediators of estrogen-induced human breast cancer proliferation. According to their hypothesis constitutional expression of these autocrine mediators could be one of the mechanisms whereby estrogen-independence is achieved in some human breast cancers[57]. Our own studies of IGF expression in many different human breast cancer cell lines have failed to identify any cell line with detectable IGF-I mRNA. This is in contrast to RNA extracted from human breast cancer biopsies where IGF-I mRNA is easily detected in a number of samples (unpublished observations).

The inability to detect IGF-I mRNA in human breast cancer cells may be a function of the relatively insensitive Northern blot technique we used or may indicate that some form of cell-cell interaction is necessary for expression in breast epithelial cells. An alternative hypothesis is that the cell type responsible for the IGF-I mRNA in human breast cancer biopsies is not present in long term human breast cancer cell cultures. A number of lines of evidence supports this latter hypothesis. Firstly, in the uterus there is little IGF-I mRNA in the epithelial cell layer. Most of the IGF-I mRNA is localized to the stroma and myometrial layers. Secondly in situ hybridization with IGF-I in various tissues localizes IGF-I mRNA to fibroblastoid like cells rather than epithelial tissues[58]. Finally, a very recent report from Lippman's laboratory now suggests that since there is no protection of IGF-I riboprobes by RNA from breast cancer cells, the IGF-I like immunoreactivity must not be authentic IGF-I[59].

There is little data which directly addresses the question of IGF-I expression in normal mammary tissue and the role of IGF-I in the hormonal response of the breast. IGF-I mRNA is detectable in rat mammary tissue and the abundance appears to decline throughout pregnancy although this may be a dilutional effect due to an increase in other abundant mRNA encoding milk proteins (unpublished observations). It is likely that IGF-I is expressed in mammary stromal tissue and has some functional role in mammary growth and development. Further studies are necessary to address this point.

Our own data suggest that IGF-I mRNA levels in the lung and kidney are not effected by estrogen administration to either immature or ovariectomized rats. There is no reported data for skin or bone which appear to be estrogen-responsive tissues.

CONCLUSIONS

The observations we have summarized here suggest that expression of IGF-I in the uterus is regulated by estrogen and that IGF-I may be an important mediator of estrogen action in the uterus in addition to its role as a mediator of GH action. If IGF-I is a necessary component of the uterine response to estrogen then local expression rather than pituitary-dependent hepatic IGF-I synthesis may be sufficient for this component of the uterine response to estrogen. Clearly IGF-I is likely to be only one of many growth factors involved in the uterine response to estrogen. Equally convincing data have been presented suggesting that epidermal growth factor may be important in mediating estrogen action in the mouse uterus. Indeed, we have hypothesized that IGF-I expression is not particularly specific for growth hormone or estrogen action but rather is an integral component of the growth response in many tissues to various hormonal and non-hormonal growth stimuli. Since we have not demonstrated a functional role for the locally synthesized IGF-I in the estrogen-induced proliferative response we can not exclude the possibility that the increase in IGF-I expression is a consequence of estrogen-induced growth rather than an important component of the growth response. In the rat, estrogen even when administered in very small physiological amounts, inhibits growth hormone induction of IGF-I expression in the liver. It remains to be determined whether this effect is seen in other tissues and in other species such as man.

REFERENCES

1.   Rinderknecht, E., and Humbel, R.E., 1978. The amino acid sequence of human insulin-like growth factor-I and its structural homology with proinsulin. J. Biol. Chem. 253:2769.

2.   Froesch, E.R., Schmid, C., Schwander, J., and Zapf, J., 1985, Action of insulin-like growth factors. Ann. Rev. Physiol. 47:443.

3.   Clemmons, D.R., and Van Wyk, J.J., 1981, Somatomedin: physiological control and effects on cell proliferation. in: Handbook of experimental pharmacology. R. Baserga, ed., Springer-Verlag, New York, Vol 57:161.

4.   D'Ercole, A.J., Stiles, A.D., and Underwood, L.E., 1984, Tissue concentrations of somatomedin C: Further evidence for multiple sites of synthesis and paracrine or autocrine mechanisms of action. Proc. Natl. Acad. Sci. USA 81:935.

5.   Lund, P.K., Moats-Staats, B.M., Hynes, M.A., Simmons, J.D., Jansen, M., D'Ercole, A.J., and Van Wyk, J.J., 1986, Somatomedin/insulin-like growth factor-I and insulin-like growth factor-II mRNAs in rat fetal and adult tissues. J. Biol. Chem. 261:14539.

6.   Murphy, L.J., Bell, G.I., Duckworth, M.L., and Friesen, H.G., 1987, Identification, characterization and regulation of a rat cDNA which encodes insulin-like growth factor-I. Endocrinology 121:684.

7.   Schoenle, E., Zapf, J., Humbel, R.E., and Froesch, E.R., 1985, Insulin-like growth factor I stimulates growth in hypophysectomized rats. Nature 296:252.

8.   Russell, W.E., D'Ercole, A.J., and Underwood, L.E., 1985, Somatomedin C/insulin like growth factor I during liver regeneration in the rat. Am. J. Physiol. 284:E618.

9. Stiles, A.D., Sosenko, I.R.S., D'Ercole, A.J., and Smith, B.T., 1985, Relationship of kidney tissue somatomedinC/ insulin-like growth factor I to post nephrectomy renal growth in the rat. Endocrinology 117:2397

10. Fagin, J.A., and Melmed. S., 1987, Relative increase in insulin-like growth factor I messenger ribonucleic acid levels in compensatory renal hypertrophy. Endocrinology 120:718

11. McMorris, F.A., Smith, T.M., DeSalvo, S., and Furlanetto, R.W., 1986, Insulin-like growth factorI/somatomedin C: A potent inducer of oligodendrocyte development. Proc. Natl. Acad. Sci. U.S.A. 83:822.

12. Ewton, D., and Florini, J.R., 1981, Effects of the somatomedins and insulin myoblast differentiation In Vitro Dev. Biol. 86: 31.

13. Zezulak, K.M., and Green, H., 1986, The generation of insulin-like growth factor-I sensitive cells by growth hormone action. Science 233:551.

14. Adashi, E.Y., Resnick, C.E., Svoboda, M.E., and Van Wyk, J.J., 1984, A noval role for somatomedin-C in the cytodifferentiation of the ovarian granulosa cells. Endocrinology 115:1227.

15. Veldhuis, J.D., and Rodgers, R.J., 1987, Mechanisms subserving the steroidogenic synergism between follicle-stimulating hormone and insulin-like growth factor I (Somatomedin C). J. Biol. Chem. 262:7658,.

16. Sirbasku, D.A., and Benson, R.H., 1979, Estrogen-inducible growth factors that may act as mediators (estromedins) of estrogen-promoted tumor cell growth. In: Hormones and Cell Culture. G.S. Sato and R. Ross, Eds., Cold Spring Harbor Laboratory, New York p 477.

17. Cooke, P.S., Uchima, F.D.A., Fujii, D.K., Bern, H.A., and Cunha, G.R., 1986, Restoration of normal morphology and estrogen responsiveness in cultured vaginal and uterine epithelia transplanted with stroma. Proc. Natl. Acad. Sci. U.S.A. 83:2109.

18. Cunha, G.R., Chung, L.Q.K., Shannon, J.M., Taguchi, O., and Fujii, H., 1983 Hormone-induced morphogenesis and growth: role of mesenchymal-epithelial interactions. Recent Prog. Horm. Res. 39:559.

19. Ikeda, T., and Sirbasku, D.A., 1984, Purification and properties of a mammary-uterine-pituitary tumor cell growth factor from pregnant sheep uterus. J. Biol. Chem. 259:4049.

20. Gonzalez, F., Lakshmanan, J., Hoath, S., and Fisher, D.A., 1984, Effect of oestradiol-17 on uterine epidermal growth factor concentration in immature mice. Acta Endocrinol. 105:425.

21. DiAuustine, R.P., Petrusz, P., Bell, G.I., Brown, C.F., Korach, K.S., MaLachlan, J.A., and Teng, C.T., 1988, Influence of estrogens on mouse uterine epidermal growth factor precursor protein and messenger ribonucleic acid. Endocrinology 122:2355.

22. Ronnstrand, L., Beckmann, M.P., Faulders, B., Ostman, A., Ek, B., and Heldin C.H., 1987, Purification of the receptor for platelet-derived growth factor from porcine uterus. J. Biol. Chem. 259:2929.

23. Murphy, L.J., Murphy, L.C., and Friesen, H.G., 1988, Estrogen induces insulin-like growth factor-I expression in the rat uterus. Mol. Endocrinol. 1:445.

24. Murphy, L.J., Bell, G.I., and Friesen, H.G., 1987, Tissue distribution of insulin-like growth factor I and II messenger ribonucleic acid in the adult rat. Endocrinology 120:1279.

25. Ghahary, A., and Murphy, L.J., 1989, Uterine insulin like growth factor-I receptors: Regulation by estrogen and variation throughout the estrous cycle. Endocrinology in press.

26. Murphy, L.J., Murphy, L.C., and Friesen, H.G., 1988, A role for the insulin-like growth factors as estromedins in the rat uterus. Trans. Ass. Amer. Physic. 99:204.

27. Hoppener, J.W.M., Mosselman, S., Roholl, P.J.M., Lambrechts, C., Slebos. R.J.C., de Pagter-Holthuizen, P., Lips, C.J.M., Jansz, H.S., and Sussenbach, J.S., 1988, Expression of insulin-like growth factor-I and -II in human smooth muscle tumors. EMBO 7:1379

28. Gardner, R.M., Kirkland, J.L., Ireland, J.S., and Stancel, G.M., 1978, Regulation of the uterine response to estrogen by thyroid hormone., Endocrinology 103:1164.

29. Murphy, L.J., and Luo, J.M., 1989, Effects of cycloheximide on hepatic and uterine insulin-like growth factor-I mRNA . Mole. Cell. Endocrinol. in press.

30. Rotwein, P., Pollock, K.M., Didier, D.K., and Krivi, G.G., 1986, Organization and sequence of the human insulin-like growth factor-I gene. J. Biol. Chem. 261:4282.

31. Shimatsu, A., and Rotwein, P., 1987, Mosaic evolution of the insulin-like growth factors> Organization, sequence and expression of the rat insulin-like growth factor-I gene. J. Biol. Chem. 262:7894.

32. Martinez, E., Givel, F., and Wahli, W., 1987, The estrogen-responsive element as an inducible enhancer: DNA sequence requirements and conversion to a glucocorticoid-responsive element. EMBO 7:3719.

33. Green, S., and Chambon, P., 1988, Nuclear receptors enhance our understanding of transcription regulation. TIG 4:309.

34. Evans, R.M., 1988, The steroid and thyroid hormone receptor superfamily. Science 240:889.

35. Rotwein, P., 1986, Two insulin-like growth factor I messenger RNAs are expressed in human liver. Proc. Natl. Acad. Sci. USA 83:77-81

36. Bell, G.I., Stempien, M.M., Fong, N.M., Rall, L.B., 1986, Sequences of liver cDNAs encoding two different mouse insulin-like growth factor I precursors. Nucleic Acid Res. 14:7873.

37. Roberts, C.T., Lasky, S.R., Lowe, W.L., LeRoith, D., 1987, Rat IGF-I cDNAs contain multiple 5'-untranslated regions. Biochem. Biophys. Res. Commun. 146, 1154.

38. Lowe, W.L., Roberts, C.T., Lasky, S.R., LeRoith, D., 1987, Differential expression of alternative 5' untranslated regions in mRNAs encoding rat insulin-like growth factor I. Proc. Natl. Acad. Sci. U.S.A. 84, 8946.

39. Tchernitchin, A., Rooryck, J., Tchernitchin, X., Vandenhende, J., and Galand, P., 1975, Effects of cortisol on uterine eosinophilia and other oestrogenic responses. Mole. Cell. Endocrinol. 2:331.

40. Luo, J.M., and Murphy, L.J., 1989, Dexamethazone inhibits growth hormone induction of insulin-like growth factor-I (IGF-I) mRNA in hypophysectomized rats and reduces IGF-I mRNA abundance in the intact rat. Endocrinology in press.

41. Murphy, L.J., 1988, Estrogen induces the expression and enhances the binding and action of insulin-like growth factor-I in the rat uterus. Clinical Res. 36:387A.

42. Rosenfeld, R.L., Furlanetto, R., and Bock, D., 1983, The relationship of somatomedin-C levels to pubertal changes. J. Pediatr. 103:723.

43. Handelsman, D.J., Spaliviero, J.A., Scott, C.D., and Baxter, R.C., 1987, Hormonal regulation of the peripubertal surge of insulin-like growth factor-I in the rat. Endocrinology 120:491.

44. Bala, R.M., Lopaka, J., Leung, A., McCoy, E., and McArthur, R.G., 1981, Serum immunoreactive somatomedin levels in normal adults, pregnant women at term, children at various ages and children with constitutionally delayed growth. J. Clin. Endocrinol. Metab., 52:508.

45. Rosenfield, R.L., and Furlanetto, R., 1985, Physiologic testosterone or estradiol induction of puberty increases plasma somatomedin-C. J Pediatr. 107:415

46. Ross, J.L., Cassorla, F.G., Skerda, M.C., Valk, I.M., Loriaux, D.L., and Cutler, G.B., 1983, A preliminary study of the effect of estrogen dose on growth in Turner's syndrome. N. Engl. J. Med. 309:1104.

47. Copeland, K.C., Johnson, D.M., Kuehl, T.J., and Castracane, V.D., 1984, Estrogen stimulates growth hormone and somatomedin-C in castrate and intact female baboons. J. Clin. Endocrinol. Metab. 58:698.

48. Liu, L., Merriam, G.R., and Sherins, R.J., 1987, Chronic sex steroid exposure increases mean plasma growth hormone concentration and

pulse amplitude in men with isolated hypogonadotropic hypogonadism. J. Clin. Endocrinol. Metab. 64:651.

49. Dean, H.J., Kellet, J.G., Bala, R.M., Guyda, H.J., Bhaumick, B., Posner, B.I., and Friesen, H.G., 1982, The effect of growth hormone treatment on somatomedin levels in growth hormone deficient children. J. Clin. Endocrinol. Metab. 55:1167.

50. Clemmons, D.R., Underwood, L.E., Ridgeway, E.C., Kliman, B., Kjellberg, RN, and Van Wyk, J.J. 1980, Estradiol treatment of acromegaly. Reduction of immunoreactive somatomedin-C and improvement in metabolic status. Am. J. Med. 69:571.

51. Freudenberger, C.B., and Hashimoto, E.I., 1937, A summary of data for the effects of ovariectomy on body growth and organ weights of the young albino rat. Am. J. Anat. 62:93.

52. Jansson, J.O., Eden, S., and Isaksson, O., 1983, Sites of action of testosterone and estradiol on longitudinal bone growth. Am. J. Physiol. 244:E135.

53. Murphy, L.J., and Friesen, H.G., 1988, Differential effects of estrogen and growth hormone on uterine and hepatic insulin-like growth factor-I gene expression in the ovariectomized, hypophysectomized rat. Endocrinology 122:325.

54. Shulman, D.I., Sweetland, M., Duckett, G., and Root, A.W., 1987, Effect of estrogen on the growth hormone (GH) secretory response to GH-releasing factor in the castrate adult female rat in vivo. Endocrinology 120:1047.

55. Huff, K.K., Kaufman, D., Gabbay, K.H., Spencer, E.M, Lippman, M.E., and Dickson, R.B, 1986, Secretion of an insulin-like growth factor-I-related protein by human breast cancer cells. Cancer Res. 46:4613.

56. Baxter, R.C., Maitland, J.E., Raisur, R.L., and Sutherland, R.L., 1983, High molecular weight somatomedin-C (IGF-I) from T-47D human mammary carcinoma cells: immunoreactivity and bioactivity. in: Insulin-like growth factors/somatomedins. E.M. Spencer, ed. Walter de Gruyter Co., Berlin. pp615.

57. Dickson, R.B., McManaway, M.E., and Lippman, M.E., 1986, Estrogen-induced factors of breast cancer cells partially replace estrogen to promote tumor growth. Science 232:1540.

58. Han, V.K.M., D'Ercole, A.J., and Lund, P.K., 1987, Cellular localization of somatomedin (insulin-like growth factor) messenger RNA in the human fetus. Science 236:193.

59.   Yee, D., Favoni, R.E., Huff, K.K., Paik, S., Dickson, R.B., Lebovic, G.S., Schwartz, A., Lippman, M.E., and Rosen, N., 1988 Insulin-like growth factor-I (IGF-I) expression and a novel IGF-I related activity in human malignancy. Proc. Amer. Ass. Cancer Res 29:216 (abstract).

REGULATION OF IGF-II GENE EXPRESSION IN HUMAN ADRENALS, GONADS AND PLACENTA

Raimo Voutilainen

Department of Pediatrics and Pathology
University of Helsinki
SF-00290 Helsinki, Finland

INTRODUCTION

Insulin-like growth factor I (IGF-I) is a 70-amino acid polypeptide mediating some of the somatotrophic effects of growth hormone. IGF-I is produced in many tissues, suggesting that it may function in an autocrine or paracrine fashion (D'Ercole et al., 1984 and 1986; Bell et al. 1985; Lund et al., 1986; Han et al. 1987a and b). Expression of IGF-I gene in the liver is regulated by growth hormone (Mathews et al., 1986; Roberts et al., 1986).

Insulin-like growth factor II (IGF-II) is a 67-amino acid polypeptide closely related to IGF-I, but its biological function is less clear. Production of rat IGF-II has been thought to be turned off soon after birth (Brown et al., 1986; Lund et al., 1986) in most tissues. Other species, including man, produce remarkable amounts of IGF-II during adulthood. Human adult serum IGF-II concentrations are three to four times higher than IGF-I, but are only partially responsive to growth hormone (Zapf et al., 1981).

It's evident that there are remarkable species differences in the expression of IGF genes and IGF secretion during different developmental stages. Therefore, we cannot draw clear conclusions to human physiology from data derived from animal experiments. This chapter reviews our experience of IGF gene expression in human fetal and some adult tissues. Our special interest is in the expression of IGF genes in human steroid producing organs, because IGFs reportedly have remarkable influences on the hormonal function of these organs in some other species (Adashi et al., 1985; Chatelain et al., 1988).

IGF-II mRNA IS ABUNDANT IN ACTIVELY STEROID PRODUCING ORGANS

IGF-II mRNA is detected in all fetal tissues examined, but in greatly varying abundances (Fig. 1). Actively steroid producing organs contain remarkable amounts of IGF-II mRNA. Human fetal adrenal contains as much or even more IGF-II mRNA than fetal liver. It's interesting to note that the relative IGF-II mRNA amounts in fetal steroid producing organs correlate well with the steroid producing capacity and the abundance of mRNAs for the steroidogenic enzymes in these organs (Voutilainen and Miller, 1986). Very limited material of adult human RNA samples was available for IGF-II mRNA analysis. IGF-II mRNA is

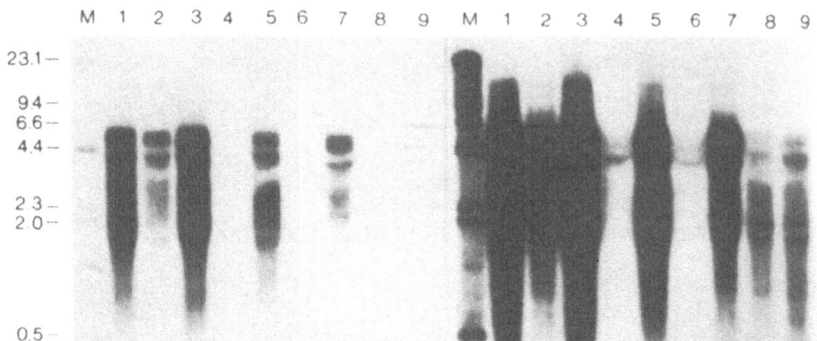

Fig. 1. Northern blot probed with ^{32}P-labelled IGF-II cDNA. Each lane was loaded with 30 ug of total RNA from the following tissues: lane 1, human fetal liver (20,4 wk gestation); lane 2, human placenta (10 wk); lane 3, human fetal adrenal (20.1 wk); lane 4, human adult adrenal; lane 5, human fetal testis (19 wk); lane 6, human testicular Leydig cell tumor; lane 7, fetal rhesus monkey testis; lane 8, human fetal ovary (20 wk); lane 9, fetal rhesus monkey ovary; lane M, molecular weight markers. A. Short autoradiographic exposure; B. Long exposure. (Reprinted from Voutilainen and Miller, 1988).

detectable also in non-fetal steroidogenic tissues (adult adrenal, childhood Leydig cell tumor, adult ovarian granulosa cells), but it is clearly less abundant than in fetal steroid producing organs (Fig. 1B, 4A).

IGF-I mRNA is clearly less abundant that IGF-II mRNA in all fetal tissues studied. We could not detect possible differences in IGF-I mRNA expression in different fetal steroidogenic tissues due to the very low IGF-I gene expression level and the limited fetal tissue material available for analysis. IGF-I and II mRNA analysis has not been done carefully in steroidogenic tissues of other mammals. Thus we don't know if the high IGF-II gene expression in fetal adrenals and testes is typical for primates only, or if it exists in other species as well.

ACTH-REGULATED EXPRESSION OF IGF-II mRNA IN HUMAN ADRENALS

In cultured human fetal adrenal cells ACTH-treatment clearly increases IGF-II mRNA accumulation (Fig. 2, 4B). IGFs added to the culture medium seem to have some inhibitory effect on IGF-II mRNA accumulation (Fig. 2). We don't know for sure what biological function IGF-II has in the adrenal. Demonstration of IGF mRNA by in situ hybridization histochemistry (Han et al., 1987a) and IGF immunoreactivity by immunohistochemistry (Han et al., 1987b) in human fetal adrenocortical cells further points out some regulatory function for IGF-II in human fetal adrenals. We know that human fetal adrenal is very big and its growth and steroidogenic function is controlled mainly by pituitary ACTH. Thus it is tempting to speculate that IGF-II regulates human fetal adrenal growth under the influence of ACTH.

In addition to the possible IGF-mediated adrenal growth regulation, IGFs may also regulate adrenocortical hormonal function. In bovine adrenocortical cells IGF-I seems to stimulate steroidogenesis (Penhoat et al., 1988). On the other hand, in our experiments IGF-I or rat IGF-II (MSA) had no effect on the mRNA levels of some steroidogenic enzymes (Fig. 2). This again suggests species differences in the effects of IGFs.

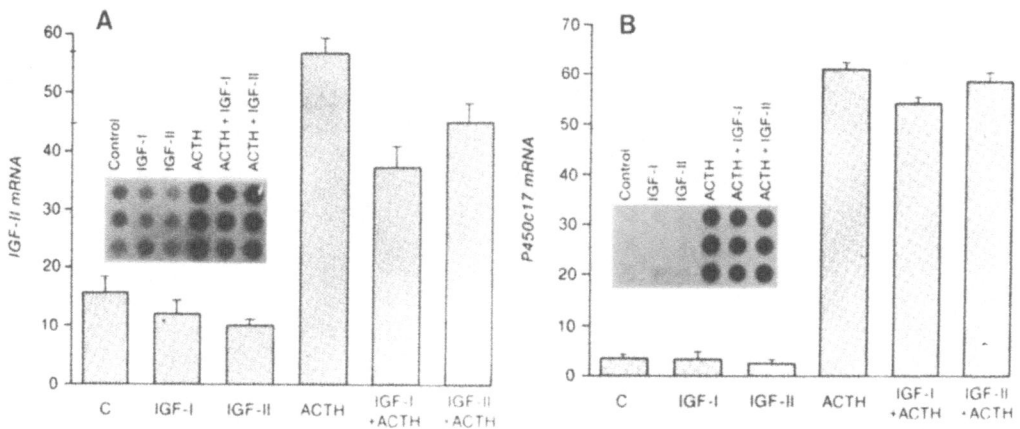

Fig. 2. Regulation of IGF-II mRNA (A) in cultured human fetal adrenal cells. The IGF-II probe was later washed away and the same filter rehybridized with P450c17 (17-hydroxylase/17, 20 lyase) probe (B) for comparison. (Modified from Voutilainen and Miller, 1988).

GONADOTROPIN-REGULATED EXPRESSION OF IGF-II mRNA IN HUMAN OVARIAN GRANULOSA CELLS

IGF-II mRNA is abundant in cultured ovarian granulosa cells, while IGF-I mRNA is barely detectable (Voutilainen and Miller, 1987a). IGF-II mRNA abundance in granulosa cells is clearly dependent on the culture stage (Fig. 3).

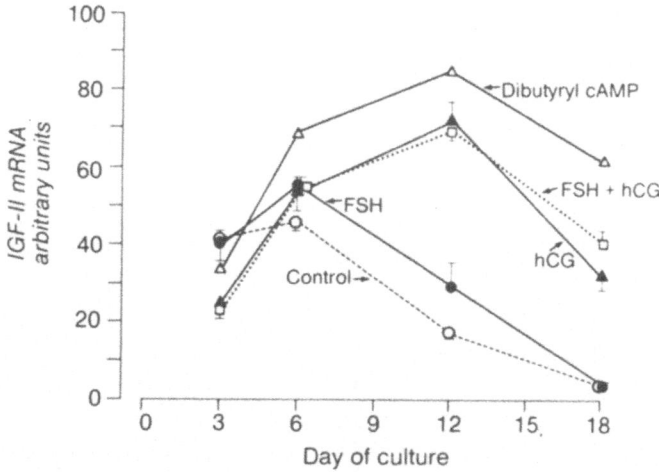

Fig. 3. Dependence of ovarian granulosa cell IGF-II mRNA expression on the culture stage. Cultures were begun on day 0, and hormonal stimuli were started 2 days before the cells were harvested on the day indicated on the x-axis. (Reprinted from Voutilainen and Miller, 1987a).

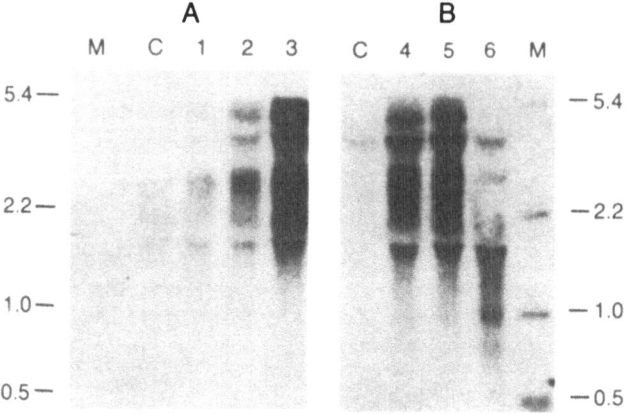

Fig. 4. Northern blots of cultured human granulosa cells (A) and cultured human fetal adrenal cells (B) probed with IGF-II cDNA. Lanes M contain molecular weight markers shown in kb. The 30 ug RNA sources analysed in A are: cultured human granulosa cells without any hormonal treatment (lane C); cultured granulosa cells treated with FSH at 100 ng/ml (lane 1), hCG at 100 ng/ml (lane 2), or 1 mM dibutyryl cAMP (lane 3). The 50 ug RNA samples analysed in B are from: cultured human fetal adrenal cells without any hormonal treatment (C), with 0.1 uM ACTH (lane 4) or 1 mM dibutyryl cAMP (lane 5), and cultured human genital skin fibroblasts without any hormonal stimulus (lane 6). (Reprinted from Voutilainen and Miller, 1987a).

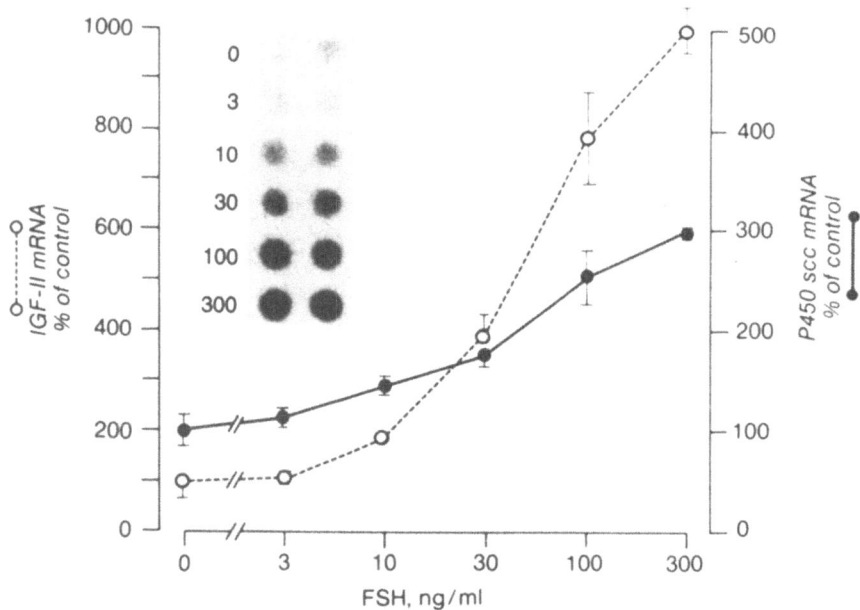

Fig. 5. Dose-dependent effects of FSH on IGF-II and P450scc mRNAs in cultured human granulosa cells. Cells were cultured 10 days without and then 2 days with FSH before harvesting. The original IGF-II dot-blot data is shown in the insert. (Reprinted from Voutilainen and Miller, 1987a).

172

We have not studied systematically the proliferation rate of granulosa cells, but as estimated by phase contrast microscopy, the time of highest IGF-II mRNA expression seems to coincide or slightly precede the most active proliferation period of the granulosa cells. This fits to the hypothesis that IGF-II is an important autocrine mitogenic agent for granulosa cells. IGF-II mRNA expression is lower in the fresh granulosa cells just harvested from in vitro fertilization patients than in cultured cells at their highest proliferation period (Voutilainen and Miller, 1989). This finding is not in contradiction with our hypothesis, because the uncultured fresh granulosa cells represent non-proliferating highly differentiated cells from mature ovarian follicles after intensive gonadotropin stimulation in vivo.

IGF-II mRNA exists in at least five size species both in granulosa cells and fetal adrenals (Fig. 4). In cultured granulosa cells follicle stimulating hormone (FSH), human chorionic gonadotropin (hCG; has the same biological activity as luteinizing hormone, LH) and dibutyryl cyclic AMP (dbcAMP) increase all the forms of IGF-II transcripts (Fig. 4). The percent increase is most remarkable in the highest molecular weight transcripts. The stimulatory effect of FSH (Fig. 5), hCG (Fig. 6) and dbcAMP (Fig. 7) on IGF-IImRNA accumulation is dose dependent, and physiological concentrations of gonadotropins increase IGF-II mRNA. The same gonadotropin doses increase progesterone production and mRNA accumulation of the cholesterol side-chain cleavage enzyme (P450scc), which is the rate limiting enzyme in progesterone production (Voutilainen et al., 1986).

Gonadotropin regulated expression of IGF-II gene in ovarian granulosa cells leads to the suggestion that gonadotropins might regulate ovarian follicular growth through this growth factor. Ramasharma and Li (1987) have shown that human granulosa cells secrete IGF-II peptide and that gonadotropins regulate IGF-II secretion in these cells. This supports our hypothesis and verifies that IGF-II transcripts are really translated to IGF-peptides in granulosa cells.

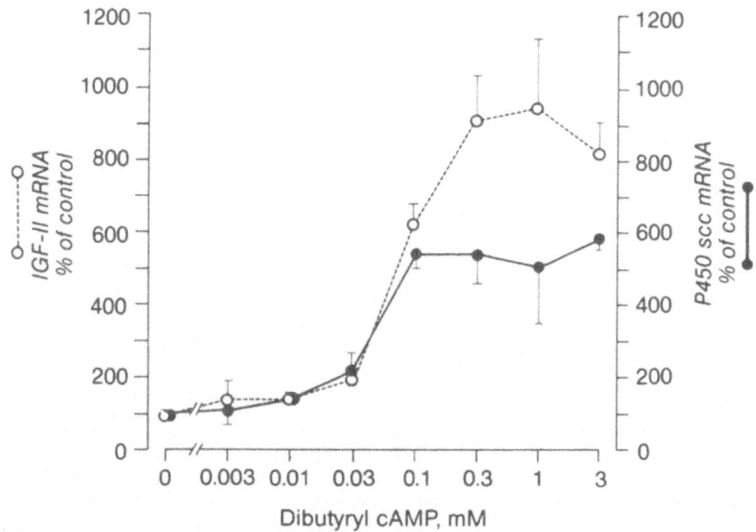

Fig. 6. Dose-dependent effects of hCG on IGF-II and P450scc mRNAs in cultured human granulosa cells. The procedures were as in Figure 5. (Reprinted from Voutilainen and Miller, 1987a).

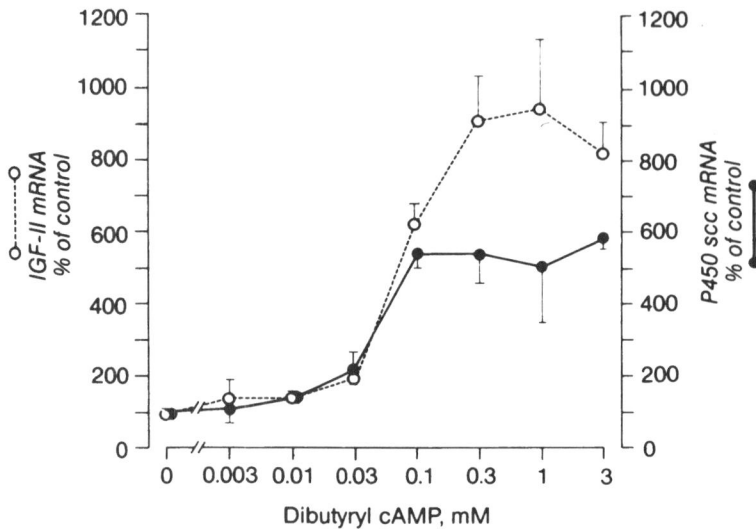

Fig. 7. Dose-dependent effects of dibutyryl cAMP on IGF-II and P450scc mRNA in cultured human granulosa cells. The procedures were as in Figure 5. (Reprinted from Voutilainen and Miller, 1987a).

Other species than man may use IGF-I instead of IGF-II to mediate the proliferative and/or differentiative effect of gonadotropins on granulosa cells. In porcine granulosa cells gonadotropins and estradiol stimulate immunoreactive IGF-I production (Hsu and Hammond, 1987). Rat granulosa cells contain both IGF-I and II type receptors, but IGF-I is more potent than IGF-II in advancing ovarian steroidogenic functions (Davoren et al., 1986). In human granulosa cell cultures IGF-I gene expression is very low, and we could see no increase of this low signal during gonadotropin or growth hormone treatments (Voutilainen and Miller, 1987a).

IGF-II mRNA EXPRESSION IN HUMAN FETAL TESTES AND OVARIES

IGF-II mRNA is much more abundant in human fetal testes than in fetal ovaries (Fig. 1). The same is true for the expression of mRNA's for steroidogenic enzymes (Voutilainen and Miller, 1986; Voutilainen et al., 1988). In fetal testes testosterone production and the level of expression of the genes needed in testosterone synthesis is highest during 12-16 weeks of gestation. Steroidogenesis in fetal testes is regulated by placental hCG. The abundance of IGF-II mRNA in fetal testes is also dependent on the gestational age, following the same pattern as steroidogenic enzyme mRNAs (Fig. 8). This suggests that expression of the IGF-II gene may be controlled by the same factors that control steroidogenesis. We don't know yet in what cell types IGF-II gene is expressed in human fetal testes. Experiments using in situ hybridization histochemistry and immunocyto-chemistry will be needed to identify the cell types producing IGF-II. In rat testes Sertoli cells seem to produce IGF-like peptides (Smith et al., 1987), but in humans we may well have a different distribution of IGF gene expression. In addition, we seem to have again clear species differences in the expression of IGF-I and II in testes. While IGF-II is highly expressed in human fetal testes, IGF-I mRNA is hardly detectable and it shows no essential change in its expression level at

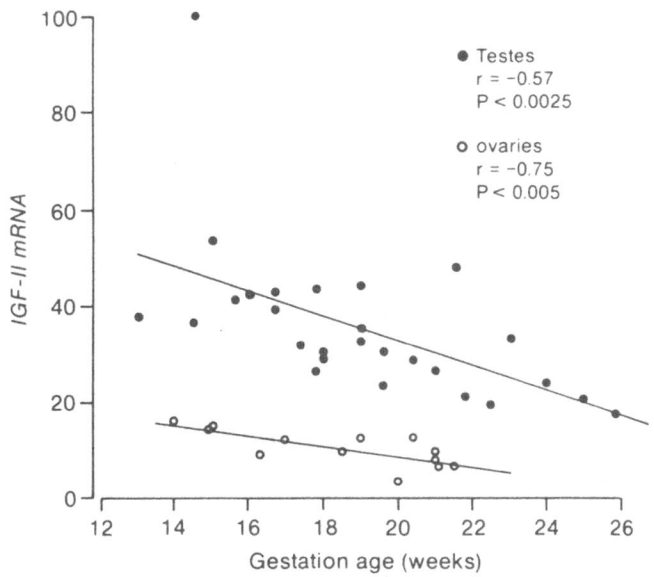

Fig. 8. IGF-II mRNA expression in human fetal testes and ovaries as a function of gestational age. The amounts of IGF-II mRNA are displayed as arbitrary units, determined by densitometric scanning of a single dot blot autoradiograph containing all the samples. The lines show the linear regression analysis of the testicular and ovarian IGF-II mRNA as a function of gestational age. (Reprinted from Voutilainen and Miller, 1988).

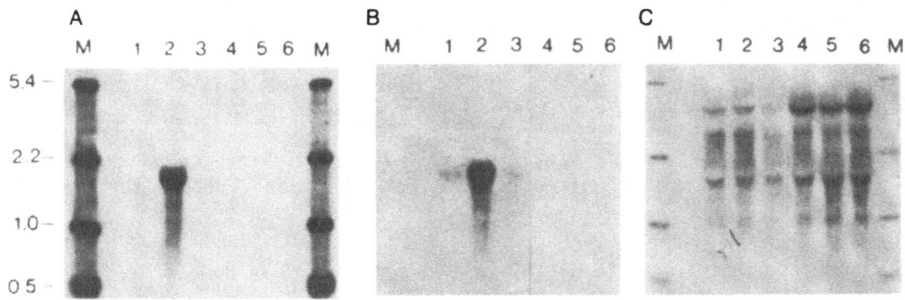

Fig. 9. Northern blot of cultured human fetal testicular cells (from a fetus of 22 wk gestation) and cultured human genital skin fibroblasts probed with P450c17 (A), P450scc (B) and IGF-II (C) cDNAs. Lanes 1-3 have 40 ug total RNA from cultured testicular cells with the following treatments: no treatment (lane 1), 0.5 mM dibutyryl cAMP for 2 days (lane 2), and 50 ng/ml hCG for 2 days (lane 3). Lanes 4-6 have 50 ug total RNA from cultured genital skin fibroblasts with the following treatments: no treatment (lane 4), human growth hormone 100 ng/ml for 2 days (lane 5), and 0.5 mM dibutyryl cAMP for 2 days (lane 6). Lanes M contain molecular weight markers shown in kb. (Reprinted from Voutilainen and Miller, 1988).

Table 1.   Expression of IGF-II, MIS (Mullerian inhibiting substance)
and P450c17 mRNAs in cultured human fetal testicular cells

Treatment	IGF-II mRNA	MIS mRNA	P450c17 mRNA
Control	100.0 ± 29.0	100.0 ± 6.5	100.0 ± 17.9
FSH (50 ng/ml)	61.3 ± 7.6	86.8 ± 17.4	80.2 ± 11.6
hCG (50 ng/ml)	62.5 ± 3.7	115.0 ± 13.0	100.0 ± 5.6
dbcAMP (0.2 mM)	73.0 ± 8.0	$829.0 \pm 17.4*$	$922.0 \pm 41.3*$

Fetal testicular cells (from a 20 wk gestation fetus) were cultured 5 days
without any treatments and then 2 additional days with the indicated
factors before RNA isolation. Data are expressed as means ± SEM of three
cultures, with he controls set at 100. *$p < 0.001$. (Modified from
Voutilainen and Miller, 1987b and 1988).

different gestational ages (Voutilainen and Miller, unpublished observation). In
any case the developmental pattern of IGF-II mRNA in human fetal testes follows
more closely the same pattern as mRNAs for steroidogenic enzymes (being Leydig
cell products) than mRNA for Mullerian inhibiting substance (MIS). MIS is a
Sertoli cell product and the expression of MIS gene is equally high from 13 to 25
weeks of gestation (Voutilainen and Miller, 1987b).

After detecting specific tropic hormone dependent regulation of IGF-II gene
in adrenals and granulosa cells, we supposed to see the same kind of regulation of
IGF-II mRNA in fetal testicular cultures by cAMP mediated mechanism. However,
IGF-II mRNA did not increase in response to gonadotropin or cAMP treatments in
our culture system, although mRNAs for steroidogenic enzymes and MIS
increased during cAMP treatments (Fig. 9, Table 1). Thus regulation of IGF-II
expression in fetal testis may be independent of gonadotropins. It is also possible
that our total testis culture system contains some inhibitors which can somehow
oppose IGF-II mRNA accumulation during cAMP stimulation.

REGULATION OF IGF-II mRNA IN HUMAN PLACENTA

Human placenta secretes both IGF-I and II. In in vitro explant culture
conditions IGF-II secretion clearly exceeds IGF-I production (Fant et al., 1986).

Table 2. Regulation of IGF-II and P450scc mRNAs in human
placental cultures

Treatment	IGF-II mRNA	P450scc mRNA
Control	100.0 ± 12.9	100.0 ± 14.5
dbcAMP (1 mM)	$190.0 \pm 11.8*$	$385.0 \pm 22.2**$
hCG (100 ng/ml)	125.0 ± 11.8	80.2 ± 11.1
hGH (100 ng/ml)	84.2 ± 1.3	84.0 ± 6.8

Placental cells were cultured for 2 days without any treatment and
then stimulated for 2 days as indicated. The values are means ± SEM
(n = 3 in each group) expressed as percentage of the control. hGH,
human growth hormone. *$p < 0.005$ compared to control, **$p < 0.001$
compared to control. (Modified from Voutilainen and Miller, 1987a).

In RNA analysis of placental tissue we can also see much more IGF-II than IGF-I mRNA. In cultured placental cells IGF-II mRNA increases during cAMP treatment (Table 2), as it does in adrenal and granulosa cell cultures. We have tried to further study the expression of IGF genes in placental trophoblast cells. We have found out that a human choriocarcinoma cell line, JEG-3, expresses IGF-II mRNA (Ritvos et al., 1988), but IGF-I mRNA is not detectable. This cell line may be very useful when studying the regulation of IGF-II gene expression and IGF-II secretion in trophoblast cells.

CONCLUSIONS

IGFs seem to have proliferative and differentiative effects in steroidogenic tissues. Local production of these peptides can regulate the growth and function of these organs by paracrine and/or autocrine mechanisms. Important species differences seem to exist in the expression of IGF-I and II genes. The very high, hormonally regulated, expression of IGF-II gene in human fetal adrenals, testes, placenta and adult ovarian granulosa cells suggests that IGF-II is an important growth factor in human steroidogenic tissues. IGF-I may have the same type of function in rat and porcine steroidogenic organs.

REFERENCES

Adashi, E.Y., Resnick, C.E., D'Ercole, A.J., Svoboda, M.E., and Van Wyk, J.J., 1985, Insulin-like growth factors as intraovarian regulators of granulosa cell growth and function, Endocr. Rev., 6:400.

Bell, G.I., Gerhard, D.S., Fong, N.M., Sanchez-Pescador, R., and Rall, L.B., 1985, Isolation of the human insulin-like growth factor genes: Insulin-like growth factor II and insulin genes are contiguous, Proc. Natl. Acad. Sci. USA, 82:6450.

Brown, A.L., Graham, D.E., Nissley, S.P., Hill, D.J., Strain, A.J., and Rechler, M.M., 1986, Developmental regulation of insulin-like growth factor II mRNA in different rat tissues, J. Biol. Chem. 261:13144.

Chatelain, P., Penhoat, A., Perrard-Sapori, M.H., Jaillard, Ch., Naville, D., and Saez, J., 1988, Maturation of Steroidogenic cells: a target for IGF-I, Acta Paediatr. Scand. (Suppl), 347:104.

Davoren, J.B., Kasson, B.G., Li, C.H., and Hsueh, A.J.W., 1986, Specific insulin-like growth factor (IGF) I- and II-binding sites on rat granulosa cells: Relation to IGF action, Endocrinology 119:2155.

D'Ercole, A.J., Stiles, A.D., and Underwood, L.E., 1984, Tissue concentrations of somatomedin C: Further evidence for multiple sites of synthesis and paracrine or autocrine mechanisms of action, Proc. Natl. Acad. Sci. USA, 81:935.

D'Ercole, A.J., Hill, D.J., Strain, A.J., and Underwood, L.E., 1986, Tissue and plasma somatomedin-C/insulin-like growth factor I concentrations in the human fetus during the first half of gestation, Pediatr. Res. 20:253.

Fant , M., Munro, H., and Moses, A.C., 1986, An autocrine/paracrine role for insulin-like growth factors in the regulation of human placental growth, J. Clin. Endocrinol. Metab., 63:499.

Han, V.K.M., D'Ercole, A.J., and Lund, P.K., 1987a, Cellular localization of somatomedin (insulin-like growth factor) messenger RNA in the human fetus, Science, 236:193.

Han, V.K.M., Hill, D.J., Strain, A.J., Towle, A.C., Lauder J.M., Underwood, L.E., and D'Ercole, A.J., 1987b, Identification of somatomedin/insulin-like growth factor immunoreactive cells in the human fetus, Pediatr. Res., 22:245.

Hsu, C.-J., and Hammond, J.M., 1987, Gonadotropins and estradiol stimulate immunoreactive insulin-like growth factor-I production by porcine granulosa cells in vitro, Endocrinology, 120:198.

Lund, P.K., Moats-Staats, B.M., Hynes, M.A., Simmons, J.G., Jansen, M., D'Ercole, A.J., and Van Wyk, J.J., 1986, Somatomedin-C/insulin-like growth factor-I and insulin-like growth factor-II mRNAs in rat fetal and adult tissues, J. Biol. Chem., 261:14539.

Mathews, L.S., Norstedt, G., Palmiter, R.D., 1986, Regulation of insulin-like growth factor I gene expression by growth hormone, Proc. Natl. Acad. Sci. USA, 83:9343.

Penhoat, A., Chatelain, P.G., Jaillard, C., Saez, J.M., 1988, Characterization of somatomedin-C/insulin-like growth factor I and insulin receptors on cultured bovine adrenal fasciculata cells. Role of these peptides on adrenal cell function, Endocrinology, 122:2518.

Ramasharma, K., and Li, C.H., 1987, Human pituitary and placental hormones control human insulin-like growth factor II secretion in human granulosa cells, Proc. Natl. Acad. Sci. USA, 84:2643.

Ritvos, O., Ranta, T., Jalkanen, J., Suikkari, A.-M., Voutilainen, R., Bohn, H., Rutanen, E.-M., 1988, Insulin-like growth factor (IGF) binding protein from human decidua inhibits the binding and biological action of IGF-I in cultured choriocarcinoma cells, Endocrinology, 122:2150.

Roberts, C.T.Jr., Brown, A.L., Graham, D.E., Seelig, S., Berry, S., Gabbay, K.H., and Rechler, M.M., 1986, Growth hormone regulates the abundance of insulin-like growth factor I RNA in adult rat liver, J. Biol. Chem., 261:10025.

Smith, E.P., Svoboda, M.E., Van Wyk, J.J., Kierszenbaum, A.L., and Tres, L.L., 1987, Partial characterization of a somatomedin-like peptide from the medium of cultured rat sertoli cells, Endocrinology, 120:186.

Voutilainen, R., Tapanainen, J., Chung, B.-C., Matteson, K.J., and Miller, W.L., 1986, Hormonal regulation of P450scc (20,22 desmolase) and P450c17 (17-hydroxylase/17,20 lyase) in cultured human granulosa cells, J. Clin. Endocrinol. Metab., 63:202.

Voutilainen, R., and Miller, W.L., 1986, Developmental expression of genes for the steroidogenic enzymes P450scc (20,22 desmolase), P450c17 (17-hydroxylase/17,20 lyase), and P450c21 (21-hydroxylase) in the human fetus, J. Clin. Endocrinol. Metab., 63:1145.

Voutilainen, R., and Miller, W.L., 1987a, Coordinate tropic hormone regulation of mRNAs for insulin-like growth factor II and the cholesterol side-chain-cleavage enzyme, P450scc, in human steroidogenic tissues, Proc. Natl. Acad. Sci. USA, 84:1590.

Voutilainen, R., and Miller, W.L., 1987b, Human Mullerian inhibitory factor messenger ribonucleic acid is hormonally regulated in the fetal testis and in adult granulosa cells, Molecular Endocrinol., 1:604.

Voutilainen, R., and Miller, W.L., 1988, Developmental and hormonal regulation of mRNAs for insulin-like growth factor II and steroidogenic enzymes in human fetal adrenals and gonads, DNA, 7:9.

Voutilainen, R., and Miller, W.L., 1989, Potential relevance of Mullerian-inhibiting substance (MIS) to ovarian physiology, Seminars in Reproductive Endocrinology, 7:87.

Voutilainen, R., Picado-Leonard, J., DiBlasio, A.M., and Miller, W.L., 1988, Hormonal and developmental regulation of adrenodoxin mRNA in steroidogenic tissues, J. Clin. Endocrinol. Metab., 66:383.

Zapf, J., Walter, H., and Froesch, E.R., 1981, Radioimmunological determination of insulin like growth factor I and II in normal subjects and in patients with growth disorders and extrapancreatic tumor hypoglycemia, J. Clin. Invest, 68:1321.

EXPRESSION OF INSULIN-LIKE GROWTH FACTORS BY THE HUMAN PLACENTA

Keith D. Boehm,[1] Michael F. Kelley,[1] and Judith Ilan[1,2,3]

Departments of [1]Reproductive Biology and [2] Developmental Genetics and Anatomy and [3]The Cancer Research Center Case Western Reserve University, Cleveland, Ohio 44106

INTRODUCTION

The placenta is an extraembryonic structure that participates in the protection and maintenance of the developing conceptus. To function in this capacity, the placenta exhibits the characteristics of many adult mammalian tissue and organ level regulatory systems. It is, therefore, not surprising that the placenta has been shown to produce effector molecules either identical or similar to those expressed by adult neural, endocrine, and immunologic cells. Furthermore, receptor molecules capable of binding and allowing cells to respond to various effector molecules (including many of those produced by the placenta) have been demonstrated to be present in placental membranes. Thus, local hormonal networks may be postulated to exist within the placenta that can regulate its various functions and its own growth and development. One such autocrine/paracrine network that has been postulated for the human placenta is the insulin-like growth factor-I and -II (IGF-I and -II) system. As we have concentrated much of our research effort in recent years on examining the expression of the IGF-I and -II genes in the human placenta, this review will provide a brief recapitulation of that work presented in conjunction with the results of others working on this placental local hormonal system. This article is not intended to be an exhaustive review of the literature, and apology is made at the outset for publications not cited.

HUMAN PLACENTAL CELL MEMBRANES POSSESS RECEPTORS FOR IGF-I and -II

The presence of receptors for IGF-I on human placental membranes was first demonstrated by Marshall et al. (1974), using ^{125}I-IGF-I, in a binding assay. Subsequent studies have further characterized placental IGF-I receptors (Type 1 IGF receptors). As shown for other tissues, the placental Type 1 IGF receptor was demonstrated to be quite similar in structure to the insulin receptor (Bhaumick et al., 1981, 1982, Chernausek et al., 1981, Massagué and Czech, 1982, Pilch et al., 1986, Feltz et al., 1988, and Sahal et al., 1988) and binding affinity studies indicated its preferential specificity for IGF-I (Bhaumick et al., 1981, Read et al., 1986, Maly and Lüthi, 1986, a,b, Feltz et al., 1988, Cara et al., 1988).

Daughaday et al. (1981) provided evidence for distinct IGF-II receptors (Type 2 IGF receptors) on human and rat placental membranes. Massagué and Czech (1982) further showed that these receptors were structurally dissimilar to Type 1 IGF and insulin receptors.

The relative amounts of both IGF receptors on placental membranes have been estimated. Massagué and Czech (1982) reported that, like the insulin receptor, the Type 1 IGF receptor was present in relatively high amount in human placental membrane preparations, whereas, the Type 2 IGF receptor was found to be in comparatively low relative amount. In contrast, Daughaday et al. (1981) and Pilistine et al. (1984a) showed that the majority of the IGF receptors in rat placental membranes are of the Type 2 variety.

Identification of both the Type 1 and 2 IGF receptors in placental membrane preparations has led to the use of these membranes for radioreceptor assays of IGF-I and -II (Marshall et al., 1974, Baxter and DeMellow, 1986).

The above mentioned placental IGF receptor studies, and others, have thus provided evidence for one part of a placental IGF local hormonal control system, the presence of specific high affinity receptors for IGF-I and -II. Furthermore, Jonas et al. (1986) provided evidence that a relatively small number of atypical insulin receptors exist in placental membranes that have moderate affinity for IGF-I and -II, and there are at least two Type 1 IGF receptors in human placental membranes (Jonas, 1988). Placental tissues, therefore, possess the membrane receptor apparatus to enable them to respond to locally-produced IGF-I and -II.

PRODUCTION OF IGF-I and -II BY THE HUMAN PLACENTA

Initial studies (D'Ercole et al., 1980, Bhaumick and Bala, 1984) suggested that the placenta did not produce IGF-I that was detectable, immunologically. Fant et al. (1986) subsequently showed that IGF-I and-II were produced by both preterm and term human placental explant cultures. IGF-I was measured by a specific radioimmunoassay and IGF-II by a rat placental radioreceptor assay. This group further measured IGF-I and -II produced by confluent monolayers of fibroblasts established from preterm human placentas. These data provided strong evidence that the human placenta is capable of producing IGF-I and-II and that this production is, in part, a consequence of synthesis and secretion of IGF-I and -II by fibroblasts resident in the human placenta. Placental cells, therefore, possess the ability to produce IGF-I and -II which, in turn, can bind to and stimulate placental cell/tissue Type 1 and 2 IGF receptors, completing the IGF-I and -II local hormonal circuit.

IGF-I AND -II GENE EXPRESSION IN THE HUMAN PLACENTA

The molecular biologic approach has also been applied in an effort to demonstrate the human placenta as a site of IGF-I and -II production. Work done in our laboratory (Liu et al., 1985) and, later, that of Younes et al.(1986) demonstrated the presence of insulin-related RNA sequences in the human placenta. A following study from our group (Mills et al., 1986) showed that a polypeptide in the spectrum of cell-free translation products of both first trimester and term human placental poly (A)+ RNAs possessed IGF-I immunoreactivity.

Reeve et al. (1985) and Scott et al. (1985) demonstrated the presence of IGF-II RNA in samples of human placental total RNA by both dot and RNA transfer blot hybridization analyses. They detected three major transcripts for placental IGF-II of 6.0, 4.9-4.8, and 2.0-1.9 kilobase (kB) in size, by RNA transfer blot hybridization. Stempien et al. (1986) isolated and sequenced a mouse placental IGF-II precursor cDNA and others have shown IGF-I and -II RNAs in placentas of rodents. Analyses of human placental poly (A)+ RNA for IGF-II transcripts from all trimesters of pregnancy were accomplished in our laboratory (Shen et al., 1986). Four major IGF-II RNA species were detected (Figure 1) by transfer blot hybridization analysis. The sizes of these transcripts are 6.0, 4.9, 3.2, and 2.2 kB. In relative abundance, the 6.0 kB placental IGF-II RNA specie is highest in amount in all trimesters of pregnancy (Figure 1, lanes 1-3). Furthermore, overall, the relative amount of IGF-II RNA is greatest in the placenta at mid gestation. This can be seen in Figure 1, lane 2 and was confirmed by densitometric analysis of RNA dot blot hybridization autoradiograms (Shen et al., 1986). Thus, the IGF-II gene is expressed in the human placenta as multiple RNA transcripts and these RNAs appear in greatest relative amount during the second trimester of pregnancy.

The detection of multiple IGF-II transcripts in the human placenta and other tissues can be ascribed to at least three possible molecular events: 1) multiple promoter utilization at the gene level, 2) alternative transcript splicing, and 3) differential transcript polyadenylation. In this regard, a recent report from our laboratory (Shen et al., 1988) described a human placental library IGF-II cDNA clone which contained a 5'-untranslated region (5'-UTR) sequence different than those described for adult human liver (Bell et al., 1984, Jansen et al., 1985) IGF-II cDNAs, yet possessed the same coding and 3'-untranslated region sequences. When this human placental IGF-II 5'UTR (only) cDNA was used as probe of human placental poly (A)+ RNAs in transfer blot hybridization experiments, only three of the four major IGF-II transcripts [previously detected using an adult human liver cDNA probe (Jansen, et al., 1985) containing coding region sequence (Shen et al., 1986)] were visualized. The sizes of the hybridizing transcripts were 6.0, 3.2, and 2.2 kB. Therefore, these three major human placental IGF-II RNA species all contain this specific IGF-II 5'UTR, whereas the 4.9 kB human placental IGF-II transcript does not. When the adult human liver IGF-II cDNA (Jansen et al., 1985) 5'-UTR was used as probe of the human placental poly (A)+ RNA population in transfer blot hybridizations, no transcripts were visualized, indicating that the 4.9 kB placental IGF-II RNA specie must contain a 5'-UTR sequence different than those described by Bell et al. (1984) and Jansen et al. (1985) for adult human liver IGF-II cDNAs, and that reported from our laboratory (Shen et al., 1988) for a human placental IGF-II cDNA. The 5'-UTR sequence of the human placental 4.9 kB IGF-II transcript is most likely than recently described by LeBouc et al. (1987) from a human placental IGF-II cDNA, although hybridization experiments using human placental RNAs and this 5'-UTR have not yet been performed.

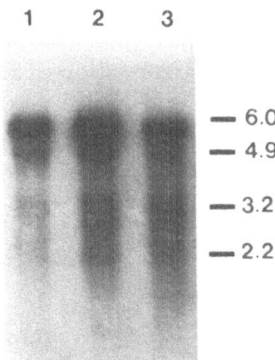
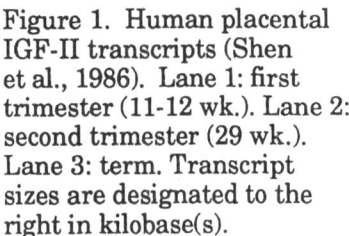
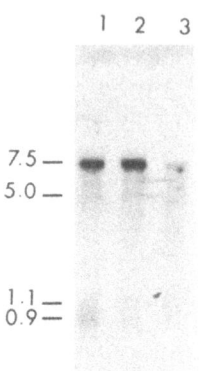

Figure 1. Human placental IGF-II transcripts (Shen et al., 1986). Lane 1: first trimester (11-12 wk.). Lane 2: second trimester (29 wk.). Lane 3: term. Transcript sizes are designated to the right in kilobase(s).

Figure 2. Human placental IGF-I transcripts (Wang et al., 1988). lane 1: first trimester (11-12 wk.). Lane 2: second trimester (29 wk.). Lane 3: term. Transcript sizes are designated to the left in kilobase(s).

Analysis of the known sequence of the human IGF-II gene indicates that the production of IGF-II transcripts with three different 5'-UTRs is most likely a consequence of the utilization of three putative promoters in the gene. Two of the promoters are apparently utilized when the IGF-II gene is expressed in the human placenta, to generate the four major IGF-II transcripts detected by transfer blot hybridization with human IGF-II coding region cDNA probe. Furthermore, as three of the four major placental IGF-II transcripts contain the same 5'-UTR, yet differ in size, alternative splicing and/or differential polyadenylation of transcript(s) [among other possibilities] must also be involved in the generation of the human placental IGF-II RNA species.

Thus, it can only be concluded that expression of the IGF-II gene in the human placenta as well as other tissues and organs is the result of complex molecular events occurring at both transcriptional and post-transcriptional levels. These events may, in turn, have effects on the ultimate production of IGF-II by the placenta. For instance, IGF-II transcripts possessing different 5'-UTRs may be translated with differing efficiencies, and IGF-II RNAs with different-sized poly (A) tails may have different stabilities and translational efficiencies. As the placenta is an extraembryonic structure, the mode of regulation of IGF-II gene transcription and IGF-II RNA processing in this organ is most likely developmental-related and/or tissue-specific.

The expression of the IGF-I gene in the human placenta appears to be as complex as that of the IGF-II gene. In our group's most recent report (Wang et al., 1988) the expression of the human placental IGF-I gene throughout gestation was described. Transfer blot hybridization analysis of first and second trimester, and term placental poly (A)+ RNAs revealed the presence of four major IGF-I transcripts (Figure 2). The sizes of these IGF-I RNAs were 7.5, 5.0, 1.1, and 0.9 kB. All four transcripts are expressed in each major stage of pregnancy. The 7.5 kB IGF-I transcript is the greatest in relative amount of the four transcripts detected, and this is the case in each trimester of gestation. The relative amount of IGF-I RNA overall, was found to be greatest in first trimester human placental poly (A)+ RNA (Figure 2, lane 1) and almost as abundant in second trimester samples (Figure 2, lane 2) with a much lower abundance in samples of term placental RNA (Figure 2, lane 3). These abundance differences were confirmed by densitometric analysis of dot blot hybridization autoradiograms. Thus, the IGF-I gene, like the IGF-II gene, is expressed in the human placenta as multiple transcripts, and, these IGF-I RNAs appear in greatest relative amount during the first trimester of pregnancy, unlike those for IGF-II which are present in highest amount during the second trimester. It is, therefore, likely that the regulation of the expression of the human placental IGF-I gene is similar to that of the human placental IGF-II gene, as both are expressed as multiple RNAs. Futher examination of human placental IGF-I transcripts is required, though, before more definitive statements can be made in this regard.

The molecular biologic studies from our laboratory and the work of others have thus provided further evidence in support of the human placenta as a site of IGF-I and -II production. The reports of Shen et al. (1986) and Wang et al. (1988) also suggest that IGF-I and -II may be of greater importance to placental growth and development during the first half of gestation as the RNAs for these two growth factors appear to be in highest relative amount in these early periods of pregnancy. In contrast, though, Fant et al. (1986) showed that preterm and term human placental explants released approximately equivalent amounts of immunoreactive IGF-I. These results may be specific to their explant system or may suggest, when viewed with the IGF-I RNA data of Wang et al (1988), that translational and/or post-translational regulatory mechanisms are differentially active in the human placental IGF-I production pathway, during gestation. The findings of Wang et al. (1988) that IGF-I RNA relative abundance is highest during the first and second trimesters of pregnancy, and lowest at term, is supported by the demonstration of Mills et al. (1986) that immunoreactive IGF-I from the cell-free translation of placental poly (A)+ RNAs is higher in relative amount from first trimester samples (although this latter study's results may only be reflecting the amounts of IGF-I RNA that are readily translatable in the rabbit reticulocyte system, in vitro). Further work is required to provide a more complete understanding of both IGF-I and -II production by the human placenta, during each major stage of gestation.

Fant et al. (1986) also provided data which indicates that both placental explants and cultured placental fibroblasts produce higher levels of IGF-II than IGF-I. These data correlate with the comparative overall relative abundances of human placental IGF-I and -II RNAs as measured by Wang et al. (1988) and Shen et al. (1986), respectively. The significance of these differences is not yet apparent, although it has been believed that IGF-II is the most important of the IGFs during the fetal period.

The studies discussed above have thus provided a solid body of evidence in support of the human placenta as an intrauterine site of IGF-I and -II production. In an effort to refine further the cellular locale of this production, Fant et al. (1986), as previously mentioned, showed that cultured human placental fibroblasts produced both IGF-I and -II. Therefore, fibroblasts resident in the placenta are a likely source of, at least a portion of, the IGF-I and -II secreted by placental explants and the IGF-I and -II RNAs detected in placental poly (A)+ RNA populations. To identify other possible placental sites of IGF-I and -II production our laboratory recently performed in situ hybridization histochemistry analyses on fixed sections of human placentas (Wang et al., 1988) to detect IGF-I and -II RNAs. Autoradiographic grains, representing IGF-I RNAs hybridized with radiolabeled IGF-I cDNA probe (Figure 3), were found to be localized predominantly in the syncytiotrophoblastic layer with a smaller number of grains scattered throughout the villus core region, which is populated with cytotrophoblasts, macrophages, and fibroblasts, among other cell types.

These data indicate that the syncytiotrophoblast is the main source of placental IGF-I and villus core cell types such as fibroblasts and cytotrophoblasts are potential minor sources of placental IGF-I. In contrast, autoradiographic grains, representing IGF-II RNAs hybridized with radiolabeled IGF-II cDNA probe (Figure 4) were found to be predominantly distributed throughout the villus core region with a minor number of grains localized to the syncytiotrophoblast layer. These results suggest that the primary source(s) of placental IGF-II are villus core cell

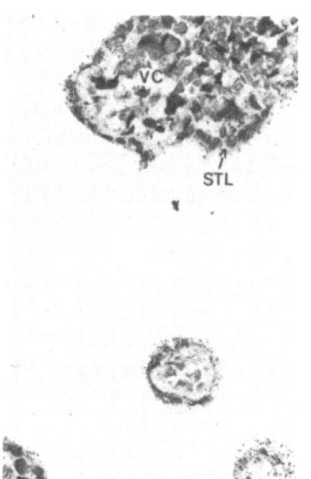

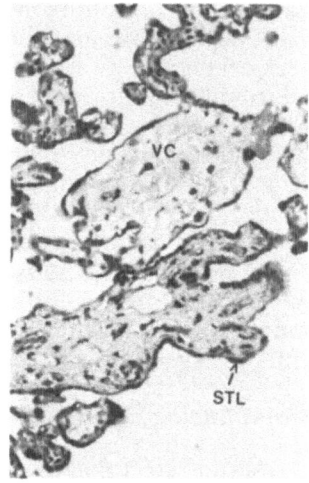

Figure 3. Human placental IGF-I transcripts in situ (Wang et al., 1988). STL: syncytiotrophoblastic layer VC: villus core region. Second trimester (100X mag.).

Figure 4. Human placental IGF-II transcripts in situ (Wang et al., 1988). VC: villus core region. STL: syncytiotrophoblastic layer. Term (100X mag.).

types such as cytotrophoblasts and fibroblasts. Further studies are required to identify placental locales of IGF-I and -II (i.e., by immunohistochemistry) and specific placental cell and layer sites of Type 1 and 2 IGF receptors for understanding, more completely, the cellular communication involved in the possible autocrine and/or paracrine mode(s) of action of IGF-I and -II during the growth and maturation of the human placenta. Fant et al. (1986) did, however, demonstrate the presence of Type 1 IGF receptors on cultured human placental fibroblasts, indicating that this placental cell type is capable of responding to IGF-I produced by itself and/or IGF-I produced by neighboring sources.

SOME THOUGHTS ON IGF(S) AND THE HUMAN PLACENTA

The previously discussed studies have provided evidence that the human placenta produces IGF-I and -II and possesses Type 1 and 2 IGF receptors that would enable it to respond to the IGFs it generates. This putative local autocrine/paracrine IGF placental circuit would most likely play a role in placental metabolism, growth, and development. As the IGFs are structurally similar to insulin they can have short-term insulin - like metabolic effects on tissues. In addition, the IGFs, can have less acute growth- and/or differentiation-promoting effects on tissues. Therefore, the local actions of IGFs in the human placenta may range from effects on tissue metabolism and proliferation of cells to cellular/tissue differentiation (/maturation). The presence of this local intrauterine placental IGF system is important, as maternal IGFs (Underwood et al., 1979) like maternal insulin (Adams et al, 1969, Kalhan et al., 1975), are apparently unable to cross the blood-placental barrier. It is not yet clear how the possible spectrum of effects elicited in response to the IGFs would be able to be accomplished. It can be speculated, though, that potential cross-reactivities of IGFs for IGF and insulin receptors (and subtypes of these receptors) is a possible mechanism, as local placental IGF concentrations may be high enough for this to occur, and although the human placenta contains a relatively high level of insulin receptors, maternal insulin is not readily available for binding and this extraembryonic organ apparently is not a source of extra-pancreatic insulin, as previously mentioned. Further complicating this putative placental IGF regulatory system is the recent determination (Fant et al., 1988) that human placental explants also produce, at least, two distinct IGF-specific binding proteins which may modulate the effects of placental-derived IGF-I and -II on tissue(s) possessing their receptors. Of special interest is the fact that the amount of these binding proteins is greatest in conditioned media from preterm (9-19 week) placental explants and these proteins are barely detectable in media conditioned by term placental explants, suggesting an important modulatory role during early gestation, if the explant system is accurately reflecting the situation in vivo.

The human placental IGF control system is also not isolated from other factors of the placental biochemical millieu. In this regard, it should be noted that IGFs have been viewed as progression growth factors, which means that they are able to support the movement of some cells through the cell cycle only after these cells have been made competent to respond to them in this fashion by prior exposure to other factor(s). One such competence growth factor is platelet-derived growth factor (PDGF). PDGF-B chain/c-sis proto-oncogene RNA has been detected in human placental cells (Goustin et al., 1985) and PDGF - A and -B chain RNA have been demonstrated to be present in the human placental total RNA population (Taylor and Williams, 1988) suggesting placental production of PDGFs. Therefore, placental IGFs most likely are able to carry out their cellular growth-promoting effects in an autocrine/paracrine fashion, as target cells have been made competent by exposure to PDGF molecules [PDGF receptors have been found in placental membranes (Taylor and Williams, 1988)]. Futher, it is logical to assume that other placental competence and progression factors are interactive in this tissue growth and differentiation regulatory system.

Other molecules derived from the human placenta may participate in the regulation of the production of IGF-I and -II (and possibly other molecules in the local IGF network, i.e., placental IGF binding proteins). Unlike adult tissues that synthesize IGFs in response to growth hormones (GH), the placental production of IGFs has been believed to be regulated by placental lactogen (PL), a hormone with structural similarity to GH. Human maternal blood PL levels have been correlated with placental weight and increasing maternal blood levels of IGF-I appear to be under the control of PL (Furlanetto et al., 1978, Merimee et al., 1982, Pilistine et al., 1984b). Somewhat recently, however, a variant of GH was shown to be produced by the human placenta (Henner et al., 1985, Frankenne et al., 1987). This GH appears at mid gestation and increases to term. Thus, at least, two placental-derived hormones, PL and GH, may be involved in the control of IGF (as well as IGF binding protein) production by this organ of pregnancy. Furthermore, Boime et al. (1982) localized PL RNAs predominantly, to the syncytiotrophoblastic layer (STL) of the human placenta, using the method of in situ hybridization (ISH). These data, in conjunction with the placental IGF ISH data of Wang et al. (1988), would allow speculation that STL-produced PL may act in an autocrine fashion to stimulate the production of STL-derived IGF-I, and the STL-produced PL may also act to stimulate villus core cell IGF-II production via a paracrine mechanism of action. In this regard, it should also be noted that IGF-I can stimulate the production of PL by human placental explants (Bhaumick et al., 1987). If these explants are mimicking the placenta in vivo, it is possible that placental-derived IGF-I (and possibly IGF-II) and PL positively feedback on one another's production. The production of the placental GH variant may also be regulated either positively or negatively by placental IGF feedback. GH variant RNAs have also been localized to the syncytiotrophoblastic epithelium of the human placenta by ISH (Liebhaber et al., 1989), thus indicating possible autocrine/paracrine interactions for STL-derived GH variant(s) and placental IGF(s), similar to those postulated for PL and placental IGF(s). Therefore, more classical endocrine hormones released by the placenta may regulate the production of IGFs by the same organ via local autocrine/paracrine action, and may, in turn, be regulated in their production by the IGFs, in a similar fashion. Perhaps this form of local control is required due to the relative self-contained and -reliant nature of the feto-placental unit within the host mother, especially in lieu of the potential immunological problems involved with pregnancy. Also, as the placenta produces many regulatory molecules which are similar or identical to those produced elsewhere in specialized adult body organs that are separated by distance and connected only by the vascular system, it (the placenta) may require complex, layered molecular control circuitry to enable it to carry out multiple functions without the benefit(s) of highly defined physical compartmentalization. In this respect, the placenta might better be viewed as a structure at the level of an organism without a circulatory system or a structure at the tissue-level of organization, which utilizes functional compartmentalization mechanisms to a greater extent than physical compartmentalization mechanisms for regulatory purposes.

Although the human placenta may be viewed by itself, it must be remembered that it is a temporary extraembryonic structure of pregnancy that resides in the maternal uterus with the developing embryo/fetus. Therefore, substances produced by the placenta may affect these structures in close apposition. The converse of this situation would also be possible.

The specialized human uterine endometrium of pregnancy, decidua, is known to produce prolactin (PRL), identical to that produced by the human pituitary gland. This hormone, like PL, is structurally similar to GH. Recently, Thrailkill et al. (1988) showed that cultured human decidual cells synthesize and release PRL in response to IGF-I administration. It is, therefore, possible that IGFs produced by the placenta may have a longer range paracrine effect on decidual PRL production, and decidual PRL may, likewise, participate in the regulation of placental IGF production. It has also been shown that human decidual tissue expresses an IGF binding protein (Koistinen et al., 1986, Ritvos et al., 1988, Pekonen et al., 1988) with which placental-derived IGFs may interact. This same binding protein is also produced by the secretory phase human endometrium (Rutanen et al., 1988). Further complicating this situation, though, are recent demonstrations that rat (Murphy et al.,1987 a,b) and porcine (Tavakkol et al., 1988) uteri, and human uterine myometrium (Höppener et al., 1988) express the IGF-I gene, and therefore, most likely produce IGF-I. Of most interest, though, is the finding that porcine uterus IGF-I RNAs and immunoreactive uterine IGF-I were highest in relative amount in early pregnancy. Therefore, uterine IGF-I may be more involved than placental IGFs in eliciting PRL production by the decidua, and decidual PRL may then affect uterine IGF production. It cannot be ruled out, though, that decidual PRL and uterine IGF may also interact with closely apposed placental tissue, and that placental GH variant, PL, and IGFs may interact with closely apposed uterine tissues, during gestation. The production of IGF binding proteins by placental and uterine tissues in close relation is likely to be affected by these hormones and growth factors also. Thus, uterine-placental IGF interactions may also occur during gestation. Further studies are required before the findings of the porcine system can be extrapolated to the human.

Placental-derived IGF-I and -II may also affect the developing embryo and fetus, especially earlier in pregnancy when placental IGF RNAs are most highly expressed. Human umbilical cord serum IGF levels correlate well with gestational age and birth size/weight (Ashton and Vesey, 1978, Bennett et al., 1983). Also, many fetal tissues possess IGF receptors and respond to IGFs (Sara et al., 1983, Grizzard et al., 1984, Bassas et al., 1985, Ashton and Francis, 1978, Kaplowitz et al., 1982, Strain et al., 1987, Swenne et al, 1987). IGFs and IGF binding protein(s) are also found in human amniotic fluid. Of special interest, is the report of Mattson et al. (1988) that insulin, IGF-I, and IGF-II can bind to early stage mouse embryos, suggesting a role for these factors very early in development. Thus, placental IGFs may act on the embryo via a paracrine mechanism very early in development and later by way of the umbilical cord through the traditional endocrine mechanism of action. Very early embryo development, though, may also be affected by uterine-derived IGF, and later fetal development is likely, more directly, affected by IGFs produced by fetal tissues, themselves (D'Ercole et al., 1980, Adams et al.,1983 a,b, Hill et al., 1985, Sara and Carlsson-Skwirut, 1986, Lund et al., 1986, Han et al., 1987).

Lastly, factors produced by the human placenta may enter the maternal circulation. As mentioned earlier, PL appears to regulate the maternal serum levels of IGFs during pregnancy, rather than GH. It is, therefore, possible that placental-derived IGF-I and -II, entering the maternal circulation in addition to maternal liver-produced IGF (as well as IGF produced by other maternal cells/tissues) contribute to the increased level of IGF observed in the human mother's serum during pregnancy. Therefore, placental IGFs may affect maternal tissues in an endocrine fashion.

ACKNOWLEDGEMENTS

The work cited from our laboratory was supported in part by NIH Grants HD-18271 and CA-43703; and grants from the American Diabetes Association; Diabetes Association of Greater Cleveland; the Kidney Foundation of Ohio; The American Heart Association, Northeast Ohio Affiliate; and The American Cancer Society, Cuyahoga County, Ohio Unit. We would also like to thank Mr. J.K. Yun and Dr. G.L. Engelmann for their assistance with aspects of the work cited from our laboratory, and Ms. M. Maurer for manuscript word processing. Dr. Boehm's present address is: Department of Dermatology, Case Western Reserve University, Cleveland, Ohio 44106. Address correspondence to Judith Ilan.

REFERENCES

Adams, P.A.J., Teramo, K., Raiha, N., Gitlin, D., and Schwartz, R., 1969, Human fetal insulin metabolism early in gestation: response to acute elevation of the fetal glucose concentration and placental transfer of human [131I] insulin, Diabetes, 18:409.

Adams, S.O., Nissley, S.P., Handwerger, S., and Rechler, M.M., 1983a, Developmental patterns of insulin-like growth factor I and II synthesis and regulation in rat fibroblasts, Nature (London), 302:150.

Adams, S.O., Nissley, S.P., Greenstein, L.A., Yang, T.W.-H., and Rechler, M.M., 1983b, Synthesis of multiplication-stimulating activity (rat insulin-like growth factor II) by rat embryo fibroblasts, Endocrinol., 112:979.

Ashton, I.K., and Francis, M.J.O., 1978, Response of chondrocytes isolated from human foetal cartilage to plasma somatomedin activity, J. Endocrinol., 76:473.

Ashton, I.K., and Vesey, J., 1978, Somatomedin activity in human cord plasma and relationship to birth size, insulin, growth hormone, and prolactin, Early Hum. Develop., 2:115.

Bassas, L., DePalo, F., Lesniak, M.A., and Roth, J., 1985, Ontogeny of receptors for insulin-like peptides in chick embryo tissues: early dominance of insulin-like growth factor over insulin receptors in brain, Endocrinol., 117:2321.

Baxter. R.C., and DeMellow, J.S.M., 1986, Measurement of insulin-like growth factor-II by radioreceptor assay using ovine placental membranes, Clin. Endocrinol., 24:267.

Bell, G.I., Merryweather, J.P., Sanchez-Pescador, R., Stempien, M.M., Priestley, L., Scott, J., and Rall, L.B., 1984, Sequence of a cDNA clone encoding human preproinsulin-like growth factor II, Nature (London), 310:775.

Bennett, A., Wilson, D.M., Liu, F., Nagashima, R., Rosenfeld, R.G., and Hintz, R.L., 1983, Levels of insulin-like growth factor I and II in human cord blood, J. Clin. Endocrinol. Metabol., 57:609.

Bhaumick, B., and Bala, R.M., 1984, Basic somatomedin (B-SM) receptors in human term placenta explants, Biochem. Biophys. Res. Comm., 122:583.

Bhaumick, B., Bala, R.M., and Hollenberg, M.D., 1981, Somatomedin receptor of human placenta: solubilization, photolabeling, partial purification, and comparison with insulin receptor, Proc. Natl. Acad. Sci. USA, 78:4279.

Bhaumick, B., Armstrong, G.D., Hollenberg, M.D., and Bala, R.M., 1982, Characterization of the human placental receptor for basic somatomedin, Can. J. Biochem., 60:923.

Bhaumick, B., Dawson, E.P., and Bala, R.M., 1987, The effects of insulin-like growth factor-I and insulin on placental lactogen production by human term placental explants, Biochem. Biophys. Res. Comm., 144:674.

Boime, I., Boothby, M., Hoshina, M., Daniels-McQueen, S., and Darnell, R., 1982, Expression and structure of human placental hormone genes as a function of placental development, Biol. Reproduct., 26:73.

Cara, J.F., Nakagawa, S.H., and Tager, H.S., 1988, Structural determinants of ligand recognition by type I insulin-like growth factor receptors: use of semisynthetic insulin analog probes, Endocrinol., 122:2881.

Chernausek, S.D., Jacobs, S., and Van Wyk, J.J., 1981, Structural similarities between human receptors for somatomedin C and insulin: analysis by affinity labeling, Biochem., 20:7345.

Daughaday, W.H., Mariz, I.K., and Trivedi, B., 1981, A preferential binding site for insulin-like growth factor II in human and rat placental membranes, J. Clin. Endocrinol. Metabol., 53:282.

D'Ercole, A.J., Applewhite, G.T., and Underwood, L.E., 1980, Evidence that somatomedin is synthesized by multiple tissues in the fetus, Develop. Biol., 75:315.

Fant, M., Munro, H., and Moses, A.C., 1986, An autocrine/paracrine role for insulin-like growth factors in the regulation of human placental growth, J. Clin. Endocrinol. Metabol., 63:499.

Fant, M., Munro, H., and Moses, A.C., 1988, Production of insulin-like growth factor binding protein(s) (IGF-BPs) by human placenta: variation with gestational age, Placenta, 9:397.

Feltz, S.M., Swanson, M.L., Wemmie, J.A., and Pessin, J.E., 1988, Functional properties of an isolated aB heterodimeric human placenta insulin-like growth factor 1 receptor complex, Biochem., 27:3234.

Frankenne, F., Rentier-Debrue, F., Scippo, M.-L., Martial, J., and Henner, G., 1987, Expression of the growth hormone variant gene in human placenta, J. Clin. Endocrinol. Metabol., 64:635.

Furlanetto, R.W., Underwood, L.E., Van Wyk, J.J., and Handwerger, S., 1978, Serum immunoreactive somatomedin-C is elevated late in pregnancy, J. Clin. Endocrinol. Metabol., 47:695.

Goustin, A.S., Betsholz, C., Pfiefer-Ohlsson, S., Persson, H., Rydnert, J., Bywater, M., Holmgren, G., Heldin, C.-H., Westermark, B., and Ohlsson, R., 1985, Coexpression of the sis and myc proto-oncogenes in developing human placenta suggests autocrine control of trophoblast growth, Cell, 41:301.

Grizzard, J.D., D'Ercole, A.J., Wilkins, J.R., Moats-Staats, B.M., and Williams, R.W., 1984, Affinity-labeled somatomedin-C receptors and binding proteins from the human fetus, J. Clin. Endocrinol. Metabol., 58:535.

Han, V.K.M., D'Ercole, A.J., and Lund, P.K., 1987, Cellular localization of somatomedin (insulin-like growth factor) messenger RNA in the human fetus, Science, 236:193.

Henner, G., Frankenne, F., Pirens, G., Gomez, F., Closset, J., Schaus, C.H., and El Khayat, N., 1985, New chorionic GH-like antigen revealed by monoclonal antibody radioimmunoassays, Lancet, 16:399.

Hill, D.J., Grace, C.J., Nissley, S.P., Morrell, D., Holder, A.T., and Milner, R.D.G., 1985, Fetal rat myoblasts release both rat somatomedin-C (SM-C)/insulin -like growth factor I (IGF I) and multiplication-stimulating activity in vitro: partial characterization and biological activity of myoblast-derived SM-C/IGF-I, Endocrinol., 117:2061.

Höppener, J.W.M., Mosselman, S., Roholl, P.J.M., Lambrechts, C., Slebos, R.J.C., de Pagter-Holthuizen, P., Lips, C.J.M., Jansz, H.S., and Sussenbach, J.S., 1988, Expression of insulin-like growth factor-I and-II genes in human smooth muscle tumours, EMBO J., 7:1379.

Jansen, M., van Schaik, F.M.A., van Tol, H., Van den Brande, J.L., and Sussenbach, J.S., 1985, Nucleotide sequences of cDNAs encoding precursors of human insulin-like growth factor II (IGF-II) and an IGF-II variant, FEBS Lett., 179:243.

Jonas, H.A., 1988, Heterogeneity of receptors for insulin and insulin-like growth factor I: evidence for receptor subtypes, in: "Insulin Receptors Part B: Clinical Assessment, Biological Responses, and Comparison to the IGF-I Receptor," C.R. Kahn and L.C. Harrison, eds., Alan R. Liss, New York.

Jonas, H.A., Newman, J.D., and Harrison, L.C., 1986, An atypical insulin receptor with high affinity for insulin-like growth factors copurified with placental insulin receptors, Proc. Natl. Acad. Sci. USA, 83:4124.

Kalhan, S.C., Schwartz, R., and Adams, P.A.J., 1975, Placental barrier to human insulin-^{125}I in insulin-dependent diabetic mothers, J. Clin. Endocrinol. Metabol. 40:139.

Kaplowitz, P.B., D'Ercole, A.J., and Underwood, L.E., 1982, Stimulation of embryonic mouse limb bud mesenchymal cell growth by peptide growth factors, J. Cell. Physiol., 112:353.

Koistinen, R., Kalkkinen, N., Huhtala, M.-L., Seppälä, M., Bohn, H., and Rutanen, E.-M., 1986, Placental protein 12 is a decidual protein that binds somatomedin and has an identical N-terminal amino acid sequence with somatomedin-binding protein from human amniotic fluid, Endocrinol., 118:1375.

LeBouc, Y., Noguiez, P., Sondermeijer, P., Dreyer, D., Girard, F., and
    Binoux, M., 1987, A new 5'-non-coding region for human placental
    insulin-like growth factor II mRNA expression, FEBS Lett., 222:181.
Liebhaber, S.A., Urbanek, M., Ray, J., Tuan, R.S., and Cooke, N.E., 1989,
    Characterization and histologic localization of human growth
    hormone-variant gene expression in the placenta, J. Clin. Invest.,
    83:1985.
Liu, K.-S., Wang, C.-Y., Mills, N., Gyves, M., and Ilan, J., 1985, Insulin-
    related genes expressed in human placenta from normal and
    diabetic pregnancies, Proc. Natl. Acad. Sci. USA, 82:3868.
Lund, P.K., Moats-Staats, B.M., Hynes, M.A., Simmons, J.G., Jansen, M.,
    D'Ercole, A.J., and Van Wyk, J.J., 1986, Somatomedin-C/insulin-
    like growth factor-I and insulin-like growth factor-II mRNAs in rat
    fetal and adult tissues, J. Biol. Chem., 261:14539.
Maly, P., and Lüthi, C., 1986a, Purification of the type I insulin-like growth
    factor receptor from human placenta, Biochem. Biophys. Res.
    Comm., 137:695.
Maly, P., and Lüthi, C., 1986b, Characterization of affinity-purified type I
    insulin-like growth factor receptor from human placenta, Biochem.
    Biophys. Res. Comm., 138:1257.
Marshall, R.N., Underwood, L.E., Voina, S.J., Foushee, D.B., and Van
    Wyk, J.J., 1974, Characterization of the insulin and somatomedin-C
    receptors in human placental cell membranes, J. Clin. Endocrinol.
    Metabol., 39:283.
Massagué, J., and Czech, M.P., 1982, The subunit structures of two
    distinct receptors for insulin-like growth factors I and II and their
    relationship to the insulin receptor, J. Biol. Chem., 257:5038.
Mattson, B.A., Rosenblum, I.Y., Smith, R.M., and Heyner, S., 1988,
    Autoradiographic evidence for insulin and insulin-like growth
    factor binding to early mouse embryos, Diabetes, 37:585.
Merimee, T.J., Zapf, J., and Froesch, E.R., 1982, Insulin-like growth factor
    in pregnancy: studies in a growth hormone-deficient dwarf, J. Clin.
    Endocrinol. Metabol., 54:1101.
Mills, N.C., D'Ercole, A.J., Underwood, L.E., and Ilan, J., 1986, Synthesis
    of somatomedin C/insulin-like growth factor I by human placenta,
    Molec. Biol. Rep., 11:231.
Murphy, L.J., Bell, G.I., and Friesen, H.G., 1987a, Tissue distribution of
    insulin-like growth factor I and II messenger ribonucleic acid in the
    adult rat, Endocrinol. 120:1279.
Murphy, L.J., Murphy, L.C., and Friesen, H.G., 1987b, Estrogen induces
    insulin-like growth factor-I expression in the rat uterus, Molec.
    Endocrinol., 1:445.
Pekonen, F., Suikkari, A.-M., Mäkinen, T., and Rutanen, E.-M., 1988,
    Different insulin-like growth factor binding species in human
    placenta and decidua, J. Clin. Endocrinol. Metabol., 67:1250.
Pilch, P.F., O'Hare, T., Rubin, J., and Boni-Schnetzler, M., 1986, The
    ligand binding subunit of the insulin-like growth factor 1 receptor
    has properties of a peripheral membrane protein, Biochem. Biophys.
    Res. Comm., 136:45.
Pilistine, S.J., Moses, A.C., and Munro, H.N., 1984a, Insulin-like growth
    factor receptors in rat placental membranes, Endocrinol., 115:1060.
Pilistine, S.J., Munro, H., and Moses, A.C., 1984b, Placental lactogen
    administration reverses the effect of low protein diet on maternal-
    fetal serum somatomedin levels in the pregnant rat, Proc. Natl.
    Acad. Sci. USA, 81:5853.
Read, L.C., Ballard, F.J., Francis, G.L., Baxter, R.C., Bagley, C.J., and

Wallace, J.C., 1986, Comparative binding of bovine, human and rat insulin-like growth factors to membrane receptors and to antibodies against human insulin-like growth factor-1, Biochem. J., 233:215.

Reeve, A.E., Eccles, M.R., Wilkins, R.J., Bell, G.I., and Millow, L.J., 1985, Expression of insulin-like growth factor-II transcripts in wilms' tumour, Nature (London), 317:258.

Ritvos, O., Ranta, T., Jalkanen, J., Suikkari, A.-M., Voutilainen, R., Bohn, H., and Rutanen, E.-M., 1988, Insulin-like growth factor (IGF) binding protein from human decidua inhibits the binding and biological action of IGF-I in cultured choriocarcinoma cells, Endocrinol., 122:2150.

Rutanen, E.-M., Pekonen, F., and Mäkinen, T., 1988, Soluble 34K binding protein inhibits the binding of insulin-like growth factor I to its cell receptors in human secretory phase endometrium: evidence for autocrine/paracrine regulation of growth factor action, J. Clin. Endocrinol. Metabol., 66:173.

Sahal, D., Ramachandran, J., and Fujita-Yamaguchi, Y., 1988, Specificity of tyrosine protein kinases of the structurally related receptors for insulin and insulin-like growth factor I: tyr-containing synthetic polymers as specific inhibitors or substrates, Arch. Biochem. Biophys., 260:416.

Sara, V.R., Hall, K., Misaki, M., Frykland, L., Christensen, N., and Wetterberg, L., 1983, Ontogenesis of somatomedin and insulin receptors in the human fetus, J Clin. Invest., 71:1084.

Sara, V.R., and Carlsson-Skwirut, C., 1986, The biosynthesis of somatomedins and their role in the fetus, Acta Endocrinol. (Copenh.), 113 (Suppl. 279):82.

Scott, J., Cowell, J., Robertson, M.E., Priestley, L.M., Wadey, R., Hopkins, B., Pritchard, J., Bell, G. I., Rall, L.B., Graham, C.F., and Knott, T.J., 1985, Insulin-like growth factor-II gene expression in wilms' tumour and embryonic tissues, Nature (London), 317:260.

Shen, S.-J., Wang, C.-Y., Nelson, K.K., Jansen, M., and Ilan, J., 1986, Expression of insulin-like growth factor II in human placentas from normal and diabetic pregnancies, Proc. Natl. Acad. Sci. USA, 83: 9179.

Shen, S.-J., Daimon, M., Wang, C.-Y., Jansen, M., and Ilan, J., 1988, Isolation of an insulin-like growth factor II cDNA with a unique 5' untranslated region from human placenta, Proc. Natl. Acad. Sci. USA, 85:1947.

Stempien, M.M., Fong, N.M., Rall, L.B., and Bell, G.I., 1986, Sequence of a placental cDNA encoding the mouse insulin-like growth factor II precursor, DNA, 5:357.

Strain, A.J., Hill, D.J., Swenne, I., and Milner, R.D.G., 1987, Regulation of DNA synthesis in human fetal hepatocytes by placental lactogen, growth hormone, and insulin-like growth factor I/somatomedin-C, J. Cell. Physiol., 132:33.

Swenne, I., Hill, D.J., Strain, A.J., and Milner, R.D.G., 1987, Growth hormone regulation of somatomedin C/insulin-like growth factor I production and DNA replication in fetal rat islets in tissue culture, Diabetes, 36:288.

Tavakkol, A., Simmen, F.A., and Simmen, R.C.M., 1988, Porcine insulin-like growth factor-I (pIGF-I): complementary deoxyribonucleic acid cloning and uterine expression of messenger ribonucleic acid encoding evolutionarily conserved IGF-I peptides, Molec. Endocrinol., 2:674.

Taylor, R.N., and Williams, L.T., 1988, Developmental expression of platelet-derived growth factor and its receptor in the human placenta, Molec. Endocrinol., 2: 627.

Thrailkill, K.M., Golander, A., Underwood, L.E., and Handwerger, S., 1988, Insulin-like growth factor I stimulates the synthesis and release of prolactin from human decidual cells, Endocrinol., 123:2930.

Underwood, L.E., D'Ercole, A.J., Furlanetto, R.W., Handwerger, S., and Hurley, T.W., 1979, Somatomedin and growth: a possible role for somatomedin-C in fetal growth, in "Somatomedin and Growth," G. Giordano, J.J. Van Wyk, and F. Minuto, eds., Academic Press, New York.

Wang, C.-Y., Daimon, M., Shen, S.-J., Engelmann, G.L., and Ilan, J., 1988, Insulin-like growth factor-I messenger ribonucleic acid in the developing human placenta and in term placenta of diabetics, Molec. Endocrinol., 2:217.

Younes, M.A., D'Agostino, J., and Besch, P.K., 1986, mRNA for insulin-like growth factor from human placenta, Alabama J. Med. Sci., 23:285.

IGFs IN PREGNANCY: DEVELOPMENTAL EXPRESSION IN UTERUS AND MAMMARY GLAND

AND PARACRINE ACTIONS DURING EMBRYONIC AND NEONATAL GROWTH

Frank A. Simmen, Rosalia C.M. Simmen, L. Roxanne Letcher
Douglas A. Schober and Yong Ko

Department of Animal Science and Laboratories of Molecular
and Developmental Biology
Ohio Agricultural Research and Development Center, The Ohio
State University, Wooster, OH 44691-4096

INTRODUCTION

The insulin-like growth factors (IGF-I, IGF-II) are structural homologs of insulin that are mitogenic in vivo and in vitro[1]. These peptides are also characterized by other actions which include the ability to alter the differentiation state of target cells by affecting specific aspects of messenger RNA (mRNA) production, protein synthesis and protein secretion. The IGF peptides are synthesized in multiple organ systems by a process that is subject to hormonal, developmental and tissue-specific regulation. The molecular mechanisms underlying the complex regulation of IGF biosynthesis remain for the most part obscure.

The widespread production of IGF mRNAs and peptides coupled with the occurrence of IGF receptors in many tissues has suggested to investigators that these peptides serve as important local (autocrine and/or paracrine) mediators of tissue growth and development. Supporting this hypothesis are the observed elevated IGF-II mRNA expression in tissues of growing fetuses[2], growth hormone induced IGF-I synthesis in proliferative zones of bone[3] and the elevated IGF-I production in hyperplastic kidney[4]. Thus, IGFs are strongly implicated in fetal, postnatal and prepubertal growth processes.

Two organs which display remarkable rates of growth in pregnant animals are the uterus and mammary glands. These organs develop in synchrony with the fetus under the influence of multiple hormonal mediators of both embryonic and maternal origins. During pregnancy, the uterus must grow and develop to accomodate the rapidly growing fetus(es) as well as provide a continued source of fetal nutrients and other factors. At late pregnancy, mammary glandular growth occurs in preparation for lactation. Failure to achieve adequate growth and development of these organs may contribute to early embryonic losses, fetal or neonatal growth retardation and pre- or post-natal death.

A number of previous reports have documented IGFs and IGF receptors in female reproductive tissues of humans and rodents[5,6]. In this paper, we describe the results of studies on pregnant and lactating pigs. This animal model is extensively characterized with respect to embryological, fetal and postnatal biology as well as provides a useful source of tissues and biological fluids. The accumulated data demonstrate the differential expression of IGFs in mammary and uterine tissues, implicate

195

uterine IGF-I as a paracrine mediator of early embryonic development and suggest a role for mammary-derived IGFs in postnatal development of the gastrointestinal tract.

RESULTS AND DISCUSSION

Uterine IGFs

Sow uterine growth initiates just prior to the passage of the fertilized ova to the uterus and continues at a relatively constant rate throughout gestation[7]. Growth is manifested as a linear increase in tissue weight. In utero, between days 10 and 14 of gestation, pig blastocysts undergo a remarkable series of morphological transitions culminating in implantation to the endometrium[8]. Placentation is epitheliochorial hence fetal and maternal blood supplies never come into direct contact[8].

During days 10 to 12 of gestation, blastocyst morphology rapidly changes from spherical to tubular to elongated filamentous forms (reviewed in ref. 8). On or around day 12, pig blastocysts release appreciable amounts of estrogens into the uterine lumen. Release of this hormone stimulates a synchronized release of secretory material (histotroph) from the uterine glandular epithelial cells. These maternal secretions include transport proteins (for iron and retinol), protease inhibitors and peptide mitogens, all of which are postulated to facilitate embryonic development in utero[8]. By day 14, these embryos reach lengths of 600-1000 mm, and begin to attach to the endometrium. Attachment is essentially complete by day 18.

The marked increase in diameter of spherical blastocysts during days 10 and 11 results from proliferation of embryonic cells[10]. In contrast, the later embryonic transitions result primarily as a consequence of cell migration and cellular remodeling (a process involving trophectoderm and endoderm)[10]. Thus, the pre-implantation period in the pregnant pig is characterized by dynamic and synchronous biochemical changes in the uterine endometrium, embryos and surrounding uterine luminal fluid (ULF), all of which are necessary for successful embryo development and implantation[8]. Our laboratories have examined porcine ULFs as potential sources of growth factors for embryos and endometrial cells[11,12,13].

The endometrial content of IGF-I around the time of implantation and in cycling nonpregnant sows is relatively high (L.R. Letcher et al., unpublished data). Levels of tissue IGF-I remain unchanged during days 8 to 14 of pregnancy. Endometria at later stages of pregnancy (day 30 and after) are characterized by a diminished IGF-I content[14]. Corresponding ULFs also contained IGF-I peptide, the levels of which were dependent upon day of gestation (Fig. 1). Total uterine luminal IGF-I content was low on day 8, increased on days 10-12 and declined by day 14

Table 1. Developmental Biology of the Pig[a]

One Cell	0-51 hr
Morula Stage	72-96 hr
Blastocyst Stage	5-6 days
Implantation	13-18 days
Gestation	115 days
Lactation	4-8 weeks

[a]Compiled from Refs. 7, 8, 9

196

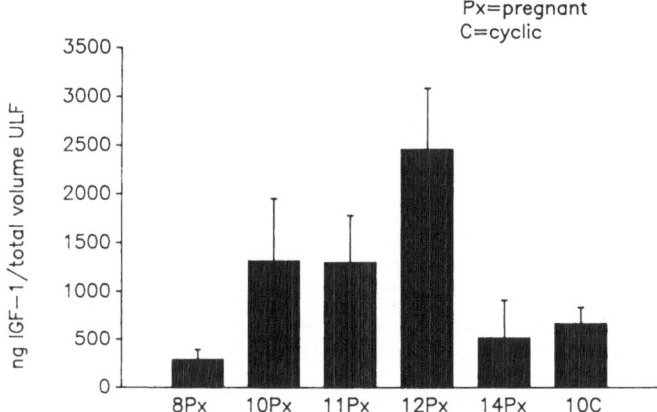

Fig. 1. IGF-I in porcine uterine luminal fluids
(ULFs). ULFs were obtained from gilts at
early pregnancy (8Px-14Px) and at estrous
cycle (10C). Each ULF was subjected to
acid-ethanol extraction and IGF-I radioimmunoassay[12]. Shown are the mean ± SEM of
IGF-I for gilts (n=3) at each day. Differences between days were highly significant (P < .001).

(Fig. 1 and ref. 12). Uterine luminal IGF-I was higher in pregnant sows
than in the nonpregnant cycling animals with comparable hormone profiles
(Fig. 1. and ref. 12). Calculations of total endometrial and embryo
IGF-I identify the maternal tissue as the major source of ULF IGF-I at
the time of blastocyst elongation (L. R. Letcher et al., unpublished
data). The idea that endometrial IGF-I is actively released into the
uterine lumen at early pregnancy is supported by the observed highest
levels in the ULF on day 12, when estrogen production by the blastocysts
and the release of endometrial proteins are maximal.

Porcine ULFs provide a growth medium for blastocysts in utero. In
addition to IGF-I, at least one other peptide mitogen has been identified, preliminarily designated uterine luminal fluid mitogen
(ULFM)[11,12]. Interestingly, ULFM is expressed in the uterine lumen
slightly earlier in pregnancy than IGF-I[12]. We believe that these peptides mediate embryo growth and/or reorganization during days 8-14 when
they are present at relatively high concentrations in ULF. The basis
for this is twofold: implantation stage pig embryos express type I IGF
receptors[15] and embryo IGF-I content varies in parallel with the corresponding ULF IGF-I concentration (L. R. Letcher et al., unpublished
data).

IGF-I, epidermal growth factor (EGF), ULFM and estrogen are in
vitro mitogens for endometrial stromal cells of day 12 pregnant pigs
(Table 2, Fig. 2). This observation raises the possibility that locally
produced IGF-I functions as an autocrine or paracrine mediator of in
vivo endometrial growth and differentiation necessary for successful
implantation[14]. Furthermore, IGF-I possibly mediates or potentiates the
mitogenic effects of estrogen and other growth factors in the endometrium of the pregnant pig and other species. The observations discussed
above are summarized in Fig. 3.

Table 2. Uterine Stromal Cell Mitogens

Growth factor	Fold-stimulation of DNA synthesis[a]
IGF-I[b]	3.3
EGF[c]	2.5
ULFM[d]	3.1

[a]Determined on density-arrested, uterine endometrial
 stromal cells previously isolated from a day 12 pregnant
 pig[11].
[b]Recombinant-derived, human IGF-I; 200ng/ml.
[c]Murine EGF; from submaxillary glands; 100 ng/ml.
[d]Porcine uterine luminal fluid mitogen; Ref. 11;60 ng/ml.

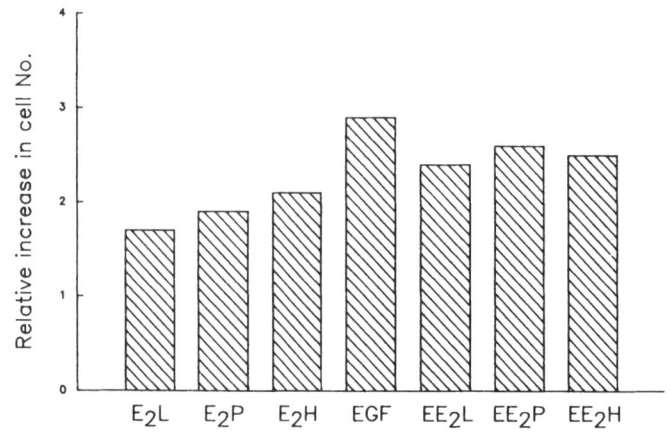

Fig. 2. Mitogenic actions of estrogen and peptide
         growth factors in the uterine stroma. Cul-
         tures of stromal fibroblasts were established
         from day 12 pregnant pig uterine endometrium
         using standard procedures. Subcultures of
         these cells were incubated in a phenol-red
         free and serum-free medium (Y. Ko et al.;
         unpublished data) supplemented with: a) low,
         b) physiological, or c) high concentrations
         of estradiol -17β , d) murine EGF
         (physiological conc.) or e-g) combinations of
         estradiol -17β and EGF. After several days
         of incubation the cells were trypsinized,
         counted, and their growth compared to control
         cells which did not receive any estradiol or
         EGF supplementation.

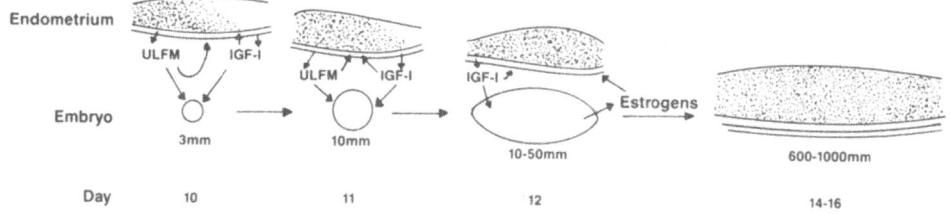

Fig. 3. Paracrine relationships of pig endometrium, embryos and uterine luminal growth
factors.

As mentioned previously, the uterine content of IGF-I declines as
pregnancy proceeds. Figs. 4 and 5 present an analysis of IGF-I mRNAs
and their corresponding quantitative variations in the pregnant pig
uterus. IGF-I mRNA abundance is highest around day 12, declines to very
low levels by day 60 and remains low to term. It is noteworthy that
IGF-I mRNA levels (per ug total RNA) in the uterus (or endometrium)
greatly exceed that for the liver, kidney, spleen or mammary glands of

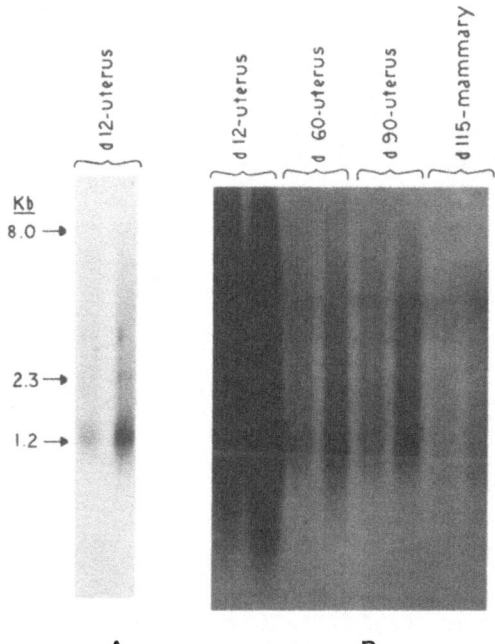

Fig. 4. IGF-I messenger RNAs (mRNAs) in pig
uteri[14]. Polyadenylated RNA was isolated from uteri of pregnant pigs
(day of pregnancy indicated) or from
mammary tissue at parturition. 10
and 20 ug aliquots of each RNA preparation were fractionated in a 1.2%
agarose-formaldehyde gel, transferred
to a nitrocellulose sheet and hybridized with ^{32}P-cDNA encoding swine
IGF-I[14]. Panel A is an autoradiogram
obtained from shorter exposure time
compared to that in panel B.

day 12 pregnant sows (F. A. Simmen et al.; unpublished data). Therefore, these data identify the uterus as a major tissue source of IGF-I
during the first trimester of pregnancy in the pig.

Fig. 5 presents a parallel analysis of uterine mRNAs encoding porcine IGF-II and IGF-binding protein (analog of the rat fetal 33 KDa
binding protein (BP-3A)). IGF-II mRNAs are undetectable in uteri until
after implantation at which time they increase and remain relatively
high until term of pregnancy. IGF binding protein mRNA is detectable in
early pregnant pig endometrium (day 12), but highest levels are observed
in uteri during later pregnancy when IGF-II expression is also found
(Fig. 5). This is similar to the situation observed in the rat where
tissue expression of the 33 KDa BP mRNA almost always parallels that of
IGF-II mRNA (M. M. Rechler, personal communication). As noted for IGF-
I, the levels of IGF-II and IGF BP mRNAs are relatively high in the pig
uterus as compared to other maternal tissues of comparable stage of
pregnancy (F.A. Simmen et al., unpublished data). Friesen and colleagues have described a relatively high level of IGF (I and II) mRNAs
in the uterus as compared to most other tissues of nonpregnant, adult
rats[16].

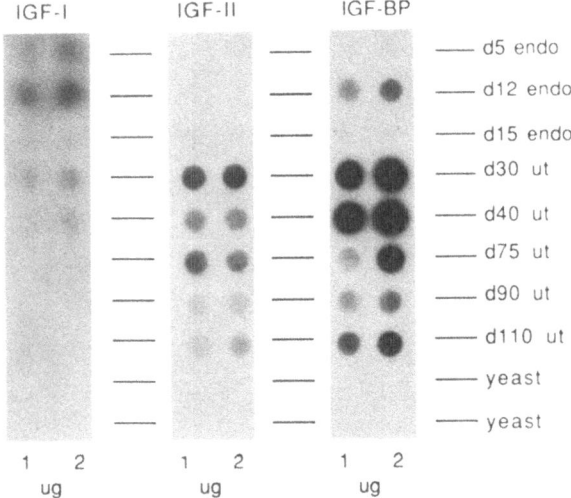

Fig. 5. Uterine expression levels during pregnancy of mRNAs
encoding IGF-I, IGF-II and an IGF binding protein
(BP-3A). Polyadenylated RNA was isolated from endometria (days 5, 12, 15) or whole uteri (days 30, 40,
75, 90, 110) of pregnant pigs. 1 and 2 ug aliquots
of each RNA preparation were denatured and immobilized (dotted) onto Biotrans nylon membranes. Rep-
lica membranes were hybridized with ^{32}P-cDNA encoding
a) swine IGF-I (sigf.3), b) rat IGF-II or c) rat (33
KDa) IGF binding protein. Yeast RNA was also dotted
onto each membrane to serve as a control for nonspecific hybridization of probes (e.g. to ribosomal
RNAs). Autoradiographic exposure times and probe
specific activities varied for the different experiments, hence quantitative comparisons of the three
panels are not possible.

It is interesting that the synthesis of IGF mRNAs in human placenta follows a developmental pattern similar to that observed for the porcine uterus. IGF-I mRNA abundance is maximal in human placenta during the first and second trimesters of pregnancy[17], whereas IGF-II mRNAs peak in concentration during the second trimester[18]. However, the magnitude of the quantitative changes observed for placental IGF mRNAs are significantly lower than that observed for porcine uterine IGF mRNAs. IGF-I mRNAs have been localized to placental syncytiotrophoblast cells whereas IGF-II mRNAs are primarily expressed in placental fibroblasts[17]. In situ information concerning uterine cell types which express IGF mRNAs is unavailable.

It is also of interest to compare the steady state levels of uterine IGF mRNAs with the endocrine status of pregnant sows. Levels of estrogens in sow plasma are relatively low until about day 75 when they sharply increase[19]. Plasma progesterone, on the other hand, is maximal during the first half of pregnancy. It has been previously noted[8] that the ratio of progesterone to estrogen is positively correlated with endometrial synthesis and secretion of proteins (e.g. porcine uteroferrin, a major uterine secretory protein in the pig). Endometrial secretory activity increases after day 30, peaks around days 60-75 and declines to parturition. This type of temporal pattern is also observed for altered levels of mRNAs encoding IGF-II and IGF BP (Fig. 5.). Prolactin and growth hormone concentrations remain fairly constant in pregnant sow plasma until around day 105 of pregnancy at which time they greatly increase. Therefore, these two hormones cannot be directly correlated with IGF-I or IGF-II gene expression in utero. However, it is possible that uterine tissues vary in numbers of prolactin and growth hormone receptors during different stages of pregnancy and as a result acquire increased sensitivity to these hormones. Friesen and colleagues have demonstrated a marked induction of uterine IGF-I mRNA synthesis in immature rats injected with physiological amounts of estradiol -17β or growth hormone[20]. However, sow uterine IGF-I mRNA levels do not similarly increase during the surges of serum estrogens and growth hormone at later pregnancy[19], although estrogen receptors are present in the pig uterus[21].

IGF binding proteins (PP12, α_1-PEG, amniotic fluid BP) have been identified as major secretory proteins of the human endometrium during the menstrual cycle and pregnancy. PP12 (probably the same protein as α_1-PEG or AFBP) is a progesterone-dependent protein produced during implantation[22,23,24]. PP12 production has been localized to syncytiotrophoblast and decidual cells and this protein is also present in amniotic fluid[22-24]. Messenger RNA encoding this IGF BP is present in human placental and decidual tissue as demonstrated by Northern hybridization with cloned cDNA[24]. PP12 is distinct in amino acid sequence from BP-3A (Fig. 5). It will be interesting to examine if mRNAs for both binding proteins are coordinately or differentially expressed in the uterus during pregnancy.

To summarize, uterine and placental growth is characterized by transitions in the tissue steady state levels of IGF mRNAs and proteins. IGF-I mRNAs predominate early in pregnancy whereas IGF-II mRNAs become more abundant at later stages. Presumably, the combined expression of both of the IGFs and their corresponding receptors and binding proteins contributes to the sustained, linear growth observed for the uterus throughout pregnancy. Other differentiative effects of uterine IGFs are also probable. It has been demonstrated for example that IGF-I stimulates decidual cell production of prolactin[25] and placental synthesis of placental lactogen in vitro[26].

Mammary IGFs

The normal pregnancy-associated development of mammary glands requires a host of essential hormones including progesterone, estrogens, growth hormone, prolactin, glucocorticoids as well as peptide growth factors. Perhaps the two best studied examples are epidermal growth factor (EGF) and IGF-I. It is pertinent that both of these growth factors are secreted at high levels in colostrum and milk of those species studied (Table 3).

IGF-I is a potent in vitro stimulator of mammary tissue growth and milk production[27,28]. Addition of IGF-I to explant cultures of bovine lactating mammary gland elicits an increased incorporation of ^{3}H-thymidine into DNA of ductal and secretory epithelial cells, myoepithelial cells, and stromal fibroblasts[28]. Growth hormone administration significantly increases milk production in the bovine by an indirect mechanism thought to involve an elevation of systemic IGF-I concentration.

IGF-I and IGF-II mRNAs are present at relatively low levels in mammary tissues (Fig. 4, refs. 14,16). However, complete studies of quantitative variations of mammary IGFs (mRNAs, proteins) during mammogenesis, lactogenesis and involution are currently lacking for any species. Thus, it is unclear whether the local production of IGFs contributes in a significant way to mammary growth and remodeling. What is evident is that mammary cells express both type I and type II receptors and that these receptors can bind plasma IGFs (triggering mitogenesis) and mediate transport of internalized IGFs to the alveolar spaces[27,28,29]. For example, bovine colostrum and milk contain a 5-6 fold molar excess of IGF-II to IGF-I (Table 3), a ratio which is also found for type II to type I IGF receptor numbers in lactating cow mammary glands[29].

Recent interest has centered on the role(s) of IGF peptides in mammary gland secretions. IGFs are present in human, bovine and porcine milk (Table 3). In general, the levels of secretory IGF-I and IGF-II are highest around the time of parturition (precolostrum and colostrum) and decline during the transition in gland production of colostrum to milk. IGF binding proteins (and perhaps receptors) are also elicited in significant quantities in milk. The bulk of secretory IGF-I is found complexed with BP complexes of 150 KDa and 40-50 KDa molecular weight. Free (unoccupied) sites are found on the 150 KDa BP complex in porcine colostrum and milk[30] and on the 40-50 KDa BP in human[31] and bovine[32] milks. No quantitative data of secretory BPs or of their (mammary) mRNAs during lactation are currently available. Data from explant culture experiments do suggest that mammary production of binding proteins occurs[33]. An IGF-I variant (destripeptide IGF-I) has recently been isolated from bovine colostrum[34]. This peptide exhibits increased bioactivity as a result of its reduced affinity for IGF binding proteins[34].

Paracrine IGFs in Neonatal Development

Early postnatal development of the pig (also of the dog, rat and mouse) is characterized by extensive growth and maturation of the gastrointestinal tract. In particular, the small and large intestines increase in length, weight, total DNA content and height of villi and microvilli. This organ response is partially dependent upon ingestion of colostrum and early milk[38]. One plausible explanation for enteric growth is that the maternal secretions are supplying a growth promoting peptide(s) to this neonatal organ. This is now well supported for the paradigm of EGF and neonatal gastrointestinal development of the murine[39]. The case for maternally-derived IGFs is less developed, although new data are emerging.

Table 3. IGF-I and IGF-II concentrations for mammary gland secretions and sera[a]

Species	IGF	Colostrum[b] (ng/ml)	Milk[c] (ng/ml)	Maternal Sera[d] (ng/ml)	Neonatal Sera[e] (ng/ml)
Human	IGF-I	7-27	6-8	400-500	40-50
	IGF-II	n.a.[f]	n.a.	1000-1200	800-900
Pig	IGF-I	67-357	4-14	20-100	11-91
	IGF-II	n.a.	n.a.	n.a.	n.a.
Cow	IGF-I	100-600	22-26	20-24	n.a.
	IGF-II	600	107-127	1000	n.a.

[a]Compiled from Refs. 30-32, 35-37.
[b]Sampled at parturition or within 2 days postpartum.
[c]Sampled between 3 and 7 days postpartum.
[d]Sampled at term or during late pregnancy.
[e]Sampled within 1 week post-parturition.
[f]Data not available.

The neonatal pig small intestine is characterized by an abundance of type I IGF receptors (Fig. 6). This receptor expression occurs in the context of low concentrations of serum IGFs and high colostral and milk IGF-I content (Table 3). The intestinal IGF-I receptors are found in both the serosal and mucosal tissue compartments and exhibit marked changes in their binding capacity during the first week of postnatal life (D. A. Schober et al., unpublished data). Work from another laboratory has documented the presence of type I and type II receptors in the intestinal epithelium of mature rats[40]. Interestingly, IGF-I and IGF-II binding to rat intestinal enterocytes decreases as the cells migrate from the crypts and onto the villus[40]. This supports the view that IGF-I promotes proliferation of crypt cells necessary for villus cell replacement. As these cells migrate onto the villus, they undergo differentiative changes in absorptive capability, membrane receptor expression and other properties.

The known stability of IGF peptides to acidic conditions and the presence in milk of IGF BPs and general protease inhibitors may allow for passage of significant amounts of IGF-I and IGF-II through the absorptive intestinal epithelium prior to gut closure. Intestinal uptake (possibly receptor-mediated) and transcellular support of such IGFs and IGF BPs could prove to be of major importance during the early neonatal period. In this regard, administration (injection) of synthetic IGF-I to neonatal rats elicits increases in body weight, precocious eye opening and increased production of erthropoietic cell precursors[41]. Similar effects can be elicited by oral administration of EGF (in artificial formula) to neonatal rats. It is also possible that IGF-I and IGF-II synergistically cooperate with other colostral growth factors (EGF, platelet-derived growth factor, colony stimulating factors, etc.) to promote early postnatal growth and development. The destripeptide IGF-I elicited by the lactating bovine may also represent a form of IGF-I evolved for just such a function as described above. Finally, we note that human neonates exhibit a transient rise in plasma IGF around days 4 to 5 of postnatal life[42]. The tissue origins and physiological actions

of this IGF remain unknown. However, its presence serves to highlight
the potential involvement of the IGFs in early neonatal growth.

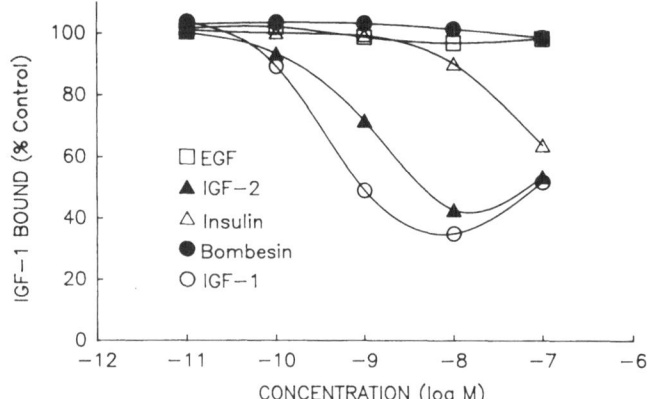

Fig. 6. Identification of type I IGF receptors in
         the intestinal mucosa of newborn pigs. Pigs
         were killed at birth, the small intestines
         removed and rinsed, and the mucosa scraped
         from the luminal surfaces. Mucosal mem-
         branes were prepared using standard proce-
         dures and the protein content was
         determined with BCA reagent (Pierce). Mem-
         branes (50ug) were incubated at 4°C for 48
         hr with 0.1 ng of ^{125}I-IGF-I and various
         amounts of unlabeled ligands. Incubations
         were terminated by centrifugation and
         membrane-associated radioactivity measured
         in a gamma counter. All peptides were pur-
         chased from commercial vendors (EGF and in-
         sulin from Collaborative Research; IGF-II
         from Bachem; bombesin from Sigma; IGF-I from
         AMGEN).

SUMMARY AND CONCLUSIONS

It is apparent that the IGFs, IGF receptors and IGF binding proteins are all intimately involved in the regulation of uterine and placental growth during pregnancy. Uterine-derived IGFs probably also directly mediate embryonic and fetal development via paracrine mechanisms.
Additional studies of uterine IGFs are clearly warranted. Areas for
further study include: characterization of the hormonal and trans-acting
factors and DNA sequences which contribute to the relatively high expression of IGF-I and IGF-II genes in utero; elucidation of the possible
role(s) of IGFs in fetal and neonatal uterine morphogenesis; and examination of IGFs as potential oncoproteins in endometrial cancer.

Current data, while not definitive, suggest that variable mammary
secretion of IGFs principally results from alterations in activity
(numbers and/or affinity) of mammary IGF receptors rather than from
changes in mammary biosynthesis of the IGFs. A large buildup of the IGF
peptides occurs in the mammary alveolar spaces just prior to parturition. It is our view that this represents a growth factor pool seques-

tered by the glands for eventual release to the newborn. This is akin to the mammary accumulation of (colostral) immunoglobulins and albumin, proteins critical to the early survival of neonates (pigs, cows). Obviously, much further work is necessary before a complete understanding of IGFs in early postnatal growth regulation is obtained. Such studies may however provide a means to ameloriate conditions of low birth weight and growth retardation of human neonates. Synthetic or recombinant IGF analogs may also someday be useful for growth promotion and early weaning of livestock species.

ACKNOWLEDGEMENTS

Research by the authors is supported by grants from the NIH (HD-22004 and HD-21961) and the USDA (87-CRCR-1-2532). Salaries and research support were also provided by State and Federal funds appropriated to the Ohio Agricultural Research and Development Center.
The authors thank F. W. Bazer (Un. Florida) for many contributions to this work which included embryos, uterine flushings, endometria and helpful discussions, M. M. Rechler (NIH) for providing rat IGF-II and IGF-binding protein cDNA clones, and C. R. Baumrucker (Penn. State Un.) for guidance and suggestions regarding the IGF-I receptor studies. Cindy Coy and Judith Riggenbach provided expert technical assistance and Norma Rickett and Beverly Fisher are thanked for manuscript preparation. We also acknowledge the other members of our laboratories for their many contributions to these studies.

References

1. J. Zapf and E. R. Froesch, Insulin-like growth factors/ somatomedins: structure, secretion, biological actions and physiological role, Hormone Res. 24:121 (1986).

2. A. L. Brown, D. E. Graham, S. P. Nissley, D. J. Hill, A. J. Strain, and M. M. Rechler, Developmental regulation of insulin-like growth factor II mRNA in different rat tissues, J. Biol. Chem. 261:13144 (1986).

3. N. L. Schlechter, S. M. Russell, E. M. Spencer, and C. S. Nicoll, Evidence suggesting that the direct growth-promoting effect of growth hormone on cartilage in vivo is mediated by local production of somatomedin, Proc. Natl. Acad. Sci. U.S.A. 83:7932 (1986).

4. J. A. Fagin and S. Melmed, Relative increase in insulin-like growth factor I messenger ribonucleic acid levels in compensatory renal hypertrophy, Endocrinology 120:718 (1987).

5. W. H. Daughaday, I. K. Mariz, and B. Trivedi, A preferential binding site for insulin-like growth factor II in human and rat placental membranes, J. Clin. Endocr. Metab. 53:282 (1981).

6. M. Fant, H. Munro and A. C. Moses, An autocrine/paracrine role for insulin-like growth factors in the regulation of human placental growth, J. Clin. Endocr. Metab. 63:499 (1986).

7. J. S. Perry and I. W. Rowlands, Early pregnancy in the pig, J. Reprod. Fertil. 4:175, (1962).

8. R. M. Roberts and F. W. Bazer, The functions of uterine secretions, J. Reprod. Fertil. 82:875 (1988).

9. R. L. Brinster, Embryo development, J. Anim. Sci. 38:1003 (1974).

10. R. D. Geisert, J. W. Brookbank, R. M. Roberts, and F. W. Bazer, Establishment of pregnancy in the pig: II. Cellular remodeling of the porcine blastocyst during elongation on day 12 of pregnancy, Biol. Reprod. 27:941 (1982).

11. R. C. M. Simmen, Y. Ko, X. H. Liu, M. H. Wilde, W. F. Pope, and F. A. Simmen, A uterine cell mitogen distinct from epidermal growth factor in porcine uterine luminal fluids: characterization and partial purification, Biol. Reprod. 38:551 (1988).

12. R. C. M. Simmen, F. A. Simmen, Y. Ko and F. W. Bazer, Differential growth factor content of uterine luminal fluids from large white and prolific Meishan pigs during the estrous cycle and early pregnancy, J. Anim. Sci. (in press).

13. F. A. Simmen, R. C. M. Simmen and F. W. Bazer, Maternal growth factors as mediators of embryonic and neonatal growth, Biochem. Soc. Transactions (in press).

14. A. Tavakkol, F. A. Simmen, and R. C. M. Simmen, Porcine insulin-like growth factor - I (pIGF-I): Complementary deoxyribonucleic acid cloning and uterine expression of messenger ribonucleic acid encoding evolutionarily conserved IGF-I peptides, Mol. Endo. 2:674 (1988).

15. A. N. Corps, C. J. Littlewood, and K. D. Brown, IGF-binding proteins from pig pre-implantation blastocysts, (abst.), International Symposium on Biotechnology in Growth Regulation (1988), Cambridge, U.K.

16. L. J. Murphy, G. I. Bell, and H. G. Friesen, Tissue distribution of insulin-like growth factor I and II messenger ribonucleic acid in the adult rat, Endocrinology 120:1279 (1987).

17. C.-Y. Wang, M. Daimon, S.-J. Shen, G. L. Engelmann, and J. Ilan, Insulin-like growth factor-I messenger ribonucleic acid in the developing human placenta and in term placenta of diabetics, Mol. Endo. 2:217 (1988).

18. S.-J. Shen, C.-Y. Wang, K. K. Nelson, M. Jansen, and J. Ilan, Expression of insulin-like growth factor II in human placentas from normal and diabetic pregnancies, Proc. Natl. Acad. Sci. U.S.A. 83:9179 (1986).

19. M. H. Dehoff, C. S. Stoner, F. W. Bazer, R. J. Collier, R. R. Kraeling, and F. C. Buonomo, Temporal changes in steroids, prolactin and growth hormone in pregnant and pseudopregnant gilts during mammogenesis and lactogenesis, Dom. An. Endo. 3:95 (1986).

20. L. J. Murphy and H. G. Friesen, Differential effects of estrogen and growth hormone on uterine and hepatic insulin-like growth factor I gene expression in the ovariectomized, hypophysectomized rat, Endocrinology 122:325 (1988).

21. A. D. Tolton, D. L. Grinwich, C. J. Belke, and M. M. Buhr, Measurement and characterization of swine uterine estradiol receptors: the effects of puberty induction on estradiol receptors and corpus luteum function, Can. J. Physiol. Pharmacol. 63:214 (1985).

22. R. Koistinen, N. Kalkkinen, M.-L. Huhtala, M. Sepala, H. Bohn and E.-M. Rutanen, Placental protein 12 is a decidual protein that binds somatomedin and has an identical N-terminal amino acid sequence with somatomedin-binding protein from human amniotic fluid, Endocrinology 118:1375 (1986).

23. S. C. Bell and J. W. Keyte, N-terminal amino acid sequence of human pregnancy-associated endometrial α-₁globulin, an endometrial insulin-like growth factor (IGF) binding protein-evidence for two small molecular weight IGF binding proteins, Endocrinology, 123:1202 (1988).

24. M. T. Brewer, G. L. Stetler, C. H. Squires, R. C. Thompson, W. H. Busby, and D. R. Clemmons, Cloning, characterization, and expression of a human insulin-like growth factor binding protein, Biochem. Biophys. Res. Commun. 152:1289 (1988).

25. K. M. Thrailkill, A. Galander, L. E. Underwood, and S. Handwerger, Insulin-like growth factor I stimulates the synthesis and release of prolactin from human decidual cells, Endocrinology, 123:2930 (1988).

26. B. Bhaumick, E. P. Dawson, and R. M. Bala, The effects of insulin-like growth factor I and insulin on placental lactogen production by human term placental explants, Biochem. Biophys. Res. Commun. 144:674 (1987).

27. A. Shamay, N. Cohen, M. Niwa, and A. Gertler, Effect of insulin-like growth factor I on deoxyribonucleic acid synthesis and galactopoiesis in bovine undifferentiated and lactating mammary tissue in vitro, Endocrinology 123:804 (1988).

28. C. R. Baumrucker and B. Stemberger, Insulin and insulin-like growth factor-I stimulates DNA synthesis in tissue slices of bovine mammary tissue, J. Anim. Sci. (in press).

29. M. H. Dehoff, R. G. Elgin, R. J. Collier, and D. R. Clemmons, Both type I and II insulin-like growth factor receptor binding increase during lactogenesis in bovine mammary tissue, Endocrinology 122:2412 (1988).

30. F. A. Simmen, R. C. M. Simmen, and G. Reinhart, Maternal and neonatal somatomedin C/insulin-like growth factor I (IGF-I) and IGF binding proteins during early lactation in the pig, Dev. Biol. 130:16 (1988).

31. R. C. Baxter, Z. Zaltsman, and J. R. Turtle, Immunoreactive somatomedin-C/insulin-like growth factor I and its binding protein in human milk, J. Clin. Endocr. Metab. 58:955 (1984).

32. P. G. Campbell and C. R. Baumrucker, Insulin-like growth factor I and its association with binding proteins in bovine milk, J. Endocr. (in press).

33. P. G. Campbell and C. R. Baumrucker, Secretion of immunoreactive insulin-like growth factor I and its binding protein from the bovine mammary gland in vitro, (abst.), Ann. Mtng. of the Endocrine Soc. (1988).

34. F. J. Ballard, G. L. Francis, M. Ross, C. J. Bagley, B. May, and J.

C. Wallace, Natural and synthetic forms of insulin-like growth factor I (IGF-I) and the potent derivative, destripeptide IGF-I: biological activities and receptor binding, Biochem. Biophys. Res. Commun. 149:398 (1987).

35. B. Bhaumick, A. D. Danilkewich, and R. M. Bala, Insulin-like growth factors (IGF) I and II in diabetic pregnancy: suppression of normal pregnancy-induced rises of IGF-I, Diabetologia 29:792 (1986).

36. P. V. Malven, H. H. Head, R. J. Collier, and F. C. Buonomo, Periparturient changes in secretion and mammary uptake of insulin and in concentrations of insulin and insulin-like growth factors in milk of dairy cows, J. Dairy Sci. 70:2254 (1987).

37. S. R. Davis, P. D. Gluckman, I. C. Hart, and H. V. Henderson, Effects of injecting growth hormone or thyroxine on milk production and blood plasma concentrations of insulin-like growth factors I and II in dairy cows, J. Endocr. 114:17 (1987).

38. C. L. Berseth, L. M. Lichtenberger, and F. H. Morriss, Comparison of the gastrointestinal growth-promoting effects of rat colostrum and mature milk in newborn rats in vivo, Am. J. Clin. Nutr. 37:52 (1983).

39. O. Tsutsumi, A. Tsutsumi, and T. Oka, A possible physiological role of milk epidermal growth factor in neonatal eyelid opening, Am. J. Physiol. 252:R376 (1987).

40. M. Laburthe, C. Rouyer-Fessard, and S. Gammeltoff, Receptors for insulin-like growth factors I and II in rat gastrointestinal epithelium, Am. J. Physiol. 254:G457 (1988).

41. A. F. Philipps, B. Persson, K. Hall, M. Lake, A. Skottner, T. Sanengen, and V. R. Sara, The effects of biosynthetic insulin-like growth factor-I supplementation on somatic growth, maturation, and erythropoiesis on the neonatal rat, Pediatr. Res. 23:298 (1988).

42. L. Tato, M. V. L. DuCaju, C. Prevot, and R. Rappaport, Early variations of plasma somatomedin activity in the newborn, J. Clin. Endocr. Metab. 40:534 (1975).

HORMONAL FEEDBACK REGULATION OF

BRAIN IGF-I AND IGF-II GENE EXPRESSION

Teresa L. Wood, Michael Berelowitz and Jeffrey F. McKelvy[*]

Department of Neurobiology and Behavior
SUNY at Stony Brook, Stony Brook, N.Y.
*Abbott Laboratories, Abbott Park, Illinois

Release of anterior pituitary hormones is mediated by peptides produced by secretory neurons in the hypothalamus or preoptic area. The releasing or inhibiting factors are stored in nerve endings that lie in the median eminence and are secreted as the result of electrical activity. The hypothalamic peptides then travel through the portal system to the anterior pituitary where they interact with specific surface receptors to trigger release of pituitary trophic hormones. These, in turn, regulate physiologic processes elsewhere in the body either directly or indirectly through release of target gland hormones.

The ultradian release of GH is coordinately controlled by two hypothalamic peptides, growth hormone releasing factor (GRF) and the inhibitory peptide, somatostatin (SRIF). Pituitary GH has direct effects on carbohydrate and lipid metabolism as well as indirect effects on growth promotion through the induction of IGF-I. The somatomedin hypothesis originally proposed that the growth-promoting influence of GH is exerted by stimulating the liver to secrete IGFs, which are the causative endocrine agents signaling peripheral tissues to grow[1]. Consistent with this hypothesis, IGF-I stimulates growth in hypophysectomized rats, and its production in liver, the major source of circulating IGF-I, is almost entirely dependent upon GH[2,3]. In man, serum concentrations of IGF-I are higher than normal in acromegaly and lower than normal in hypopituitary individuals[4-6].

In addition to the hepatic production of IGF-I, immunoreactive IGF-I and IGF-I mRNA have been detected in a number of other tissues though at lower levels than in liver[7-10]. IGF-I peptide concentrations in many of these tissues also are dependent on the growth hormone status of the organism and, when growth hormone is administered, exhibit changes greater than those that occur in serum[11,12]. In addition, IGF-I mRNAs in these tissues also demonstrate at least some degree of GH dependence[9,10,13].

IGF-II, although structurally related to IGF-I, does not respond as dramatically as IGF-I to changes in the GH status of the organism. In man, serum IGF-II levels remain normal in acromegaly although some hypopituitary individuals do have lower serum IGF-II concentrations[14]. Also in contrast to IGF-I, there is no evidence for GH regulation of IGF-II secretion in cultured rat fibroblasts[15].

In addition to growth hormone, other hormones including corticosteroids, thyroid hormone and gonadotropins also regulate the localized production of IGF-I and IGF-II. Glucocorticoids both reduce IGF-I peptide levels in fetal rat plasma and reduce IGF-I peptide release from liver explants[16,17]. Dexamethasone decreases IGF-I mRNA levels in rat neuronal and glial cells in primary culture[18]. These results have led to the suggestion that the growth-inhibitory effects of glucocorticoids may be mediated by this local reduction of IGF-I production[18]. Thyroid hormone also may exert some of its effects by altering production of the IGFs either directly or indirectly. TH administration increases plasma IGF-I levels in hypothyroid patients possibly by regulating hypothalamic GRF and thus pituitary GH secretion[19,20]. In addition, TH directly increases the release of IGFs from dissociated cultures of fetal rat hypothalami[21]. Finally, gonadotropins and estradiol enhance the secretion of immunoreactive IGF-I from ovarian granulosa cells in vitro[22].

IGF-I FEEDBACK REGULATION OF GH

Classically, target-gland hormones and pituitary hormones are known to function through feedback mechanisms to regulate their own release. Target gland hormones can regulate their own release through "long-loop" feedback mechanisms to either the hypothalamus or the pituitary. In addition, pituitary hormones can regulate the hypothalamic secretion of releasing factors directly through "short-loop" feedback. The cellular actions of these feedback mechanisms can include: 1) alteration of electrical activity impinging on brain cells that release hypothalamic-releasing factors, 2) modification of the synthesis of releasing factors, 3) modification of the receptor system for the releasing factor in pituitary cells, and 4) modification of the rate of synthesis of the pituitary trophic hormones.

Thus, according to the classic endocrine feedback loops described above, if IGF-I mediates the growth-promoting actions of GH, then IGF-I would be expected to inhibit pituitary GH secretion. Recent studies demonstrate that IGF-I regulates GH and are consistent with the existence of IGF-I feedback loops at the level of both the brain and the pituitary. Membrane receptors for both IGF-I and IGF-II have been characterized from pituitary membranes and have been localized throughout the brain, including the hypothalamus, and on neurons and glia in culture[23-25]. Intracerebroventricular administration of IGF-I reduces the in vivo release of GH from the anterior pituitary[26,27]. This may result from stimulation of somatostatin release from the hypothalamus[28]. Evidence for feedback at the level of the pituitary comes from studies demonstrating that IGF-I inhibits GH mRNA levels and GH synthesis as well as GRF-elicited GH release in pituitary cell cultures[29,30]. These results suggest a potential role for the IGFs in pituitary GH regulation.

BRAIN SITES OF IGF-I AND IGF-II SYNTHESIS

For either IGF-I or IGF-II to interact with the hypothalamic-somatotroph axis in vivo, the peptides must be present in the central nervous system and/or the hypophyseal portal circulation. Immunoreactive IGF-I has been detected in the adult brain though at much lower levels than in liver[11,31]. Immunoreactive IGF-II also has been detected in extracts of human brain and anterior pituitary and in human cerebrospinal fluid[32]. It is unclear from these studies however, whether the immunoreactive IGFs detected in brain tissues are actually synthesized there or whether their presence there is the result of peripheral IGFs which have been transported into the CSF and brain tissue.

Initial reports of IGF-I and IGF-II mRNAs in specific brain regions by nuclease protection assays indicated that both IGFs are synthesized throughout the CNS of adult rats including the hypothalamus but are differentially expressed between brain regions[33]. More recently, IGF-II mRNA has been localized by in situ hybridization to the choroid plexus and leptomeninges of the brain[34,35]. These findings strongly suggest that the choroid plexus is a primary site of synthesis of IGF-II and a probable source of IGF-II in cerebrospinal fluid. The detection of IGF-II mRNA in specific brain regions, then, may be the result of this expression. The in situ hybridization results, however, do not rule out the existence of low levels of IGF-II expression in neurons and/or glia in the brain.

Although detectable by nuclease protection assays, IGF-I mRNA has not been detected by in situ hybridizations (unpublished data). This indicates that its low level of expression in the adult brain is not due to its presence in only a few cell types but rather its presence at low levels in many cell types. To the extent that in vitro cultures reflect normal in vivo synthesis, the observations of IGF-I mRNA in rat brain neuronal and glial cultures are consistent with low level IGF-I expression in multiple cell types in the brain[18,33,36]

TECHNIQUES FOR INVESTIGATING mRNA REGULATION

The precise physiological roles for either IGF-I or IGF-II in the brain are unknown. The identification of IGF-I mRNA in the hypothalamus and of high levels of IGF-II production by the choroid plexus and leptomeninges suggest possible roles in pituitary GH regulation operating at the level of either the pituitary or the hypothalamus. One approach to investigate the role of the IGFs in the brain is to study the regulation of their synthesis. If either IGF-I or IGF-II are involved in the regulation of pituitary GH for example, then their production should be altered by the removal of GH. This regulation could occur at many levels including transcriptional controls, mRNA processing and stability, mRNA transport from nucleus to cytoplasm, translational controls and post-translational modifications. Many peptides studied to date are regulated at least partially at the level of transcription. Additionally, IGF-I mRNAs in peripheral tissues have been shown to respond to GH[9,10,13]. In the liver, this GH dependence is at the level of gene transcription[37]. The measurement of steady-state RNA levels is a standard mechanism to investigate regulation of mRNA and reflects transcriptional regulation as well as changes in mRNA stability, processing or transport.

The two most commonly used techniques for investigating steady-state RNA levels are Northern or RNA blot hybridizations and nuclease protection assays. The primary difference between the two techniques is whether the hybridization of the radiolabeled probe for the sequence of interest is hybridized to the tissue RNA in solution or after the RNA has been bound to a membrane. Hybridization in solution for nuclease protection assays is more efficient and thus gives greater sensitivity. This is useful when isolating RNA from small pieces of tissue where the yields are low or when the RNA species of interest is relatively rare. For accurate quantitation using either of these procedures, it is important to express the number of hybridized CPMs in the unknown samples into a mass amount by comparison to a standard curve with known amounts of unlabeled "sense" RNA corresponding to the labeled antisense probe[38].

REGULATION OF BRAIN IGF-I AND IGF-II mRNAs

Multiple RNA transcripts have been identified for both the IGF-I and IGF-II genes. Rat liver IGF-I mRNAs (estimated size classes of 0.8-1.2kb, 1.5-1.9kb and 7.0-7.5kb) each are comprised of multiple closely migrating IGF-I mRNA species containing alternate 5'-sequences and either IA or IB type E domain coding sequences[10,39-41]. These multiple transcripts are detected in other tissues as well including the brain. Similarly, multiple transcripts (from 4.6 to 1.0kb) have been described for the single-copy IGF-II gene. In both rats and humans, part of this heterogeneity is generated by the presence of two different promoters, each transcribing alternative 5'-noncoding regions which are spliced to common coding exons[42-46].

Studies of the regulation of IGF-I and IGF-II mRNAs by Northern blot hybridization or by nuclease protection assays using probes specific to the various RNA species indicate that these multiple RNA transcripts are regulated in a developmental and tissue-specific manner. The size classes of mRNAs for either IGF-I or IGF-II have not been characterized completely; thus the best method to determine differential regulation of these species is by Northern blot hybridizations. This is difficult for the investigation of IGF-I RNAs in specific brain regions however due to its low abundance in the adult brain. We have found that nuclease protection assays are extremely sensitive for the detection of small amounts of RNA and will give information on the overall regulation of IGF-I mRNAs in adult brain regions.

Initial evidence for hormonal feedback regulation of brain IGF-I and IGF-II mRNAs came from studies by Hynes et al.[13]. Both IGF-I and IGF-II mRNA levels decreased to 25% and 20% of sham levels respectively in whole rat brain following hypophysectomy (hypox). In addition, intracerebroventricular injection of human GH into hypox rats increased whole

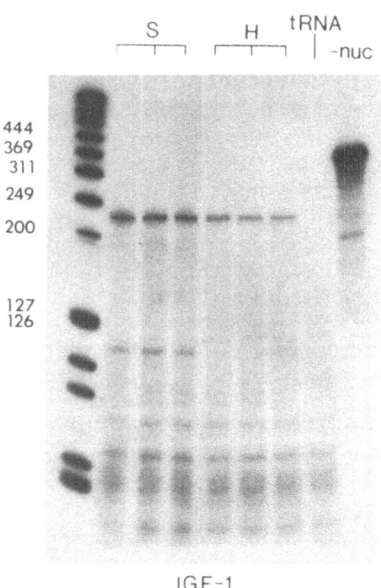

IGF-1

Fig. 1. Nuclease protection assay showing IGF-I mRNA from adult male rat hypothalami 15 days after hypox (H) or sham operation (S). 5ug of total RNA from individual hypothalami were hybridized to a ^{32}P-labeled antisense riboprobe transcribed from a rat IGF-I cDNA. The protected fragments were treated with RNase A and T$_1$, separated by polyacrylamide gel electrophoresis and visualized by autoradiography.

brain IGF-I mRNAs to 80% of normal levels. In contrast, IGF-II mRNAs from whole brain did not change significantly in response to ip injection of hGH and increased to only 50% of control values as the result of icv injection of hGH. The results of these studies in whole brain indicate that GH may be a major regulator of IGF-I but not IGF-II mRNA levels.

As discussed previously, brain IGFs may be regulated by several different hormones including GH, glucocorticoids and TH. This may depend on their location and function in a particular cell type in the brain. Investigating the regulation of IGFs in whole brain therefore may not accurately reflect the regulation of IGFs in specific regions or cell types. Since the known GH-modulators GRF and SRIF, as well as the release and inhibiting factors for other pituitary hormones, are located in specific neurons of the hypothalamus, we specifically investigated whether hormonal manipulations regulated hypothalamic IGF-I and IGF-II mRNAs. Using nuclease protection assays, we determined that hypothalamic IGF-I mRNA levels declined progressively from 58% to 38% of sham levels by 8 and 15 days after hypox in male rats (Figure 1). IGF-II mRNA levels, determined by Northern blot hybridizations, declined to 38% of sham values by 8 days and remained at this level 15 days after hypox (Figure 2). Thus, the decreases in hypothalamic IGF mRNAs paralleled the decreases observed in whole brain. Further, hypothalamic IGF-I mRNA levels were restored in hypox rats given chronic GH replacement but were not restored in hypox rats given T4, corticosterone and testosterone without GH (unpublished data). In contrast, there was no effect of any of these replacement paradigms on hypothalamic IGF-II mRNA levels.

The finding that IGF-I mRNA levels in hypox rat hypothalami are restored by physiological levels of growth hormone given sub-cutaneously suggests that hypothalamic IGF-I regulates pituitary GH in vivo. Hypothalamic IGF-I may

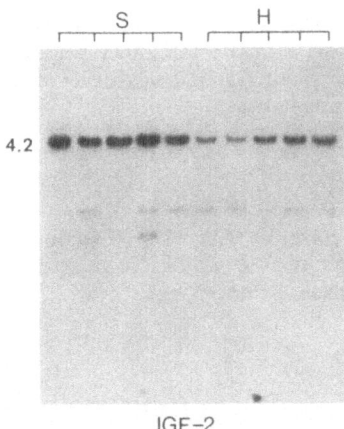

IGF-2

Fig. 2. Northern blot hybridization showing IGF-II mRNA from adult male rat hypothalami 15 days after hypox (H) or sham operation (S). Each lane represents RNA from individual hypothalami. Following electrophoresis, RNAs were transferred to a nylon membrane (Biodyne, ICN) and hybridized to a ^{32}P-labeled DNA probe random primed from a rat IGF-II cDNA insert.

be released from terminals in the median eminence similar to GRF and SRIF and exert its effects directly on the pituitary somatotrophs. Alternatively, hypothalamic IGF-I may act to alter the levels of GRF or SRIF in hypophysiotropic neurons.

In contrast, since hypothalamic IGF-II mRNA levels decrease following hypox and cannot be restored by GH, corticosterone, TH or testosterone replacement, it appears that that there is a yet unidentified pituitary hormone or pituitary target hormone which is responsible for the maintenance of normal IGF-II mRNA levels in the hypothalamic region. It is clear from in situ hybridization results that the IGF-II mRNA in our hypothalamic sections is derived primarily from the leptomeninges (unpublished data). These results do not rule out the possibility, however, that IGF-II is expressed in hypothalamic neurons or glia although at a lower level than in the choroid plexus and leptomeninges. If so, its function and regulation in those cells may be entirely different from that in the choroid plexus and leptomeninges of the brain. Finally, the IGFs may act as localized growth factors in the brain both during development and in the adult: 1) nanomolar levels of IGF-I or insulin are able to induce adrenergic expression in cultures of dorsal root ganglia from embryonic quail and to stimulate the development of oligodendrocytes in culture[47-49], 2) IGF-I and IGF-II stimulate DNA synthesis in fetal rat brain cultures[50,51], 3) insulin and IGF-II enhance neurite outgrowth in cultures of human neuroblastoma cells[52], and 4) immunoreactive IGF-I and IGF-I mRNA are elevated in the rat cerebellum during a limited time of cerebellar differentiation[53].

In conclusion, it is likely that the IGFs have multiple functions in the brain depending on their temporal expression as well as their location in specific cell types. In the developing brain, both IGF-I and IGF-II may act as localized growth factors for neuronal and glial differentiation. In the adult brain, IGF-I may function in the hypothalamus as a modulator of pituitary GH either directly as a hypophysiotropin, or indirectly by regulating the activity of peptidergic neurons, and in other brain regions as a neurohormone or growth factor. The identity of the IGF-II receptor with the mannose-6-phosphate receptor, suggests that IGF-II may function in lysozomal enzyme targetting which could be important in restructuring and plasticity in the adult brain[54]. Further studies on the regulation of brain IGFs, the localization of the individual cells producing IGFs in the CNS, as well as similar studies on the IGF receptors and binding proteins will provide further clues to their functions in both the developing and adult brain.

ACKNOWLEDGEMENTS

The authors gratefully acknowledge Dr. Charles Roberts and Dr. Derek LeRoith for providing the rat IGF-I cDNA clone and Dr. Mathew Rechler for providing the rat IGF-II cDNA clone.

REFERENCES

1. W. H. Daughaday, K. Hall, M. S. Raben, W. D. Salmon, Jr., J. L. van der Brande, and J. J. van Wyk, Somatomedin: Proposed designation for sulphation factor, Nature 235:107 (1972).
2. E. Schoenle, J. Zapf, R. E. Humbel, and E. R. Froesch, Insulin-like growth factor I stimulates growth in hypophysectomized rats, Nature 296:252 (1982).
3. C. D. Scott, J. L. Martin, and R. C. Baxter, Rat hepatocyte insulin-like growth factor-I and binding protein: Effect of growth hormone in vitro and in vivo, Endocrinology 116:1102 (1985).

4. D. R. Clemmons, J. J. van Wyk, E. C. Ridgway, B. Kliman, R. N. Kjellberg, and L.E. Underwood, Evaluation of acromegaly by radioimmunoassay of somatomedin-C, New Engl. J. Med. 301:1138 (1979).

5. R. C. Baxter, The somatomedins: Insulin-like growth factors, Adv. Clin. Chem. 25:49 (1986).

6. G. Oppizzi, M. M. Petroncini, D. Dallabonzana, R. Cozzi, G. Verde, P. G. Chiodini, and A. Liuzzi, Relationship between somatomedin-C and growth hormone levels in acromegaly: Basal and dynamic evaluation, J. Clin. Endocrin. Metab. 63:1348 (1986).

7. J. C. Schwander, C. Hauri, J. Zapf, and E. R. Froesch, Synthesis and secretion of insulin-like growth factor and its binding protein by the perfused rat liver: Dependence on growth hormone status, Endocrinology 113:297 (1983).

8. L. J. Murphy, G. I. Bell, and H. G. Friesen, Tissue distribution of insulin-like growth factor I and II messenger ribonucleic acid in the adult rat, Endocrinology 120:1279 (1987).

9. L. J. Murphy, G. I. Bell, M. L. Duckworth, and H. G. Friesen, Identification, characterization, and regulation of a rat complementary deoxyribonucleic acid which encodes insulin-like growth factor-I, Endocrinology 121:684 (1987).

10. C. T. Roberts, Jr., S. R. Lasky, W. L. Lowe, Jr., W. T. Seaman, and D. LeRoith, Molecular cloning of rat insulin-like growth factor I complementary deoxyribonucleic acids: differential messenger ribonucleic acid processing and regulation by growth hormone in extrahepatic tissues, Mol. Endocrinol. 1:243 (1987).

11. A. J. D'Ercole, A. D. Stiles, and L. E. Underwood, Tissue concentrations of somatomedin C: Further evidence for multiple sites of synthesis and paracrine or autocrine mechanisms of action, Proc. Natl. Acad. Sci. USA 81:935 (1984).

12. C. C. Orlowski and S. D. Chernausek, Discordance of serum and tissue somatomedin levels in growth hormone-stimulated growth in the rat, Endocrinology 123:44 (1988).

13. M. A. Hynes, J. J. van Wyk, P. J. Brooks, A. J. D'ercole, M. Jansen, and P. K. Lund, Growth hormone dependence of somatomedin-C/insulin-like growth factor-I and insulin-like growth factor-II messenger ribonucleic acids, Mol. Endocrinol. 1:233 (1987).

14. J. Zapf, H. Walters, and E. R. Froesch, Radioimmunological determination of insulinlike growth factors I and II in normal subjects and in patients with growth disorders and extra-pancreatic tumor hypoglycemia, J. Clin. Invest. 68:1321 (1981)

15. S. O. Adams, S. P. Nissley, M. Kasuga, T. P. Foley, and M. M. Rechler, Receptors for insulin-like growth factors and growth effects of multiplication-stimulating activity (rat IGF-II) in rat embryo fibroblasts, Endocrinology 112:971 (1983).

16. H. D. Mosier, Jr., E. M. Spencer, L. C. Dearden, and R. A. Jansons, The effect of glucocorticoids on plasma insulin-like growth factor I concentration in the rat fetus, Pediatr. Res. 22:92 (1987).

17. K. Pavelic, D. Vrbanec, S. Marusic, S. Levanat, and T. Cabrijan, Autocrine tumor growth regulation by somatomedin C: An in vitro model, J. Endocrinol. 109:233 (1986).

18. M. Adamo, H. Werner, W. Farnsworth, C. T. Roberts, Jr., M. Raizada, and D. LeRoith, Dexamethasone reduces steady state insulin-like growth factor I messenger ribonucleic acid levels in rat neuronal and glial cells in primary culture, Endocrinology 123:2565 (1988).

19. H. Katakami, T. R. Downs, and L. A. Frohman, Decreased hypothalamic growth hormone-releasing hormone content and pituitary responsiveness in hypothyroidism, J. Clin. Invest. 77:1704 (1986).

20. R. Valcavi, C. Dieguez, M. Preece, A. Taylor, I. Portioli, and M. F. Scanlon, Effect of thyroxine replacement therapy on plasma insulin-like growth factor 1 levels and growth hormone responses to growth hormone releasing factor in hypothyroid patients, Clin. Endocrinol. 27:85 (1987).

21. M. Binoux, A. Faivre-Bauman, C. Lassarre, A. Barret, and A. Tixier-Vidal, Triiodothyronine stimulates the production of insulin-like growth factor (IGF) by fetal hypothalamus cells cultured in serum-free medium, Dev. Brain Res. 21:319 (1985).

22. C.-J. Hsu and J. M. Hammond, Gonadotropins and estradiol stimulate immunoreactive insulin-like growth factor-I production by porcine granulosa cells in vitro, Endocrinology 120:198 (1987).

23. V. K. M. Han, J. M. Lauder, and A. J. D'Ercole, Characterization of somatomedin/insulin-like growth factor receptors and correlation with biologic action in cultured neonatal rat astroglial cells, J. Neurosci. 7:501 (1987).

24. N. J. Bohannon, E. S. Corp, B. J. Wilcox, D. P. Figlewicz, D. M. Dorsa, and D. G. Baskin, Localization of binding sites for insulin-like growth factor-I (IGF-I) in the rat brain by quantitative autoradiography, Brain Res. 444:205 (1988).

25. M. Smith, J. Clemens, G. A. Kerchner, and L. G. Mendelsohn, The insulin-like growth factor-II (IGF-II) receptor of rat brain: regional distribution visualized by autoradiography, Brain Res. 445:241 (1988).

26. H. Abe, M. E. Molitch, J. J. van Wyk, and L. E. Underwood, Human growth hormone and somatomedin C suppress the spontaneous release of growth hormone in unanesthetized rats, Endocrinology 113:1319 (1983).

27. G. S. Tannenbaum, H. J. Guyda, and B. I. Posner, Insulin-like growth factors: a role in growth hormone negative feedback and body weight regulation via brain, Science 220:77 (1983).

28. M. Berelowitz, S. L. Firestone, and L. A. Frohman, Effects of growth hormone excess and deficiency on hypothalamic somatostatin content and release and on tissue somatostatin distribution, Endocrinology 109:714 (1981).

29. S. Yamashita and S. Melmed, Insulin-like growth factor I action on rat anterior pituitary cells: Suppression of growth hormone secretion and messenger ribonucleic acid levels, Endocrinology 118:176 (1986).

30. G. P. Ceda, R. G. Davis, R. G. Rosenfeld, and A. R. Hoffman, The growth hormone (GH)-releasing hormone (GHRH)-GH-somatomedin axis: Evidence for rapid inhibition of GHRH-elicited GH release by insulin-like growth factors I and II, Endocrinology 120:1658 (1987).

31. T. Noguchi, L. M. Kurata, and T. Sugisaki, Presence of a somatomedin-C-immunoreactive substance in the central nervous system: Immunohistochemical mapping studies, Neuroendocrinology 46:277 (1987).

32. G. K. Haselbacher, M. E. Schwab, A. Pasi, and R. E. Humbel, Insulin-like growth factor II (IGF II) in human brain: Regional distribution of IGF II and of higher molecular mass forms, Proc. Natl. Acad. Sci. USA 82:2153 (1985).

33. P. Rotwein, S. K. Bergess, J. D. Milbrandt, and J. E. Krause, Differential expression of insulin-like growth factor genes in rat central nervous system, Proc. Natl. Acad. Sci. USA 85:265 (1988).

34. F. Stylianopoulou, J. Herbert, M. B. Soares, and A. Efstratiadis, Expression of the insulin-like growth factor II gene in the choroid plexus and the leptomeninges of the adult rat central nervous system, Proc. Natl. Acad. Sci. USA 85:141 (1988).

35. M. A. Hynes, P. J. Brooks, J. J. van Wyk, and P. K. Lund, Insulin-like growth factor II messenger ribonucleic acids are synthesized in the choroid plexus of the rat brain, Mol. Endocrinol. 2:47 (1988).

36. R. Ballotti, F. C. Nielson, N. Pringle, A. Kowalski, W. D. Richardson, E. van Obberghen, and S. Gammeltoft, IGF-I in cultured rat astrocytes, expression of the gene and receptor tyrosine kinase, EMBO J. 6:3633 (1987).

37. L. S. Mathews, G. Norstedt, and R. D. Palmiter, Regulation of insulin-like growth factor I gene expression by growth hormone, Proc. Natl. Acad. Sci. USA 83:9343 (1986).

38. M. Blum, Regulation of neuroendocrine peptide gene expression, Meth. Enzymol. 168:3 (1988).

39. W. L. Lowe, Jr., C. T. Roberts, Jr., S. R. Lasky, and D. LeRoith, Differential expression of alternative 5' untranslated regions in mRNAs encoding rat insulin-like growth factor I, Proc. Natl. Acad. Sci. USA 84:8946 (1987).

40. W. L. Lowe, Jr., S. R. Lasky, D. LeRoith, and C. T. Roberts, Jr., Distribution and regulation of rat insulin-like growth factor I messenger ribonucleic acids encoding alternative carboxyterminal E-peptides: Evidence for differential processing and regulation in liver, Mol. Endocrinol. 2: 528 (1988).

41. E. C. Hoyt, J. J. van Wyk, and P. K. Lund, Tissue and development specific regulation of a complex family of rat insulin-like growth factor I messenger ribonucleic acids, Mol. Endocrinol. 2:1077 (1988).

42. M. B. Soares, A. Turken, D. Ishii, L. Mills, L. V. Episkopou, S. Cotter, S. Zeitlin and A. Efstratiadis, Rat insulin-like growth factor II gene, J. Mol. Biol. 192, 737 (1986).

43. A. Gray, A. W. Tam, T. J. Dull, J. Hayflick, J. Pintar, W. K. Cavenee, A. Koufos, and A. Ullrich, Tissue-specific and developmentally regulated transcription of the insulin-like growth factor 2 gene, DNA 6:283 (1987).

44. J.-C. Irminger, K. M. Rosen, R. E. Humbel, and L. Villa-Komaroff, Tissue-specific expression of insulin-like growth factor II mRNAs with distinct 5' untranslated regions, Proc. Natl. Acad. Sci. USA 84:6330 (1987).

45. P. N. Schofield and V. E. Tate, Regulation of human IGF-II transcription in fetal and adult tissues, Development 101:793 (1987).

46. L. Chiariotti, A. L. Brown, R. Frunzio, D. R. Clemmons, M. M. Rechler, and C. B. Bruni, Structure of the rat insulin-like growth factor II transcriptional unit: Heterogeneous transcripts are generated from two promoters by use of multiple polyadenylation sites and differential ribonucleic acid splicing, Mol. Endocrinol. 2:1115 (1988).

47. F. A. McMorris, T. M. Smith, S. DeSalvo, and R. W. Furlanetto, Insulin-like growth factor I/somatomedin C: A potent inducer of oligodendrocyte development, Proc. Natl. Acad. Sci. USA 83:822 (1986).

48. R. H. M. van der Pal, J. W. Koper, L. M. G. van Golde, and M. Lopes-Cardozo, Effects of insulin and insulin-like growth factor (IGF-I) on oligodendrocyte-enriched glial cultures, J. Neurosci. Res. 19:483 (1988).

49. Z. G. Xue, N. M. Le Douarin, and J. Smith, Insulin and insulin-like growth factor-I can trigger the differentiation of catecholaminergic precursors in cultures of dorsal root ganglia, Cell Diff. Dev. 25:1 (1988).

50. D. Lenoir and P. Honegger, Insulin-like growth factor I (IGF I) stimulates DNA synthesis in fetal rat brain cell cultures, Dev. Brain Res. 7:205 (1983).

51. E. DiCicco-Bloom and I. B. Black, Insulin growth factors regulate the mitotic cycle in cultured rat sympathetic neuroblasts, Proc. Natl. Acad. Sci. USA 85:4066 (1988).

52. E. Recio-Pinto and D. Ishi, Effects of insulin, insulin-like growth factor-II and nerve growth factor on neurite outgrowth in cultured human neuroblastoma cells, Brain Res. 302:323 (1984).

53. I. K. Andersson, D. Edwall, G. Norstedt, B. Rozell, A. Slottner and H.-A. Hansson, Differing expression of insulin-like growth factor I in the developing and in the adult rat cerebellum, Acta Physiol Scand 132: 167 (1988).

54. D. O. Morgan, J. C. Edman, D. N. Standring, V. A. Fried, M. C. Smith, R. A. Roth, and W. J. Rutter, Insulin-like growth factor II receptor as a multifunctional binding protein, Nature 329:301 (1987).

EXPRESSION OF THE IGF-II GENE IN BRAIN AND MUSCLE

Kenneth M. Rosen[1], Bruce M. Wentworth[2],
Edward D. Lamperti[1,3], Stanislaus Kinota[1],
Richard O'Brien[1], Nadia Rosenthal[2], Bruce Yankner[1]
and Lydia Villa-Komaroff[1]

[1]Department of Neurology, The Children's Hospital and Harvard
Medical School, 300 Longwood Ave., Boston, MA 02115
[2]Department of Biochemistry, Boston University Medical School
and [3]Department of Cell Biology, University of Massachusetts
Medical School

INTRODUCTION

The insulin-like growth factors (IGFs) have been implicated in many
processes related to growth and development. Despite a growing interest
by many investigators into the biological role of the insulin-like growth
factors, their function remains poorly understood. Advances in molecular
biological techniques have allowed for the characterization of the genes for
these factors, and provided tools with which to study their expression.
Using these tools, investigators have been elucidating the pattern of IGF
gene expression in various tissues throughout a wide range of species. We
have examined the expression of insulin-like growth factor II (IGF-II)
during development in the nervous system and in muscle. IGF-II gene
expression is developmentally regulated in both sites, suggesting that it
plays an important role during the developmental program.

IGF-II is a member of the family of peptides called somatomedins, or
insulin-like growth factors, which are related to insulin by structure and
function (for review see Blundell & Humbel, 1980; Froesch, et al., 1985;
Phillips & Vassilipolou-Sellin, 1980). IGF-I and IGF-II are 62% homologous
to each other at the amino acid level (Rinderknecht & Humbel, 1978). The
A and B chains of these two peptides have about 50% homology to the
corresponding chains of insulin. Both IGF-I and II are found in human
serum bound to carrier proteins and these forms are probably synthesized in
the liver (Acquaviva, et al., 1982). Both peptides are potent mitogens for
cultured cells (Froesch, et al., 1985) and both peptides, in the unbound
form, can mimic the metabolic functions of insulin (Froesch, et al., 1985).
Despite their similarities, evidence has begun to accumulate that these two
peptides have distinct and different roles.

There is now general agreement that IGF-I mediates the growth effects
of growth hormone in childhood and adolescence, but the role of IGF-II in
growth and development has not been determined, although there are
several indirect lines of evidence which implicate IGF-II as a growth factor

during fetal development. IGF-II can be synthesized by a number of fetal tissues (D'Ercole, et al., 1980), and levels of IGF-II are high in fetal serum and decline postnatally (Daughaday, et al., 1982; Moses, et al., 1980; Wilson, et al., 1982). Rat fibroblasts in culture mimic this developmental switch. The synthesis of IGF-II by cultured rat embryo fibroblasts increases in response to placental lactogen but not to growth hormone. Both growth hormone and placental lactogen stimulate IGF-I synthesis in fibroblasts from older rats but have no effect on IGF-II synthesis (Adams, et al., 1983). However, it should be noted that human fetal fibroblasts produce 100 to 1000 fold less IGF-II than rat fibroblasts (Gaynes, et al., 1985). Since IGF-II continues to be synthesized by some adult tissues, including the brain, and is present at four times the level of IGF-I in human adult serum, (Haselbacher, et al., 1985), and since IGF-II is a less potent mitogen than IGF-I in several assays (Froesch, et al., 1985) it is likely that IGF-II serves a function both in development and in the adult.

The role of peptide growth factors in the regulation of the development and differentiation of various tissues is poorly understood at present, despite a general consensus as to their importance. In most studies, *in vitro* systems have been used to identify and assay factors that affect the survival or differentiation of specific cell types. Nerve growth factor (NGF) is the best characterized of these factors and provides a paradigm for how such factors can be studied (for review see Calisano, et al., 1984; Berg, 1984). The characterization of factors other than NGF has been difficult because sources rich in the factors, analogous to the submaxillary gland of mouse in the case of NGF, have not been found, making it difficult to obtain large enough amounts of peptides pure enough for characterization. Nevertheless, a number of factors which influence cellular growth and differentiation have been identified and partially characterized. The role such factors play during the embryonic period is difficult to assess because of continuously evolving temporal changes in cell number, type, and position. The advent of techniques which allow the isolation of specific genes and the introduction of these genes into cultured cells or intact organisms promises to provide greater insights into the process of development.

IGF-II Expression in the Nervous System

The biological role of IGF-II in the central nervous system is currently unknown. Two very preliminary studies have demonstrated the existence of abnormal levels of IGF-II in cerebral spinal fluid and serum in two neuropathologic conditions, raising additional questions as to the function of the peptide. In children with Down's syndrome, the serum concentration of IGF-II is at adult levels and the concentration of IGF-I does not rise to normal (Sara, et al., 1983a; Anneren, et al., 1984). In the single case examined, levels of IGF-II were below the level of detectability in a fetus with Down's syndrome, while normal fetuses have high, easily detectable levels of IGF-II (Sara, et al., 1981). These results imply that the developmental expression of IGF-II is impaired in children with Down's syndrome. This is obviously not a direct effect on the IGF-II gene, since IGF-II has been mapped to chromosome 11 (Tricoli, et al., 1984) and Down's syndrome is a trisomy of chromosome 21. In individuals with Alzheimer's disease, serum levels of IGF-II were also elevated relative to normal adults. Levels of IGF-II in cerebral spinal fluid were measured in two individuals with Alzheimer's and were also elevated (Sara, et al., 1982). It must be pointed out that measurements of IGF-II in cerebral spinal fluid or serum may not reflect the amount of active peptide. Tissues may synthesize the peptide as an autocrine or paracrine factor, and peptide bound to carrier protein may not have the same activity as peptide produced locally and not bound to carrier protein.

The most dramatic evidence that altered levels of IGF-II might have a deleterious effect is the finding of very high levels of IGF-II in a case of megalencephaly (Schoenle et al., 1985). In this case IGF-II levels were investigated in a case where the head circumference at birth was very enlarged (40.5 cm). The child died at 19 weeks with a head circumference of 51.5 cm. At autopsy, the brain weight was 1450g; the normal weight is about 567g. IGF-I and II levels in serum were within the normal range, but IGF-II levels in cerebral spinal fluid were about twice the IGF-II levels in a pool of CSF from controls of the same age. IGF-II levels in the frontal cortex were 10 fold higher than levels found in controls of similar age.

Receptors which bind IGF-II, IGF-I and insulin are distributed throughout the CNS (Havrankova & Roth, 1978; Pacold & Blackard, 1979; van Houten et al., 1980; Sara et al., 1983b,c). The insulin receptor binds IGF-II and IGF-I as well as insulin. Two subtypes of IGF receptor have been characterized. Type I binds IGF-I better than IGF-II and also binds insulin. Type II binds IGF-II better than IGF-I and does not appear to bind insulin (Rechler et al, 1980; Nissley and Rechler, 1984). The type II receptor has been shown to be identical to the mannose-6-phosphate receptor (MacDonald et al, 1988). It is not clear if brain type I and insulin receptors are identical to those found in other tissues. IGF-II, like other growth factors, may either interfere with or potentiate the growth or differentiation of cells and the interactions between them. Several lines of evidence suggest that IGF-II plays a role in CNS development and maintenance. In general, high levels of insulin are required for the culture of neuronal cells in defined media. Because IGF-II has not been readily available, very few studies have been done to determine whether physiological levels of IGF-II will substitute for insulin. In a pure neuronal culture grown in defined serum-free medium, insulin was the only hormone found to enhance the growth of neurons obtained during their early proliferative stage (Aizenman et al, 1986). Physiological levels of IGF-II enhanced the survival of hippocampal and septal neurons in serum-free medium (Onifer et al, 1987).

In serum-free medium, SH-SY5Y human neuroblastoma cells lose the capacity to bind NGF and do not grow neurites when NGF is added to the medium. NGF binding and neurite outgrowth are regained when insulin or IGF-II are added to the serum-free medium. These peptides also have a direct effect on neurite outgrowth (Recio-Pinto et al, 1984). The levels of α and β tubulin mRNAs transiently increase in the SH-SY5Y cells in response to IGF-II and insulin (Mill et al, 1985). Actin mRNA levels do not increase in these cells in response to IGF-II or insulin. IGF-II and insulin also effect neurite outgrowth and neuronal survival in peripheral ganglion cell cultures from chick embryos (Bothwell, 1982; Recio-Pinto et al, 1986).

The 7.5 kDa form of IGF-II, as well as larger forms of the peptide are present both in human cerebral spinal fluid (Haselbacher and Humbel, 1982) and in several regions of the adult brain (Haselbacher, et al., 1985). In these studies IGF-I was not detected although isolation of a truncated form of IGF-I has been reported (Sara, et al., 1987). At least some of these larger forms are probably prohormones since the sequence of cDNA clones indicated that IGF-II is contained within a larger protein (Bell, et al., 1984; Whitfield, et al., 1984; Soares, et al., 1985; Stempien, et al., 1986). We have shown that the 7.5 kDa form found in brain is identical to the originally characterized major 7.5 kDa form found in plasma (Irminger, et al., 1987) (see below). Whether the larger proteins have unique biological properties or simply represent long-lived precursors to the small form is not known.

RESULTS

Genomic Clones Encoding Murine IGF-II

We have isolated recombinant phages containing the murine IGF-II gene. Recombinant libraries were constructed by inserting DNA which had been partially digested with Sau3A into the BamHI site of EMBL4 (Wentworth et al, 1986). Three overlapping phages were isolated that contain the IGF-II gene and span about 40 kb of DNA. A restriction map of the phages has been obtained and fragments containing exons corresponding to the cDNA have been identified. The phage containing the 5'-most region of the IGF-II gene overlaps with one of the previously isolated phages containing the murine insulin II gene.

Isolation of a cDNA from Brain Encoding Human IGF-II

We have isolated cDNAs encoding IGF-II from recombinant libraries made using human hypothalamus RNA as template for the synthesis of double-stranded cDNA (Irminger, et al., 1987). These results provide definitive evidence that IGF-II found in the brain is synthesized by cells in the brain and is not accumulated from the serum. The sequence of the clones shows that at least one species of the IGF-II found in the brain is identical to the major 7.5 kDa form found in serum; however, we find that the 5' untranslated sequence of the brain cDNA has no homology to the 5' untranslated sequence of the previously reported liver cDNAs. We and others have isolated RNA from a variety of tissues and find that IGF-II RNA varies in size and abundance in different tissues. A 6.0 kb transcript is present in placenta, hypothalamus, adrenal gland, kidney, Wilms' tumors and also in a pheochromocytoma which contains an additional 5.0 kb transcript. A probe specific to the 5' untranslated sequence of the brain cDNA hybridizes to the 6 kb transcripts, but neither to the liver 5.3 kb transcript nor to the pheochromocytoma 5.0 kb transcript. The lengths of the 5' and 3' untranslated sequence of human IGF-II RNAs have been deduced by utilizing the specificity of RNAse H. This experiment indicated that the human IGF-II RNAs differ only at the 5' end and have a common 3' end.

IGF-II Expression in Murine Fetal Brain

In preliminary experiments, no IGF-II RNA was detected in RNA isolated from whole brain of a first trimester human abortus. Because the mouse represents a far more convenient organism for developmental studies, we have begun characterization of expression of IGF-II in the developing mouse brain. Although the **timing** of brain development differs in mouse and in human, the general **pattern** of development is very similar in all mammals, so that stages of development can be meaningfully compared (Dobbing, 1975). Since we detected no IGF-II RNA in first trimester human brain, we began our study by examining murine brain RNA from stages of embryonic development which correspond to later stages of human brain development. RNA was isolated from murine brains at embryonic days 14 through 18 (E14-E18), P0 (neonates within 24 hours after birth), P2, P3, P4, P6, P8 and P10 (the gestational period for mice is 18 days). The level of IGF-II mRNA increases slightly between E14 and E15 and then remains constant through P0. The level of IGF-II then drops after birth but remains detectable through P8. In contrast, when we examine RNA from whole fetal bodies or from fetal heads, we find, as have others, that IGF-II RNA levels are high throughout embryonic development and fall abruptly at birth. We also examined RNA isolated from eyes, since the retina represents an easily accessible portion of the CNS with a well-defined anatomy, and the levels of IGF-II mRNA remain high through P4.

The major IGF-II transcripts in mouse are 3.75 kb and 1.15 kb in size, although a number of other transcripts can be visualized. There are also multiple transcripts in rats (Murphy et al, 1987). Our results indicate that the expression of IGF-II is regulated during development such that it is expressed at high levels during embryogenesis and then expression declines postnatally. This gene is regulated differently in different tissues: it is shut off almost completely in fibroblasts at birth, but continues to be expressed in brain and eye. Our results further indicate that IGF-II mRNA levels are highest **before** the central nervous system growth spurt, when the brain undergoes a rapid increase in size. This growth spurt occurs between 5 and 10 days after birth in mice and occurs just before birth in humans (Davison and Dobbing, 1968). Since it is during the growth spurt that glial cells undergo a rapid increase in number and differentiate into myelinating cells, and since IGF-II mRNA levels decline before this time, it is tempting to speculate that IGF-II might affect neuronal growth, migration or differentiation rather than the growth or differentiation of glial cells. On the other hand, IGF-II might affect both neuronal cells and early progenitors of the glial cells.

In human adult brain, the anterior pituitary has the largest amount of immunoreactive 7.5 kDa IGF-II, with 6 times more than any other region and 50 to 100 times more than pons or cerebellum (Haselbacher et al., 1985). To get an idea of the sites of synthesis of IGF-II in the murine brain, we dissected brains from adult (P60) mice into cerebrum, cerebellum, midbrain, pons and medulla, thalamus and hypothalamus and prepared RNA using guanidine thiocyanate. We cloned the sense strand of the 307 bp Pvu II-Bal I fragment from the human brain cDNA (Irminger et al, 1987) into M13 to generate single-stranded probe of high specific activity. Ten μg of RNA from each region was examined by northern blot analysis. No bands were detected after a 70 hour exposure at -70° with a screen. We then analyzed 5 μg of RNA from each region of adult brain and from E18 brain by S1 nuclease protection assay. Hypothalamus contained about 50 fold less protected material than E18 brain and about 2 fold more material than thalamus. Midbrain contained about as much material as thalamus. Cerebrum and cerebellum each contained about 1/5 the material found in thalamus. Pituitary contained the least amount of material – 5 to 10 fold less than cerebrum or 1/50 to 1/100 the amount in hypothalamus. These results indicate that the pituitary either stores the IGF-II it synthesizes or receives it from some other source. Alternatively, the distribution of immunoreactive IGF-II might be different in mouse and human (Rosen and Villa-Komaroff, 1987). In rat brain, the expression of IGF-II mRNA is somewhat different (Rotwein et al, 1988): it is highest in pons, medulla and cerebellum, while hypothalamus contains about 0.4 of the amount in these regions. Furthermore, IGF-II sequences were found in glial RNA but not in neuronal RNA prepared from primary cultures of E17 rat brain. By contrast, *in situ* hybridization indicated that IGF-II mRNA is expressed in epithelium of the choroid plexus and not in either glial or neuronal cells of adult rat (Stylianopoulou et al, 1988). However, immunoreactive IGF-II can be found in neurons in the ventral hypothalamus, dorsal to the optic tract, with dense staining of fiber bundles in the arcuate nucleus and retrochiasmatic areas of the hypothalamus (Lauterio et al, 1987). We are currently undertaking experiments to definitively determine if the IGF-II gene is active in neuronal cell types in the developing brain.

Expression of IGF-II in Muscle

Myogenesis is controlled by the interaction of numerous factors, the program ultimately overseen by the regulated expression of many genes (Gunning et al., 1987; Blau, 1988). Compounds ranging from peptide growth factors to anabolic steroids have been implicated as modulators of muscle growth and development (Florini, 1987). IGF-II mRNA is present in large amounts in mammalian embryonic smooth, skeletal, and cardiac muscle (Scott, et al., 1985). The level of IGF-II mRNA declines postnatally, becoming undetectable in adult muscle (Soares et al., 1985,1986; Brown et al., 1986; Gray et al., 1987). Ishii (1987) has reported that the decline in IGF-II mRNA levels parallels the time course for the elimination of superfluous synapses, and the selection of the ultimate muscle endplate. It has been known for some time that media conditioned by muscle cells or myogenic cell lines contains factors which influence both neuronal survival and neurite outgrowth (Heaton and Kemperman, 1987). Perhaps IGF-II plays a role in the interaction between muscle and nerve.

Cultured myogenic cell lines have been an important tool in defining the steps involved in myogenesis, and in identifying factors that regulate the process. The rat skeletal muscle cell line L_6, and subclones derived from it, has been used extensively for this purpose (Yaffe, 1968). Using L_6 derivatives, several investigators have shown that IGF-I has a potent stimulatory effect upon the growth of myoblasts (Ewton and Florini, 1980, 1981; Ballard et al., 1986). At present, little is known about what role IGF-II plays in muscle either *in vivo* or *in vitro*.

We have examined the expression of IGF-II mRNA in mouse skeletal muscle by northern blot analysis. As is seen in other tissues, the gene is active during late embryonic and early postnatal development, but it is not expressed in adult muscle. The observation is in agreement with the results of other investigators who have been studying the expression of IGF-II in rat muscle (see above). To better understand the timing of IGF-II expression in muscle cells, we turned to the use of cultured myogenic cell lines and examined the expression of IGF-II in the mouse myogenic cell line C_2C_{12} (Blau et al, 1983). The C_2C_{12} cell line was derived from the adult mouse skeletal muscle cell line C_2 (Yaffe and Saxel, 1977). When grown in DMEM medium supplemented with 20% fetal calf serum and 0.5% chick embryo extract, the cells remain myoblasts. When the medium is changed to DMEM plus 2% horse serum, the cells differentiate and fuse to form multinucleated myotubes and express a group of muscle-specific genes (Blau et al, 1983; Chiu and Blau, 1984). By comparison with L_6 cells, which express embryonic isoforms of the contractile apparatus proteins, C_2C_{12} cells express adult isoforms of these proteins, indicating that this cell line may be representative of a different stage in muscle development. This line has been used to examine the events occurring during differentiation and to identify a muscle-specific enhancer (Donoghue et al., 1988).

We isolated cytoplasmic RNA from both myoblasts and myotubes and, after electrophoresis and transfer, hybridized to a murine IGF-II probe. We detected no IGF-II mRNA in myoblasts, but found high levels of IGF-II mRNA in myotubes. We examined RNA isolated at varying times after the addition of differentiation medium to myoblasts and find that expression of IGF-II occurs within one hour, making induction of IGF-II mRNA a very early event in the differentiation of these myoblasts. The expression of IGF-II precedes the expression of muscle specific sequences since mRNA for myosin light chain 1/3 appears 8 to 10 hours after transfer of the cells to differentiation medium. This is in stark contrast to the L_6E_9 cell line, which we found takes as long as 48 to 72 hours to activate IGF-II mRNA production. These results may indicate another important difference between C_2 derivatives and L_6 derivatives. The question that immediately arises is whether IGF-II expression is essential for the later expression of muscle specific genes in C_2C_{12} cells, or whether it has some other role.

SUMMARY

The regulation of the IGF-II gene is complex and interesting. It is expressed ubiquitously at high levels in fetal tissues, and it is expressed either not at all or at very low levels in most adult tissues. Both differential splicing and transcription initiation at different promoters occur. While there is tissue-specific expression of different transcripts in humans, there does not appear to be differential expression of any of the transcripts during development or in different tissues in the mouse, rather all of the transcripts appear to be regulated in a similar manner. It has been reported that at least one promoter in the rat IGF-II gene consists of "minimal" control elements (Evans et al, 1988) and the authors speculate that this gene does not contain tissue-specific or stage specific enhancers.

Expression of IGF-II in the central and peripheral nervous systems remains an area of great interest and intense investigation. The brain is one of the few tissues where expression of IGF-II persists beyond early postnatal times in both rodents and humans. Several investigators have shown that there is a very high level of expression of IGF-II mRNA in the choroid plexus and in meningeal cells. Experiments by others have also indicated that expression occurs in glial cell populations. Immunohisto-chemical localizations suggest that neuronal expression of IGF-II takes place, but no definitive data are yet available.

Muscle provides a useful, relatively well defined system in which to study the role of IGF-II and the pattern of expression of IGF-II in developing muscle is quite similar to that of other tissues. Large amounts of IGF-II transcripts are present during late embryonic development and the transcripts decrease to undetectable levels a short time after birth. The time course of disappearance of the transcripts parallels the cessation of polyneuronal innervation, and the ultimate selection of the motor endplate (Ishii, 1987). Experiments utilizing continuously cultured muscle cell lines show that IGF-II expression occurs in the myotube and not in its precursors, the myoblasts. The availability of a cell system where expression of the IGF-II gene is regulated is critical for the identification of controlling elements and trans-acting factors that act upon the gene. Our results indicate that the myogenic cell line C_2C_{12} may be such a model system since the expression of the IGF-II gene is dramatically induced very soon after the cells have been given the signal to differentiate. These data indicate the existence of an enhancer associated with the IGF-II gene or, at the very least, specific promoter associated positive or negative regulatory elements. We are continuing our studies on this cell line in the hope of identifying these elements.

Ultimately, we hope to be able to better define the exact role of IGF-II during development. Its dramatic pattern of expression suggests that it plays an important role in the developmental program. Until recently, the lack of both purified peptide and high affinity antibodies against the peptide have limited our capacity to examine the biological role of IGF-II. The recent increase in availability of these reagents promises to aid in our understanding of IGF-II, and its specific roles in biological processes. Work by many investigators on the peptide, its binding proteins, its receptor, and its gene will surely rapidly advance our knowledge of this enigmatic protein.

LITERATURE CITED

Acquaviva AM, Bruni CB, Nissley SP and Rechler MM, 1982, Cell-free synthesis of rat insulin-like growth factor II. Diabetes 31:656-658.
Adams SO, Nissley SP, Handwerger S and Rechler MM, 1983, Developmental patterns of insulin-like growth factor-I and II synthesis and regulation in rat fibroblasts. Nature 302:150-153.
Aizenman Y, Weichsel ME, and deVellis J, 1986, Changes in insulin and

transferrin requirements of pure brain neuronal cultures during embryonic development. Proc. Natl. Acad. Sci. USA 83:2263-2266.

Anneren G, Enberg G and Sara VR, 1984, The presence of normal levels of serum immunoreactive insulin-like growth factor 2 (IGF-2) in patients with Down's syndrome. Ups. J. Med. Sci. 89:274-278.

Ballard FJ, Read LC, Francis GL, Bagley CJ, and Wallace JC, 1986, Binding properties and biological potencies of insulin-like growth factors in L6 myoblasts. Biochem. J., 233:223-230.

Bell GI, Merryweather JP, Sanchez-Pescador R, Stempien MM, Priestly L, Scott J and Rall LB, 1984, Sequence of a cDNA clone encoding human preproinsulin-like growth factor genes: the insulin-like growth factor II and insulin genes are contiguous. Proc. Natl. Acad. Sci. USA 82:6450-6454.

Berg DK, 1984, New neuronal growth factors. Ann. Rev. Neuro. 7:149-170.

Blau HM, 1988, Hierarchies of regulatory genes may specify mammalian development. Cell, 53:673-674

Blau HM, Chiu C-P, and Webster C, 1983, Cytoplasmic activation of human nuclear genes in stable heterocaryons. Cell 32: 1171-1180

Blundell TL and Humbel RE, 1980, Hormone families: pancreatic hormones and homologous growth factors. Nature 287:781-787.

Bothwell M, 1982, Insulin and somatomedin MSA promote the nerve growth factor independent neurite formation by cultured chick dorsal root ganglionic sensory neurons. J. Neurosci. Res. 8:225-231.

Brown AL, Graham DE, Nissley SP, Hill DJ, Strain AJ, and Rechler MM, 1986, Developmental regulation of insulin-like growth factor II mRNA in different rat tissues. J. Biol. Chem. 261:13144-13150

Calisano P, Cattaneo A, Aloe L, and Levi-Montalcini R, 1984, The nerve growth factor (NGF), Hormonal Proteins and Peptides 12:1-56.

Chiu C-P and Blau HM, 1984, Reprogramming cell differentiation in the absence of DNA synthesis. Cell 37:879-887.

Daughaday WH, Parker KA, Borowsky S, Trivedi B, and Kapadia M, 1982, Measurement of somatomedin-related petides in fetal, neonatal, and maternal rat serum by insulin-like growth factor (IGF) I radioimmunoassay, IGF-II radioreceptor assay (RRA) and multiplication stimulating activity RRA after acid-ethanol extraction. Endocrinology 110:575-581.

Davison AN, Dobbing J, 1968, The Developing Brain, in Applied Neurochemistry ed AN Davison, J Dobbing, FA. Davis, pp 253-285.

D'Ercole AJ, Applewhite GT and Underwood LE, 1980, Evidence that somatomedin is synthesized by multiple tissues in the fetus. Devel. Biol. 75:313-328.

Dobbing J, 1975, Prenatal nutrition and neurological development. In Mental Retardation ed. NA Buchwald and MAB Brazier. Academic Press pp. 401-420.

Donoghue MJ, Ernst H, Wentworth BM, Nadal-Ginard B, and Rosenthal N, 1988, A muscle-specific enhancer is located at the 3' end of the myosin light-chain 1/3 gene locus. Genes and Development 2:1779-1790.

Evans T, DeChiara T, and Efstratiadis A, 1988, A promoter of the rat insulin-like growth factor II gene consists of minimal control elements. J. Mol. Biol. 199:61-81.

Ewton DZ and Florini JR, 1980, Relative effects of the somatomedins, multiplication-stimulating activity, and growth hormone on myoblasts and myotubes in culture. Endocrinology 106:577-583

Ewton DZ and Florini JR, 1981, Effects of somatomedins and insulin on myoblast differentiation in vitro. Dev. Biol., 86:31-39

Florini JR, 1987, Hormonal control of muscle growth. Muscle & Nerve 10:577-598.

Froesch ER, Schmid C, Schwander J and Zapf J, 1985, Actions of
   insulin-like growth factors. Ann. Rev. Physiol. 47:443-467.
Gaynes LA, Nissley SP, Greenstein LA and Lee L, 1985, Insulin-like growth
   factors (IGF) and IGF binding proteins in conditioned media from fetal
   and postnatal human fibroblasts. The Endocrine Society, Abstract 188.
Gray A, Tam AW, Dull TJ, Hayflick J, Pintar J, Cavenee WK, Koufos A, and
   Ullrich A, 1987, Tissue-specific and developmentally regulated
   transcription of the insulin-like growth factor 2 gene. DNA 6:283-295.
Gunning P, Hardeman E, Wade R, Ponte P, Bains W, Blau HM, and Kedes L,
   1987, Differential patterns of transcript accumulation during human
   myogenesis. Mol. Cell. Biol. 7:4100-4114.
Haselbacher G and Humbel RE, 1982, Evidence for two species of insulin-like
   growth factor II (IGF-II and "Big" IGF-II) in human spinal fluid.
   Endocrinology 110:1822-1824.
Haselbacher G, Schwab ME, Pasi A and Humbel RE, 1985, Insulin-like growth
   factor II (IGF-II) in human brain: regional distribution of IGF-II and of
   higher molecular mass forms. Proc. Natl. Acad. Sci. USA 82: 2153-2157.
Havrankova J and Roth J, 1978, Insulin receptors are widely distributed in
   the central nervous system of the rat. Nature 272:827-829.
Heaton MB and Kemperman H, 1987, The influence of muscle-conditioned
   media on chick embryo brainstem neurons in culture. J. Neurosci. Res.
   17:384-390
Irminger JC, Rosen KM, Humbel RE and Villa-Komaroff L, 1987, Tissue
   specific expression of IGF II mRNAs with distinct 5' untranslated
   regions. Proc. Natl. Acad. Sci. USA 84:6330-6334.
Ishii D, 1987, Insulin-like growth factor II gene expression in muscle:
   relationship to synapse elimination and nerve regeneration. Soc. for
   Neuroscience 13:1211.
Lauterio TJ, Marson L, Della-Fera MA, Baile CA, 1987, Insulin-like growth
   factor II (IGF-II) in rat brain: Distribution of IGF-II neurons and fibers
   and peptide concentrations changes with fed state. Soc. for
   Neuroscience 13:611.
MacDonald RG, Pfeffer SR, Coussens L, Tepper MA, Brocklebank CM, Mole
   JE, Anderson JK, Chen E, Czech MP, and Ullrich A, 1988, A single
   receptor binds both insulin-like growth factor II and
   mannose-6-phosphate. Science 239:1134-7.
Mill JF, Chao MW, and Ishii DN, 1985, Insulin, insulin-like growth factor
   and nerve growth factor effects on tubulin mRNA levels and neurite
   formation. Proc. Natl. Acad. Sci. USA 85:7126-7130.
Moses AC, Nissley SP, Short PA, Rechler MM, White RM, Knight AB and
   Higa OZ, 1980, Increased levels of multiplication-stimulating activity, an
   insulin-like growth factor, in fetal rat serum. Proc. Natl. Acad. Sci.
   USA 77:3649-3653.
Murphy LJ, Bell GI, and Friesen HG, 1987, Tissue distribution of insulin-like
   growth factor I and II messenger ribonucleic acid in the adult rat.
   Endocrinology 120:1279-82.
Nissley SP and Rechler M, 1984, Insulin-like growth factors: biosynthesis,
   receptors and carrier proteins. Hormonal Peptides and Hormones
   12:127-203.
Onifer SM, Faber SD, Murphy SH, Kaseda Y, and Lau WC, 1987, Effects of
   insulin-like growth factor II (IGF-II) on hippocampal and septal neurons
   maintained in tissue culture. Neuro. Sci. Abs. 13:1615.
Pacold ST and Blackard WG, 1979, Central nervous system insulin receptors
   in normal and diabetic rats. Endocrinology 105:1452-1457.
Phillips LS and Vassilipoulou-Sellin R, 1980, Somatomedins. New Engl. J.
   Med. 302:371-446.
Rechler MM, Zapf J, Nissley SP, Froesch ER, Moses AC, Podskalay JM,

Schilling EE and Humbel RE, 1980, Interactions of insulin-like growth factors I and II and multiplication-stimulating activity with receptors and serum carrier proteins. Endocrinology 107:1451-1459.

Recio-Pinto E, Lang FF and Ishii DN, 1984, Insulin and insulin-like growth factor II permit nerve growth factor binding and the neurite formation response in cultured human neuroblastoma cells. Proc. Natl. Acad. Sci. USA 81:2562-2566.

Recio-Pinto E, Rechler M and Ishii DN, 1986, Effects of insulin, insulin-like growth factor and nerve growth factor on neurite formation and survival in cultured sympathetic and sensory neurons. J. Neur. 6:1211-1219.

Rinderknecht E and Humbel RE, 1978, Primary structure of human insulin-like growth factor II, FEBS Letts. 89:283-286.

Rosen KM and Villa-Komaroff L, 1987, Distribution of insulin-like growth factor II mRNA in the central nervous system. Soc. Neuro. Abs. 13:1706.

Rotwein P, Burgess SK, Milbrandt JD, and Krause JE, 1988, Differential expression of insulin-like growth factor genes in rat central nervous system. Proc. Natl. Acad. Sci. USA 85:265-269.

Sara VR, Hall K, Enzell K, Gardner A, Morowski R and Wetterberg L, 1982, Somatomedins in aging and dementia disorders of Alzheimer type. Neurobiol. Aging 3:117-120.

Sara VR, Hall K, Rodeck CH and Wetterberg L, 1981, Human embryonic somatomedins. Proc. Natl. Acad. Sci. USA 78:3175-3179.

Sara VR, Gustavsson KH, Anneren G, Hall K and Wetterberg L, 1983a, Somatomedins in Downs syndrome. Biol. Psychiat. 18:803-811.

Sara VR, Hall K, Holtz HV, Humbel R, Sjogren B and Wetterberg L, 1983b, Evidence for the presence of specific receptors for insulin-like growth factors 1 (IGF-1) and 2 (IGF-2) and insulin throughout the adult human brain, Neurosci. Letts 34:39-44.

Sara VR, Hall K, Misaki M, Fryklund L, Christensen N and Wetterberg L, 1983c, Ontogenesis of somatomedin and insulin receptors in the human fetus, J. Clin. Invest. 71:1084-1094.

Sara VR, Carlsson-Skwirut C, Andersson C, Hall E, Sjoegren B, Homgren A, and Joernvall H, 1987, Characterization of somatomedins from human fetal brain: Identification of a variant form of insulin-like growth factor 1. Proc. Natl. Acad. Sci. USA 83:4904-4907.

Schoenle EJ, Haselbacher GK, Briner J, Janzer RC, Gammeltoft S, Humbel RE, and Prader A, 1986, Elevated concentration of IGF II in brain tissue from an infant with macrencephaly. J. Pediatr. 108:737-40.

Scott J, Cowell J, Robertson ME, Priestly LM, Wadey R, Hopkins B, Pritchard J, Bell GI, Rall LB, Graham CF, and Knott TJ, 1985, Insulin-like growth factor-II gene expression in Wilms' tumour and embryonic tissues. Nature 317:260-262.

Soares MB, Ishii DN and Efstratiadis AE, 1985, Developmental and tissue specific expression of a family of transcripts related to rat insulin-like growth factor II mRNA. Nucl. Acids Res. 13:1119-1134.

Soares MB, Turken A, Ishii D, Mills L, Episkopou V, Cotter S, Zeitlin S, and Efstratiadis A, 1986, Rat insulin-like growth factor II gene: A single gene with two promoters expressing a multitranscript family. J. Mol. Biol., 192:737-752

Stempien MM, Fong NM, Rall LB, and Bell GI, 1986, Sequence of a placental cDNA encoding the mouse insulin-like growth factor II precursor. DNA 5:357-361.

Stylianopoulou F, Herbert J, Soares MB, and Efstratiadis A, 1988, Expression of the insulin-like growth factor II gene in the choroid plexus and the leptomeninges of the adult rat central nervous system. Proc. Natl. Acad. Sci. USA 85:141-145.

Tricoli JV, Rall LB, Scott J, Bell GI and Shows TB, 1984, Localization of insulin-like growth factors to human chromosomes 11 and 12 Nature 310:784-786.

van Houton M, Posner BI, Kopriwa BM, and Brawer JR, 1980. Insulin binding sites localized to nerve terminals in rat median eminence and arcuate nucleus. Science 207:1081-1083.

Wentworth BM, Schaefer IM, Villa-Komaroff L, and Chirgwin JM. Characterization of the two nonallelic genes encoding mouse preproinsulin. J. Mol. Evol. 1986; 23:305-312.

Whitfield HJ, Bruni CB, Frunzio R, Terrell JE, Nissley SP and Rechler MM, 1984, Isolation of a cDNA clone encoding rat insulin-like growth factor-II precursor. Nature 312:277-280.

Wilson DM, Bennett A, Adamson GD, Nagashima RJ, Liu F, DeNatale ML, Hintz RL and Rosenfeld RG, 1982, Somatomedins in pregnancy: a cross sectional study of insulin-like growth factors I and II and somatomedin peptide content in normal human pregnancies. J. Clin. Endocrinol. Metab. 55:858-861.

Yaffe D, 1968, Retention of differentiation potentialities during prolonged cultivation of myogenic cells. Proc. Nat. Acad. Sci. USA 61:477-483.

Yaffe D and Saxel O, 1977, Serial passaging and differentiation of myogenic cells isolated from dystrophic mouse muscle. Nature 270:725-727

REGULATION OF THE NEURONAL INSULIN-LIKE PEPTIDE

S. Devaskar and H. F. Sadiq

Division of Neonatology, Department of Pediatrics
St. Louis University, and
The Pediatric Research Institute
Cardinal Glennon Children's Hospital, St. Louis, MO

Insulin or an insulin-like peptide has been demonstrated within the brains of various animal species[1] and more importantly insulin immunoreactivity has been observed within a few neurons.[2,3] In an attempt to determine whether this peptide is synthesized locally within the CNS, we have previously employed the technique of in-situ hybridization and a biotinylated rat insulin I cDNA to detect insulin mRNA within isolated and cultured 10 day old rabbit neurons. Neurons were characterized and detected by the neuron-specific enolase staining and glial cells contaminating these cultures were detected by the glial fibrillary acidic protein staining. We observed the presence of insulin transcripts within 3 to 5% of neurons (75% pure cultures) that were maintained in a serum-containing medium (Figure 1a). Interestingly, the same percentage of cells was positive for insulin immunoreactivity. Under similar hybridization conditions utilized for insulin, IGF I and II cDNAs (human) failed to cross-hybridize with these "insulin" transcripts.[3]

We have preliminary evidence that the expression of this neuronal insulin-like peptide can be modulated by in-vitro culture conditions. When neuronal cultures were maintained in a serum-free defined medium containing 5 μg/ml of insulin,[4] we observed that although there was some glial contamination (15%), an absence of the flat astrocytic and epithelial cells present in serum containing cultures was noted. Omission of insulin from the defined medium resulted in some neuronal loss but more distinctly the neurons appeared to have lost their neurites. Similar to the serum-containing media, insulin transcripts along with insulin immunoreactivity was observed in 3 to 5% of neurons maintained in an insulin-containing defined medium. On the other hand, despite the loss of neuronal well-being and a diminution in the total number of cells, the number of insulin-immunoreactive as well as the insulin-transcript-containing (Figure 1b) neurons increased to 15-20%. The preservation of insulin-synthesizing neurons with a loss of other non-insulin-synthesizing neurons is a distinct possibility. Additionally we observed the presence of insulin immunoreactivity within the conditioned media of the neurons grown under insulin-free conditions. The insulin secreted by neurons into the insulin-free media was further characterized by high-performance liquid chromatography to co-migrate with a porcine insulin standard. These observations are highly suggestive of insulin synthesis and secretion by neurons in culture. Further there appears to be some degree of autoregulation by exogenous insulin on the synthesis of the neuronal insulin-like peptide.

One can argue that the observations made in cultured neurons do not reflect the events within the whole brain and may have no physiologic relevance to the whole animal. In an attempt to pursue the fact that insulin is synthesized within the CNS, we performed in-situ hybridization with vibratome sections of fetal and neonatal rabbit brains. Employing the antisense (T7) and sense (T3) (control) biotinylated rat insulin I riboprobes, we demonstrated insulin transcripts that were specifically localized to neurons within the paraventricular structures alone. More importantly,

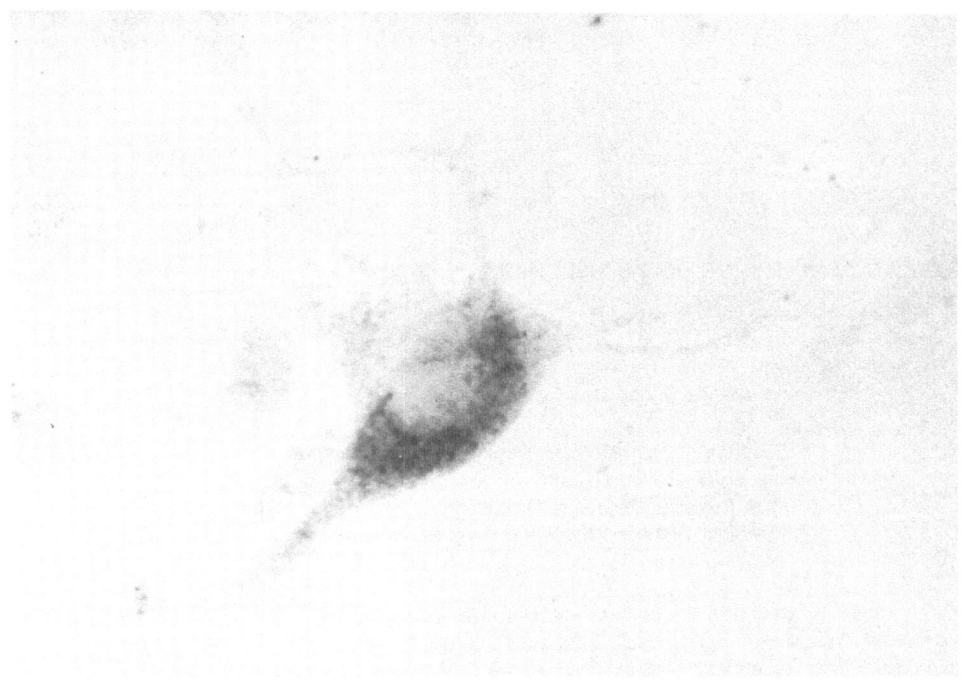

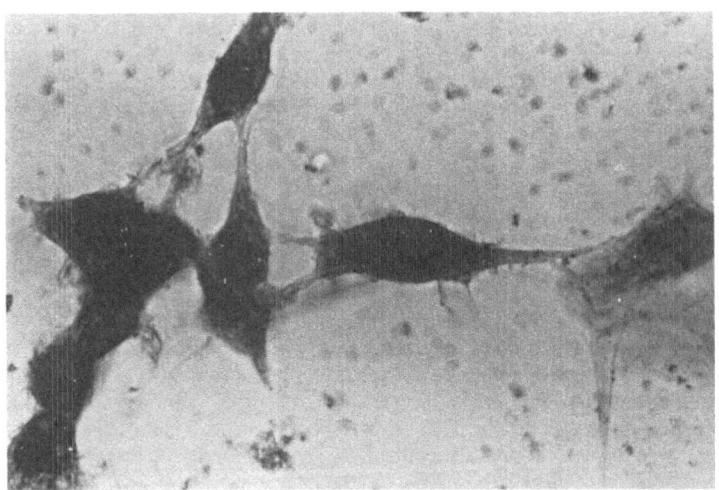

Figure 1. *In-situ* hybridization in neuronal cultures. This method has been described previously.[3] Briefly, the cells were fixed with 4% paraformaldehyde and permeabilized with Triton-X-100 (0.05%, 5 mM MgCl$_2$ in PBS). The cells were prehybridized in hybridization buffer for 1 hour at 37°C. Hybridization buffer contained 50% formamide, 10% dextrose sulfate, 0.5% BSA, 10 mM vanadyl ribonucleoside, 10 μg/ml ssDNA and 20 μg/ml yeast tRNA. The cells were then hybridized overnight at 37°C with a biotin (UTP)-labeled Nick-translated probe in hybridization buffer. The cells were washed with 50% formamide, 2 × SSC and then with 50% formamide, 1 × SSC (30 minutes each at 37°C). After strepavidin-peroxidase and 3',3'-diamino-benzadine treatments, the signal was intensified by silver staining.

Upper panel: Serum-containing media. A single neuron demonstrating the insulin transcript (×1000).

Lower panel: Insulin-free media. Neurons with the insulin transcripts (×400).

232

BRAIN

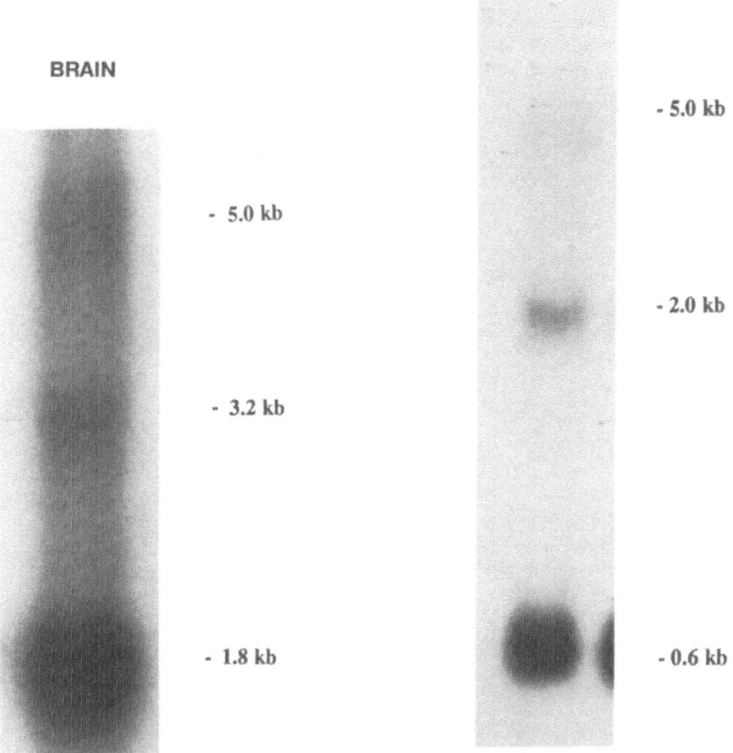

Figure 2.   *Northern Blots:* 50 μg of rabbit whole brain poly (A+) RNA was separated on a 1.5% agarose-formaldehyde gel and transferred to a Nytran filter. The filter was prehybridized (4 hrs, 42°C) followed by hybridization (1.0 × 10^6 cpm/ml of (^{32}P) GTP-labeled rat insulin cDNA probe (Nick-translated), 50% formamide, 5 × Denhardt's, 1 mM EDTA, 50 mM NaPO$_4$, 2.5 mM HEPES, 200 μg/ml ssDNA) for 48 hours at 37°C.   The filters were washed with 2 × SSC 0.1% SDS for 5 minutes at room temperature × 2, and 1 × SSC, 1% SDS for 30 minutes at 42°C, × 1, followed by autoradiography.
Left panel: Brain. The insulin transcript is seen in the center as a 3.2 kb band. Ribosomal RNA is seen as a 5.0 kb band, and actin mRNA (a control) as a 1.8 kb band.
Right panel: Pancreas. The insulin transcript is seen as a 0.6 kb band, while the 2.0 and 5.0 kb bands represent the ribosomal RNA.

Northern blot analysis performed with a ^{32}P-labeled rat insulin I cDNA and fetal rabbit whole brain poly (A+) RNA demonstrated that the insulin mRNA had a different size (3.2 kb) (Figure 2a) when compared to the pancreatic insulin transcripts (0.6 kb) (Figure 2b). This observation in the brain is similar to that seen with other extrapancreatic insulin transcripts, namely the embryonic extrafetal membranes[5] and rat testicular tissue.[6]   Solution hybridization of rabbit brain, rabbit and rat pancreatic poly (A+) RNA with the antisense (T7) rat insulin I cRNA (310 bp) was next performed. The rat pancreas revealed a single prominent band (310 bp) (Figure 3) that corresponded in size with the rat insulin I cRNA along with other smaller but minor bands, suggesting complete homology with the probe.   The rabbit pancreatic transcripts on the other hand demonstrated an absence of the 310 bp band, but revealed multiple smaller bands (109, 103 and 101 bp) indicative of non-homologous regions.   The brain mRNA band migrated as a smaller (than the probe) single

233

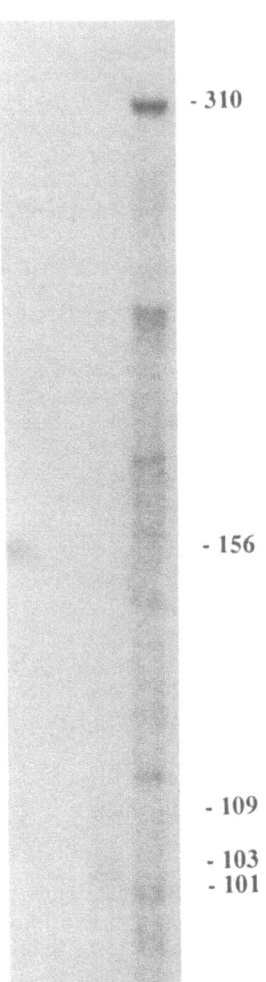

Figure 3. *Solution Hybridization:* 50 μg of rabbit whole brain poly (A+) RNA and 4 μg of pancreatic poly (A+) RNA (rat or rabbit) were hybridized (1.0 × 10^6 (^{32}P) GTP-labeled rI cRNA probe (T7 riboprobe transcript), 80% formamide, 0.4 M NaCl, 0.04 M PIPES, 0.001 M EDTA) overnight at 42°C. This was followed by RNase digestion (40 μg/ml RNase T) at 37°C for 30 minutes. Digestion was terminated by addition of 0.1 vol 10% SDS and 2.5 mg/ml proteinase K and incubation for 15 minutes at 37°C. Following ethanol precipitation, the samples were separated on a sequencing gel (5% polyacrylamide-urea) and then subjected to autoradiography.
(L to R): *First lane:* brain mRNA.
              Second lane: rabbit pancreas mRNA.
              Third lane: rat pancreas mRNA.

156 bp band which in addition to being different from that of the rat was distinct from the rabbit pancreatic insulin mRNA as well. These observations are consistent with the presence of a brain transcript(s) for an insulin-like peptide that is non-homologous to the rat and rabbit pancreatic insulin mRNAs.

In summary, we have demonstrated the presence of insulin (-like) transcripts within the CNS that are distinct from those present within the pancreas. Additionally, in isolated and cultured neurons, the number of cells containing insulin transcripts is autoregulated by the presence or absence of exogenous insulin. Thus, these observations confirm the presence of insulin within the CNS, but more importantly, demonstrate a local source for insulin synthesis, namely the neurons.

ACKNOWLEDGEMENTS

The authors acknowledge Drs. A. Permutt and P. Rotwein, Washington University, St. Louis, Missouri, for their generous gifts of the insulin and IGF I and II cDNAs.

Part of this work was supported by the Fleur-de-Lis Foundation, and the Monsanto-Searle Research Funds, St. Louis, MO.

REFERENCES

1. D. LeRoith, S. A. Hendricks, M. A. Lesniak, S. Rishi, K. L. Becker, J. Havrankova, J. L. Rosenweig, M. J. Brownstein and J. Roth, Insulin in brain and other extrapancreatic tissues of vertebrates and non-vertebrates, *Adv. Metab. Disorders* 10:303-340 (1983).
2. M. K. Raizada, Localization of insulin-like immunoreactivity in the neurons from primary cultures of rat brain, *Exp. Cell Res.* 143:351-357 (1983).
3. R. Schechter, L. Holtzclaw, F. Sadiq, A. Kahn and S. Devaskar, Insulin synthesis by isolated rabbit neurons, *Endocrinology* 123:505-513 (1988).
4. J. E. Bottenstein, Growth and differentiation of neural cells in defined media, *in:* "Cell Culture in the Neurosciences" (eds. J. E. Bottenstein and G. Sato), Plenum Press, New York (1985).
5. K. Rau, L. Muglia and J. Locker, Insulin-gene expression in extrafetal membranes of rats, *Diabetes* 38:39-43 (1989).
6. A. Efstratiadis, Unpublished data presented at the Symposium on Insulin, IGFs and Their Receptors, University of Florida College of Medicine, Gainesville, FL, January, 1989.

GROWTH REGULATION OF HUMAN BREAST CANCER BY INSULIN-LIKE GROWTH FACTORS

Douglas Yee, Kevin J. Cullen, Soonmyoung Paik,
Marc E. Lippman, and Neal Rosen

Lombardi Cancer Research Center
Georgetown University Medical Center
Washington, D.C. 20007

In the United States, breast cancer is the commonest malignancy of women and accounts for more deaths than any other tumor (1). Nearly 100 years ago, Beatson (2) demonstrated that oophorectomy could induce remission in some women with metastatic breast cancer. This was the first clinical suggestion that breast cancer growth could be regulated by the hormone estrogen. Since that time, manipulations aimed at reducing the estrogen available to breast cancer tissues have included surgical oophorectomy, adrenalectomy and hypophysectomy. Additionally, pharmacologic agents that inhibit estrogen synthesis, estrogen binding to its receptor and pituitary releasing factor analogs have been developed. These therapies are successful in only a minority of patients with breast cancer (3). Many breast cancers are not hormonally sensitive at the time of diagnosis and most of those cancers that are initially sensitive will eventually escape their dependence on estrogen.

In addition to estrogen, breast cancer cells in vitro are responsive to a wide variety of other trophic substances and growth factors such as glucocorticoids, androgens, progestins, retinoids, epidermal growth factor, calcitonin, thyroid hormone, prolactin, and vitamin D (4, 5). A better understanding of the role of these factors in the regulation of breast cancer growth could conceivably lead to clinically useful therapeutic strategies analogous to those available for estrogen. Moreover, it has been suggested that unregulated expression of growth factors may stimulate autocrine growth of human malignancies (6). Identification of these autocrine factors could be important in understanding the biology of tumor growth.

The insulin family of growth factors have important stimulatory effects on breast cancer cells. Osborne et al. (7) has reported that physiologic concentrations of insulin can stimulate the net synthesis of DNA, RNA, and protein in breast cancer cells. However, in breast cancer epithelial insulin-like growth factor I (IGF-I) has been shown to be 10 to 100 fold more potent than insulin as a mitogen for these cells (8). Although stimulation of both the insulin receptor and type I IGF receptor can induce mitogenesis in fibroblasts (9), the greater potency of IGF-I compared to insulin in breast cancer cells suggests that the binding of IGF-I to its receptor may mediate a relatively more important mitogenic response than stimulation of the insulin receptor.

237

Several laboratories have found that IGF-I is a mitogen at nanomolar concentrations (8, 10). Previous work has shown that rat MSA III (insulin-like growth factor II, IGF-II) is also a potent mitogen for breast cancer cells (11). We and others have recently found that human IGF-II is only slightly less potent than IGF-I in stimulating breast cancer cell growth (12, 13). These observations suggest that both IGF-I and -II are both potentially important mitogens for breast cancer.

Although insulin has important effects on the growth of breast cancer cells in vitro, it is primarily expressed by the beta cells of the pancreas (14) and it is an unlikely candidate for an autocrine or paracrine growth regulator of breast cancer cells. Furthermore, reduction of circulating insulin levels would likely lead to prohibitive physiologic side effects in vivo. In contrast, IGF-I and IGF-II expression has been described in many different tissues (15), and these factors could potentially regulate the growth of breast cancer cells in a fashion that is more amenable to therapeutic manipulation.

The suggestion that an IGF may be an autocrine growth factor for breast cancer came from the observation that media conditioned by breast cancer cells contained an immunoreactive IGF-I related protein. Huff et al. (10) found that conditioned media could displace authentic IGF-I in the radioimmunoassay (RIA). This material was secreted in association with an IGF binding activity and acid conditions separated the binding activity from the immunoreactive IGF-I related protein. This protein had a Mr of 13 kilodaltons. Although this is larger than the 7.6 kD circulating form of IGF-I, it is compatible with the size of the preprohormone. Minuto et al. (16) demonstrated similar results. Acid gel filtration of media conditioned by breast cancer cells yielded two peaks of IGF-I related activity by RIA. The 35-45 kD peak displayed IGF binding activity while the smaller peak co-migrated with labelled IGF-I. Therefore, the larger peak may be explained entirely or in part by interference in the RIA by a binding protein while the smaller peak may represent an IGF-I related peptide.

Using an IGF-I cDNA probe in a Northern analysis of breast cancer cell line RNA, several hybridizing RNA species were found (10). However the size of most of these bands were different from the transcript sizes seen in human liver. This result has also been found by other groups (17). Since the entire IGF-I gene has not been cloned, the RNA species seen in breast cancer could potentially represent products of alternative splicing of the IGF-I gene or related RNA species that have nucleotide sequence homology to IGF-I.

To further investigate these possibilities we used a ribonuclease (RNAse) protection assay with an IGF-IA cDNA probe to detect authentic IGF-I transcripts (18). The probe we used contains the entire coding portion of the IGF-I gene, therefore if the IGF-I related protein detected in breast cancer conditioned media represents translation of an alternatively spliced RNA, these transcripts would protect all or part of this probe. Moreover, this technique is more sensitive than Northern analysis for detecting IGF-I transcripts. Rotwein et al. (19) have shown that RNAse protection is 100 fold more sensitive than Northern analysis in detecting IGF-I transcripts in mouse liver. We have found that this technique was able to detect 1 picogram of IGF-I transcript (18).

RNAse protection assays were able to detect IGF-I transcripts in 30 micrograms of total liver RNA or in as little as 0.5 micrograms of poly(A)+ liver RNA. In contrast, no IGF-I transcripts were found in breast cancer cell line RNAs (0/11). We have altered technical conditions to increase the sensitivity of the assay and still were unable to detect any full length or partially protected transcripts. Therefore, it is unlikely that the immunoreactive IGF-I related protein is transcribed from the coding portions of the IGF-I gene.

Because IGF-I and IGF-II have a high degree of protein homology, we examined the breast cancer cell lines for IGF-II mRNA to determine if the immunoreactive IGF material found in conditioned media could be IGF-II (12). Northern analysis using and IGF-II cDNA probe and poly(A)+ selected breast cancer cell line mRNA yielded 6.0, 4.8, and 0.5 kb bands in only the breast cancer cell line T-47D. When other breast cancer cell lines were examined for IGF-II gene transcripts by RNAse protection, only T-47D consistently expressed IGF-II mRNA. Although some subclones of late passage MCF-7 contained IGF-II mRNA, early passage cells were always negative. These results supported those of Peres et al. (20) who found that IGF-II mRNA was produced by T-47D.

Since T-47D is an estrogen responsive cell line, we investigated the effects of estrogen on IGF-II mRNA synthesis. We have found that 1 nM estradiol increases steady state levels of IGF-II mRNA within 1 hour of treatment. This estrogen induced increase in IGF-II mRNA can be inhibited by the anti-estrogen 4-0H tamoxifen (12). Therefore, in the cell line T-47D, IGF-II could potentially be an estrogen induced autocrine growth factor. We are further examining the response of IGF-II protein production to estrogen stimulation in T-47D.

We have been unable to find any breast cancer cell lines that express IGF-I and only T-47D and late passage MCF-7 express IGF-II. Therefore, most breast cancer cell lines do not express authentic IGF-I or -II mRNA. Since large amounts of the immunoreactive IGF-I related protein was found in the media of all the breast cancer cell lines examined (10), it seems unlikely that this activity represents authentic IGF-I or -II protein. The identity of the immunoreactivity is unclear. The IGFs are associated with binding proteins in the serum and extracellular fluid (21) and it is known that these proteins can interfere with the IGF radioimmunoassay (22). We and others have shown that the immunoreactive IGF-I is associated with a binding activity (10, 16). DeLeon et al. (23) have cross linked radiolabelled IGF-I to conditioned media and demonstrated binding activities with Mr of 45, 36, and 29 kD. We incubated media conditioned by the breast cancer cell line MDA-MB 231 with radiolabelled IGF-I and determined the size of the resulting complex by gel filtration chromatography (24). Under neutral conditions, the labelled IGF-I migrated with an Mr of 35-40 kD and incubation with excess unlabelled IGF-I partially displaced the radiolabelled IGF-I to its appropriate 7.5 kD size. This suggested that the IGF-I was associated with a binding protein of 28-33 kD. This observation was supported by analysis of the size fractionated conditioned media with the IGF-charcoal separation method to assay for binding activity. This assay revealed a broad range of binding activity from 67 to 25 kD.

Since a low molecular weight IGF binding protein of this approximate size, BP-25, had been cloned from the hepatocellular carcinoma cell line HepG2 (25), we examined breast cancer cell lines for production of this binding protein (24). Northern analysis and RNAse protection analysis revealed that two breast cancer cell lines, Hs578T and MDA-MB 231, expressed BP-25 mRNA. However, other breast cancer cell lines (MCF-7, T-47D) that have been reported to express IGF binding activity, did not express BP- 25. BP-25 protein was also detected by radioimmunoassay of conditioned media of these cell lines with a polyclonal antiserum directed against purified BP-25. In addition, this antiserum immunoprecipitated the 30 kD species in the media conditioned by metabolically labelled MDA-MB 231 cells. The 30 kD M.W. is consistent with the size of BP-25 after post-translational modification (25).

It has been reported that the expression of BP-25 is limited to the liver (26), however we have shown that it is clearly expressed by some breast cancer cell lines. Busby et al. (27) have suggested that BP-25 can either augment or inhibit the mitogenicity of IGF-I depending on its post-translational modifications, and expression of BP-25 by these breast cancer cells is likely to modulate the effects of IGF-I. Furthermore, other low molecular weight IGF binding proteins are also produced by breast cancer cells (23). For example, MCF-7 does not express BP-25, yet this cell line clearly produces binding activity (10, 23). The physiologic function of these other low molecular weight binding proteins has not been defined, further investigation will clarify the effects of IGF binding protein production on the growth regulation of breast cancer.

The immunoreactive IGF-I related protein detected in breast cancer conditioned media may represent interference of binding protein with the RIA. Purified BP-25 can bind IGF-I or IGF-II tracer in a dose dependent fashion that results in a false positive RIA displacement curve (K. Cullen, unpublished data). However, since IGF-I Northern analysis of breast cancer RNAs will detect some RNA species, this suggests that breast cancer cells produce some RNAs that share nucleotide homology with authentic IGF-I. Additionally, Minuto et al. (16) demonstrated that the low molecular weight IGF-I immunoreactive protein did not have binding activity. We are actively investigating the nature of the radioimmunoactive IGF material produced by breast cancer cells since it could potentially represent a novel IGF species.

Since IGF-I and -II are mitogens for breast cancer cells, this suggests that these cells express IGF receptors. Several groups have found that breast cancer cells have both type I and type II IGF receptor. Cross linking studies using radiolabelled IGF-I have detected the presence of type I receptors in breast cancer cells (8, 23, 28). We have used RNAse protection with the cDNAs of the type I and II receptor to demonstrate the presence of these receptors in almost all breast cancer cell lines (D. Yee, K. Cullen, unpublished observations).

The physiologic effects of blocking the type I receptor has been demonstrated by in vitro and in vivo model systems. Rohlik et al. (29) has shown that blockade of the type I receptor with a specific monoclonal antibody, αIR3, inhibited the binding of labelled IGF-I to the breast cancer cell line MCF-7. αIR3 also inhibited the growth of these cells in the presence of serum. In this experiment it was not clear if the antibody inhibited the mitogenic effects of serum derived IGF-I or an IGF-like material produced by the cells themselves. However, αIR3 decreased the growth of these cells induced by 1 nM estradiol suggesting that the mitogenic effects of estradiol may partially be mediated by an endogenously produced IGF-like material. We have shown that IGF-II mRNA was induced by estradiol in another hormone dependent cell line, T-47D, and some late passage MCF-7 cells produced IGF-II mRNA (12). Since the mitogenic effects of IGF-II are most likely mediated through the type I IGF receptor (30), these data would suggest that αIR3 may inhibit the mitogenic effects of estradiol in these MCF-7 by interfering with an IGF-II autocrine loop. Although we have not found any subclones of MCF-7 to produce authentic IGF-I, it is also possible that the antibody blocks the effects of another IGF related protein.

Arteaga et al. (31) have demonstrated that in vivo blockade of the type I IGF receptor can also inhibit the growth of breast cancer cells. In nude mice, intraperitoneal injections of the antibody inhibited the tumor formation of the estrogen independent cell line MDA-MB 231. In contrast to the in vitro studies of Rohlik et al. (29), MCF-7 tumor formation was not inhibited by the antibody. In this system it is not clear if the antibody inhibited the growth of MDA-MB 231 by interfering with an autocrine, paracrine, or endocrine pathway. We have been unable to demonstrate IGF-I mRNA in breast cancer tumors in vitro, however it is possible that conditions that are not met in vitro are required for IGF-I or IGF-II expression.

In order to evaluate this possibility, we have examined RNA extracted from breast cancer tissue samples for IGF-I mRNA expression (18). Unlike breast cancer cell lines, approximately 60% of these RNAs contained IGF-I message. However when RNA was extracted from non-pathologic areas adjacent to the breast cancer or from benign fibroadenomas, IGF-I mRNA was found at levels higher or equal to most breast cancers. One possible interpretation of this finding is that the stromal cells of the breast, and not the malignant epithelial cells, expressed IGF-I mRNA. Clemmons and Van Wyk (32) have shown that skin fibroblasts produced IGF-I protein in vitro and we have found that skin fibroblasts had easily detectable levels of IGF-I mRNA (18). The stromal origin of IGF-I has also been noted in fetal tissues. Han et al. (15) found that in situ hybridization localized IGF-I and IGF-II mRNA to stromal elements.

When tissues obtained from breast cancer specimens were examined by in situ hybridization, no IGF-I was contained in malignant epithelial cells; rather, the stromal areas of the non-malignant lobules had the highest signal (18; S. Paik, manuscript in preparation). Multiple samples were examined and we have not found evidence of IGF-I mRNA expression by malignant epithelial cells. Since no breast cancer cells we have examined produce IGF-I mRNA in vitro, these finding support the observation that authentic IGF-I is not likely to be an autocrine growth factor for breast cancer.

The pattern of expression suggests that IGF-I may be an important paracrine stimulator of breast cancer cells. Furthermore, the level of IGF-I expression in non-pathologic breast tissue was generally higher than that seen in the liver. Since breast cancer cells are responsive to nanomolar concentrations of IGF-I and blockade of the type I receptor inhibits tumor formation, the local production of IGF-I may influence the growth of malignant epithelial cells. Of note is that breast cancer cells produce platelet derived growth factor (PDGF) yet have no PDGF receptor (33). Since PDGF enhances the production of IGF-I by fibroblasts (34), PDGF production by breast cancer cells may increase the stromal production of IGF-I which in turn may stimulate the growth of the breast cancer cells. We are currently examining in vitro and in vivo methods of interfering with this putative dual paracrine loop.

Examination of breast cancer tissue RNAs demonstrated that most breast cancers had detectable levels of IGF-II mRNA when examined by RNAse protection (12). Northern analysis of total RNA derived from breast cancer tissue revealed a 6.0 and 4.8kb band which suggested that the IGF-II transcripts detected in breast cancer were transcribed from the fetal promoter (35). Furthermore, the level of expression measured by RNAse protection varied approximately 20 fold between samples. Interestingly, the tumor that expressed the most IGF-II mRNA contained only small amounts of stromal cells when examined by routine histologic techniques. This sample did not contain IGF-I mRNA, suggesting that this specimen did not contain a large amount of fibroblasts. Since we had found that some malignant breast cancer cells in vitro expressed IGF-II, this was supportive evidence that some breast cancer cells in vivo also expressed IGF-II.

However, we and others also demonstrated detectable levels of IGF-II mRNA in benign breast tumors and non-pathologic breast tissues. Travers et al. (36)

demonstrated that dot blot hybridization was able to detect IGF-II transcripts in 15/15 RNAs derived from non-malignant tumors and normal breast. In contrast, only 11/21 breast carcinomas had detectable IGF-II. Like IGF-I, this suggests that IGF-II may be expressed by stromal components. However unlike IGF-I, some breast cancer cell lines express IGF-II mRNA, suggesting that malignant breast cells potentially could maintain this capacity in vivo. To fully evaluate the source of IGF-II mRNA in breast tissues, we have used in situ hybridization.

Preliminary IGF-II in situ hybridizations have demonstrated that some breast cancer samples have intense IGF-II signal in stromal elements surrounding normal ductules (S. Paik, manuscript in preparation). This pattern of expression is similar to that seen in IGF-I in situ studies. However, one specimen which expressed high levels of IGF-II mRNA on RNAse protection also demonstrated IGF-II mRNA in malignant epithelial cells when studied by in situ hybridization. Thus the in vitro observation that some malignant breast epithelial cell lines expressed IGF-II has also been demonstrated in patient samples. Therefore, IGF-II could function as an autocrine growth factor for some tumors.

The mechanism that accounts for the expression of IGF-II by breast cancer cells has not been defined. Ali et al. (42) has found that 60% of breast cancer tissue specimens have a deletion in the short arm of chromosome 11 suggesting the existence of a recessive breast cancer gene. Of note is that this deletion is near the Wilms' tumor gene locus and these tumors are known to express IGF-II (37,38). Therefore, it is possible that a genetic mechanism may account for IGF-II expression by some breast cancer cells. Additionally, there may be other hormones or factors present in vivo that stimulate IGF-II expression. For example it is known that FSH, LH, and hCG, stimulate the expression of IGF-II by ovarian granulosa cells (39). Since breast epithelial cells also respond to pituitary factors (40, 41), it is possible that some of these hormones in vivo also stimulate breast cancer growth. We are currently examining the possibility that IGF-II production by breast cancer cells is regulated by endocrine pathways.

In summary, work done by several laboratories has shown that IGFs are potentially important growth regulatory polypeptides in breast cancer. Virtually all breast tumors and breast cancer cell lines have receptors for the IGFs and both IGF-I and IGF- II are mitogenic for these cells. In several systems, blockade of the type I IGF receptor inhibits the growth of breast cancer cells or tumor, supporting the concept that IGFs can effect breast cancer growth.

Examination of mRNA expression by breast cancer cell lines and tissues has revealed that IGF-I and -II are probably paracrine stimulators of growth. However, IGF-II may also be an autocrine growth factor for some breast cancers. Furthermore breast cancer cells in vitro produce IGF binding proteins that could modulate the interaction between the IGFs and their receptors. Further evaluation of the role of the IGFs, their receptors, and binding proteins in breast cancer growth and development may yield new therapeutic modalities.

References

1. D. M. Parkin, J. Stjernsward and C. S. Muir, Estimates of the world-wide frequency of twelve major cancers, Bull. WHO, 62:163 (1984).

2. G. T. Beatson, On the treatment of inoperable cases of carcinoma of the mamma. Suggestions for a new method of treatment with illustrative cases, Lancet, 2:104 (1896).

3. I. C. Henderson, Endocrine therapy in metastatic breast cancer, in: Breast Diseases, J. R. Harris, S. Hellman, I. C. Henderson, D. W. Kinne, eds., J. B. Lippincott Co., Philadelphia, (1987).

4. M. E. Lippman, Definition of hormones and growth factors required for optimal proliferation and expression of phenotypic responses in human breast cancer cells, in: Cell Culture Methods for Molecular and Cell Biology, Vol.2 , D. W. Barnes, D. A. Sirbasku and G. H. Sato, ed., Alan R. Liss, New York, (1984).

5. J. A. Eisman, T. J. Martin, I. MacIntyre, R. J. Framptin, J. M. Moseley and R. Whitehead, 1,25-dihydroxyvitamin D3 receptor in a cultured breast cancer cell line (MCF-7 cells), Biochem. Biophys. Res. Comm., 93:9 (1980).

6. M. B. Sporn and A. B. Roberts, Autocrine growth factors and cancer, Nature, 313:745 (1985).

7. C. K. Osborne, G. Bolan, M. E. Monaco and M. E. Lippman, Hormone responsive human breast cancer in long-term tissue culture: effect of insulin, Proc. Natl. Acad. Sci., 73:4536 (1976).

8. R. W. Furlanetto and J. N. DiCarlo, Somatomedin-C receptors and growth effects in human breast cells maintained in long term tissue culture, Canc. Res., 44:2122 (1984).

9. J. S. Flier, P. Usher and A. C. Moses, Monoclonal antibody to the type I insulin-like growth factor (IGF-I) receptor blocks IGF-I receptor-mediated DNA synthesis: clarification of the mitogenic mechanisms of IGF-I and insulin in human skin fibroblasts, Proc. Natl. Acad. Sci., 83:664 (1986).

10. K. K. Huff, D. Kaufman, K. H. Gabbay, E. M. Spencer, M. E. Lippman and R. B. Dickson, Secretion of an insulin-like growth factor-I-related protein by human breast cancer cells, Canc. Res., 46:4613 (1986).

11. Y. Myal, R. P. C. Shiu, B. Bhaumick and M. Bala, Receptor binding and growth-promoting activity of insulin-like growth factors in human breast cancer cells (T-47D) in culture, Canc. Res., 44:5486 (1984).

12. D. Yee, K. J. Cullen, S. Paik, J. F. Perdue, B. Hampton, A. Schwartz, M.E. Lippman and N. Rosen, Insulin-like growth factor II mRNA expression in human breast cancer, Canc. Res., 48:6691 (1988).

13. K. P. Karey and D. A. Sirbasku, Differential responsiveness of human breast cancer cell lines MCF-7 and T47D to growth factors and 17ß-estradiol, Canc. Res., 48:4083 (1988).

14. P. E. Cryer, Glucose homeostasis and hypoglycemia, in: Williams Textbook of Endocrinology, J. D. Wilson and D. W. Foster, eds. W.B. Saunders Co., Philadelphia (1985).

15. V. K. M. Han, A. J. D'Ercole and P. K. Lund, Cellular localization of somatomedin messenger RNA in the human fetus, Science, 236:193 (1987).

16. F. Minuto, P. DelMonte, A. Barreca, A. Nicolin and G. Giordano, Partial characterization of somatomedin C-like immunoreactivity secreted by breast cancer cells in vitro, Mol. Cell. Endo., 54:179 (1987).

17. C. K. Osborne, C. R. Ross, E. B. Coronado, S. A. Fuqua and L. J. Kitten, Secreted growth factors from estrogen receptor-negative human breast cancer do not support growth of estrogen receptor-positive breast cancer in the nude mouse model, Breast Canc. Res. Treat., 11:211 (1988).

18. D. Yee, S. Paik, G. S. Lebovic, R. R. Marcus, R. E. Favoni, K. J. Cullen, M. E. Lippman, and N. Rosen, Analysis of IGF-I gene expression in malignancy, evidence for a paracrine role in human breast cancer, Mol. Endo., in press (1989).

19. P. Rotwein, K. M. Pollock, N. Watson and D. T. Milbrandt, Insulin-like growth factor gene expression during rat embryonic development, Endo., 121:2141 (1987).

20. R. Peres, C. Betsholtz, B. Westermark and C.-H. Heldin, Frequent expression of growth factors for mesenchymal cells in human mammary carcinoma cell lines, Canc. Res., 47:3425 (1987).

21. M. Binoux, P. Hossenlopp, S. Hardouin, D. Seurin, C. Lassarre and M. Gourmelen, Somatomedin (insulin-like growth factors)-binding proteins, molecular forms and regulation, Hormone Res., 24:141 (1986).

22. W. H. Daughaday, M. Kapadia and I. Mariz, Serum somatomedin binding proteins: physiologic significance and interference in radioligand assay, J. Lab. Clin. Med., 109:355 (1987).

23. D. De Leon, B. Bakker, D. M. Wilson, R. L. Hintz and R. G. Rosenfeld, Demonstration of insulin-like growth factor (IGF-I and -II) receptors and binding protein in human breast cancer cell lines, Biochem. Biophys. Res. Comm., 152:398 (1988).

24. D. Yee, R. E. Favoni, R. Lupu, K. J. Cullen, G. S. Lebovic, K. K. Huff, P. D. K. Lee, Y. L. Lee, D. R. Powell, R. B. Dickson, N. Rosen, and M. E. Lippman, The insulin-like growth factor binding protein BP-25 is expressed by human breast cancer cells, Biochem. Biophys. Res. Comm., in press, (1989)

25. Y.-L. Lee, R. L. Hintz, P. M. James, P. D. K. Lee, J. E. Shively and D. R. Powell, Insulin-like growth factor (IGF) binding protein cDNA from human HEP G2 hepatoma cells: predicted protein sequence suggests an IGF binding domain different from those of the IGF-I and IGF-II receptors, Mol. Endo., 2:404 (1988).

26. A. Brinkman, C. Groffen, D. J. Kortleve, A. Geurts van Kessel and S. L. S. Drop, Isolation and characterization of a cDNA encoding the low molecular weight insulin-like growth factor binding protein (IBP-1), EMBO J., 7:2417 (1988).

27. W. H. Busby Jr., D. G. Klapper and D. R. Clemmons, Purification of a 31,000-dalton insulin-like growth factor binding protein from human amniotic fluid, isolation of two forms with different biological actions, J. Biol. Chem., 26:14203 (1988).

28. J.-P. Peyrat, J. Bonneterre, J. Beuscart, J. Dijane and A. Demaille, Insulin-like growth factor 1 receptors in human breast cancer and their relation to estradiol and progesterone receptors, Canc. Res., 48:6429 (1988).

29. Q. T. Rohlik, D. Adams, F. C. Kull Jr. and S. Jacobs, An antibody to the receptor for insulin-like growth factor I inhibits the growth of MCF-7 cells in tissue culture, Biochem. Biophys. Res. Comm., 149:276 (1987).

30. R. A. Roth, Structure of the receptor for insulin-like growth factor II: the puzzle amplified, Science, 239:1269 (1988).

31. C. L. Arteaga, L. Kitten, E. Coronado, S. Jacobs, F. Kull and C. K. Osborne, Blockade of the type I somatomedin receptor inhibits growth of estrogen receptor negative human breast cancer cells in athymic mice, Proc. Endo. Soc. 70th Annual Meeting, A683 (1988).

32. D. R. Clemmons and J. J. Van Wyk, Evidence for a functional role of endogenously produced somatomedinlike peptides in the regulation of DNA synthesis in cultured human fibroblasts and porcine smooth muscle cells, J. Clin. Invest., 75:1914 (1985).

33. D. A. Bronzert, P. Pantazis, H. N. Antoniades, A. Kasid, N. Davidson , R. B. Dickson, and M. E. Lippman, Synthesis and secretion of platelet-derived growth factor by human breast cancer cell lines, Proc. Natl. Acad. Sci., 84:5763 (1987).

34. D. R. Clemmons, Multiple hormones stimulate the production of somatomedin by cultured human fibroblasts, J. Clin. Endo. Metab., 58:850 (1984).

35. P. de Pagter-Holthuizen, M. Jansen, F. M. A. van Schaik, R. vanderKammen, C. Oosterwijk, J. L. Van den Brande, and J. S. Sussenbach, The human insulin-like growth factor II gene contains two development-specific promoters, FEBS letters, 214:259 (1987).

36. M. T. Travers, P. J. Barrett-Lee, U. Berger, Y. A. Luqmani, J.-C. Gazet,T. J. Powles, and R. C. Coombes, Growth factor expression in normal, benign, and malignant breast tissue, Br. Med. J., 296:1621 (1988).

37. J. Scott, J. Cowell, M. E. Robertson, L. M. Priestley, R. Wadey, B. Hopkins, J. Pritchard, G. I. Bell, L. B. Rall, C. F. Graham, and T. J. Knott, Insulin-like growth factor-II gene expression in Wilms'tumour and embryonic tissues, Nature, 317:260 (1985).

38. A. E. Reeve, M. R. Eccles, R. J. Wilkins, G. I. Bell and L. J. Millow, Expression of insulin-like growth factor-II transcripts in Wilms' tumour, Nature, 317:258 (1985).

39. K. Ramasharma and C. H. Li, Human pituitary and placental hormones control human insulin-like growth factor II secretion in human granulosa cells, Proc. Natl. Acad. Sci., 84:2643 (1987).

40. R. P. C. Shiu, L. C. Murphy, Y. Myal, T. C. Dembinski, D. Tsuyuki and B. M. Iwasiow, Actions of pituitary prolactin and insulin-like growth factor II in human breast cancer, in: Breast Cancer: Cellular and Molecular Biology, M.E. Lippman, R.B. Dickson, eds. Kluwer Academic Publishers, Boston, (1988).

41. C. B. Newman, H. Cosby, H. G. Friesen, M. Feldman, P. Cooper, V. DeCrescito, M. Pilon, and D. L. Kleinberg, Evidence for a nonprolactin, non-growth-hormone mammary mitogen in the human pituitary gland, Proc. Natl. Acad. Sci., 84:8110 (1987).

42. I. Ali, R. Lidereau, C. Theillet, R. Callahan, Reduction to homozygosity of genes on chromosome 11 in human breast neoplasia, Science, 238:185 (1987).

INSULIN-LIKE GROWTH FACTORS IN LUNG

CANCER CELL LINES

Gabriele Jaques and Klaus Havemann

Dept. Internal Medicine, Div. Hematology/Oncology, Philipps-University of Marburg, Baldingerstrasse, 3550 Marburg, F.R.G.

INTRODUCTION

Lung cancer is the most common type of cancer in men all over the world and is the leading cause of cancer mortality in males in more than 35 countries. For females, lung cancer is expected to become more common as the percentage of woman smokers continues to increase. In 1980 the World Health Organisation (WHO) estimated that there will be 660,500 new cases of lung cancer worldwide. It is also expected that lung cancer will surpass gastric cancer as the most common tumor globally by the end of the 1980's.[1,2]
Human lung cancer can be subgrouped into two major categories, one being the small-(SCLC)- and the other the nonsmall (NSCLC) -cell carcinoma. Small cell carcinoma (SCLC) represents about 25-30% of all pulmonary carcinomas.

Clinically, SCLC differs from NSCLC by its characteristic cellular morphology, by a tendency to metastasize rapidly and widely, by its responsiveness to chemotherapy and by its frequent association with various paraneoplastic syndroms.[3,4]

We and others have developed in vitro growth of permanent cell lines of SCLC and NSCLC from patients with this disease, in order to better understand the origin and biology of this tumor.[5-9] Most SCLC cell lines, express in contrast to NSCLC cell lines, properties of the neuroendocrine system including L-DOPA decarboxylase, bombesin-like immunoreactivity, neuron-specific enolase and the BB isoenzyme of creatine kinase. SCLC cell lines lacking these markers are called variant subtype in contrast to the classic subtype of SCLC.[10] These cell lines have provided material for studying the origin and growth requirements of these tumors, the regulation of secretory products and the mechanism of hormone synthesis.

It has been proposed that transformed cells may express, maintain or enhance their malignant phenotype by the secretion of autocrine growth factors, thus enabling a positive feedback loop, whereby a cell may both produce and respond to the same growth factor.[11] Besides bombesin[12], an amphibian analogue to the gastrin-releasing peptide (GRP), recent data also indicate the family of insulin-like growth factors (IGF) as important growth factors for lung cancer.

This chapter will review recent data concerning production, secretion of IGF-related material by lung cancer cell lines, binding sites on these cells and the growth-promoting effects of these IGFs in in vitro systems.

IGF-immunoreactive material in cell extracts

Normal lung tissue contains immunoreactive IGF-I.[13] In a previous report it could be demonstrated that tissue extracts from malignant lung tumors, obtained after surgery, contain a higher amount of immunoreactive IGF-I than normal lung tissue. The concentration of IGF-I related material in tissue extracts from SCLC and NSCLC primary tumors, however, showed no difference.[14] So far, it is unclear whether the increased IGF-I content in these tumor samples was due to the production of the neoplastic cells or the interstitial cells in this tumor. The examination of cell lines derived from these tumors however, can help to answer this question in vitro. We have examined several SCLC cell lines for the presence of IGF-I immunoreactivity in the cell pellets (Tab. 1).[15] IGF-I immunoreactivity was detected in 11/14 pellets of SCLC cell lines. The levels ranged from 12-76 mIU/mg soluble protein. There was no correlation between the IGF-I levels and the classic or variant subtype of the SCLC cell lines.[10] Macaulay et al.[16] reported the detection of immunoreactive somatomedin C in 4/5 SCLC cell homogenates. Somatomedin C was also present in cell preparations from 2/3 NSCLC cell lines established from squamous- and adenocarcinoma. These results were proven not to be false positive since the sample IGF-I concentrations diluted in parallel with standard IGF-I.

A prerequisite of a IGF-I production in lung cancer cell lines is the expression of IGF-I mRNA. Only one report deals with the expression of IGF-I and IGF-II mRNA in lung tumor cell lines. Betsholz et al.[17] examined 10 lung cancer cell lines (4 SCLC, 6 NSCLC) and found only one of the NSCLC cell lines positive for IGF-II mRNA. None of the other investigated cell lines, belonging to both subclasses expressed detect-

Table 1.   IGF-I immunoreactivity in cells and conditioned media (CM) of SCLC cell lines

Cell line	Type	IGF-I immuno-reactivity in cells (mIU/mg)	IGF-I immuno-reactivity in CM (mIU/mg)
SCLC-16HC	c	30 + 19	15 + 1
SCLC-22H	c	17 + 4	13 + 2
SCLC-24H	c	23 + 8	5 + 1
SCLC-86M1	c	< 0.5	< 0.5
NCI-H69	c	39 + 22	35 + 7
NCI-H146	c	22 + 4	18 + 2
NCI-N592	c	35 + 9	50 + 4
SW-201 5	c	76 + 3	n.d.
SCLC-16HV	v	47 + 17	118 + 10
SCLC-21H	v	20 + 9	n.d.
NCI-H82	v	< 0.5	< 0.5
NCI-N417	v	12 + 5	n.d.
NCI-H526	v	< 0.5	< 0.5
DMS-79	v	43 + 13	n.d.

able levels of specific mRNA, neither for IGF-I nor for IGF-II. Preliminary data obtained in our laboratory with a human IGF-I cDNA[18] probe
indicated that the SCLC cell lines were all negative for IGF-I expression. These data are confirmed by a report on IGF-I immunreactive
material in breast cancer cell lines.[19] IGF-I immunreactive material
secreted by breast cancer cell lines could be detected by a radioimmunoassay. Northern analysis using poly A+RNA extracted from these
cell lines revealed several band, but a ribonuclease protection assay
showed that no breast cancer cell line produced authentic IGF-I mRNA.
The authors conclude that the immunreactive material might be a novel
IGF-I related protein.

IGF-I immunoreactivity in conditioned media (CM)

In a series of preliminary experiments, we found that conditioned
media (CM) prepared from SCLC cell lines stimulated clonogenic growth
of the same or other SCLC cell lines. Previously Cuttitta et al.[20]
observed that CM - used as a growth supplement for culturing lung
tumors - contained IGF-I related material. To identify and later
characterize the components secreted by SCLC cells required for auto-
stimulation, it was necessary to develop a serum-free culture system
which is not influenced by the presence of growth factor containing
fetal calf serum. We therefore grew SCLC cell lines in RPMI 1640 medium
supplemented with 30 nM sodium selenite and 10 µg/ml 98% purified human
transferrin designated as ST medium.[5] First the secretion of IGF-I
immunreactive material into culture medium was investigated as a function of time. Different SCLC cell lines were washed and plated in ST,
and daily samples were collected for radioimmunological determination.

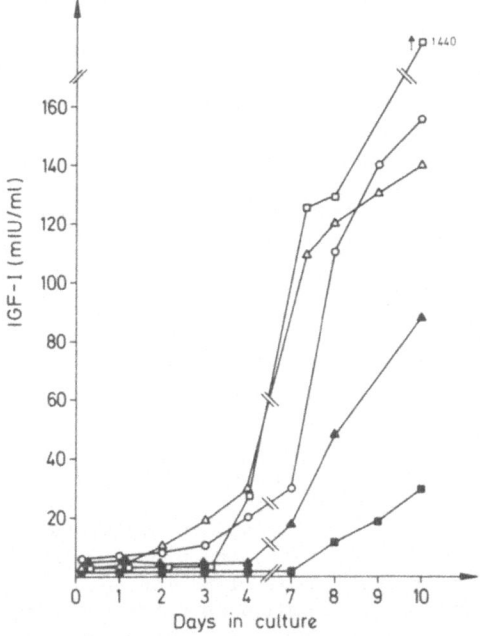

Fig. 1. Production of immunoreactive IGF-I by SCLC-16HC (■), SCLC-16HV
(Δ), NCI-H69 (o), NCI-N592 (□) and NCI-H146 (▲). The increase of IGF-I immunoreactivity in the medium (mIU/ml) is
shown as a function of days in culture.

As shown in Fig. 1 IGF-I immunoreactivity increased after an initial lag phase in a linear manner. Individual cell lines expressed different patterns of production and showed a correlation to the corresponding cell pellets. Interestingly, one SCLC cell line, which converted in vitro from the classic to the variant subtype (SCLC-16HC to SCLC-16HV) started to produce more IGF-I related material. The production of IGF-I immunoreactivity was prevented by the addition of the protein synthesis-inhibitor cycloheximide (10^{-5} M), thus indicating the involvement of protein synthesis. RIA displacement experiments of acid/ethanol extracts[21] of the conditioned medium necessary to remove IGF-binding protein (see below), compared with untreated medium, showed that displacement curves of the CM were parallel to authentic IGF-I. These results suggest that SCLC cell lines secrete a substance which interacts in the radioimmunological determination of IGF-I. At the moment it is not clear, whether this substance is IGF-I, IGF-II, or even another molecule which interacts in the radioimmunoassay. Minuto[22] confirmed our observations showing that the CM from the NSCLC cell line CALU released high amounts of IGF-I immunoreactivity measured by the same radioimmunoassay.

Gel filtration studies were performed to describe the molecular weight distribution of the immunreactive IGF-I in the conditioned media. Minuto et al.[22] applied a method under acidic conditions, which is suggested to separate IGF-I from its binding protein.[23] He achieved a chromatographic profile with peaks of immunreactivity in a high molecular weight region and in the elution volume corresponding to the molecular weight of authentic IGF-I. Detailed information as to the molecular weight of the high molecular fraction were not given. We also found two peaks of immunreactivity under neutral or acidic conditions (Fig. 2), one in the 40-60 KD and the other in the 6-7 KD molecular weight range. In contrast to Minuto et al.[22] we observed a delayed elution of the low molecular weight as compared to native IGF-I. This smaller IGF-I might have arisen by limited proteolysis of the IGF-I molecule. A report on truncated IGF-I isolated from uterus, pituitary and platelets is given recently by Ogasawara et al.[24] On the other hand the high molecular weight immunreactive IGF-I material could be a prohormone. However, since after acidic treatment only a part of the high molecular activity was shifted to the low molecular weight range we investigated whether the high molecular immunreactive material might be an IGF binding protein.[25] Fig. 3 shows radioactivity profiles of gel-filtrated ^{125}I-IGF-I, incubated with CM of SCLC-16HV. The profile clearly shows (broken line) a significant binding of ^{125}I-IGF-I to the higher molecular range. The addition of unlabeled IGF-I in the incubation mixture resulted in a competitive loss of radioactivity in the M_r 40-60 KD, demonstrating that the binding of ^{125}I-IGF-I was specific. Further data obtained from binding studies with ^{125}I-IGF-I revealed that this high molecular weight complex could be detected by a charcoal assay[26] or by a polyethylenglycol precipitation method.[27] The above-mentioned parallelism of the RIA displacement curves can therefore be explained by the presence of IGF-binding proteins in samples interfering with our radioimmunoassay. Thus, IGF-binding proteins may compete with the antiserum in the radioimmunoassay for the binding of added ^{125}I-IGF-I. Crosslinking experiments with CM of different cell lines confirmed this assumption. As shown in Fig. 4 IGF-binding proteins could be demonstrated applying an affinity-labeling technique similar to that described by D'Ercole.[28] 6/9 CM from SCLC cell lines produced this binding proteins. The addition of unlabeled IGF-I but not insulin

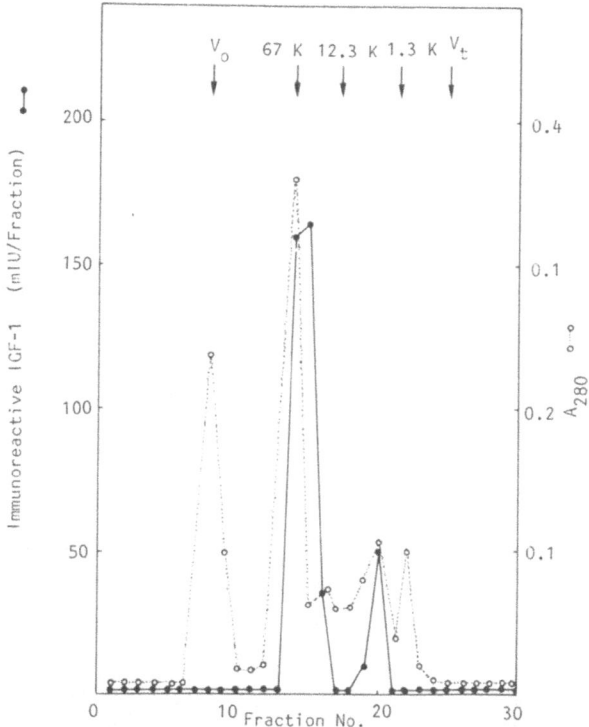

Fig. 2. The elution profile of conditioned medium of SCLC-16HV on a
Superose 12^{TM} column (Pharmacia, Fine Chemicals, Uppsala, Swe-
den) in 0.15 M NH_4HCO_3 after high-performance gel chromatography (broken line, optical density at 280 nm). The 1.0 ml fractions were collected at a flow rate of 0.2 ml per min. The
fractions were lyophilized, redissolved in RIA buffer and
assayed for IGF-I (solid line). Arrows indicate the elution
position of bovine serum albumin (67 k), cytochrom C (12.3 k),
cyanocobalmin (1.3 k), and the void volume (V_o).

inhibited this reaction. Insulin cross reacts with IGF-I in its binding
to the type I receptor, but does not bind to IGF-binding proteins. It
remains to be seen which biological role these IGF-binding proteins
play in the postulated autocrine/paracrine regulation in lung cancer.
Recent investigations indicate that they can modulate the biological
activity of the IGFs.[29] At the cellular level the IGF binding proteins
have been described to have both inhibitory and stimulatory effects on
the action of IGFs.[30,31]

IGF-I binding sites in lung cancer cell lines

Receptors for IGF-I have been already described on a variety of
other cells.[32] We, however, were the first to examine IGF-I binding
sites in several lung carcinoma cell lines, SCLC and NSCLC cell lines.
[15,33] They have been characterized by competition binding experiments
with ^{125}I-IGF-I and unlabeled IGF-I. The binding parameters (B_{max} and
K_D) for IGF-I were determined by Scatchard blot analysis. Nonspecific
binding was calculated from the difference between samples with or

251

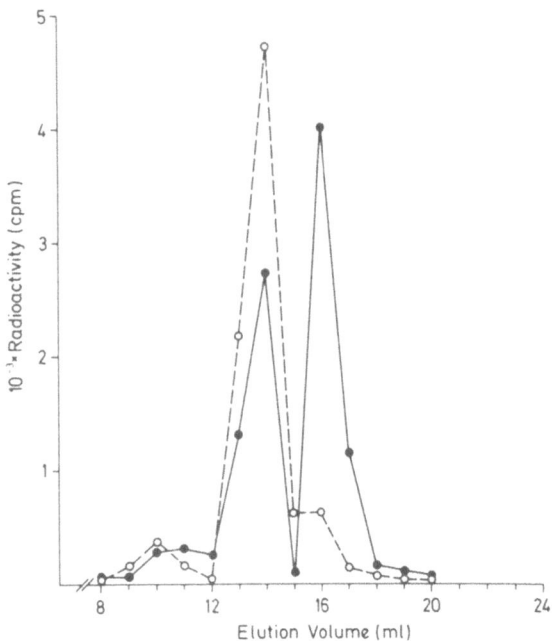

Fig. 3. Elution profile of radioactivity (cpm) on a Superose 12 column
when the conditioned medium of SCLC-16HV was incubated for 2 h
at 22°C with ^{125}I-IGF-I in the presence of 1 μg (●—●) or without (o---o) unlabelled IGF-I. The unlabelled IGF-I was added to
the incubation medium simultaneously with the labelled IGF-I
and the CM-16HV.

without an excess of unlabeled IGF-I. Fig. 5 shows the displacement of
^{125}I-IGF-I by increasing concentrations of unlabeled IGF-I for SCLC-
24H. The Scatchard analysis is consistent with one class of binding
sites and assuming a molecular mass of 7649 for IGF-I, we calculated a
KD of 0.89 nM with a B_{max} of 317 fmol/mg. The dissociation constants
(K_D) were similar for both small cell and non small cell lung cancer
cell lines. The amount of maximally bound IGF-I (B_{max}) ranged from 110
fmol/mg to 1230 fmol/mg. There were no characteristic differences
between small cell and non small cell lung cancer cell lines observed.
To our knowledge comparable datas are not reported concerning IGF
receptors on lung tumor cell lines. To investigate the type of receptor
present on the cell lines - type I or type II receptor - we performed
competition experiments with porcine insulin.[15] A 50% displacement of
^{125}I-IGF-I was achieved with 1 nM unlabeled IGF-I, compared to 100 nM
insulin, a 100-fold higher concentration. From these data we conclude
that the investigated cell lines possess the type I receptor for IGF-I.
Further studies in particular crosslinking and hybridization experiments are necessary to characterize these binding sites more precisely.

Mitogenic activity of IGF

One important action of the IGFs, apart from the anabolic effect
on cartilage cell and insulin-like effects on fat cells, is their
ability to stimulate DNA-synthesis and cell proliferation in fibroblasts and other cell types. This activity is termed multiplication-
stimulating activity.[34] In order to test whether the production and/or

252

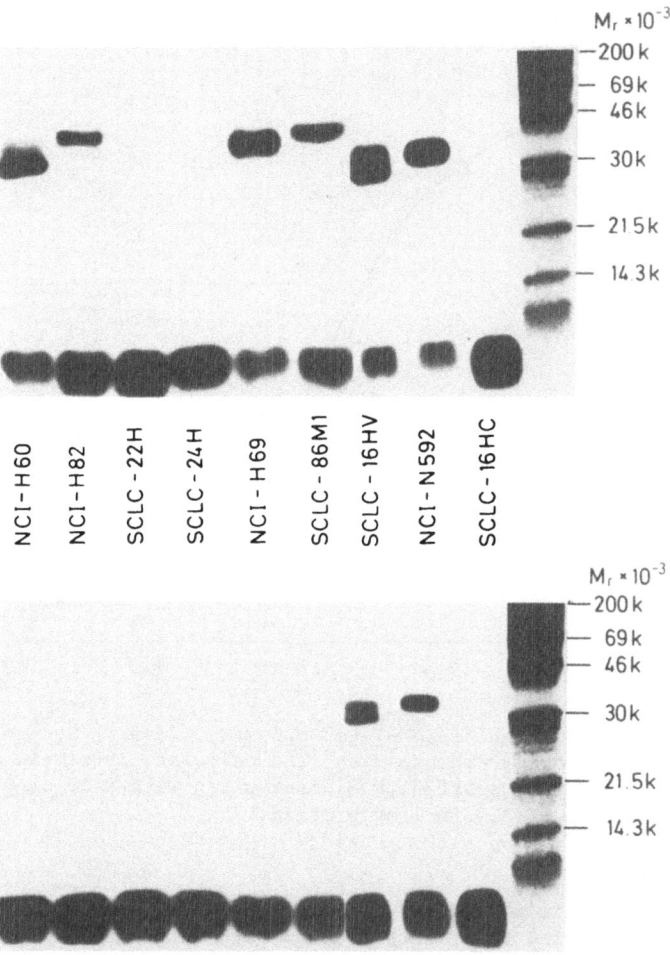

Fig. 4. Electrophoretic analysis (SDS-PAGE) and autoradiography of ^{125}I-IGF-I binding protein complexes in conditioned media of SCLC cell lines cross linked in the absence (upper gel) and in the presence of unlabelled IGF-I (1 µg) (lower gel) under non-reducing conditions. The right lane shows the positions of the mol wt markers M_r (x10^{-3}).

binding of the IGFs have effects on the proliferation on lung cancer cell lines, the mitogenic response to exogenously added and cell-derived IGF-I was studied by [³H] thymidine uptake.[15,16,22] We and others found a dose-related increase of [³H] thymidine uptake, reaching a maximum between 1-100 nM IGF-I, depending on the individual cell lines (Fig. 6). The stimulatory effect reached a factor of 1.2-4.0 and was comparable to the effect gained by cell-derived, partially purified IGF-related material.

This enhancement of DNA synthesis was seen in cell lines of the classic or variant subtypes of SCLC cultures as well as in NSCLC cell lines. The response to IGF-I, however, did not correlate to the

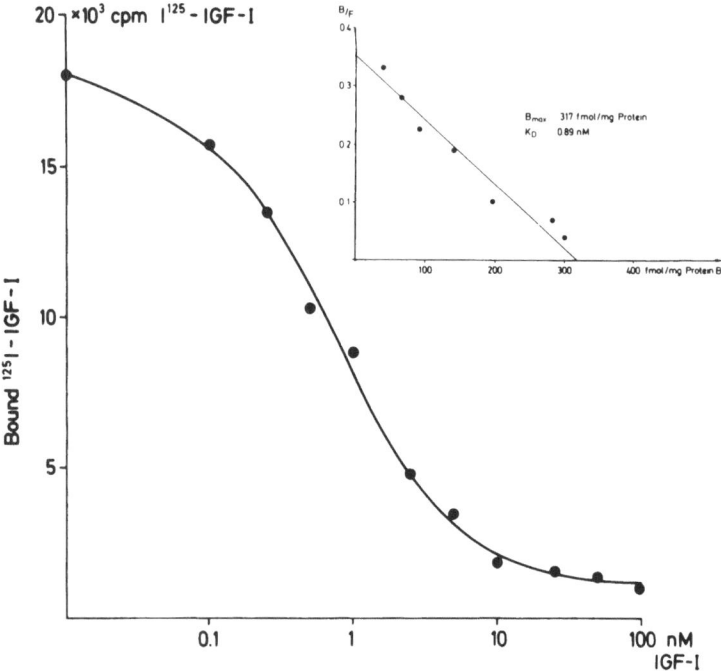

Fig. 5. Binding of ^{125}I-IGF-I to SCLC-24H. (Insert) Scatchard plot of
IGF-I dose-response data. The calculations revealed the presence of high-affinity binding sites with a K_D of 0.89 nM and
a B_{max} of 317 fmol/mg protein.

cellular synthesis of the IGF-I immunoreactive material since cell
lines not producing the factor reacted with exogenously added IGF-I or
cell lines producing the factor did not respond in the growth assay.

This discrepancy could be explained by cessation of production,
saturation of binding sites with endogenously produced factor or by
binding of authentic IGF-I to its binding protein. Fig. 7 demonstrates
that only a much higher insulin concentration (10 µg/ml) is necessary
to obtain the same half-maximal-growth stimulation with IGF-I. This
result is in agreement with our finding (see above) that a 50% displacement of ^{125}I-IGF-I from its receptor needs a 100-fold higher
insulin than IGF-I concentration.[15] The stimulatory effect of IGF-I was
reproduced by experiments showing a marked increase in clonal growth or
by a recently described semi-automated colorimetric assay.[15,36] The
addition of monoclonal antibodies against IGF-I[35] prevented the enhanced DNA-synthesis as shown in Tab.2.[37] Minuto additionally
demonstrated that a similar monoclonal antibody abolished the
mitogenic effect of the cell derived IGF-I.[22]

Preliminary data obtained in our laboratory indicate a growth
stimulation with the rat-derived (MSA, multiplication stimulating activity) and recombinant human IGF-II in SCLC cell lines. Interestingly,
the same cell line stimulated with IGF-I reacted to MSA (Fig. 8).

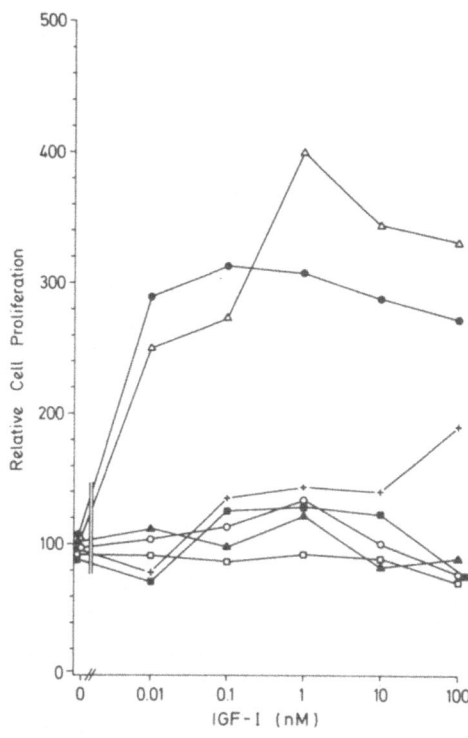

Fig. 6. Effect of IGF-I (recombinant analog of human IGF-I, Amersham International, UK) in different concentrations on proliferation of 7 SCLC cell lines. Cell lines are denoted by symbols as follows: SCLC-16HC (■), SCLC-16HV (Δ), SCLC-24H (●), NCI-H69 (o), NCI-H82 (+), NCI-H146 (▲), NCI-N592 (□). Relative cell proliferation represents the ratio (in cpm) of IGF-I treated versus not treated (control) cells measured by [^{3}H] thymidine uptake.

Table 2.   Inhibition of mitogenic effect by m AB 3 D1/2[35] in the presence of added IGF-I for SCLC-24H

conditions	cpm $\pm$ S
Antibody + 10^{-8} M IGF-I	
1: 5,000	132 $\pm$ 80
1: 10,000	415 $\pm$ 240
1: 100,000	1149 $\pm$ 199
Control	
10^{-8} M IGF-I	1659 $\pm$ 124

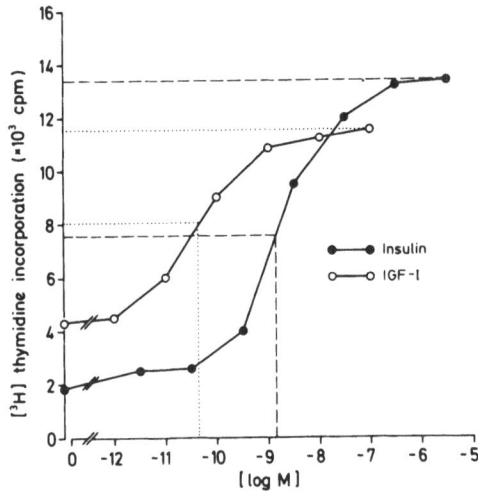

Fig. 7. Comparison of proliferative effects of insulin and IGF-I on
SCLC-22H. Broken lines show half maximal effects on [³H]
thymidine uptake (ordinate) and the corresponding concentration
(abscissa).

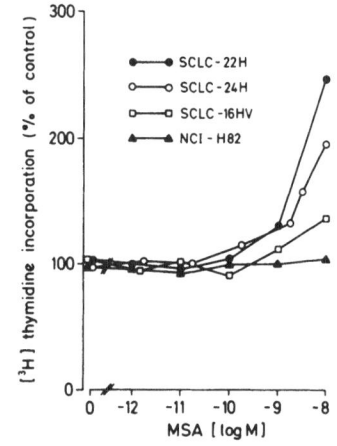

Fig. 8. Effect of MSA (multiplication stimulating activity) in
different concentrations on the proliferation of four SCLC
cell lines measured by [³H] thymidine uptake.

SUMMARY

The observation that malignant cells require less exogenous growth factors than normal cells, led to the hypothesis of an autocrine mechanism of growth control, i.e., tumor cells produce and respond to their own growth factors.[38] Gastrin-releasing peptide (GRP, mammalian equivalent of bombesin) was suggested to be an autocrine growth factor for SCLC. Most SCLC cell lines of the classic subtype produce GRP[39-42], about 20% of the cell lines tested bind detectable GRP[43] and some SCLC cell lines have been reported to respond to GRP.[10,44,45] The IGFs are shown to be another group of growth factors in lung cancer, which at least in SCLC fulfill all criteria of the autocrine/paracrine model. This report has shown production of immunreactive IGF-I, but so far no corresponding gene expression in lung cancer cell lines. Another important finding is the production of IGF binding proteins by most of the SCLC cell lines, a binding protein which might be different from the binding protein produced in liver and placenta. This binding protein could be important in the paracrine pathway by supporting the tumor cells with IGFs produced in the surrounding normal tissue. One has to keep in mind, however, that this binding protein might interfere with the radioimmunoassay for IGF-I and with the receptor binding assays. Most cell lines express IGF-binding sites on their cell surfaces, but the interpretation of these data may be difficult if IGF binding proteins are attached to the cell membrane. Authentic and cell derived IGFs stimulate growth of lung cancer cell lines and here again it has to be investigated, whether the IGF-binding proteins enhance or inhibit the effects of the IGFs on the lung cancer cell lines.

ACKNOWLEDGEMENTS

The work was supported by the Deutsche Forschungsgemeinschaft. We wish to thank C. Born for her secretarial assistance, K. Soltz and Dr. A. Rodden for correcting the manuscript.

REFERENCES

1. D. M. Parkin, E. Läärä, and C. S. Muir, Estimates of the worldwide frequency of sixteen major cancers in 1980, Int. J. Cancer 41 (1988 in press).
2. World Health Organization, Female lung cancer increases in developed countries, Weekl. epidem. rec. 61:297 (1986).
3. J. D. Minna, G. A. Higgins, and E. J. Glatsten, Cancer of the lung, in: "Principles and Practice of Oncology," V. T. de Vita, ed., J. B. Lippincott, Philadelphia (1981).
4. A. F. Greco, "Small cell lung cancer," Grune and Stratton, New York (1981).
5. G. Bepler, G. Jaques, K. Neumann, G. Aumüller, C. Gropp, and K.Havemann, Establishment, growth properties, and morphological characteristics of permanent human small cell lung cancer cell lines, J. Cancer Res. Clin. Oncol. 113:31 (1987).
6. A. F. Gazdar, D. N. Carney, E. K. Russell, H. L. Sims, S. B. Baylin, P. A. Bunn, J. G. Guccion, and J. G. Minna, Establishment of continuous clonable cultures of small cell carcinoma of the lung which have amine-precursor uptake and decarboxylation properties, Cancer Res. 40:3502 (1980).

7. O. S. Pettengill, G. D. Sorenson, D. H. Wurster-Hill, T. J. Curphey, W. W. Noll, C. C. Cate, and L. H. Maurer, Isolation and growth characteristics of continuous cell lines from small cell carcinoma of the lung, Cancer 45:906 (1980).

8. J. Bergh, K. Nilsson, R. Ekman, and B. Giovanella, Establishment and characterization of cell lines from human small and large cell carcinomas of the lung, Acta path. microbiol immun. Scand., Sect. A 93:133 (1985).

9. H. Baillie-Johnson, P. R. Twentyman, N. E. Fox, G. A. Walls, P. Workman, J. V. Watson, N. Johnson, J. G. Reeve, and N. M. Bleehen, Establishment and characterization of cell lines from patients with lung cancer (predominantly) small cell carcinoma, Br. J. Cancer 52:495 (1985).

10. D. N. Carney, A. F. Gazdar, G. Bepler, J. G. Guccion, P. J. Marangos, T. W. Moody, M. H. G. Zweig, and J. D. Minna, Establishment and identification of small cell lung cancer cell lines having classic and variant features, Cancer Res. 45:2913 (1985).

11. P. L. Kaplan, and B. Ozaune, Cellular responsiveness to growth factors correlates with a cell's ability to express the transformed phenotype, Cell 33:931 (1983).

12. F. U. Cuttitta, D. N. Carney, J. Mulshine, T. W. Moody, J. Fedorko, A. Fischler, and J. D. Minna, Bombesin-like peptides can function as autocrine growth factors in human small cell lung cancer, Nature 316:823 (1985).

13. A. J. D'Ercole, A. D. Stiles, and L. E. Underwood, Tissue concentration of somatomedin-C: further evidence for multiple sites of synthesis and paracrine or autocrine mechanism of action, Proc. Natl. Acad. Sci. USA 81:935 (1984).

14. F. Minuto, P. Del Monte, A. Barreca, P. Fortini, G. Cariola, G. Catrambone, and G. Giordano, Evidence for an increased somatomedin-C/insulin-like growth factor I content in primary human lung tumors, Cancer Res. 46:985 (1986).

15. G. Jaques, M. Rotsch, C. Wegmann, U. Worsch, M. Maasberg, and K. Havemann, Production of immunoreactive insulin-like growth factor I and response to exogenous IGF-I in small cell lung cancer cell lines, Exp. Cell Res. 176:336 (1988).

16. V. M. Macauly, J. D. Teale, M. J. Everard, G. P. Joshi, I. E. Smith, and J. L. Millar, Somatomedin-C/insulin-like growth factor-I is a mitogen for human small cell lung cancer, Br. J. Cancer 57:91 (1988).

17. C. Betsholtz, J. Bergh, M. Bywater, M. Petterson, A. Johnson, C.-H. Heldin, R. Ohlsson, T. J. Knott, J. Scott, G. I. Bell, and B. Westermark, Expression of multiple growth factors in a human lung cancer cell line, Int. J. Cancer 39:502 (1987).

18. G. I. Bell, J. P. Merryweather, R. Sanchez-Pescator, M. M. Stempien, L. Preistly, J. Scott, and L. B. Rall, Sequence of a cDNA clone encoding human preproinsulin-like growth factor II, Nature 310:775 (1984).

19. D. Yee, R. E. Favoni, K. K. Huff, S. Paik, R. B. Dickson, G. S. Lebovic, A. Schwartz, M. E. Lippman, and N. Rosen, Insulin-like growth factor I (IGF-I) expression and a novel IGF-I related activity in human malignancy, Proc. Am. Assoc. Cancer Res. 29:54 (1988).

20. F. Cuttitta, M. L. Levitt, J. G. Park, P. Kasprzyk, Y. Nakanishi, J. Reeve, J. Walsh, J. Mulshine, A. F. Gazdar, and J. D. Minna, Growth of human cancer cell lines in unsupple-

mented basal media as a means of identifying autocrine growth factors, Proc. Am. Assoc. Cancer Res. 28:27 (1987).

21. W. H. Daughaday, I. K. Mariz, and S. L. Blethen, Inhibition of bound somatomedin to membrane receptor and immunobinding sites: a comparison of radioreceptor and radioimmunoassay of somatomedin in native and acid-ethanol-extracted serum, J. Clin. Endocrinol. Metab. 51:781 (1980).

22. T. Minuto, P. Del Monte, A. Barreca, A. Alama, G. Cariola, and G. Giordano, Evidence for autocrine mitogenic stimulation by somatomedin C/insulin-like growth factor I on an establish human lung cancer cell line, Cancer Res. 48:3716 (1988).

23. J. M. Horner, F. Liu, and Hintz, Comparison of $[^{125}I]$ somatomedin A and $[^{125}I]$ somatomedin C radioreceptor assays for somatomedin peptide content in whole and acid chromatographed plasma, J. Clin. Endocrinol. Metab. 47:1287 (1978).

24. M. Ogasawara, K. P. Karey, and D. A. Sirbasku, Insulin-like growth factor I: relationship between plasma and tissue sources, biological potency and N α truncation, Proc. Am. Assoc. Cancer Res. 29:52 (1988).

25. G. Jaques, and K. Havemann, IGF-binding protein in small cell lung cancer cell lines, Exp. Cell Res. (1988 submitted).

26. C. D. Scott, J. L. Martin, and R. C. Baxter, Production of insulin-like growth factor I and its binding protein by adult rat hepatocytes in primary culture, Endocrinology 116:1094 (1985).

27. M. A. De Vroede, L. Y. H. Tseng, G. Pamayotis, S. Katsoyannis, S. P. Nissley, and M. M. Rechler, J. Clin. Invest. 77:602 (1986).

28. A. J. D'Ercole, and J. R. Wilkins, Affinity labeled somatomedin C binding proteins in rat sera, Endocrinology 114:1141 (1984).

29. D. R. Clemmons, R. G. Elgin, V. K. M. Han, S. J. Casella, A. J. D'Ercole, and J. J. Van Wyk, Cultured fibroblast monolyers secrete a protein that alters the cellular binding of somatomedin C-insulin-like growth factor I, J. Clin. Invest. 77:1548 (1986).

30. O. Ritvos, T. Ranta, J. Jalkanen, A. H. Suikkari, R. Voutilainen, H. Bohn, and E. M. Rutanen, Insulin-like growth factor (IGF) binding protein from human decidua inhibits the binding and biological action of IGF-I in cultured choriocarcinoma cells, Endocrinology 122:2150 (1988)

31. R. G. Elgin, W. H. Busby, and D. R. Clemmons, An insulin-like growth factor (IGF) binding protein enhances the biologic response to IGF-I, Proc. Natl. Acad. Sci. 84:3254 (1987).

32. P. S. Nissley, and M. M. Rechler, Insulin-like growth factors: biosynthe is, receptors, and carrier proteins, in: "Hormonal proteins and peptides," Li Ch., ed., Academic Press, Orlando (1984).

33. M. Rotsch, U. Worsch, G. Jaques, M. Maasberg, and K. Havemann, Insulin-like growth factor I binding sites in non small cell lung cancer cell lines, J. Cancer Res. Clin. Oncol. 114:542 (1988).

34. W. D. Salmon, Jr., and M. R. DuVall, A serum fraction with "sulfation factor activity" stimulates in vitro incorporation of leucine and sulfate into protein-polysaccharide complexes, uridine into RNA and thymidine into DNA of costal cartilage from hypophysectomized rats, Endocrinology 86:721 (1970).

35. R. C. Baxter, S. Axiak, and R. L. Raison, Monoclonal antibody against human somatomedin-C/insulin-like growth factor I, J. Clin. Endocrinol. Metab. 54:474 (1980).

36. Y. Nakanishi, F. Cuttitta, P. Kasprzyk, I. Avis, S. Steinberg, A. Gazdar, and J. Mulshine, Valiation of a rapid colorimetric assay demonstrating growth stimulatory effects including insulin-like growth factor-I (IGF-I) on small cell lung cancer cell lines (SCLC-CL), Proc. Am. Assoc. Cancer Res. 28:53 (1987).

37. G. Jaques, C. Wegmann, M. Rotsch, M. Maasberg, and K.Havemann, Production and secretion of immunreactive insulin-like growth factor I, J. Cancer Res. Clin. Oncol. 114:S154 (1988).

38. M. B. Sporn, and G. J. Todaro, Autocrine secretion and malignant transformation of cells, N. Engl. J. Med. 303:878 (1980).

39. S. M. Wood, J. R. Wood, M. A. Glatei, Y. C. Lee, D. O'Shanghnessy, and S. R. Bloom, Bombesin, somatostatin and neurotensin-like immunreactivity in bronchial carcinoma, J. Clin. Endocrinol. Metab. 53:1310 (1981).

40. W. Luster, C. Gropp, H. F. Kern, and K. Havemann, Lung tumour cell lines synthesizing peptide hormones established from tumours of four histological types: characterization of the cell lines and analysis of their peptrole hormone production, Br. J. Cancer 51:865 (1985).

41. T. W. Moody, C. P. Pert, A. F. Gazdar, D. N. Carney, and J. D. Minna, High levels of intracellular bombesin characterize human small cell lung carcinoma, Science 214:1246 (1981).

42. G. D. Sorenson, S. R. Bloom, M. A. Ghateis, S. A. Del Prete, C. C. Cate, and O. S. Pettengill, Bombesin production by human small cell carcinoma of the lung, Regul. Pept. 4:59 (1982).

43. T. W. Moody, V. Bertness, and D. N. Carney, Bombesin-like peptides and receptors in human tumor cell lines, Peptides 4:683 (1983).

44. S. Weber, J. E. Zuckermann, D. G. Bostwick, K. G. Bensch, B. I. Sikic, and T. A. Raffin, Gastrin releasing peptide is a selective mitogen for small cell lung carcinoma in vitro, J. Clin. Invest. 75:306 (1985).

45. D. N. Carney, F. Cuttitta, T. W. Moody, and J. D. Minna, Selective stimulation of small cell lung cancer clonal growth by bombesin and gastrin-releasing peptide, Cancer Res. 47:821 (1987).

STRUCTURAL BASIS FOR LIGAND-DEPENDENT TRANSMEMBRANE SIGNALLING OF THE INSULIN AND IGF-1 RECEPTOR KINASES

Jeffrey E. Pessin and Judith L. Treadway

Department of Physiology and Biophysics
The University of Iowa
Iowa City, IA

Introduction

Insulin and IGF-1 modulate a number of acute metabolic and long term growth-related responses in target cells (for several comprehensive reviews see Refs. 1-11). The initial requisite for these insulin and IGF-1 dependent biological responses is the presence of specific high affinity receptor proteins capable of transducing the cell surface binding events into an intracellular signal. A key issue fundamental to all plasma membrane receptor-mediated events is the understanding of the molecular mechanisms responsible for the flow of information across the phospholipid bilayer. An implicit assumption is that the receptor complexes themselves contain the necessary structural information to transmit the extracellular binding event to a recognizable intracellular signal, namely activation of tyrosine-specific protein kinase activity. Thus, a detailed understanding of the molecular properties of these receptor species is central in elucidating the ligand-dependent transmembrane signalling mechanism responsible for the pleiotrophic responses to hormone binding. We emphasize that although many investigators use the term transmembrane signalling in reference to the generation of second messengers and different receptor-mediated biological responses, we will use this term exclusively to refer to the ability of ligand binding to propagate an intramolecular transmembrane signal within the holoreceptor molecule itself. It is this primary event which results in the activation of the receptor intracellular tyrosine-specific protein kinase domain.

Insulin and IGF-1 Receptor Structure

The insulin and IGF-1 receptors share a large degree of structural as well as functional properties. Both these mature holoreceptor molecules are minimally composed of two identical α subunits (Mr $\sim$ 135,000) and two identical β subunits (Mr ~ 95,000) disulfide-linked into an $\alpha_2\beta_2$ heterotetrameric complex. Small differences in the Mr between the IGF-1 and insulin receptor α and β subunits in various tissues reflects alterations in post-translation modification since both receptor species are synthesized from essentially identical α-β fusion protein precursors of Mr = 151,869 and Mr = 152,784 for the IGF-1 and insulin receptor, respectively.[12-14] The insulin receptor precursor is initially synthesized in the endoplasmic reticulum where it is cotranslationally acylated with myristic and palmitic acids and glycosylated into a high mannose form.[15-17] The receptor is then transferred to the Golgi apparatus where it undergoes additional processing, including removal of mannose residues, complex addition of other monosaccharides, proteolytic cleavage of the $\alpha\beta$ precursor, and assembly into an $\alpha_2\beta_2$ heterotetrameric complex.[15-21]

The IGF-1 and insulin receptor α subunits reside exclusively on the extracellular face of the plasma membrane and encode for the high affinity ligand binding domain.[22,23] In analogy with the low density lipoprotein receptor,[24] the α subunits of the insulin and IGF-1 receptors contain a highly related N-terminal cysteine rich domain thought to be involved in the high affinity ligand binding activity displayed by these receptors.[25] The α subunits are generally thought to be disulfide-linked to each other through what has operationally been defined as Class I disulfide bonds.[26,27] The α subunits are also anchored to the plasma membrane by disulfide linkages to the extracellular portion of the transmembrane β subunits (Class II disulfide bonds). The intracellular region of the β subunits contains the ATP binding domain,[28-30] tyrosine-specific autophosphorylation acceptor sites,[31-33] the intrinsic substrate protein kinase activity of the holoreceptors[34-42] and serine and threonine phosphoacceptor sites for other endogenous protein kinases.[43-45] The deduced sequences of the IGF-1 and insulin receptors predict an overall amino acid similarity of approximately 50% with the highest degree of sequence identity (84%) occurring in kinase domain.[12-14] The largest degree of sequence divergence between these receptors occurs in the α subunit, extracellular portion of the β subunit, and C-terminal domain (approximate 45% identity). A schematic model depicting the major structural features of the insulin and IGF-1 receptors is presented in Figure 1.

Insulin and IGF-1 Receptor Ligand Binding Properties

Insulin binding to cells, membranes, detergent soluble preparations, and purified insulin receptors has been well established to display curvilinear properties when analyzed by the method of Scatchard.[46] This nonlinear binding characteristic has been interpreted to suggest the presence of at least two different insulin receptor affinity states[47-49] or negative cooperativity of insulin binding.[50-52] Evidence for the negative cooperativity of insulin binding has been substantiated by observations that the addition of unlabeled insulin to receptors prebound with ^{125}I-insulin enhances the rate of insulin dissociation.[50-55] In addition, the insulin

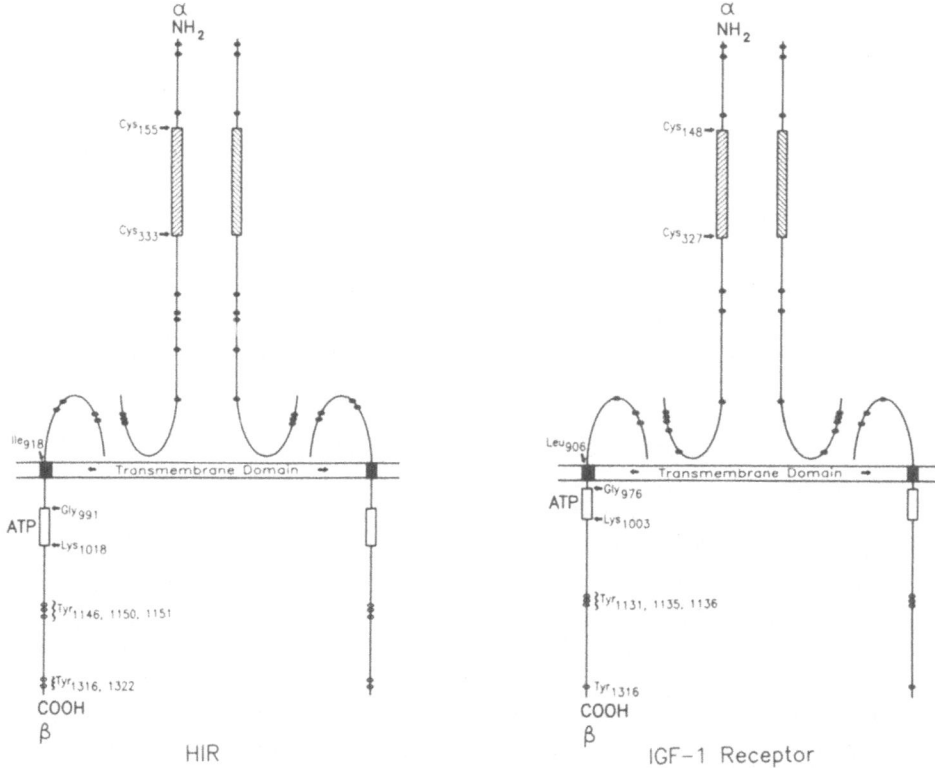

Figure 1. Schematic representation of the major structural
domains of the human insulin and IGF-1 receptors. The hatched region in the α subunits signifies a cysteine rich domain;
individual cysteine residues are denoted by a dark circle.
The looped region between the α and β subunits represents a
proposed protected hydrophobic pocket in which the class II
disulfides are thought to reside. The transmembrane domain
is represented by the dark rectangle and the intracellular
ATP binding pocket is denoted by the open box. The intracellular tyrosine-specific autophosphorylation acceptor sites
are indicated by sequence position and with an open circle.

domains responsible for the negative cooperative interactions within the
insulin receptor have been identified.[56] More recent data has identified
unique site(s) within the insulin receptor α subunit that are involved in
the complex nature of insulin binding.[57]

In contrast to the insulin receptor, IGF-1 binding to the IGF-1
receptor (cell, membrane, detergent soluble preparations and purified
receptors) displays relatively homogeneous ^{125}I-IGF-1 binding properties
with no indication of negative cooperative interactions.[2,58,59] Although
^{125}I-IGF-1 binding has often been observed to generate Scatchard plots
which slightly deviate from linearity,[60-62] they are never as concave as
those typically observed for ^{125}I-insulin binding to the insulin receptor.
The non-linear binding of IGF-1 may reflect the presence of multiple IGF-1

receptor forms which have recently been resolved by immunoprecipitation[60] and differential recovery from insulin and IGF-1 affinity chromatography columns.[62]

To examine the role of subunit interactions in the ligand binding properties of the insulin and IGF-1 receptors, methods have recently been developed to obtain functional $\alpha\beta$ heterodimeric receptor complexes from native $\alpha_2\beta_2$ heterotetrameric holoreceptors.[61,63-67] This procedure involves treatment of membranes or soluble receptors with low concentrations of DTT (1-2mM) under alkaline conditions (pH ~8.5) to selectively reduce the Class I disulfide bonds responsible for the covalent linkage between $\alpha\beta$-$\alpha\beta$ heterodimeric subunits. Since prolonged exposure to DTT and alkaline pH can denature the receptor and impair function, the receptors are incubated for a short period of time (5-30 min) followed by rapid removal of DTT and restoration to neutral pH by either centrifugation and washing for membranes or rapid gel filtration in the case of soluble receptors. A flow chart of the procedure for isolating $\alpha\beta$ heterodimeric insulin and IGF-1 receptor complexes from membranes or soluble $\alpha_2\beta_2$ heterotetrameric receptor preparations is outlined in Figure 2.

In contrast to the typical curvilinear Scatchard plot of insulin binding to the $\alpha_2\beta_2$ heterotetrameric insulin holoreceptor complex, isolated detergent soluble $\alpha\beta$ heterodimers generate a linear binding isotherm with no evidence for negative cooperative interactions.[64-66] The affinity for insulin binding to the purified human placenta $\alpha\beta$ heterodimeric complex (~1.0nM) was found to be intermediate to the high affinity (~0.2nM) and low affinity (~40nM) binding components found for the $\alpha_2\beta_2$ heterotetrameric

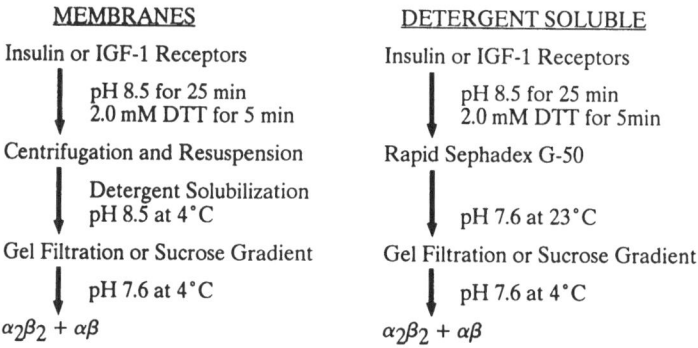

Figure 2. Flow diagram for the isolation of $\alpha\beta$ hetero-dimeric insulin or IGF-1 receptor complexes from the native $\alpha_2\beta_2$ heterotetrameric state. This method is applicable to both membrane-associated receptors as well as detergent soluble receptors.[63,65]

insulin receptor complex.[65] These data collectively indicate that $\alpha\beta$-$\alpha\beta$ insulin receptor subunit interactions are necessary for the high affinity curvilinear binding of insulin to native $\alpha_2\beta_2$ heterotetrameric receptor complexes.

This conclusion is further supported by the observation that alkaline pH and DTT treated placental membrane receptors (nondissociated) display curvilinear insulin binding despite the fact that the Class I disulfide bonds have been reduced.[64] However, subsequent detergent solubilization and dissociation of the alkaline pH and DTT treated membrane receptors results in the linear insulin binding typical of the $\alpha\beta$ heterodimeric state (Figure 3). Thus, when the Class I disulfide bonds between the $\alpha\beta$ heterodimeric insulin receptor subunits are reduced in the placental membrane these subunits remain associated through noncovalent interactions and still display complex insulin binding behavior. Once these interactions are disrupted by the physical separation of the $\alpha\beta$ heterodimers upon detergent solubilization and dilution, homogeneous insulin binding is obtained.

It has been postulated that insulin receptors exist in different subunit association states in the cell surface membranes of target cells. Solubilized membranes from cultured hepatocytes, rat liver, turkey erythrocytes and human placenta have all been observed to contain varying proportions of both the $\alpha_2\beta_2$ heterotetrameric and $\alpha\beta$ heterodimeric insulin recep-

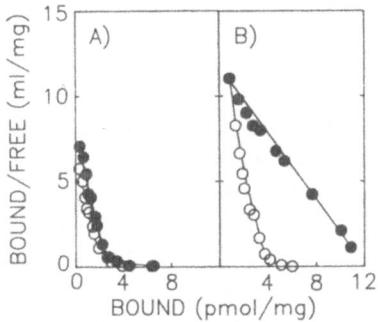

Figure 3. Scatchard analysis of insulin binding to the alkaline pH and DTT treated placental membranes and detergent soluble insulin receptors. A) The placental membranes were incubated at pH 8.5 for 25 min followed by an additional 5 min in the absence (o) or presence (•) of 2mM DTT. The membranes were washed by centrifugation, resuspended and ^{125}I-insulin binding was determined. B) The identical placental membranes treated above were solublized in Triton X-100 and partially purified by Bio-Gel A-1.5m gel filtration chromatography. The isolated $\alpha\beta$ heterodimeric and $\alpha_2\beta_2$ heterotetrameric insulin receptors were then subjected to Scatchard analysis of ^{125}I-insulin binding.

tor complexes.[68-73] However, recent data have indicated that these $\alpha\beta$ heterodimeric insulin receptor complexes do not exist in intact cells, but rather are generated by reduction of the insulin receptor during the preparation of crude membranes. This conclusion is based on the absence of $\alpha\beta$ heterodimeric insulin receptor complexes in samples from membranes prepared in the presence of specific sulfhydryl or oxidizing agents (unpublished results). In addition, ^{125}I-insulin crosslinking to intact cells uniquely identifies the $\alpha_2\beta_2$ heterotetrameric receptor complex.[74]

The role for $\alpha\beta$-$\alpha\beta$ subunit interactions in the regulation of ligand binding by IGF-1 receptors is less well understood, due in large part to the existing variability of results for Scatchard binding to the control $\alpha_2\beta_2$ heterotetrameric IGF-1 receptors. Results from our lab have consistently demonstrated that the partially purified IGF-1 receptor yields a homogeneous ^{125}I-IGF-1 binding plot. We have also observed that IGF-1 binding to the partially purified $\alpha\beta$ heterodimeric IGF-1 receptor[61] generates a linear Scatchard plot with ligand binding affinity similar to that found for the $\alpha_2\beta_2$ heterotetrameric IGF-1 receptor.[59,61] These data differ from the results of others reporting a more complex binding isotherm for the $\alpha_2\beta_2$ heterotetrameric form of the IGF-1 receptor,[62] and reduced binding affinity by the isolated purified human placenta $\alpha\beta$ heterodimeric IGF-1 receptor complex.[66] The reason for this discrepancy in ^{125}I-IGF-1 binding properties is not yet apparent. Nevertheless, these data do collectively support a role for $\alpha\beta$-$\alpha\beta$ heterodimeric subunit interaction in the ligand stimulation of protein kinase activity (see below).

Structural Properties of the Receptor Kinases

The intrinsic tyrosine-specific protein kinase activity of the insulin and IGF-1 holoreceptors is similar to many other protein kinases in that it displays the ability to autophosphorylate as well as to phosphorylate exogenous substrates. Although the role for autophosphorylation in other receptors with intrinsic protein tyrosine kinase activity (i.e. the EGF receptor kinase) has been controversial,[75-77] it is clear that β subunit autophosphorylation of both the insulin and IGF-1 receptors results in ligand-independent protein kinase activation.[41,78-83] The major *in vivo* and *in vitro* tyrosine autophosphorylation acceptor sites of the insulin receptor have been recently identified.[31-33] Using the numbering system of Ullrich et al.,[12] autophosphorylation occurs predominantly at tyrosine residues 1146, 1151, 1152, 1316 and 1322. The precise order of autophosphorylation and correlation of these autophosphorylation sites with protein kinase activation has not been clearly established to date. It has been suggested that autophosphorylation is a highly concerted reaction in which each phosphotyrosine residue contributes proportionately to kinase activation.[84] However, another study has suggested the protein kinase activation tracks specifically with the autophosphorylation of tyrosine residues 1151 and 1152.[33] Consistent with the notion that the 1146-1152 domain is critical for insulin receptor kinase function, truncation of the 1316-1322 tyrosine-containing domain from the insulin receptor has no effect on insulin-stimulated protein kinase activity[85] although this mutant is defective in insulin stimulation of biological responsiveness.[86] Further, substitution of phenylalanine for the 1146-1152 tyrosine residues greatly reduces the biological response of cells to insulin as well as rendering these insulin receptor mutants relatively kinase inactive.[87]

Differences in the number and combination of phosphotyrosine acceptor sites between *in vivo* and *in vitro* autophosphorylated insulin receptors have recently been reported.[33] It has been observed that the phosphopeptide containing tyrosine residues 1146, 1151 and 1152 undergoes bis-phosphorylation (i.e. either 1146 plus 1151 or 1152) in response to insulin *in vivo*, but undergoes tris-phosphorylation *in vitro* (1146, 1151, and 1152). This is in contrast to another study in which the tris-phosphorylated form of the 1146-1152 domain of the insulin receptor was not observed.[31] Interestingly, since there are apparently four or five major autophosphorylation acceptor sites per β subunit, the maximal stoichiometry of autophosphorylation in the $\alpha_2\beta_2$ heterotetrameric holoreceptor complex would be expected to be either 8 or 10. However, the maximal extent of insulin receptor autophosphorylation reported has not exceeded 4 mol phosphate/mol holoreceptor.[88] Whether this represents an asymetric pattern of insulin receptor autophosphorylation remains to be determined. Due to the highly conserved sequence similarity between the insulin and IGF-1 receptors, particularly in the tyrosine kinase domain, it is anticipated that relatively equivalent tyrosine sites of autophosphorylation are present in the IGF-1 receptor. However, this will also require direct experimental verification.

In the purified state, the ligand stimulation of receptor protein kinase activity occurs exclusively on tyrosine residues due to an intramolecular autophosphorylation reaction.[29,89,90] However, ligand stimulation of both the insulin and IGF-1 receptors *in vivo* results in a rapid tyrosine autophosphorylation (<1 min) followed by a slower (>10 min) increase in phosphoserine and phosphothreonine content.[35,45,91,92] The increase in phosphoserine/threonine results from either an insulin-dependent activation of intracellular serine/threonine specific protein kinases and/or inhibition of serine/threonine specific protein phosphatases. Recently, several reports have indicated that a serine/threonine protein kinase activity may be co-purified in association with the insulin receptor.[93-95] The functional consequence of the insulin-dependent serine/threonine phosphorylation of the insulin receptor has not yet been established. However, in analogy with phosphorylation of the insulin receptor by protein kinase C and protein kinase A,[96-101] it is thought that serine/threonine phosphorylation results in an attenuation of the tyrosine-specific insulin and IGF-1 receptor protein kinase activities (ie. desensitization).

Several potential *in vivo* target substrate proteins for the insulin and IGF-1 receptor kinases have been suggested.[102-108] Like most protein kinases, these receptor kinases are very prolific *in vitro* and phosphorylate a variety of proteins, albeit with different efficacies. Differences in the *in vitro* substrate specificities between the insulin and IGF-1 receptors have recently been observed.[109] Insulin stimulation of the insulin receptor kinase primarily results in an increase in the Vmax of substrate phosphorylation with little or no change in Km.[34,110] The Vmax for ADP formation from ATP has not been directly determined but presumably reflects the Vmax of substrate phosphorylation since these reactions occur on an equimolar basis. The Km for ATP has been much more difficult to determine. Several studies have suggested that insulin decreases the Km for ATP[111] whereas other reports indicate no effect.[110,112] However, interpretation of many of these studies is problematic due to the increase in the receptor protein kinase activity by increasing the extent of receptor autophosphorylation under the assay conditions employed. Thus, only the Km for ATP of the maximally activated (autophosphorylated) receptors can be reliably estimated.[113] In addition, the Vmax and Km for autophosphorylation cannot be determined since the reaction occurs in an intramolecular manner such that the concentrations of substrate and catalytic site remain constant.[29,89,90]

The insulin and IGF-1 receptors also share a common divalent metal ion dependence for Mn as well as MgATP in order to display ligand-stimulated protein kinase activity *in vitro*.[34,110,112,114] Mn appears to decrease the insulin receptor kinase Km for ATP in the standard *in vitro* protein kinase assays. Typical ligand-dependent stimulation requires the presence of supraphysiological concentrations of Mn (2-10mM) for both autophosphorylation and substrate phosphorylation. Recently, basic proteins such as protamine sulfate, histone H2fb and polylysine have been observed to markedly activate the insulin-dependent insulin receptor protein kinase activity nearly 100 fold.[115,116] The stimulatory effect of these agents occurs only at reduced divalent metal ion concentrations (0.1mM) similar to that found *in vivo*.[115,117] Surprisingly, although polylysine was found to stimulate the insulin-dependent insulin receptor kinase it was completely ineffective in altering the IGF-1 stimulation of the IGF-1 receptor kinase (Figure 4). These data suggest that specific basic intracellular proteins may regulate the metal ion dependence, substrate specificity, and enzymatic activity of the insulin and IGF-1 receptor kinases *in vivo*. Direct support for this hypothesis will require the identification and ultimate purification of such putative regulatory molecules.

Receptor Kinase and Biological Responsiveness

Since the original observation that the insulin receptor is an

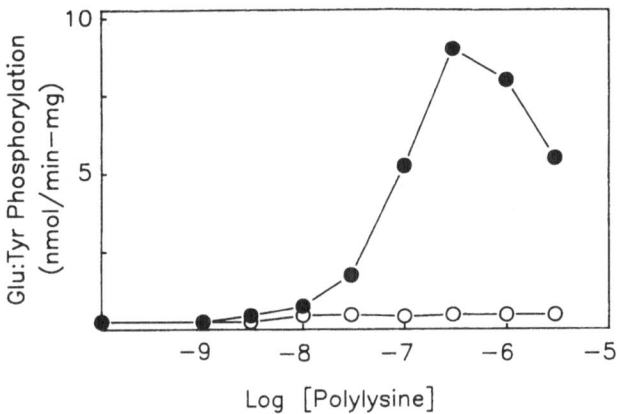

Figure 4. Polylysine concentration dependence of the
insulin and IGF-1 receptor protein kinase activity.
The partially purified insulin (●) and IGF-1 (o) receptors were incubated with 0.1mM Mg/Mn and various concentrations of polylysine in the presence of 100nM insulin
or 50nM IGF-1, respectively. The initial rate of
Glu:Tyr phosphorylation was determined.

insulin-stimulated tyrosine-specific protein kinase,[33,118] the data supporting the requirement of the kinase in the insulin stimulation of biological responsiveness has vacillated. Several lines of evidence have
strongly supported the requirement for insulin stimulation of the protein
kinase activity in insulin action. Cells transfected with the human
insulin and IGF-1 receptors display enhanced biological sensitivity with no
significant effect on responsiveness whereas cells containing kinase
deficient mutant insulin receptors have reduced biological sensitivity to
insulin.[87,119-121] Further, specific anti-insulin receptor or phosphotyrosine antibodies which block the insulin receptor protein kinase activity
inhibit insulin action.[122-124] In contrast, several studies have suggested
that the insulin receptor protein kinase activity may not be required for
insulin action. Specific anti-insulin receptor monoclonal antibodies have
been observed to stimulate glucose transport activity without activating
the insulin receptor protein kinase.[125] In addition, expression of insulin
receptors containing a point mutation that converts tyrosine 960 to phenylalanine does not alter the insulin-dependent β subunit autophosphorylation
but does inhibit insulin-stimulated biological responsiveness as well as *in
vivo* substrate phosphorylation.[126]

Obviously a key issue which remains to be resolved is the precise
role for the insulin receptor protein kinase activity in biological responsiveness. Based upon the current data, we postulate that insulin-stimu-

lated autophosphorylation alters the conformation of the insulin receptor such that it becomes capable of interacting with some as yet undefined effector molecule(s). For some of the insulin-dependent responses this post-receptor effector system may involve the stimulation of a phosphorylation cascade or alternatively the generation of an intracellular insulin mediator.[127] Agents which elicit insulin responses independent of insulin receptor kinase activation may do so by inducing a conformational change similar to that induced by autophosphorylation, and thereby coupling with the same intermediate effector. The activation of this postulated effector then propagates the chain of events necessary for cellular responsiveness to insulin and IGF-1. In either case, it is clear that under normal conditions the ligand-stimulated activation of the tyrosine-specific protein kinase of these receptors is initially required to induce the conformational changes necessary for activation of the intrinsic substrate protein kinase activity and/or interaction with a putative effector system.

A Molecular Hypothesis for Transmembrane Signalling

Over the past five years, our primary effort has been to address the molecular basis of the insulin and IGF-1 receptor transmembrane signalling event. As originally stated in the introduction of this chapter, in this context transmembrane signalling is referred to as the ability of extracellular ligand binding to activate the intracellular protein kinase domain. Any model which attempts to address this question must take into account that ligand binding information is propagated across the plasma membrane phospholipid bilayer through a single spanning transmembrane segment in each β subunit. Numerous studies have demonstrated that ligand-stimulated protein kinase activity is observed in detergent-solubilized insulin and IGF-1 receptor preparations,[1-11] as well as receptors reconstituted into phospholipid vesicles.[90,128,129] These data demonstrate that domain separation by the phospholipid membrane environment is not essential for receptor function.

Recently, several laboratories have taken advantage of the methods to isolate functional $\alpha\beta$ heterodimeric receptor complexes from the native $\alpha_2\beta_2$ heterotetrameric holoreceptors to investigate the role of subunit interaction in the transmembrane signalling process. Initial studies indicated that the isolated $\alpha\beta$ heterodimeric insulin receptor complex is devoid of kinase activity and that insulin-stimulated protein kinase activity requires the $\alpha_2\beta_2$ heterotetrameric form of the receptor.[63] Interestingly, insulin was observed to induce the covalent association of the $\alpha\beta$ heterodimeric receptor into a kinase active $\alpha_2\beta_2$ heterotetrameric state.[63,67,130] The time course and insulin concentration dependence of the formation of the $\alpha_2\beta_2$ heterotetrameric state from the isolated $\alpha\beta$ heterodimeric complexes was found to directly correlate with the activation of substrate protein kinase activity.[130,131] In addition, treatment of the $\alpha\beta$ heterodimeric insulin receptor complexes with Mn/MgATP was observed to induce a reversible noncovalent association into an $\alpha_2\beta_2$ heterotetrameric state (Figure 5). This reassociation was the direct consequence of Mn/MgATP binding and not autophosphorylation, since Mn/MgADP and Mn/MgAMPPCP also induced the noncovalent interaction between the $\alpha\beta$ heterodimeric insulin receptor complexes.

The role of $\alpha\beta$-$\alpha\beta$ interactions in activation of the insulin receptor kinase has been further examined by dilution experiments. We and others have found that autophosphorylation of the $\alpha_2\beta_2$ heterotetrameric insulin receptors occurs in a dilution-independent manner (intramolecular), whereas β subunit autophosphorylation of the isolated $\alpha\beta$ heterodimeric insulin receptor preparations occurred in a dilution-dependent fashion (intermolecular).[67] To determine whether this interaction was covalent or noncovalent in nature, two experimental approaches were employed. Treat-

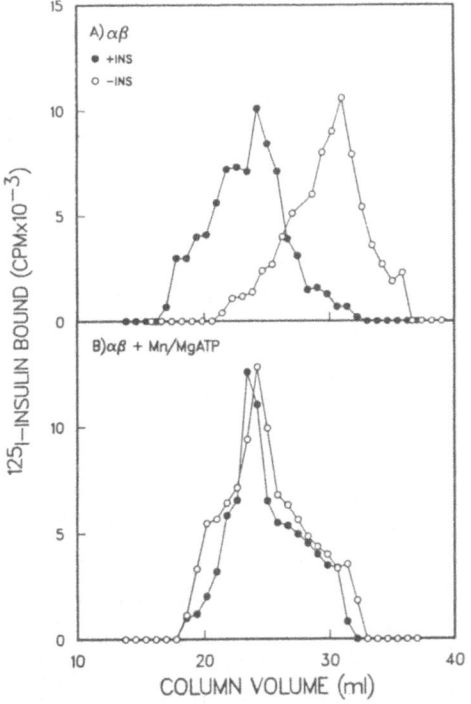

Figure 5. Mg/MnATP-induced noncovalent association of $\alpha\beta$ heterodimeric insulin receptors into an $\alpha_2\beta_2$ heterotetrameric state. A) Isolated $\alpha\beta$ heterodimeric insulin receptors were incubated with 2mM Mg and 10mM Mn plus 100μM ATP in the presence ($\bullet$) or absence (o) of 200nM insulin. The samples were then subjected to Bio-Gel A-1.5m gel filtration chromatography in the absence of Mg/MnATP and ^{125}I-insulin binding was determined. B) The $\alpha\beta$ heterodimeric insulin receptors incubated in the presence ($\bullet$) or absence (o) of 200nM insulin plus Mg/MnATP were subjected to gel filtration chromatography in columns equilibrated with Mg/MnATP.

ment of the $\alpha\beta$ heterodimeric insulin receptors with insulin prior to immobilization on wheat germ agglutinin (WGA)-Sepharose resulted in an insulin-stimulated protein kinase activity.[131] However, immobilization of the $\alpha\beta$ heterodimers prior to the addition of insulin completely prevented the activation of the insulin receptor protein kinase activity. In addition, we have reported that the specific sulfhydryl reagent iodoacetamide inhibits the insulin-dependent covalent association of the $\alpha\beta$ heterodimers but not the noncovalent association induced by Mg/MnATP.[132] Under conditions in which the covalent association of the $\alpha\beta$ heterodimers was prevented, the noncovalently associated $(\alpha\beta)_2$ heterotetrameric complex displayed similar insulin-stimulated protein kinase activity compared to the control disulfide-linked $\alpha_2\beta_2$ heterotetrameric insulin holoreceptor. These results support the role of both covalent and noncovalent $\alpha\beta$-$\alpha\beta$ heterodimeric subunit interactions in the regulation of insulin receptor protein kinase activity.

Several previous studies have addressed the role of receptor aggregation in the insulin-dependent activation of biological responsiveness.[133] These investigators have employed bivalent agents (anti-receptor antibodies) which are known to mimic insulin action *in vivo*.[134-136] These experiments have demonstrated that bivalent anti-insulin antibodies elicit the same responses in target cells as insulin, whereas monovalent Fab fragments do not. However, biological action of these Fab fragments is restored upon crosslinking the Fab-receptor complex with anti-Fab bivalent secondary antibodies.[137] These data coupled with electron microscopy studies of insulin-ferritin conjugates[138,139] have been interpreted to indicate that insulin induces microclustering (intermolecular crosslinking) of insulin receptors. Whether this insulin-induced aggregation is necessary for insulin responsiveness is unclear, as distinctions between the effects of intermolecular versus intramolecular crosslinking in the activation of the insulin receptor protein kinase activity have not been determined.

We have recently addressed this issue using WGA, a bifunctional lectin that mimics insulin action in adipocytes.[140] WGA was found to stimulate the insulin receptor protein kinase activity of the $\alpha_2\beta_2$ heterotetrameric complex in the basal state (in the absence of insulin) without affecting the insulin-stimulated protein kinase activity (Figure 6). Under these conditions, the insulin holoreceptor mobility in Bio-Gel A-1.5m gel filtration columns was consistent with the size of the $\alpha_2\beta_2$ heterotetrameric complex. WGA also stimulated the basal insulin receptor protein kinase activity of the isolated $\alpha\beta$ heterodimeric complex. Under these conditions the $\alpha\beta$ heterodimeric complex migrated with the identical mobility as the $\alpha_2\beta_2$ heterotetrameric complex by Bio-Gel A-1.5m gel filtration chromatography. However, the WGA-induced reassociation into the $(\alpha\beta)_2$ complex appears to be noncovalent, since only the $\alpha\beta$ heterodimeric species was observed upon ^{125}I-insulin affinity crosslinking and SDS-polyacrylamide gel electrophoresis in the presence of iodoacetamide (Figure 7). In the absence of iodoacetamide, the appearance of the Mr = 400,000 $\alpha_2\beta_2$ heterotetrameric band is due to the insulin-induced covalent association of the $\alpha\beta$ heterodimeric insulin receptors.[130,132,140] Further, monomeric WGA was incapable of stimulating the insulin receptor protein kinase activity of either the $\alpha_2\beta_2$ heterotetrameric or $\alpha\beta$ heterodimeric insulin receptor complexes.[140] These data demonstrate that bivalent agents, such as WGA, can induce the appropriate intramolecular conformational changes within the native $\alpha_2\beta_2$ heterotetrameric structure sufficient to activate the insulin receptor kinase domain.

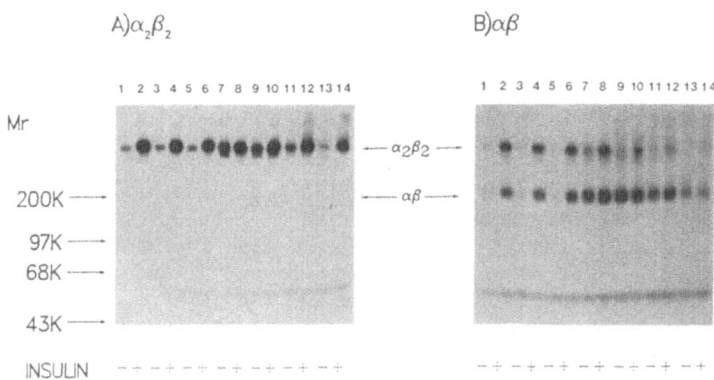

Figure 6. WGA stimulation of β subunit autophosphorylation in the $\alpha_2\beta_2$ heterotetrameric and $\alpha\beta$ heterodimeric insulin receptor complexes. A) The native $\alpha_2\beta_2$ heterotetrameric insulin receptors were pretreated in the absence (lanes 1, 3,5,7,9,11 and 13) or presence (lanes 2,4,6,8,10,12 and 14) of 200nM insulin. The samples were then incubated for an additional 1h in the absence (lanes 1 and 2) and presence of 2.5ng/ml (lanes 3 and 4), 25ng/ml (lanes 5 and 6), 250ng/ml (lanes 7 and 8), 2.5μg/ml (lanes 9 and 10), 25μg/ml (lanes 11 and 12) or 250μg/ml WGA (lanes 13 and 14). The insulin receptors were then autophosphorylated for 5 min and subjected to SDS-polyacrylamide gel electrophoresis and autoradiography. B) An identical protocol was utilized for the isolated $\alpha\beta$ heterodimeric insulin receptor complex.

Based upon these data we propose the following hypothesis for the mechanism of insulin-stimulated intramolecular transmembrane signalling of the insulin receptor. Since the isolated individual $\alpha\beta$ heterodimeric complex is protein kinase inactive, we postulate that the kinase activity of this species is tonically inhibited in an intramolecular fashion. If this

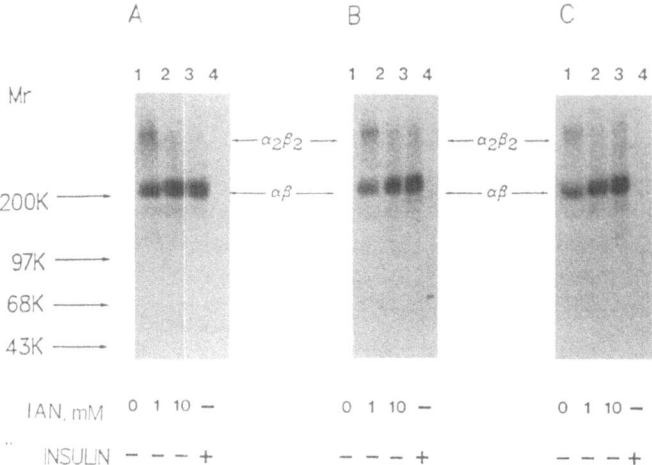

Figure 7. ^{125}I-insulin crosslinking of WGA-treated $\alpha\beta$ heterodimeric insulin receptors. A) The isolated $\alpha\beta$ heterodimeric insulin receptors were incubated in the absence (lane 1) or presence of 1mM iodoacetamide (lane 2), 10mM iodoacetamide (lane 3) or 1μM unlabeled insulin (lane 4) prior to affinity crosslinking with ^{125}I-insulin. The $\alpha\beta$ heterodimeric insulin receptors were also incubated with 200ng/ml (B) or 800ng/ml (C) WGA before ^{125}I-insulin affinity crosslinking as described in A. The $\alpha\beta$ heterodimeric insulin receptors in (A) migrated as $\alpha\beta$ complexes on Bio-Gel A-1.5m gel filtration chromatography, whereas samples in (B) and (C) migrated with the identical mobility as the native $\alpha_2\beta_2$ heterotetrameric insulin receptor complex.

were not the case, the isolated $\alpha\beta$ heterodimeric insulin receptor complex would be an activated insulin-independent protein kinase, similar to that observed for the baculovirus expressed truncated β subunit.[141] The tonic intramolecular inhibition of the $\alpha\beta$ heterodimeric insulin receptor protein kinase is released by either interaction between the two $\alpha\beta$ heterodimeric complexes or by proteolysis of the native holoreceptor complex.[142] The productive interactions between the $\alpha\beta$ heterodimeric complexes occur by either insulin binding, Mn/MgATP binding followed by autophosphorylation, or by crosslinking with bifunctional agents such as WGA and anti-insulin receptor antibodies.

Since the individual pairs of α and β subunits must necessarily orient in the same direction with respect to the plasma membrane (ie: both α subunits located on the extracellular side), there are only two ways in which the native $\alpha_2\beta_2$ heterotetrameric insulin receptor complex can be assembled. The identical $\alpha\beta$ heterodimeric subunits could assemble in a head to tail arrangement similar to that of several cytoskeletal proteins. However, this is highly unlikely since the insulin receptor apparently does not exist in a long filamentous polymerized state. Alternatively, we hypothesize that the $\alpha\beta$ heterodimeric subunits are assembled within the native $\alpha_2\beta_2$ heterotetrameric complex in a mirage image that is 180° out of phase with respect to each other. Under such physical constraints and to accomodate the established binding and kinase properties of the insulin receptor, we predict that the extracellular binding of insulin results in a rotation of the $\alpha\beta$ heterodimeric insulin receptor halves within the plane of the phospholipid bilayer. This extracellular rotation would be transferred through the transmembrane segment to the intracellular region of the β subunit, and subsequently would alter the spatial relationship between the opposing β subunits. Thus, the energy required for the release of the tonic inhibition imposed on the kinase domain would be provided by the energy released upon conformational change following insulin binding. This rotational activation model can readily account for the kinase activation properties of bifunctional agents, autophosphorylation, and rapid reversibility upon removal of a given stimulus (Figure 8).

This model of insulin receptor transmembrane signalling may also be generally applicable for several other ligand-stimulated tyrosine-specific protein kinase receptors. Although not studied in as much detail as the insulin receptor, the analogous IGF-1 receptor also apparently requires $\alpha\beta$ heterodimeric subunit association in order to display IGF-1 dependent protein kinase activation.[66,143] The EGF receptor tyrosine-specific protein kinase exists as a Mr = 170,000 polypeptide chain containing an extracellular EGF binding domain, a single transmembrane domain and an intracellular tyrosine kinase domain.[144] The overall structural features of the EGF receptor are highly related to the $\alpha\beta$ heterodimeric insulin and IGF-1 receptor complexes. Several studies have indicated that the EGF receptor exists as a monomer in detergent solution, but rapidly forms a kinase active noncovalent homodimer in response to EGF binding.[145-147]

In summary, the current information available suggests that ligand-dependent regulation of the intracellular protein kinase activity of the insulin and IGF-1 receptors requires complex interactions between homologous receptor $\alpha\beta$ heterodimeric subunits. Although the rotational model for transmembrane signalling is consistent with the experimental data to date, direct demonstration of this hypothesis is currently lacking. However, this model does provide a working framework for systematically addressing several predicted consequences of the transmembrane signalling mechanism and the role of $\alpha\beta$-$\alpha\beta$ heterodimeric subunit interactions in the ligand binding properties of these receptor species.

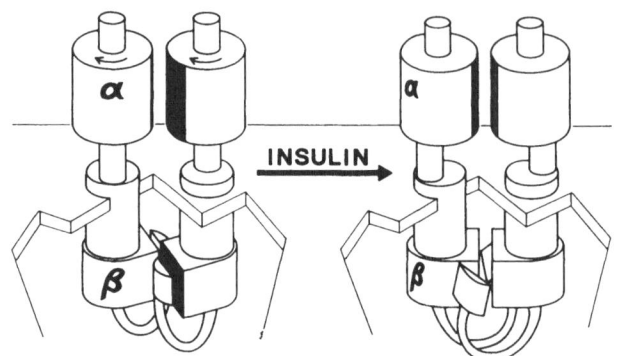

INSULIN

Figure 8. Schematic representation of the postulated
intramolecular subunit rotational model for the ligand-
dependent transmembrane signalling mechanism. The shaded
regions of the α subunit represent the insulin binding
domains. On the left, the unbound $\alpha_2\beta_2$ insulin receptor
complex is hypothesized to exist with opposing α subunit
insulin binding sites facing forward and behind the plane
of representation. The shaded portions of the β subunit
represent the intracellular ATP/substrate domains, each
with a complimentary intramolecular inhibitory domain.
The hypothesized insulin-induced rotation (right) is predicted to sterically relieve this intramolecular inhibition. This drawing represents the major salient features
of the proposed rotational model, but is not meant to
infer the location of the interacting domains based upon
the linear representation of the insulin receptor presented in Figure 1.

Literature cited

1. S. P. Nissley and M. M. Rechler, Somatomedian/Insulin-like growth
   factor tissue receptors, Clin. Endo. Metab. 13:43 (1984).
2. R. M. Bala, B. Baumick, G. H. Armstrong and M. D. Hollenberg, Receptors for insulin-like growth factors: basic somatomedin receptors
   in human and rodent tissues, in: "Insulin-like Growth Factors/
   Somatomedins," E. M. Spencer, ed., Walter de Gruyter, Berlin
   (1983).
3. E. R. Froesch, C. Schmid, J. Schwander and J. Zapf, Actions of
   insulin-like growth factors, Annu. Rev. Physiol. 47:443 (1985).
4. S. Jacobs and P. Cuatrecasas, Insulin receptor: structure and
   function, Endocr. Rev. 2:251 (1981).
5. C. R. Kahn, The molecular mechanism of insulin action, Annu. Rev.
   Med. 36:429 (1985).
6. M. P. Czech, Insulin action and regulation of hexose transport,
   Diabetes 29:399 (1980).
7. I. D. Goldfine, Effects of insulin on intracellular functions, in:
   "Biochemical Actions of Hormones," G. Litwack ed., Vol 8, Academic
   Press, NY (1981).
8. M. P. Czech, The nature and regulation of the insulin receptor:
   structure and function, Annu. Rev. Physiol. 47:357 (1985).

9. I. D. Goldfine, The insulin receptor: molecular biology and transmembrane signaling, Endo. Rev. 8:235 (1987).
10. S. Jacobs and P. Cuatrecasas, Insulin receptors, Annu. Rev. Pharmacol. Toxicol. 23:461 (1983).
11. M. M. Rechler and S. P. Nissley, The nature and regulation of the receptors for insulin-like growth factors, Annu. Rev. Physiol. 47:425 (1985).
12. A. Ullrich, J. R. Bell, E. Y. Chen, R. Herrera, L. M. Petruzzelli, T. J. Dull, A. Gray, L. Coussens, Y.-C. Liao, M. Tsubokawa, A. Mason, P. H. Seeburg, C. Grunfeld, O. M. Rosen and J. Ramachandran, Human insulin receptor and its relationship to the tyrosine kinase family of oncogenes, Nature (London) 313:756 (1985).
13. A. Ullrich, A. Gray, A. W. Tam, T. Yang-Fang, M. Tsubokawa, C. Collins, W. Henzel, T. R. LeBon, S. Kathuria, E. Chen, S. Jacobs, U. Francke, J. Ramachandran and Y. Fugita-Yamaguchi, Insulin-like growth factor I receptor primary structure: comparison with insulin receptor suggests structural determinants that define functional specificity, EMBO J. 5:2503 (1986).
14. Y. Ebina, L. Ellis, K. Jarnagin, M. Edery, L. Graf, E. Clauser, J. H. Ou, F. Maslarz, Y. W. Kan, I. D. Goldfine, R.A. Roth and W. J. Rutter, The human insulin receptor cDNA: the structural basis for hormone-activated transmembrane signalling, Cell 40:747 (1985).
15. E. Van Obberghen, M. Kasuga, M. LeCam, J. A. Hedo, A. Itin and L. C. Harrison, Biosynthetic labeling of insulin receptor studies of subunits in cultured human IM-9 lymphocytes, Proc. Natl. Acad. Sci. USA 78:1052 (1981).
16. J. Forsayeth, B. Maddux and I. D. Goldfine, Biosynthesis and processing of the human insulin receptor, Diabetes 35:837 (1986).
17. J. A. Hedo, M. Kasuga, E. Van Obberghen, J. Roth and C. R. Kahn, Direct demonstration of glycosylation of insulin receptor subunits by biosynthetic and external labeling evidence for heterogeneity, Proc. Natl. Acad. Sci. USA 78:4791 (1981).
18. G. V. Ronnett, V. P. Knutson, R. A. Kohanski, T. L. Simpson and M. D. Lane, Role of glycosylation in the processing of newly translated inulsin proreceptor in 3T3-Ll adipocytes, J. Biol. Chem. 259:4566 (1984).
19. S. Jacobs, F. C. Kull, Jr. and P. Cuatrecasas, Monensin blocks the maturation of receptors for insulin and somatomedin C: identification of receptor precursors, Proc. Natl. Acad. Sci. USA 80:1228 (1983).
20. P. J. Deutsch, C. F. Wan, O. M. Rosen and C. S. Rubin, Latent insulin receptors and possible receptor precursors in 3T3L-1 adipocytes, Proc. Natl. Acad. Sci. USA 80:133 (1983).
21. J. A. Hedo, E. Collier and A. Watkinson, Myristyl and palmityl acylation of the insulin receptor, J. Biol. Chem. 262:954 (1987).
22. P. F. Pilch, T. O'Hare, J. Rubin and M. Boni-Schnetzler, The ligand binding subunit of the insulin-like growth factor 1 receptor has properties of a peripheral membrane protein, Biochem. Biophys. Res. Commun. 136:45 (1986).
23. C. Grunfeld, J. K. Shigenaga and J. Ramachandran, Urea treatment allows dithiothreitol to release the binding subunit of the insulin receptor from the cell membrane: implications for the structural organization of the insulin receptor, Biochem. Biophys. Res. Commun. 113:389 (1985).
24. J. L. Goldstein, M. S. Brown, R. G. Anderson, D. W. Russel and W. J. Schneider, Receptor-mediated endocytosis: concepts emerging from the LDL receptor system, in: "Annual Review of Cell Biology," G. E. Palade, ed., Vol 1, Annual Reviews Inc., Palo Alto (1985).
25. S. M. Waugh, E. E. DiBella and P. F. Pilch, Isolation of a proteolytically derived domain of the insulin receptor containing

the major site of crosslinking/binding, Biochemistry in press (1989).

26. J. Massague and M. P. Czech, Role of disulfides in the subunit structure of the insulin receptor, J. Biol. Chem. 257:6729 (1982).

27. L. J. Sweet, P. A. Wilden and J. E. Pessin, Dithiothreitol activation of the insulin receptor/kinase does not involve subunit dissociation of the native $\alpha_2\beta_2$ insulin receptor subunit complex, Biochemistry 25:7068 (1986).

28. R. A. Roth and D. J. Cassell, Insulin receptor: evidence that it is a protein kinase, Science (Washington, D.C.) 219:299 (1983).

29. M. A. Shia and P. F. Pilch, The β subunit of the insulin receptor is an insulin-activated protein kinase, Biochemistry 22:717 (1983).

30. E. Van Obberghen, B. Rossi, A. Kowalski, H. Gazzano and G. Ponzio, Receptor-mediated phosphorylation o fthe hepatic insulin receptor: evidence that the Mr 95,000 receptor subunit is its own kinase, Proc. Natl. Acad. Sci. USA 80:945 (1983).

31. H. E. Tornqvist, M. W. Pierce, A. R. Frackelton, R. A. Nemenoff and J. Avruch, Identification of insulin receptor tyrosine residues phosphorylated in vitro, J. Biol. Chem. 262:10212 (1987).

32. H. E. Tornqvist, J. R. Gunsalus, R. A. Nemenoff, A. R. Frackelton, M. W. Pierce and J. Avruch, Identification of the insulin receptor residues undergoing insulin-stimulated phosphorylation in intact rat hepatoma cells, J. Biol. Chem. 263:350 (1988).

33. M. F. White, S. E. Shoelson, H. Keutman and C. R. Kahn, A cascade if tyrosine autophosphorylation in the β subunit acitvates the phosphotransferase of the insulin receptor, J. Biol. Chem. 263:2969 (1988).

34. J. Avruch, R. A. Nemehoff, P. J. Blackshear, M. W. Pierce and R. Osathandonh, Insulin-stimulated tyrosine phosphorylation of the insulin receptor in detergent extracts of human placental membranes, J. Biol. Chem. 257:15162 (1982).

35. M. Kasuga, Y. Zick, D. L. Blithe, F. A. Karlsson, H. U. Haring and C. R. Kahn, Insulin stimulation of phosphorylation of the β subunit of the insulin receptor, J. Biol. Chem. 257:9891 (1982).

36. M. Kasuga, Y. Zick, D. L. Blithe, M. Crettaz and C. R. Kahn, Insulin stimulates tyrosine phosphorylation of the insulin receptor in a cell-free system, Nature (London) 298:667 (1982).

37. S. Jacobs, F. C. Kull, Jr., H. S. Earp, M. E. Svoboda, J. J. Van Wyk and P. Cuatrecasas, Somatomedin-C stimulates the phosphorylation of the β-subunit of its own receptor, J. Biol. Chem. 258:9581 (1983).

38. J. B. Rubin, M. A. Shia and P. F. Pilch, Stimulation of tyrosine-specific phosphorylation in vitro by insulin-like growth factor I, Nature (London) 305:438 (1983).

39. J. J. Van Wyk, M. E. Svodoba and L. E. Underwood, Evidence from radioligand assays that somatomedin-C and insulin-like growth factor-I are similar to each other and different from other somatomedins, J. Clin. Endocrinol. Metab. 50:206 (1980).

40. Y. Zick, N. Sasaki, R. W. Rees-Jones, G. Grunberger, S. P. Nissely and M. M. Rechler, Insulin-like growth factor-I (IGF-I) stimulates tyrosine kinase activity in purified receptors from a rat liver cell line, Biochem. Biophys. Res. Commun. 119:6 (1984).

41. N. Sasaki, R. W. Rees-Jones, Y. Zick, S. P. Nissely and N. M. Rechler, Characterization of insulin-like growth factor-I stimulated tyrosine kinase activity associated with the β-subunit of Type I insulin-like growth factor receptors of rat liver cells, J. Biol. Chem. 260:9793.

42. Y. Fujita-Yamaguchi, T. R. LeBon, M. Tsubokawa, W. Henzel, S. Kathuria, D. Koyal and J. Ramachandran, Comparison of insulin-like growth factor I receptor and insulin receptor purified from human placenta membranes, J. Biol. Chem. 261:16727 (1986).

43. K.-T. Yu, M. A. Peters and M. P. Czech, Similar control mechanisms

regulate the insulin and Type I insulin-like growth factor receptor kinases, J. Biol. Chem. 261:11341 (1986).

44. H.-U. Haring, M. Kasuga, M. F. White, M. Crettaz and C. R. Kahn, Phosphorylation and dephosphorylation of the insulin receptor: evidence against an intrinsic phosphatase activity, Biochemistry 23:3298 (1984).

45. H. Gazzano, A. Kowalski, M. Fehlman and E. Van Obberghen, Two different protein kinase activities are associated with the insulin receptor, Biochem. J. 216:575 (1983).

46. G. Scatchard, The attractions of proteins for small molecules and ions, Ann. N.Y. Acad. Sci. 51:660 (1949).

47. J. M. Hammond, L. Jarett, I. K. Mariz and W. H. Daughaday, Heterogeneity of insulin receptors on fat cell membranes, Biochem. Biophys. Res. Commun. 49:1122 (1972).

48. C. R. Kahn, P. Freychet, J. Roth and D. M. Neville, Jr., Quantitative aspects of the insulin-receptor interaction in liver plasma membrane, J. Biol. Chem. 249:2249 (1974).

49. R. J. Pollet, M. L. Standaert and B. A. Haase, Insulin binding to the human lymphocyte receptor, J. Biol. Chem. 252:5828 (1977).

50. P. De Meyts, J. Roth, D. M. Neville, Jr., J. R. Gavin III and M. A. Lesniak, Insulin interaction with its receptors: experimental evidence for negative cooperativity, Biochem. Biophys. Res. Commun. 55:154 (1973).

51. P. De Meyts, A. R. Bianco and J. Roth, Site-site interactions among insulin receptors, J. Biol. Chem. 251:1877 (1976).

52. B. H. Ginsberg, C. R. Kahn, J. Roth and P. De Meyts, Insulin-induced dissociation of its receptor into subunits: possible molecular concomitant of negative cooperativity. Biochem. Biophys. Res. Commun. 73:1068 (1976).

53. P. De Meyts, Cooperative properties of hormone receptors in cell membranes, J. Supramolec. Struct. 4:241 (1976).

54. J. R. Gavin III, P. Gorden, J. Roth, J. A. Archer and D. N. Buell, Characteristics of the human lymphocyte insulin receptor, J. Biol. Chem. 248:2202 (1973).

55. P. De Meyts, Insulin and growth hormone receptors in human cultured lymphocytes and peripheral blood monocytes, in: "Methods in Receptor Research", M. Blecher, ed., Vol 9, Part 1, M. Decker, NY (1976).

56. P. De Meyts, E. Van Obberghen, J. Roth, A. Wollmer and D. Brandenburg, Mapping the residues in the receptor-binding region of insulin responsible for the negative cooperativity, Nature 273:504 (1978).

57. P. De Meyts, J.-L. Gu, R.M. Shymko, B. Kaplan, G. Bell and J. Whittaker. Insulin receptor α-subunit sequence 83-103 contains a binding domain complementary to the cooperative site of insulin. in: "The action of insulin and related growth factors in diabetes mellitus," Joslin Diabetes Center Symposium, Harvard Medical School (1988).

58. R. C. Baxter and P. F. Williams, Reciprocal modulation of insulin and insulin-like growth factor-I receptor affinity by calcium, Biochem. Biophys. Res. Commun. 116:62 (1983).

59. S. J. Casella, V. K. Han, A. J. D'Ercole, M. E. Svoboda and J. J. Van Wyk, Insulin-like growth factor II binding to the type I somatomedin receptor, J. Biol. Chem. 261:9268 (1986).

60. H. A. Jonas and L. C. Harrison, The human placenta contains two distinct binding and immunoreactive species of insulin-like growth factor-I receptors, J. Biol. Chem. 260:2288 (1985).

61. H. A. Jonas and L. C. Harrison, Disulphide reduction alters the immunoreactivity and increases the affinity of insulin-like growth-factor-I receptors in human placenta, Biochem. J. 236:417 (1986).

62. S. E. Tollefsen, K. Thompson and D. J. Petersen, Separation of the

high affinity insulin-like growth factor I receptor from low affinity binding sites by affinity chromatography, J. Biol. Chem. 262:16461 (1987).

63. M. Boni-Schnetzler, J. B. Rubin and P. F. Pilch, Structural requirements for the transmembrane activation of the inuslin receptor kinase, J. Biol. Chem. 261:15281 (1986).

64. M. Boni-Schnetzler, W. Scott, S. M. Waugh, E. DiBella and P. F. Pilch, The insulin recptor, J. Biol. Chem. 262:8395 (1987).

65. L. J. Sweet, B. D. Morrison and J. E. Pessin, Isolation of functional $\alpha\beta$ heterodimers from the purified human placental $\alpha_2\beta_2$ heterotetrameric insulin receptor complex, J. Biol. Chem. 262:6939 (1987).

66. S. E. Tollefsen and K. Thompson, The structural basis for insulin-like growth factor I receptor high affinity binding, J. Biol. Chem. 263:16267 (1988).

67. L. J. Sweet, B. D. Morrison, P. A. Wilden and J. E. Pessin, Insulin-dependent intermolecular subunit communication between isolated $\alpha\beta$ heterodimeric insulin receptor complexes, J. Biol. Chem. 262:16730 (1987).

68. C. C. Yip and M. L. Moule, Structure of the insulin receptor of rat adipocytes: the 3 interconvertible redox forms, Diabetes 32:760 (1983).

69. M. Crettaz, I. Jialal, M. Kasuga and C.R. Kahn, Insulin receptor regulation and desensitization in rat hepatoma cells. J. Biol. Chem. 259:11543 (1984).

70. R. A. Aiyer, Structural characterization of insulin receptors. I. Hydrodynamic properties of receptors from turkey erythrocytes, J. Biol. Chem. 258:14992 (1983).

71. R. A. Aiyer, Structural characterization of insulin receptors. II. Subunit composition of receptors from turkey erythrocytes, J. Biol. Chem. 258:15000 (1983).

72. A. Deger, H. Kramer, R. Rapp, R. Koch and U. Weber, The nonclassical insulin binding of insulin receptors from rat liver is due to the presence of two interacting α-subunits in the recptor complex, Biochem. Biophys. Res. Commun. 135:458 (1986).

73. R. Koch, A. Deger, H. M. Jack, K. N. Klotz, D. Schenzle, H. Kramer, S. Kelm, G. Muller, R. Rapp and U. Weber, Characterization of solubilized insulin receptors from rat liver microsomes, Eur. J. Biochem. 154:281 (1986).

74. J. Massague and M. P. Czech, Unique proteolytic cleavage site on the β subunit of the insulin receptor, J. Biol. Chem. 256:3182 (1981).

75. P. J. Bertics and G. N. Gill, Self-phosphorylation enhances the protein kinase activity of the epidermal growth factor receptor, J. Biol. Chem. 260:14642 (1985).

76. C. Cassel, L. J. Pike, G. A. Grant, E. G. Krebs and L. Glaser, Interaction of epidermal growth factor-dependent protein kinase with endogenous membrane proteins and soluble peptide substrate, J. Biol. Chem. 258:2945 (1983).

77. W. H. Moolenaar, A. J. Bierman, B. C. Tilly, I. Verlaan, L. H. K. Defize, A. M. Honegger, A. Ullrich and J. Schlessinger, A point mutation at the ATP-binding site of the EGF receptor abolishes signal transduction, EMBO J. 7:707 (1988).

78. O. M. Rosen, R. Herrera, Y. Olowe, L. M. Petruzzelli and M. M. Cobb, Phosphorylation activates the insulin receptor tyrosine protein kinase, Proc. Natl. Acad. Sci. USA 80:3237 (1983).

79. H. K. Klein, G. Friedenberg, M. Kladde and J. M. Olefsky, Insulin activation of insulin receptor tyrosine kinase in intact rat adipocytes, J. Biol. Chem. 261:4691 (1986).

80. K.-T. Yu and M. P. Czech, Tyrosine phosphorylation of the insulin receptor β subunit activates the receptor-associated tyrosine kinase activity, J. Biol. Chem. 259:5277 (1984).

81. K.-T. Yu and M. P. Czech, Tyrosine phosphorylation of insulin recep-

tor β subunit activates the receptor tyrosine kinase in intact H-35 hepatoma cells, J. Biol. Chem. 261:4715 (1986).

82. B. D. Morrison and J. E. Pessin, Insulin stimulation of the insulin receptor kinase can occur in the complete absence of β subunit autophosphorylation, J. Biol. Chem. 262:2861 (1987).

83. R. A. Kohanski and M. D. Lane, Kinetic evidence for activating and non-activating components of autophosphorylation of the insulin receptor protein kinase, Biochem. Biophys. Res. Commun. 134:1312 (1986).

84. H. E. Tornqvist and J. Avruch, Relationship of site-specific β subunit tyrosine autophosphorylation to insulin activation of the insulin receptor (tyrosine) protein kinase activity, J. Biol. Chem. 263:2969 (1988).

85. D. A. McClain, H. Maegawa, J. Levy, T. Huecksteadt, T. J. Dull, J. Lee, A. Ullrich and J. M. Olefsky, Properties of a human insulin receptor with a COOH-terminal truncation. I. Insulin binding, autophosphorylation and endocytosis, J. Biol. Chem. 263:8904 (1988).

86. H. Maegawa, D. A. McCalin, G. Friedenberg, J. M. Olefsky, M. Napier, T. Lipari, T. J. Dull, J. Lee and A. Ullrich, Properties of human insulin receptor with a COOH-terminal truncation. II. Truncated receptors have normal kinase but are defective in signaling metabolic effects, J. Biol. Chem. 263:8912 (1988).

87. L. Ellis, E. Clauser, D. O. Morgan, M. Edery, R. A. Roth and W. J. Rutter, Replacement of insulin receptor tyrosine residues 1162 and 1163 compromises insulin-stimulated kinase activity and uptake of 2-deoxyglucose, Cell 45:721 (1986).

88. P. A. Wilden, T. R. Boyle, M. L. Swanson, L. J. Sweet and J. E. Pessin, Alteration of intramolecular disulfides in the insulin receptor/kinase by insulin and dithiothreitol: insulin potentiates the apparent dithiothreitol-dependent subunit reduction of the insulin receptor, Biochemistry 25:4381 (1986).

89. L. Petruzzelli, R. Herrera and O. M. Rosen, Insulin receptor is an insulin-dependent tyrosine protein kinase: copurification of insulin-binding activity and protein kinase activity to homogeneity from human placenta, Proc. Natl. Acad. Sci. USA 81:3327 (1984).

90. L. J. Sweet, P. A. Wilden, A. A. Spector and J. E. Pessin, Incorporation of the purified human placental insulin receptor into phospholipid vesicles, Biochemistry 24:6571 (1985).

91. M. F. White, S. Takayama and C. R. Kahn, Differences in the sites of phosphorylation of the insulin receptor in vivo and in vitro, J. Biol. Chem. 260:9470 (1985).

92. D. Pang, B. Sharma, J. Schafer, M. White and C. R. Kahn, Predominance of tyrosine phosphorylation of insulin receptors during the initial response of intact cells to insulin, J. Biol. Chem. 260:7131 (1985).

93. H. Gazzano, A. Kowalski, M. Fehlmann and E. Van Obberghen, Two different protein kinase activities are associated with the insulin receptor, Biochem. J. 216:575 (1983).

94. K.-T. Yu, N. Khalaf and M.P. Czech, Insulin stimulates a membrane-bound serine kinase that may be phosphorylated on tyrosine. Proc. Natl. Acad. Sci. USA 84:3972 (1987).

95. D. M. Smith, M. J. King and G. J. Sale, Two systems in vitro that show insulin-stimulated serine kinase activity towards the insulin receptor, Biochem. J. 250:509 (1988).

96. S. Takayama, M. F. White, V. Lauris and C. R. Kahn, Phobol esters modulate insulin receptor phosphorylation and insulin action in cultured hepatoma cells, Proc. Natl. Acad. Sci. USA 81:7797 (1984).

97. G.E. Bollag, R.A. Roth, J. Beaudoin, D. Mochly-Rosen and D. E.

Kosland, Jr. Protein kinase C directly phosphorylates the insulin receptor *in vitro* and reduces its protein-tyrosine kinase activity, Proc. Natl. Acad. Sci. USA 83:5822 (1986).

98. R.A. Roth and J. Beaudoin, Phosphorylation of purified insulin receptor by cAMP kinase, Diabetes 36:123 (1987).

99. H. Haring, D. Kirsch, B. Obermaier, B. Ermel and F. Machicao, Decreased tyrosine kinase activity of insulin receptor isolated from rat adipocytes rendered insulin-resistant by catecholamine treatment *in vitro*, Biochem. J. 234:59 (1986).

100. L. Stadtmauer and O.M. Rosen. Increasing the cAMP content of IM-9 cells alters the phosphorylation state and protein kinase activity of the insulin receptor, J. Biol. Chem. 261:3402 (1986).

101. S. Jacobs, N. E. Sahyoun, A. R. Saltiel and P. Cuatrecasas, Phorbol esters stimulate the phosphorylation of receptors for insulin and somatomedin, Proc. Natl. Acad. Sci. USA 80:6211 (1983).

102. M. F. White, R. Maron, C. R. Kahn, Insulin rapidly stimulates tyrosine phosphorylation of a Mr 185,000 protein in intact cells, Nature 318:183 (1985).

103. K. T. Yu, N. Khalaf and M. P. Czech, Insulin stimulates the tyrosine phosphorylation of a Mr = 160,000 glycoprotein in rat adipocyte plasma membranes, J. Biol. Chem. 262:7865 (1987).

104. R. A. Kohanski, S. C. Frost and M. D. Lane, Insulin-dependent phosphorylation of the insulin receptor-protein kinase and activation of glucose transport in 3T3L-1 adipocytes, J. Biol. Chem. 261:12272 (1986).

105. R. W. Rees-Jones and S. I. Taylor, An endogenous substrate for the insulin receptor-associated tyrosine kinase, J. Biol. Chem. 260:4461 (1985).

106. T. Izumi, M. F. White, T. Kadowaki, F. Takaku, Y. Akanuma and M. Kasuga, Insulin-like growth factor I rapidly stimulates tyrosine phosphorylation of a Mr 185,000 protein in intact cells, J. Biol. Chem. 262:1282 (1987).

107. H.-U. Haring, M. F. White, F. Machicas, B. Ermel, E. Schleicher and B. Obermaier, Insulin rapidly stimulates phsophorylation of a 46-kDa membrane protein on tyrosine residues as well as phosphorylation of several soluble proteins in intact fat cells, Proc. Natl. Acad. Sci. USA 84:113 (1987).

108. R. C. Hresko, M. Bernier, R. D. Hoffman, J. R. Flores-Riveros, K. Liao, D. M. Laird and D. M. Lane, Identification of phosphorylated 422(aP2) protein as pp15, the 15-kilodalton target of the insulin receptor tyrosine kinase in 3T3L-1 adipocytes, Proc. Natl. Acad. Sci. USA 85:8835 (1988).

109. D. Sahal, J. Ramachandran and Y. Fugita-Yamaguchi, Specificity of tyrosine protein kinase of the structurally related receptors for insulin and insulin-like growth factor I: Tyr-containing synthetic polymers as specific inhibitors or substrates, Arch. Biochem. Biophys. 260:416 (1988).

110. Y. C. Kwok, R. A. Nemenoff, A. C. Powers and J. Avruch, Kinetic properties of the insulin receptor tyrosine protein kinase: activation through an insulin-stimulated tyrosine specific, intramolecular autophosphorylation, Arch. Biochem. Biophys. 244:102 (1986).

111. G. Arsenis and J. N. Livingston, Alterations in the tyrosine kinase activity of the insulin receptor produced by *in vitro* hyperinsulinemia, J. Biol. Chem. 261:147 (1986).

112. M. F. White, H.-U. Haring, M. Kasuga and C. R. Kahn, Kinetic properties and sites of autophosphorylation of the partially purified insulin receptor from hepatoma cells, J. Biol. Chem. 259:255 (1984).

113. D. H. Walker, D. Kuppuswamy, A. Visvanathan and L. J. Pike, Substrate specificity and kinetic mechanism of human placental insulin receptor/kinase, Biochemistry 26:1428 (1987).

114. Y. Zick, M. Kasuga, C. R. Kahn and J. Roth, Characterization of insulin-mediated phosphorylation of the insulin receptor in a cell-free system, J. Biol. Chem. 258:75 (1983).

115. O. M. Rosen and D. E. Lebwohl, Polylysine activates and alters the divalent cation requirements of the insulin receptor protein tyrosine kinase, FEBS Lett. 231:397 (1988).

116. D. B. Sacks and J. M. McDonald, Insulin-stimulated phosphorylation of calmodulin by rat liver insulin receptor preparations, J. Biol. Chem. 263:2377 (1988).

117. B. D. Morrison and J. E. Pessin, Polylysine specifically activates the insulin-dependent insulin receptor protein kinase, J. Biol. Chem. submitted.

118. M. Kasuga, F. A. Karlsson and C. R. Kahn, Insulin stimulates the phosphorylation of the 95,000-dalton subunit of its own receptor, Science 215:185 (1982).

119. C. K. Chou, T. J. Cull, D. S. Russell, R. Gherzi, D. Lebwohl, A. Ullrich and O. M. Rosen, Human insulin receptos mutated at the ATP-binding site lack protein tyrosine kinase activity and fail to mediate postreceptor effects of insulin, J. Biol. Chem. 262:1842 (1986).

120. G. Steele-Perkins, J. Turner, J. C. Edman, J. Hari, S. B. Pierce, C. Stover, W. J. Rutter and R. A. Roth, Expression and characterization of a functional human insulin-like growth factor I receptor, J. Biol. Chem. 263:11486 (1988).

121. Y. Ebina, E. Araki, M. Taira, F. Shimada, M. Mori, C. Craik, K. Siddle and S. Pierce, Replacement of lysine residue 1030 in the putative ATP-binding region of the insulin receptor abolishes insulin- and antibody-stimulated glucose uptake and receptor kinase activity, Proc. Natl. Acad. Sci. USA 83:704 (1987).

122. D. O. Morgan, L. Ho, L. J. Korn and R. A. Roth, Insulin action is blocked by monoclonal antibody that inhibits the insulin receptor kinase, Proc. Natl. Acad. Sci. USA 83:328 (1986).

123. D. O. Morgan and R. A. Roth, Acute insulin action reguires insulin receptor kinase activity: introduction of an inhibitory monoclonal antibody into mammalian cells blocks the rapid effects of insulin, Proc. Natl. Acad. Sci. USA 84:41 (1987).

124. C. R. Kahn, R. Maron and M. S. White, Modulation fo growth of antibodies to phosphotyrosine containing proteins, Clin. Res. 34:684 (1986).

125. J. R. Forsayeth, J. F. Caro, M. K. Sinha, B. A. Maddux and I. D. Goldfine, Monoclonal antibodies to the human insulin receptor that activate glucose transport but not insulin receptor kinase activity, Proc. Natl. Acad. Sci. USA 84:3448 (1987).

126. M. F. White, J. N. Livingston, J. M. Backer, V. Lauris, T. J. Dull, A. Ullrich and C. R. Kahn, Mutation of the insulin receptor of tyrosine 960 inhibits signal transmission but does not affect its tyrosine kinase activity, Cell 54:641 (1988).

127. O. M. Rosen, After insulin binds, Science 237:1452 (1987).

128. G. N. Panoyotou, A. I. Magee and M. J. Geisow, Title, FEBS Lett. 183:321 (1985).

129. R. E. Lewis and M. P. Czech, Phospholipid environment alters hormone-sensitivity of the purified insulin receptor kinase, Biochem. J. 248:829 (1987).

130. B. D. Morrison, M. L. Swanson, L. J. Sweet and J. E. Pessin, Insulin-dependent covalent reassociation of isolated $\alpha\beta$ heterodimeric

insulin receptors into an $\alpha_2\beta_2$ heterotetrameric disulfide-linked complex, J. Biol. Chem. 263:7806 (1988).

131. M. Boni-Schnetzler, A. Kaligian, R. Del Vecchio and P. F. Pilch, Ligand-dependent intersubunit association within the insulin receptor complex activates its intrinsic kinase activity, J. Biol. Chem. 263:6822 (1988).

132. P. A. Wilden, B. D. Morrison and J. E. Pessin, Relationship between insulin receptor subunit association and protein kinase activation, Biochemistry, in press, (1988).

133. J. E. Pessin and M. P. Czech, Hexose transport and its regulation in mammalian cells, in: "The Enzymes of Biological Membranes," A. N. Martonosi, ed., Vol. 3, Plenum Press, NY (1985).

134. D. Heffetz and Y. Zich, Receptor aggregation is necessary for activation of the soluble insulin receptor kinase, J. Biol. Chem. 261:889 (1986).

135. C. R. Kahn, K. Baird, J. S. Flier and D. B. Jarrett, Effects of autoantibodies to the insulin receptor on isolated adipocytes: studies of insulin binding and insulin action, J. Clin. Invest. 60:1094 (1977).

136. J. S. Flier, C. R. Kahn, D. B. Jarrett and J. Roth, Characterization of antibodies to the insulin receptor: a cause of insulin-resistant diabetes in man, J. Clin. Invest. 58:1442 (1976).

137. C. R. Kahn, K. Baird, R. Baird, D. B. Jarrett and J. S. Flier, Direct demonstration that receptor cross-linking or aggregation is important in insulin action, Proc. Natl. Acad. Sci. USA 75:4209 (1978).

138. L. Jarett and R. M. Smith, Electron microscopic demonstration of insulin receptors on adipocyte plasma membranes utilizing a ferritin-insulin conjugate, J. Biol. Chem. 249:7024 (1974).

139. L. Jarett and R. M. Smith, Effect of cytochalasin B and D on groups of insulin receptors and on insulin action in rat adipocytes, Clin. Invest. 6:571 (1979).

140. P. A. Wilden, B. D. Morrison and J. E. Pessin, Wheat germ agglutinin stimulation of $\alpha\beta$ heterodimeric insulin receptor β subunit autophosphorylation by noncovalent association into an $\alpha_2\beta_2$ heterotetrameric state, Endocrinology 124: in press, (1988).

141. R. Herrera, D. Lebwohl, A. Garcia de Herreros, R. G. Kallen and O. M. Rosen, Synthesis, purification, and characterization of the cytoplasmic domain of the human insulin receptor using a baculovirus expression system, J. Biol. Chem. 263:5560 (1988).

142. S. E. Shoelson, M. F. White and C. R. Kahn, Tryptic activation of the insulin receptor, J. Biol. Chem. 263:4852 (1988).

143. S. M. Feltz, M. L. Swanson, J. A. Wemmie and J. E. Pessin, Functional properties of an isolated $\alpha\beta$ heterodimeric human placenta insulin-like growth factor 1 receptor complex, Biochemistry 27:3234 (1988).

144. A. Ullrich, L. Coussens, J. S. Hayflick, T. J. Dull, A. Gray, A. W. Tam, J. Lee, Y. Yarden, T. A. Libermann, J. Schlessinger, J. Downward, E. L. V. Mayes, N. Whittle, M. D. Waterfield and P. H. Seeburg, Human epidermal growth factor receptor cDNA sequence and aberrant expression of the amplified gene in A431 epidermoid carcinoma cells, Nature (London) 309:418 (1984).

145. Y. Yarden and J. Schlessinger, Self-phosphorylation of epidermal growth factor receptor: evidence for a model of intermolecular allosteric activation, Biochemistry 26:1434 (1987).

146. Y. Yarden and J. Schlessinger, Epidermal growth factor induces rapid, reversible aggregation of the purified epidermal growth factor receptor, Biochemistry 26:1443 (1987).

147. M. Boni-Schnetzler and P. F. Pilch, Mechanism of epidermal growth factor receptor autophosphorylation and high-affinity binding, Proc. Natl. Acad. Sci. USA 84:7832 (1987).

IDENTIFICATION OF THE DOMAINS OF IGF I WHICH INTERACT

WITH THE IGF RECEPTORS AND BINDING PROTEINS

Margaret A. Cascieri and Marvin L. Bayne

Departments of Biochemical Endocrinology and
Growth Factor Research
Merck Sharp & Dohme Research Laboratories
Rahway, N.J.

INTRODUCTION

The elucidation of the primary sequence of human insulin-like growth factor I (IGF I) (1) and the subsequent modeling of its secondary and tertiary structure based on its homology with insulin (2,3) have made it possible to predict which structural features of this peptide are involved in its binding to the types 1 and 2 IGF receptors and to soluble IGF binding proteins. The amino terminal 29 amino acids of IGF I (B-region) are homologous with the B-chain of insulin. A 12 amino acid linking sequence (C-region) joins the B-region with a 21 amino acid A-region which is homologous to the A-chain of insulin. The molecule terminates with an 8 amino acid sequence termed the D-region (Figure 1).

The aromatic residues 23-25 of IGF I are highly homologous to a sequence in the B-chain of insulin (residues 24-26) in which naturally occurring mutations produce modified insulins which bind poorly to the insulin receptor (4,5). Thus, it was predicted that residues 23-25 of IGF I are important determinants of the interaction of IGF I with both the type 1 IGF receptor and the insulin receptor. The predicted tertiary structure of the C- and D-regions of IGF I suggests that these regions provide partial sheltering of residues 23-25 which may modulate binding of IGF I to the insulin receptor and/or the IGF I receptor (3).

Since insulin has weak affinity for the type 1 IGF receptor and does not bind at all to the type 2 IGF receptor or to soluble IGF binding proteins, consideration of the regions of IGF I in which significant divergence from the primary structure of insulin occur may lead to the discovery of those determinants important for modulating binding affinity for these proteins. This approach has been taken by Katsoyannis, Rechler and their colleagues (6-9). They have produced a series of molecules in which the A- or B-regions of IGF I were hybridized with the A- or B-chains of insulin. In contrast to insulin, which has no measurable affinity for soluble IGF binding proteins, an insulin A-chain/IGF I B-region hybrid binds to serum IGF binding proteins (6). This suggests that major determinants for binding to these proteins are present in the B-region of IGF I. This approach has not provided strong evidence for structural determinants involved in receptor binding. These hybrids have much lower affinity than insulin for the insulin receptor and much lower affinity than IGF I for the type 1 IGF receptor.

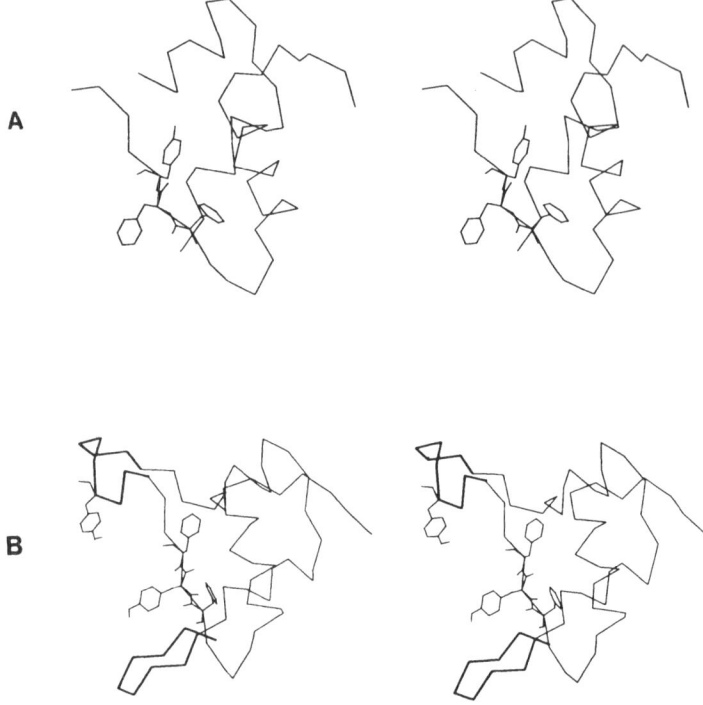

Figure 1. Stereo comparison of the crystal structure
of porcine insulin (A) with the proposed model of IGF
I (2,3) (B). In the insulin structure, residues Phe
24, Phe 25, and Tyr 26 of the B-chain are shown. In
the IGF I structure, residues Phe 23, Tyr 24, Phe 25,
and Tyr 31 are shown. The C- and D-regions of IGF I
are highlighted.

     Our approach has been to prepare a series of IGF I analogs by site-
directed mutagenesis of a synthetic gene encoding human IGF I (10). We have
shown that distinct domains of IGF I are involved in binding to each of the
receptors and the soluble binding proteins. In addition, these analogs have
proven to be useful tools to study the regulation of the biological activity
of IGF I by each of these classes of proteins.

CHARACTERIZATION OF RECOMBINANTLY-PRODUCED IGF I

     We have expressed the synthetic gene for IGF I in two different sys-
tems. The protein can be expressed in murine L cells under the control of
the human cytomegalovirus major immediate early transcriptional regulatory
region. The expression vector also contains the gene sequence of the bovine
growth hormone signal peptide which directs the secretion of IGF I into the
conditioned medium (10,11). Amino terminal sequencing of the purified, secreted peptide shows that proteolytic processing has occurred between the
penultimate and ultimate residues of the growth hormone signal peptide
sequence so that the product is [Ala 0]-IGF I (11).

     We have also expressed the gene in yeast utilizing an expression
vector containing the MFα1 promoter and pre-pro leader peptide (Figure

2). Transformation of a protease-deficient strain of Saccharomyces cerevisiae (BJ1995) with this vector results in secretion of active IGF I into the media (12). Amino and carboxyl terminal sequencing of the purified, secreted peptide shows that proteolytic processing of the pre-pro sequence occurs to give intact IGF I (12).

The activity of these two recombinantly-derived IGF I peptides is shown in Table 1. Receptor and binding protein assays were performed as previously described (12). Briefly, type 1 IGF receptor affinity and insulin receptor affinity were determined by measuring the ability of peptides to inhibit the specific binding of ^{125}I-IGF I or ^{125}I-insulin, respectively, to human placental membranes. Type 2 receptor affinity was measured by determining the ability of peptides to inhibit the specific binding of ^{125}I-MSA to rat liver membranes. Serum binding protein affinity was determined by measuring the ability of peptides to inhibit the specific binding of ^{125}I-IGF I to acid treated human serum.

Both [Ala 0]-IGF I and the yeast-derived IGF I are as potent as serum-purified IGF I (from Dr. Rene Humbel) at the human placental type 1 IGF receptor (Table 1). This potency is reflected in their ability to stimulate DNA synthesis in the rat clonal smooth muscle cell line, A10 (11,12). Both peptides have high affinity for a crude preparation of human serum binding proteins (Table 1). However, as has been previously noted with other recombinantly-produced and synthetic IGF I (13-15), neither [Ala 0]-IGF I nor yeast-derived IGF I bind with high affinity to type 2 IGF receptors. This discrepancy has been attributed to the potential contamination of serum-purified IGF I with IGF II or post-translationally altered forms of IGF I (14).

Thus, we have been able to recombinantly produce fully active IGF I using two different expression systems. Because of the ease with which large quantities of yeast conditioned media can be obtained, we have chosen this system to produce the analogs of IGF I which are described below. The peptides can be purified in three steps using BioRex-70, Biogel P10, and reverse-phase HPLC. The purity of all peptides has been confirmed using SDS-PAGE on 17-27% gradient gels followed by silver staining. The yields of purified IGF I and its analogs in this system have varied from 100-800 µg/L. This variation is caused by differences in the expression levels of individual isolated yeast transformants as well as by differences in the expression levels achieved in individual fermentations.

Table 1. Relative Potency[1] of Serum-Purified IGF I and Recombinantly-Derived IGF I at IGF Receptors and Soluble IGF Binding Proteins.

PEPTIDE	HUMAN PLACENTA TYPE 1 IGF RECEPTOR	RAT LIVER TYPE 2 IGF RECEPTOR	HUMAN SERUM BINDING PROTEINS
Serum-purified IGF I[2]	1	1	1
[Thr 59]-IGF I[2]	0.9	0.08	-
[Ala 0]-IGF I	1.1	0.04	0.9
Yeast-derived IGF I	0.8	0.05	0.6

[1] Relative potency is the ratio of IC_{50} of serum-purified IGF I/IC_{50} of recombinant IGF I.

[2] Serum-purified IGF I was the kind gift of Dr. Rene Humbel and [Thr 59]-IGF I was from Amgen.

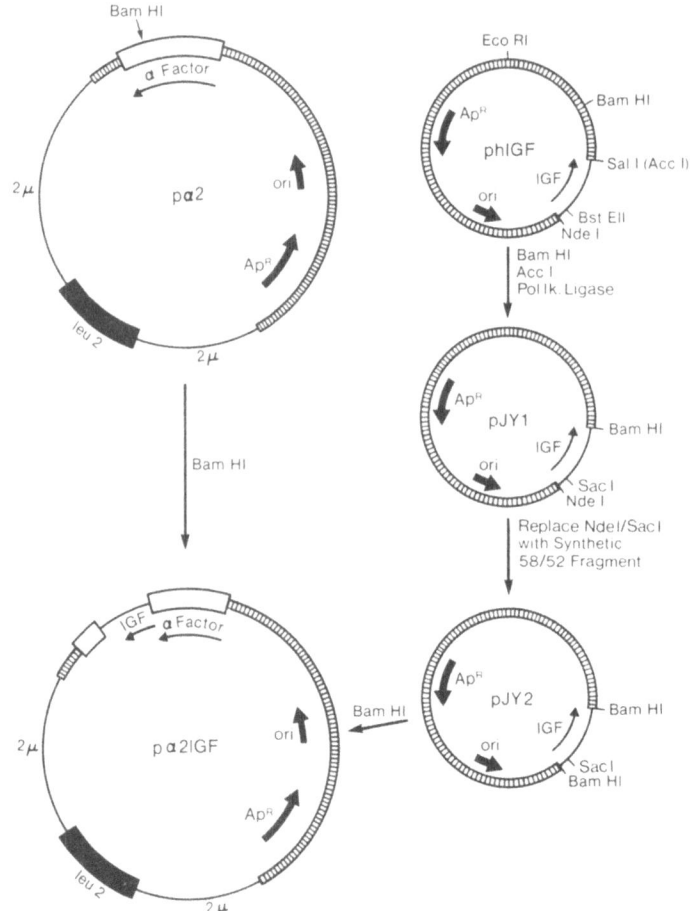

Figure 2. Construction of expression vector p α 2 IGF.
Plasmid pα 2 contains the ampicillin-resistance gene (ApR)
and the origin of DNA replication (ori) derived from pBR322
(striped bars), yeast LEU2 gene (blackened bars), yeast 2μ
sequences, and the MF α1 promoter and pre-pro polypeptide.
Plasmid phIGF, a pBR322 derivative, contains the chemically
synthesized human IGF I gene (10) inserted between the NdeI
and SalI restriction sites. Plasmid phIGF was digested
with BamHI+AccI, the ends were blunted with PolIk and then
ligated to form pJY1. The 5' end of the synthetic IGF I
gene was modified for expression in p α2 by replacing the
NdeI-SacI fragment of pJY1 with a synthetic fragment formed
by the annealing of oligodeoxynucleotides 5'-TATGCCGGATCCT-
TTCCTTGGATAAAAGAGGTCCGCAAGCTTTGTGTGGTGCTGAGCT-3' AND 5'-
CAGCACCACACAAAGTTTCCGGACCTCTTTTATCCAAGGAAAGGATCCGGCA-3' as
previously described (10). After two rounds of transformation into E. coli DH5, a clone designated pJY2, was identified by HpaII digestion and DNA sequencing. The BamHI
cassette from pJY2 was ligated into BamHI-digested p α2 by
standard procedures; proper orientation for p α2IGF was
determined by PstI mapping. Reprinted with permission from
reference 12.

IDENTIFICATION OF REGIONS INVOLVED IN TYPE 1 IGF RECEPTOR BINDING

We have produced and characterized fourteen analogs of IGF I which were designed to substitute regions of IGF I with analogous regions of insulin; delete or replace regions of IGF I which are absent in insulin; or replace specific residues of IGF I which other evidence suggested might be important. The relative potencies of these analogs for the types 1 and 2 IGF receptors, the insulin receptor, and human serum binding proteins are summarized in Table 2. Substitution of Tyr-Phe at residues 24 and 25 of IGF I with the corresponding residues in insulin, Phe-Tyr, ([F23,F24,Y25]-IGF I) has little effect on the affinity for either the type 1 IGF receptor or the insulin receptor (16). However, substitution of residue 24 with leucine or serine results in a 30-fold and a 10-fold loss in affinity for the type 1 IGF receptor, respectively (16). These analogs also have 10-fold and 2-fold reduced affinity for the insulin receptor but relatively normal affinity for the type 2 IGF receptor and human serum binding protein (16). Thus, residue 24 in IGF I, which is analogous to the residue shown to be important in maintaining high affinity binding of insulin to the insulin receptor (i.e., tyrosine 25 in the B-chain of insulin) (4,5), is very important in maintaining high affinity binding of IGF I to the type 1 IGF receptor.

We predicted that replacement of residues 28-37 of the C-region of IGF I with a four glycine bridge would allow the analog to fold into a tertiary structure similar to IGF I since the glycine tetramer should easily span the distance between residues 27 and 38 determined for insulin by X-ray crystallography (17). This analog, [1-27, Gly$_4$, 38-70]-IGF I, has 30-fold reduced affinity for the type 1 IGF receptor, but has nearly normal affinity for the insulin receptor, the type 2 IGF receptor, and human serum binding proteins (18). Thus, IGF I contains residues within its C-region which are important for maintaining high affinity binding to the type 1 IGF receptor. We have unpublished data which suggests that the tyrosine residue at position 31 of IGF I is one of the residues which are important.

[1-62]-IGF I, in which the eight amino acid D-region of IGF I is deleted, has normal affinity for the type 1 IGF receptor and 2-fold increased potency for the insulin receptor (18). [L24, 1-62]-IGF I has 100-fold reduced affinity for the type 1 IGF receptor as would be expected from the low potency of [L24]-IGF I described above (16). [1-27, Gly$_4$, 38-62]-IGF I, in which the D-region deletion and the C-region replacment are combined, has a 50-fold loss in affinity for the type 1 IGF receptor (18). Our data suggest that all of the analogs described above have tertiary structure similar to IGF I, since their binding affinities for the type 2 IGF receptor and human serum binding proteins are maintained or even increased (Table 2). However, at present we have not analyzed the physical characteristics of these analogs to directly address this question.

Substitution of residues 1-16 of IGF I with residues 1-17 of the B-chain of insulin (B-chain mutant) or of residues 42-56 of IGF I with residues 1-15 of the A-chain of insulin (A-chain mutant) has very little effect on the ability to interact with the type 1 IGF receptor (19,20). These data strongly suggest that most of the approximately 100-fold difference in affinity of IGF I and insulin for the type 1 IGF receptor is due to insulin's lack of specific binding determinants in the region 28-37 of IGF I.

In contrast, [1-27, Gly$_4$, 38-62]-IGF I has a 5-fold increased affinity for the insulin receptor (18). Also, any analog we have made in which residues Gln 15-Phe 16 of IGF I have been changed to the corresponding residues in insulin (ie., the B-chain mutant, [Y15,L16]-IGF I and [Q3,A4,Y15,L16]-IGF I) has increased affinity for the insulin receptor relative to IGF I (19).

These data suggest that the approximately 500-fold difference in affinity of insulin and IGF I for the insulin receptor is due, in part, to the presence of the C- and D-regions of IGF I and differences in their sequences at the positions correlating to 15 and 16 in IGF I.

Since substitution of residues 15 and 16 has no effect on type 1 IGF receptor binding, these data imply that the peptide binding pockets of the insulin receptor and the type 1 IGF receptor are not completely analogous. The stereo diagram in Figure 1 highlights the predicted structures of the C- and D-regions of IGF I as well as the aromatic residues at positions 23, 24, 25, and 31. Our data suggest that the type 1 IGF receptor recognizes the aromatic residues at positions 24 and 31 as well as other determinants in the C-region of IGF I, and that binding to this receptor is not affected by the presence of the D-region. The insulin receptor recognizes the aromatic residue at position 24 and one or both of the residues 15 and 16 (within the helical region including residues 8-18 of IGF I). The presence of the D- and/or C-regions of IGF I impairs the binding of the peptide to the receptor.

IDENTIFICATION OF REGIONS INVOLVED IN TYPE 2 IGF RECEPTOR BINDING

Three of the IGF I analogs described in Table 2 have reduced binding affinity relative to IGF I for the type 2 IGF receptor. The B-chain mutant has >> 20-fold lower affinity than IGF I (19). The three other analogs in this series, [Y15,L16]-IGF I, [Q3,A4]-IGF I, and [Q3,A4,Y15,L16]-IGF I have normal affinity for the type 2 receptor suggesting that these regions are not important for binding or that the more dramatic total change is required.

The A-chain mutant also has >20-fold reduced affinity for the type 2 IGF receptor (20). The four areas of divergence between IGF I and the A-chain mutant are Asp 45-Glu 46 to Glu 45-Gln 46, Phe 49-Arg 50-Ser 51 to Thr 49-Ser 50-Ile 51, Arg 55-Arg 56 to Tyr 55-Gln 56 and Asp 53 to Ser 53. We have directly investigated the effects of changes in two of these areas on type 2 IGF receptor binding. [T49,S50,I51]-IGF I has even lower affinity for the type 2 IGF receptor than does the A-chain mutant (20). In contrast, [Y55,Q56]-IGF I has 7-fold higher affinity for the type 2 IGF receptor than does IGF I. Interestingly, IGF II also has neutral amino acids at these positions (i.e., Ala-Leu) suggesting that the presence of the double argi- nine at positions 55 and 56 of IGF I contributes to the relatively lower affinity of IGF I for the type 2 IGF receptor. All three of these analogs have normal affinity for the type 1 IGF receptor, the insulin receptor and serum binding proteins. These data suggest that the tertiary structure of these analogs is not drastically different from that of IGF I.

In summary, these data suggest that undetermined features within residues 1-17 of the B-region and residues 49-51 and 55-56 of the A-region are important for recognition by the type 2 IGF receptor. Residues 49-51 are at the end of the first helical region of the A-region and residues 55 and 56 are at the beginning of the second helical region. The model des- cribing the tertiary structure of IGF I predicts that all of three of these regions are on the same face of the molecule (see Figure 1). This is the opposite face of the molecule from that which we have determined is important for maintaining type 1 IGF receptor binding.

IDENTIFICATION OF REGIONS INVOLVED IN SERUM BINDING PROTEIN BINDING

The B-chain mutant has >2000-fold lower affinity than IGF I for the acid-stable IGF binding protein in human serum (19). The three areas of divergence between IGF I and the B-chain mutant are Gly 1-Pro 2-Glu 3-Thr 4 to Phe -1-Val 1-Asn 2-Gln 3-His 4, Ala 8-Glu 9 to Ser 8-His 9 and Gln 15-Phe 16 to Tyr 15-Leu 16 (Figure 3). Both [Q3,A4]-IGF I and [Y15,L16]-IGF I have

Table 2. Relative Potency[1] of IGF I and Analogs at IGF Receptors, Insulin Receptors, and Soluble IGF Binding Proteins.

PEPTIDE	HUMAN PLACENTA TYPE 1 IGF RECEPTOR	RAT LIVER TYPE 2 IGF RECEPTOR	HUMAN SERUM BINDING PROTEINS	HUMAN PLACENTA INSULIN RECEPTOR
IGF I	1	1	1	1
IGF I-B Chain Mutant[2]	0.46	< 0.06	< 0.0005	4
[Y15,L16]-IGF I	1.2	3.75	0.23	9.3
[Q3,A4]-IGF I	1.1	0.3	0.26	2.0
[Q3,A4,Y15,L16]-IGF I	1.1	2.7	0.002	14
IGF I-A Chain Mutant[2]	1.5	< 0.08	1.6	0.8
[Y55,Q56]-IGF I	1.0	6.7	2.3	1.3
[T49,S50,I51]-IGF I	0.7	< 0.04	0.3	1.3
[F23,F24,Y25]-IGF I	0.7	3.0	0.4	1.9
[L24]-IGF I	0.03	0.3	1.1	0.1
[S24]-IGF I	0.06	0.5	1.1	0.5
[L24,1-62]-IGF I	0.01	1.0	1.8	0.2
[1-62]-IGF I	0.8	1.0	4.7	1.8
[1-27,Gly$_4$,38-70]-IGF I	0.03	0.4	2.8	1.8
[1-27,Gly$_4$,38-62]-IGF I	0.02	0.4	4.0	4.7

---

[1] Relative potency is the ratio of IC_{50} of IGF I/IC_{50} of analog. The IC_{50} of IGF I at the types 1 and 2 IGF receptors, the serum binding proteins and the insulin receptor are 4.9-5.6 nM, 300-400 nM, 0.5-0.8 nM and 1400-2800 nM, respectively.

[2] IGFI-B Chain mutant is [F-1,V1,N2,Q3,H4,S8,H9,E12,Y15,L16]-IGF I
IGFI-A Chain mutant is [I41,E45,Q46,T49,S50,I51,S53,Y55,Q56]-IGF I

4-fold lower affinity than IGF I (19). The analog, [Q3,A4,Y15,L16]-IGF I has 600-fold lower affinity than IGF I for acid-stable human serum binding protein (19). Thus, the effect of changing all four residues is much greater than would be predicted if it was additive.

Szabo et al. have purified a destripeptide truncated form of bovine IGF I and have shown that it has reduced binding affinity for a soluble binding protein in the conditioned medium of bovine kidney cells (21). We have prepared both destripeptide and destetrapeptide IGF I and have found that they have 4-fold and 24-fold reduced affinity for human serum binding proteins (unpublished data). Thus, these data confirm that determinants in the amino terminal segment of IGF I are important for binding to soluble binding proteins.

Figure 3 compares the predicted structure for this region of IGF I (2,3) with the structure of insulin determined by X-ray crystallography (17). The model predicts that residues 1-6 have an extended structure and that residues 8-17 form an alpha-helix. The phenylalanine at residue 16 of IGF I faces into the hydrophobic core of the molecule whereas the model predicts a shift of the position of the aromatic residue so it now faces outward from the helix in those analogs which have tyrosine at residue 15. This shift is clearly unfavorable for binding to serum IGF binding proteins.

[T49,S50,I51]-IGF I also has 3-fold reduced binding affinity to acid-stable human serum IGF binding proteins and [Y55,Q56]-IGF I has a 2-fold increased affinity. These data suggest that the serum IGF binding proteins bind to the same face of the molecule as does the type 2 IGF receptor, but that the specific residues involved in the binding are somewhat different since [Q3,A4,Y15,L16]-IGF I has normal affinity for the type 2 IGF receptor. Interestingly, those analogs which have the D-region deleted or which have residues 28-37 of the C-region replaced with a four glycine bridge have increased affinity for the IGF binding protein suggesting that these regions may negatively affect the ability of the binding protein to interact with IGF I.

The major IGF binding component of our acid-stable human serum is the 40 kD protein which is thought to be part of the 150 kD growth hormone-dependent IGF binding protein (22). We have shown that the B-chain mutant and [Q3,A4,Y15,L16]-IGF I also have reduced affinity for the native 150 kD IGF binding protein in rat serum (23). Several groups have identified and cloned a smaller, 25 kD soluble binding protein which is immunologically distinct from the 150 kD serum protein (24,25). We have identified such binding proteins in the conditioned medium from rat A10 cells and murine Balb/c 3T3 cells, and have shown that the B-chain mutant and [Q3,A4,Y15,L16] IGF I have > 1000-fold and 100-fold reduced affinity for these proteins, respectively (26).

CHARACTERIZATION OF THE BIOLOGICAL ACTIVITY OF IGF I ANALOGS WITH REDUCED AFFINITY FOR SOLUBLE IGF BINDING PROTEINS

^{125}I-IGF I, ^{125}I-B-chain mutant, and [125]I-[Q3,A4,Y15,L16]-IGF I have in vivo serum half-lives of 100, 27.5, and 26.9 min, respectively, after i.v. injection into 300 g, male rats (23). These data suggest that one function of the serum IGF binding proteins is to slow the clearance of IGF I from the circulation. However, despite the increased rate of clearance of these two IGF I analogs, both are more active relative to IGF I in at least one type of in vivo assay system (23).

Both insulin and IGF I stimulate the incorporation of ^{14}C-glucose into diaphragm glycogen in rats injected intraperitoneally with labeled glucose and peptide two hours before sacrifice. The doses of IGF I and insulin required to see significant stimulation are similar suggesting that this effect of IGF I is mediated by the type 1 IGF receptor (23). In the same experiment, insulin also stimulates incorporation of the labelled glucose into the total lipid fraction of epididymal fat tissue. However, at the same doses of IGF I which cause 30-fold increases in glucose incorporation into diaphragm glycogen, no significant accumulation of glucose into adipose lipid occurs (23). This can be explained by the observations of King et al.

A

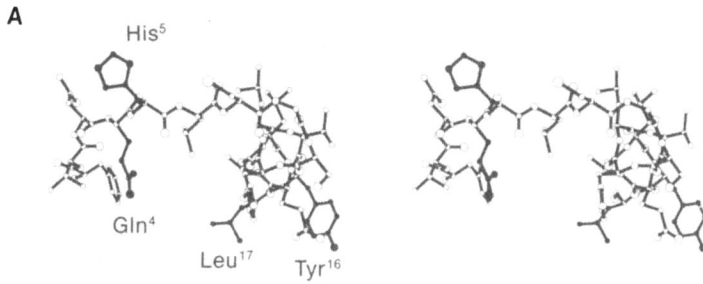

His⁵

Gln⁴

Leu¹⁷   Tyr¹⁶

B

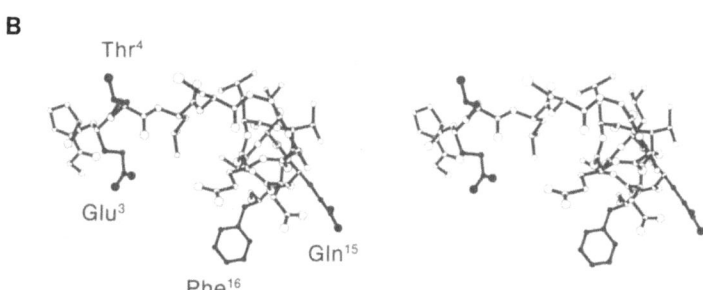

Thr⁴

Glu³

Gln¹⁵

Phe¹⁶

Figure 3. Stereo comparison of the crystal structure
of the B-chain of insulin (A) with the proposed structural
model of IGF I (B) (2,3). Reprinted with permission from
reference 19.

(27) that adipose tissue does not contain type 1 IGF receptors and that IGF
I action in this tissue is mediated by its interaction with the insulin
receptor. Thus, this is stronger evidence that the IGF I effect in diaphragm is mediated by the type 1 IGF receptor.

[Q3,A4,Y15,L16]-IGF I is at least 2 times more potent than IGF I in
stimulating the incorporation of glucose into diaphragm glycogen (Figure
4). A similar type of experiment shows that the B-chain mutant is 4 times
as potent as IGF I (23). These data strongly suggest that the binding of
IGF I to serum IGF binding proteins in the rat impairs the ability of IGF I
to interact with the type 1 IGF receptor in diaphragm.

Busby et al. (28) have described the purification of two species of
soluble IGF binding proteins from human amniotic fluid which have distinct
functional properties. Peak C inhibits the IGF I-mediated stimulation of
DNA synthesis in smooth muscle cells in a dose-dependent fashion, while peak
B causes a 4-fold enhancement of the IGF I-mediated stimulation. We have
shown that the B-chain mutant and [Q3,A4,Y15,L16]- IGF I have 5-10 fold
increased potency relative to IGF I in stimulating DNA synthesis in Balb/c
3T3 cells, but are equipotent to IGF I in rat A10 cells (26). This difference in relative potency in Balb/c 3T3 cells is due to a decrease in the
sensitivity to IGF I (26). Since these analogs are equipotent to IGF I at
the type 1 IGF receptors on these cell types, this loss in sensitivity to
IGF I may be due to the properties of the binding protein which is secreted

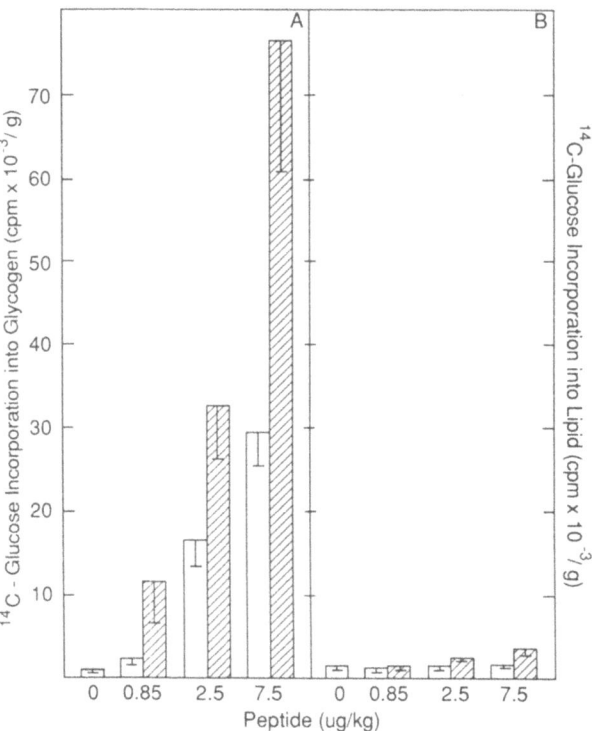

Figure 4. Stimulation of ^{14}C-glucose incorporation into
diaphragm glycogen (A) and epididymal fat lipid (B) by
IGF I (open bars) and [Q3,A4,Y15,L16]- IGF I (striped
bars). ^{14}C-glucose (1.8 μCi) and peptides were injected
i.p. into 200 g fasted rats. After 2 h, the animals
were sacrificed, diaphragm and fat were removed and
weighed, and glycogen and total lipid, respectively,
were extracted. Data are expressed as the mean + SEM
(n=5) counts per min incorporated into glycogen (A) or
lipid (B)/g wet tissue wt at each dose of peptide. The
stimulation of glycogen synthesis by both peptides is
significant by t test (P < 0.005 at 2.5 μg/kg; P < 0.001
at 7.5 μg/kg). There are no significant changes in the
level of lipid synthesis at any dose. Reprinted with
permission from reference 23.

by Balb/c 3T3 cells. In support of this, Balb/c 3T3 cell-conditioned media reduces the sensitivity of A10 cells to IGF I but not to [Q3,A4,Y15,L16]-IGF I (26). Thus, the IGF binding protein secreted by Balb/c 3T3 cells inhibits the activity of IGF I, but the protein secreted by A10 cells does not. These data suggest that soluble IGF binding proteins will play a complex role in the regulation of the biological effects of IGF I.

CONCLUSION

The binding data which we have generated with these analogs have broadly defined the regions involved in IGF I binding to its receptors and soluble binding proteins. These analogs will be valuable in studying potential species- and tissue-specific differences in the binding specificities of these proteins. They will also prove to be useful in defining the relative importance of the receptor subtypes and soluble binding protein subtypes in regulating the biological activity of IGF I.

ACKNOWLEDGEMENTS

We gratefully acknowledge the assistance of our associates in these studies, Mr. Gary G. Chicchi, Ms. Joy Applebaum, Ms. Nancy S. Hayes and Ms. Barbara G. Green. We would also like to thank Drs. Rene Humbel, James Florini and Elizabeth Jones for their kind gifts of serum-purified IGF I, MSA, and S. cerevisiae strain BJ1995, respectively.

REFERENCES

1. Rinderknecht, E. and Humbel, R.E. (1978) J. Biol. Chem. 253, 2769-2766.
2. Blundell, T.L., Bedarker, S., and Humbel, R.E. (1983) Fed. Proc. 42, 2592-2597.
3. Blundell, T.L., Bedarker, S., Rinderknecht, E., and Humbel, R.E. (1978) Proc. Natl. Acad. Sci. USA 75, 180-184.
4. Tager, H., Thomas, N., Assoian, R., Rubenstein, A., Saekow, M., Olefsky, J., and Kaiser, E.T. (1980) Proc. Natl. Acad. Sci. USA 77, 3181-3185.
5. Kobayashi, M., Ohgaku, S., Iwasaki, M., Maegawa, H., Shigeta, Y, and Inouye, K. (1982) Biochem. J. 206, 597-603.
6. DeVroede, M.A., Rechler, M.M., Nissley, S.P., Joshi, S., Thompson Burke, G., and Katsoyannis, P.G. (1985) Proc. Natl. Acad. Sci USA 82, 3010-3014.
7. Joshi, S., Burke, G.T., and Katsoyannis, P.G. (1985) Biochemistry 24, 4208-4214.
8. Joshi, S., Ogawa, H., Thompson Burke, G., Tseng, L.Y-H., Rechler, M.M., and Katsoyannis, P.G. (1985) Biochem. Biophys. Res. Commun. 133, 423-429.
9. Tseng, L. Y-H., Schwartz, G.P., Sheikh, M., Chen, Z.Z., Joshi S., Wang, J-F., Nissley, S.P., Thompson Burke, G., Katsoyannis, P.G., and Rechler, M.M. (1987) Bioch. Biophys. Res. Commun. 149, 672-679.
10. Bayne, M.L., Cascieri, M.A., Kelder, B., Applebaum, J., Chicchi, G., Shapiro, J.A., Pasleau, F., and Kopchick, J.J. (1987) Proc. Natl. Acad. Sci. USA 84, 2638-2642.
11. Cascieri, M.A., Hayes, N.S., Kelder, B., Kopchick, J.J., Chicchi, G.G., Slater, E.E., and Bayne, L.L. (1988) Endocrinology 122, 1314-1320.
12. Bayne, M.L., Applebaum, J., Chicchi, G.G., Hayes, N.S., Green, B.G., and Cascieri, M.A. (1988) Gene (Amstr.) 66, 235-244.
13. Rosenfeld, R.G., Conover, C.A., Hodges, D., Lee, P.D.K., Misra, P., Hintz, R.L., and Li, C.H. (1987) Biochem. Biophys. Res. Commun. 143, 199-205.
14. Tally, M., Enberg, G., Li, C.H., and Hall, K. (1987) Biochem. Biophys. Res. Commun. 147, 1206-1212.

15. Ewton, D.Z., Falen, S.L., and Florini, J.R. (1987) Endocrinology 120, 115-123.
16. Cascieri, M.A., Chicchi, G.G., Applebaum, J., Hayes, N.S., Green, B.G., and Bayne, M.L. (1988) Biochemistry 27, 3229-3233.
17. Dodson, E.J., Dodson, G.G., Hodgkin, D.C., and Reynolds, C.D. (1974) Can. J. Biochem. 57, 469-479.
18. Bayne, M.L., Applebaum, J., Underwood, D., Chicchi, G.G., Green, B.G., Hayes, N.S., and Cascieri, M.A. (1989) J. Biol. Chem., in press.
19. Bayne, M.L., Applebaum, J., Chicchi, G.G., Hayes, N.S., Green, B.G., and Cascieri, M.A. (1988) J. Biol. Chem. 263, 6233-6239.
20. Cascieri, M.A., Chicchi, G.G., Applebaum, J., Green, B.G., Hayes, N.S., and Bayne, M.L. (1989) J. Biol. Chem., in press.
21. Szabo, L., Mottershead, D.G., Ballard, F.J., and Wallace, J.C. (1988) Biochem. Biophys. Res. Commun. 151, 207-214.
22. Martin, J.L. and Baxter, R.C. (1986) J. Biol. Chem. 261, 8754-8760.
23. Cascieri, M.A., Saperstein, R., Hayes, N.S., Green, B.G., Chicchi, G.G., Applebaum, J., and Bayne, M.L. (1988) Endocrinology, 123, 373-381.
24. Brewer, M.T., Stetler, G.L., Squires, C.H., Thompson, R.C., Busby, W.H., and Clemmons, D.R. (1988) Biochem. Biophys. Res. Commun. 152, 1289-1297.
25. Lee, Y-L., Hintz, R.L., James, P.M., Lee, P.D.K., Shively, J.E., and Powell, D.R. (1988) Mol. Endocrinology 2, 404-411.
26. Cascieri, M.A., Hayes, N.S., and Bayne, M.L. (1989) J. Cell. Physiol., in press.
27. King, G.L., Kahn, C.R., Rechler, M.M., Nissley, S.P. (1980) J. Clin. Invest. 66, 130-140.
28. Busby, Jr., W.H., Klapper, D.G., and Clemmons, D.R. (1988) J. Biol. Chem. 263, 14203-14210.

COMPARISON OF INSULIN-LIKE GROWTH FACTOR RECEPTORS

IN HUMAN RETINAL CELLS

Joyce F. Haskell[1], Linda E. Haws[1], Alberta Davis[2], and
Richard Hunt[2]

[1]Research Service, Veterans' Administration Medical Center
and Department of Obstetrics and Gynecology, University of
Alabama at Birmingham, Birmingham, Alabama
[2]Department of Microbiology and Immunology, University of
South Carolina

INTRODUCTION

Specific receptors for insulin and insulin-like growth factors have
been identified in a variety of tissues, including brain and retinal cells.
Covalent crosslinking analysis using disuccinimidyl suberate has enabled us
to determine altered molecular weight forms of both the insulin and Type I
IGF receptors in human retinal tissue. Autopsy eye donations were frozen
and collected until adequate tissue was available for tissue to be
solubilized using Triton-X-100. Retinal pigment epithelial cells were also
identified and similarly treated. In addition, retinal pigment epithelial
cells were cultured following transformation with plasmid pSV2Neo. Using
these cells or tissues, we have analyzed the binding of [^{125}I] insulin, IGF
I and IGF II, and have determined the molecular weight for the insulin and
IGF receptors. We have also examined the effect of mannose-6-phosphate
(M6P) on the binding of IGF I to adult rat tissue. M6P had a dramatic
effect on the crosslinking of IGF I binding to the Type II IGF/M6P receptor
in detergent solubilized tissue preparations; however, no effect of M6P has
been seen on autoradiographs of ^{125}I-IGF I crosslinked to solubilized
human retinal pigment epithelial cells or human neural retinal membranes.
This agrees with our finding that human retinal cells have few Type II
IGF/M6P receptors.

DIABETIC RETINOPATHY

In diabetic retinopathy, patients experience a decrease in visual
acuity due to micro-aneurysms and hemorrage within the retina. These
pathological complications appear to be preceded by an increase in
permeability of the retinal capillaries, a decrease in the number of
pericytes, vascular supporting cells, and a proliferation of endothelial
cells which line the arteries and capillaries.[1] Despite several years of
intensive investigation, it is not known whether these cellular responses
in diabetics are due to the hyperglycemia or the deficiency of insulin.

Endothelial cells and pericytes have been shown by King and others[1] to possess insulin receptors and to be responsive to insulin stimulation. With the discovery of somatomedins, identified as a family of insulin-like growth factors (IGFs), several groups have examined the possibility of a role of these insulin related peptides in the pathophysiology observed in diabetic retinopathy. [2-4] Salarde et al. [2] found no correlation between serum IGF-1 levels in diabetic versus compared to normal children, although nocturnal secretion of GH was elevated in diabetics. Hyers et al.[3] similarly reports no significant differences in diabetic versus nondiabetic patients, however patients with proteinuria had markedly higher IGF I levels than those without proteinuria. Serum values for IGFs are complicated by the presence of serum binding proteins, and serum values may not represent tissue levels of insulin like growth factors. Elevated levels of IGF-1 have been recently reported in the vitreus of human diabetics by Grant.[4] We would like to propose that IGFs and their receptors may be involved in the pathological damage observed in diabetics, especially in the retina.

COMPARISON OF RETINAL INSULIN AND IGF RECEPTORS

An lower molecular weight form of the insulin receptor has been reported in brain homogenates[5-15]. Brain tissue contains receptors with a lower molecular weight than those reported in liver and other nonneural tissue when covalently crosslinked to [^{125}I]-insulin and examined using SDS-gel electrophoresis followed by autoradiography. In this chapter, we will present some new information based on studies examining human retinal tissue for insulin-like growth factor receptors by covalent crosslinking analysis. We report a lower molecular weight form of the insulin and Type I insulin-like growth factor receptors in human neural retinal homogenates, while retinal pigment epithelial cells contain insulin and Type I insulin-like growth factor receptors of normal molecular weight.

Human neural retinal tissue was obtained from six adult males (33-65 years of age). Eyes, removed post mortem from human donors, were obtained from the South Carolina Lions' Eye banks. Retinas were dissected and neural retinal membrane was frozen in Tris-EDTA-sucrose buffer, pH 7.6, until adequate tissue obtained. Samples were thawed, pooled, and gently homogenized using a glass homogenizer with loose fitting teflon pestle. Different particulate samples were obtained by filteration using various pore sizes of nylon sieving material in an attempt to isolate a vascular component according to the methods of Meezan et al[5]. However, no enrichment of blood vessels was detectable in samples retained on a 220 μm pore size nylon sieve or samples isolated using a 153 μm pore size sieve. An aliquot of unhomogenized human neural retinal membrane was also detergent solubilized and examined using affinity crosslinking techniques.

Retinal pigment epithelial membranes from a 65 year old caucasian male were frozen at -70C until solubilized and compared with other retinal samples (Fig. 1-3). Neonatal retinal pigment epithelial membranes (RPE) were cultured on laminin-coated tissue culture plates until confluent. In order to obtain an RPE cell line, primary cultures of human RPE cells were transfected with plasmid pSV40neo and several clones selected using Geneticin. The clone used in these experiments consisted of cells which were columnar in appearance, rapid growing and positive for keratins. These transformed RPE cells, identified as D-99t, also express SV40 T antigen, have transferrin receptors (M_r=93,000 dimers). They internalize [^{125}I]-transferrin and recycle it to the surrounding medium.(Hunt, et. al., manuscript in preparation).

Fig. 1 illustrates the difference in the molecular weight of the subunit of the insulin receptor in human neural retinal homogenates (lanes 1-3) compared to purified rat Leydig cells (lane 4) and rat testicular microvessels (lane 5). The neural retinal insulin receptor α-subunit (Mr= 115,000) is distinctly lower molecular weight than that seen in the testicular samples and the retinal pigment epithelial cell membranes (data not shown), Mr= 130,000. These result agree with the results that we have previously shown for rat and neonatal pig brain insulin receptors[5]. Brain microvessels have the higher molecular weight insulin receptor α-subunit, while brain homogenates have the lower molecular weight subunit. Many investigators[6-16] have shown similar results for brain and neuronal cells. It is not yet known whether this alteration in molecular weight is physiologically important.

The lower molecular weight may be due to a underglycosylation of the insulin receptor or an variant form of the receptor. Heidenrich and others[10-16] have used enzymatic degradation of the receptor using endo-glycosidases and neuraminidase and shown that enzymatic treatment removes sugar residues from the various forms of the insulin and IGF-I receptors. Although the digested receptors migrate at the lower molecular weight, these experiments do not conclusively show that the lower molecular weight brain and retinal receptors are due to underglycosylation.

Gammeltoft et al.[11] report an abnormal form of the Type I IGF receptor in human glial brain tumor cells. These cells contains an IGF I receptor which binds IGF I and IGF II with equal affinity. Both IGF I and II stimulate the tyrosine kinase activity and autophosphorylation in these glial astrocytes. The Type I receptor α-subunit has a M_r= 130,000 rather than the M_r= 115,000 reported for adult rat brain cortex. A curvilinear Scatchard plot is obtained for both IGF I and IGF II, whereas most investigators report linear Scatchard plots for both Type I and Type II IGF receptors. Whereas others find that most regions of adult rat brain bind more IGF I than IGF II, Gammeltoft[11] report that human glial asteocytoma cells bind more IGF II than IGF I.

Recently, several groups[15,16] have identified specific IGF-I receptors in retinal tissue preparations. Waldbillig and Chader[15] report an anomalous form of the insulin receptor in neural retinal homogenates. They also report that IGF I bound to both forms of the α-subunit (Mr=120,000 and 130,000). We found similar results for the neonatal pig brain insulin receptor (M_r= 120,000) which copurified with the cerebral microvascular (Mr= 130,000) insulin receptor[5].

In Fig.2, an autoradiogram compares the Type I IGF receptors in human RPE cell membranes (lane 1), neural retinal homogenates (lanes 2-4) and purified rat Leydig cells (lane 5). The reduced crosslinked IGF-I binding migrates corresponding to a distinctly lower molecular weight in lanes 2-4, while the RPE and testicular cells both have the higher molecular forms of the IGF-I receptor. It is not yet clear whether the upper bands (approximately Mr= 255,000) represent dimers of the α-subunit of the Type I receptor or IGF-1 crossreacting with the Type II IGF receptor. Using higher concentrations of dithiothreitol (>50 mM) does not diminish any of the upper band intensity, leading us to believe that this species does not represent disulfide linked dimers of the Type I α-subunit.

Lower molecular weight crosslinked bands are observed in a variety of tissues incubated with either [^{125}I] IGF I or IGF II. These bands (Mr

18,000-65,000) are usually not seen for [125] I-insulin unless albumin is present during incubation conditions. Several investigators[17,18] have shown that some of these bands may correspond to IGF binding proteins that are present in the serum and secreted by cells. Binding proteins have high affinity binding specifications and usually prefer IGFs to insulin. In Fig. 1 and 2, we were careful to avoid any presence of albumin in the binding and crosslinking conditions. Multiple low molecular weight bands (<72,000) are noted in both autoradiograms. We have no explanation for these multiple bands other than possible fragments of the concentrated receptor or an experimental artifact due to the large amount of unbound radiolabeled hormone present during crosslinking. No washing was attempted until solubilized tissue preparations were crosslinked and quenched with tris-EDTA buffer and samples concentrated using Centricon filter-concentrators (Amicon Corporation).

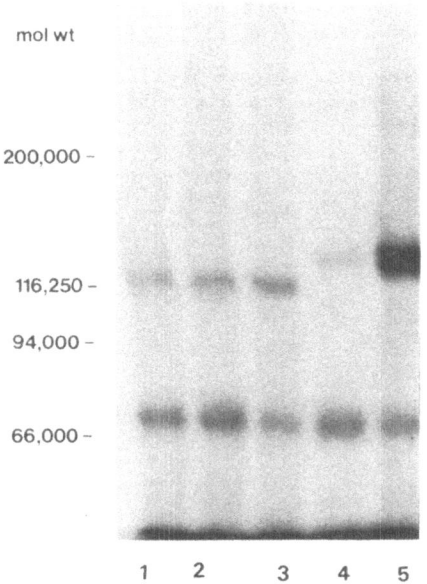

Fig. 1 Autoradiogram of [125I] -insulin cross-linked to detergent solubilized tissue samples from (1) human retinal homogenate on 220 μm pore size nylon sieve; (2) human retinal homogenate retained on 153 μm pore size nylon sieve; (3) human neural retinal membrane; (4) rat testicular microvessels.

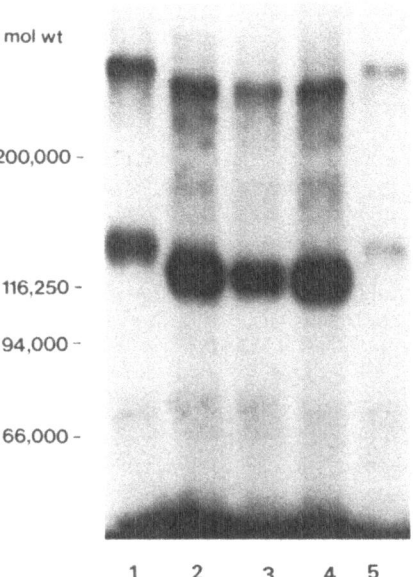

Fig. 2 Autoradiogram of [125I] -IGF I crosslinked to detergent solubilized tissue samples from (1) human retinal pigment epithelial membranes; (2) human retinal membranes from 220 μm pore size sieve; (3) human retinal homogenate retained on 153 μm pore size sieve; (4) human neural retinal membranes; compared with (5) rat testicular Leydig cells.

Several groups of investigators have demonstrated that insulin and insulin-like growth factors can interact with neuronal and vascular tissue to produce a wide variety of cellular responses. Insulin has been shown to be a potent growth regulator in arterial smooth muscle cells[19] and may be important for maintaining cell integrity in addition to the well-noted effects on glucose metabolism[18-21]. Insulin and IGFs appear to have many similar effects on cell metabolism and growth. Both IGF I and IGF II stimulate proteoglycan formation in endothelial cells from large and small blood vessels[20]. More recently, IGF I has been shown to direct the growth of retinal endothelial cells through a chemotactic mechanism[21].

One possible need for an alternate form of the insulin and IGF I receptors would be the directional transport of insulin-like peptides from the vascular system to the tissues or vice versa. In the case of brain, retina and to a lesser extent, the gonads, these tissues are protected by a blood-tissue barrier. Retinal pigment epithelial cells are part of the blood-retinal barrier. Several groups of investigators[5,14,20-24] have been interested in determining whether the cerebral or retinal microvessels have specific receptors and whether these receptors may be involved with processing or translocation of the insulin-like peptides. Brain and retina appear also to be immunoprivileged sites and therefore cells within these tissues may not need the terminal glycosylation of the insulin and IGF I receptor to protect the receptors from immunological degradation (Pillion, manuscript in preparation). Diabetic complications may also be a result of pulsatile injections of long acting insulin which may bind to IGF I receptors. Blood glucose levels may effect the amount of nonenzymatic glycosylation which may also be a critical factor in the prevention of diabetic complications.[25] Although differences have been observed in the molecular weights of brain and retinal receptors, little or no effect has been observed on the tyrosine kinase activity or autophosphorylation of either the insulin or IGF I receptor.[6-16]

Recently, anti-Type II receptor blocking antibodies have been developed and found to have no effect on IGF I stimulated biological responses[26]. A unique role for the Type II receptor has yet to be demonstrated. The Type II IGF receptor gene was finally sequenced and found to be homologous to the cation independent form of the mannose-6-phosphate receptor[27]. Rat IGF-II, formerly called "multiplication stimulating activity", MSA, closely resembles proinsulin and IGF I with discrete amino acid differences. We have obtained radioiodinated MSA (peak III-2) from S.P. Nissley, Metabolism Branch, National Cancer Institute, NIH, and used this preparation to identify the Type II IGF/M6P receptor in isolated adult rat glomeruli[28,29]. Using covalent crosslinking techniques, the Type II IGF/M6P receptor band has a M_r=220,000 under nonreduced conditions, and a M_r=255,000 under reduced conditions. This increase in apparent molecular weight may be explained by the unfolding of an single chain glycoprotein with multiple disulfide linkages.

In Fig. 3, we show an autoradiogram of $[^{125}I]$-r IGF II crosslinked to detergent solubilized tissue samples from (1) human retinal pigment epithelial cell membrane; (2) human retinal homogenate from a 220 μm pore size nylon sieve; (3) human retinal homogenate from 153 μm pore size nylon sieve, compared with rat testicular artery (lane 4). Rat Leydig cells and other testicular samples on the same gel gave similar results to lane 4. Samples in Fig. 1 and Fig. 2 contained identical tissue samples of approximately 100-200 ug membrane protein, assayed by the Bradford method using bovine serum albumin as a standard. All three 6% polyacrylamide gels

(represented in Fig.1-3) were analyzed by electrophoresis under the same
experimental conditions (reduced with 50 mM dithiothreitol). The absence
of a reduced crosslinkable M$_r$=255,000 band using retinal preparations may
be due to species variability. We are currently attempting to confirm
these results using recombinant human IGF-II. The lower intensely labeled
bands seen only in retinal preparations (Mr= 115,000) do not correspond to
the α-subunit of the Type I IGF receptor, based on comparison with IGF I
binding to purified rat Leydig cells on an unshown lane of the
autoradiogram. Whether these bands represent IGF II binding to a degraded
receptor fragment or the insulin receptor is not known. The RPE membranes
used in Fig. 1-3 were from a single donor but similar results have been
obtained with several different RPE membranes and cultured RPE cells. These
RPE cells have been shown to possess both transferrin and insulin-like
growth factor receptors[30].

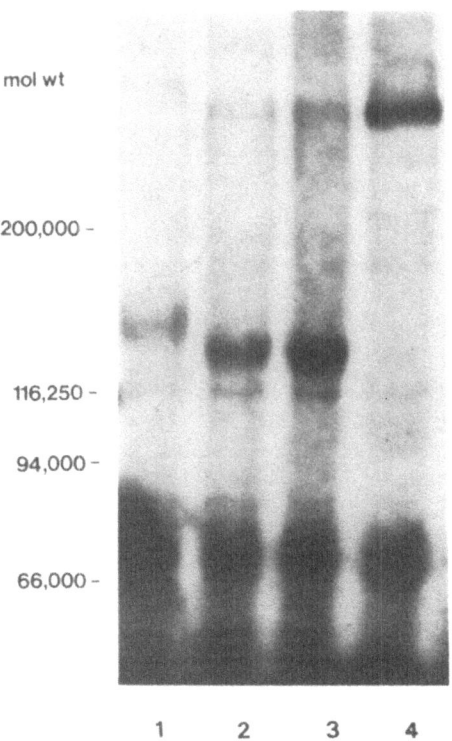

Fig. 3. Autoradiogram of [125]-I-rIGF II crosslinked to detergent-
solubilized human RPE (lane 1), human neural retinal homogenates (lanes
2,3) and rat testicular (lane 4) tissue.

Since both IGF II and mannose-6-phosphate bind to the same
glycoprotein[27], several investigators[31-37] have examined the effect of

mannose-6-phosphate on IGF II binding to the Type II IGF/M6P receptor. Polychronakos and Piscina[31] found that M6P enhances the endocytosis of IGF II in IM9 cells. Results from Braulke et al.[33] indicate that the binding sites for M6P and IGF II are present on different portions of the IGF II/M6P receptor. They find no enhancement of IGF II binding to human skin fibroblasts by pentamannosylphosphate-substituted bovine serum albumin. Roth et al[35] found that M6P increases binding of IGF II to human fetal brain membranes, although Tong et al.[36] found that while M6P and IGF II both bind to the same protein, M6P had no effect on IGF II binding to solubilized bovine liver or testes. Kiess and others[37] have demonstated that both M6P and IGF II affinity columns bind and can be used to purify the same protein in detergent extracts. Todderud and Carpenter[38] recently have detected mannose-6-phosphate on the epidermal growth factor receptor in A431, indicating a possible mechanism for interaction between these two growth factor receptors.

As shown in Fig. 4., we used covalent crosslinking techniques to examine the IGF-I receptor in solubilized vascular homogenates incubated with mannose-6-phosphate (M6P). Solubilized tissue samples were incubated with 0.5 uCi of [^{125}I]-IGF-1 in 25 mM Hepes, pH 7.6, with (lanes 1, 3, 5, 7, 9) or without (lanes 2, 4, 6, 8, 10) 1 mM M6P overnight at 4 C. Samples were crosslinked with 0.25 mM disuccinimidyl suberate, quenched with five volumes of 10 mM Tris- 1 mM EDTA, pH 7.4 and concentrated using Amicon Centricon-30 disposable filter-concentrators. The entire sample was loaded onto each lane on the 6% polyacrylamide gel. Lanes 1, 2 represent rat testicular artery, lanes 3, 4 represent rat aorta, lanes 5, 6 represent rat superior vena cava, lanes 7,8 contained inferior vena cava, and lanes 9, 10 contained renal artery. M6P enhanced [^{125}I]-IGF-I binding to the M_r=255,000 band in all rat vascular tissues studied to date. The rat testicular artery preparations (lanes 1,2) have little Type I IGF binding compared with ^{125}I-IGF-II binding; mannose-6-phosphate clearly enhanced the amount of ^{125}I-IGF-I labeling to this detergent-solubilized tissue. The majority of the binding occurs in the higher molecular weight band (M_r=255,000) with very little effect seen on the M_r= 130,000 band or the lower binding protein-sized bands (M_r< 66,000). Anti-type II IGF receptor antibody 3637[26] specifically displaces the M_r= 255,000 bands in similar experiments using rat neonatal cardiac myocytes.

As shown in Fig. 5, we have also examined the effect of mannose-6-phosphate on detergent solubilized human neural retinal tissue and transformed human RPE cells. In these experiments, detergent solubilized tissue was incubated with 0.5 uCi [^{125}I] IGF-1 for 18 h at 4 C in the presence(lanes 2-4,6-8) or absence (lanes 1,5) of 1 mM mannose-6-phosphate. ^{125}I-IGF-II does not bind to these cells to any appreciable degree. Fig. 5 clearly shows a distinct difference in molecular weight of the reduced IGF-1 receptor, between human neural retinal tissue (lanes 1,2) and transformed RPE cells (lanes 5,6). The retinal pigment epithelial cells form the retinal-blood barrier and may be involved in the transport of nutrients into the eye. Retinal insulin receptor in isolated microvessels also have the higher molecular weight form observed in RPE cells (Fig.1) and most non-neuronal tissue[4-16]. In a similar experiment, neonatal rat brain homogenates had a similar molecular size to that observed for the detergent solubilized neural retina.

Both 1 ug/ml IGF-1 (lanes 4,8) and 1 ug/ml IGF-II (lanes 3,7; recombinant human IGF-II from M. Smith, Eli Lilly, Indianapolis, IN) displaced ^{125}I-IGF-I crosslinking to detergent solubilized tissue. Binding to both upper (M_r=230,000) and lower (M_r= 125,000, for lanes 1,2 or 135,000 band, seen in lanes 5,6) are blocked by excess unlabeled IGFs in either detergent solubilized transformed RPE cells or neural retinal tissue. No enhancement of binding to the Mr= 255,000 band is seen in the presence of M6P (all lanes except 1,5). The excess IGFs do not appear to displace all of the M_r= 125,000 band observed in neural retinal membrane samples, leading us to speculate some of the binding may be due to the insulin receptor. The addition of a specific anti-insulin receptor antibody (alphaIR-1, generously given by S. Jacobs, Chapel Hill, NC) did not inhibit radiolabeled insulin-like growth factor binding, consistent with a specific Type I IGF receptor, rather than the insulin receptor.

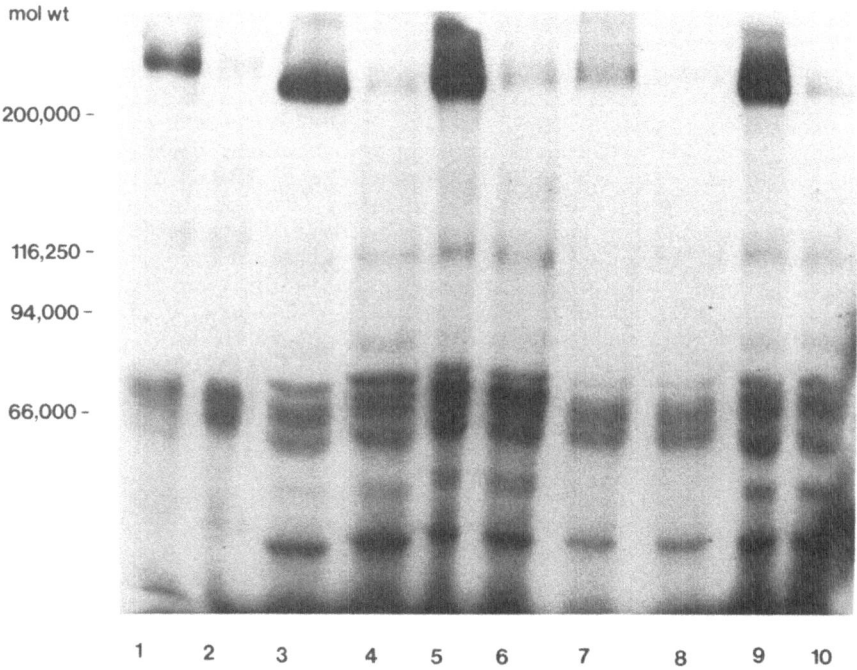

Fig. 4. Detergent solubilized rat vascular tissue covalently crosslinked to (^{125}I)-IGF I: Effect of mannose-6-phosphate. Samples were incubated with (odd-numbered lanes) or without (even-numbered lanes) 1 mM M6P.

SUMMARY

We have demonstrated rIGF-II binding to a variety of tissues, yet we cannot demonstrate specific IGF II binding to retinal preparations under conditions in which IGF I binding occurs. In this report, we find that M6P does not stimulate IGF I binding in human retinal tissues and transformed

mol wt

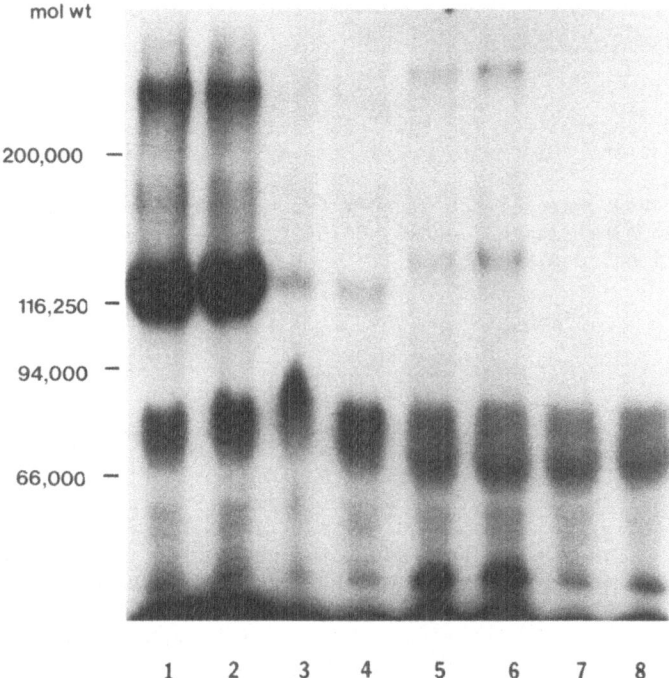

Figure 5. Crosslinked [^{125}I]-IGF I to detergent-solubilized human retinal tissue (lanes 1-4) or transformed RPE cells (lanes 5-8): effect of mannose-6-phosphate (1 mM, all lanes except 1,5). Nonspecific binding represented in lanes 3,7 (excess IGF II), lanes 4,8 (excess IGF-I).

RPE cells, although IGF I is crosslinked to a M_r=255,000 band. Binding of [^{125}I]-IGF I in the presence of mannose-6-phosphate is enhanced by 1mM M6P in all vascular and non-vascular tissues studied to date, except human retinal preparations, leading us to conclude that very little IGF-II/M6P receptor is present in retinal tissues.

Clearly, there are two definite forms of the type I IGF receptor in neuronal tissue.[5-15] As described in numerous publications and shown here for human neural retinal cells versus RPE cells, a lower molecular weight variant of the insulin and IGF-I receptor exists. Whether this alteration in molecular weight is due to a variant transcription of the insulin or IGF I receptor gene or a modification of post-translational processing is not yet known, although other investigators have shown that treatment of insulin and IGF-I receptors from a variety of cells or tissues with endoglycosidases results in a lowering of the molecular weight determined on SDS-gels.

ACKNOWLEDGEMENTS

We would like to acknowledge the excellent technical assistance of Joe Modzelewski, David Eddie Hunt, and Denise Whidby.

Our gratitude to Dr. S.P. Nissley. for providing [^{125}I]rat IGF II (MSA.

peak III-2) and the South Carolina Lions Club Eye Bank for supplying donor eyes.

We are also grateful for the critical review of this manuscript by several colleagues at the University of Alabama at Birmingham.

The authors are grateful for support from the National Institute of Health (HL 38442 to J.F. Haskell, and EY 06164 to R.C. Hunt) and the Veterans Administration (to J.F. Haskell).

REFERENCES

1. G.L. King, Cell Biology as an approach to the study of the vascular complications of diabetes, Metabolism 34:17 (1985).
2. S. Salardi, E. Cacciari, D. Ballardini, F. Righetti, M. Capello, A. Cicognani, S. Zucchini, G. Natali, and D. Tassinari, Relationships between growth factors (somatomedin-C and growth hormones) and body development, metabolic control, and retinal changes in children and adolescents with IDDM. Diabetes 35:832 (1986).
3. S.L. Hyer, P.S. Sharp, R.A. Brooks, J.M. Burrin, and E.M. Kohner, Serum IGF-1 concentration in diabetic retinopathy, Diabetic Med. 5:356 (1986).
4. M. Grant, B. Russell, C. Fitzgerald, and T.J. Merimee, Insulin-like growth factors in vitreous. Studies in control and diabetic subjects with neovascularization, Diabetes 35:416 (1986).
5. J.F. Haskell, E. Meezan, and D.J. Pillion, Identification of the insulin receptor of cerebral microvessels, Amer. J. Physiol. 248:E115 (1985).
6. A. McElduff, P. Poronnik, R.C. Baxter, and P. Williams, A comparison of the insulin-like growth factor 1 receptors from rat brain and liver, Endocrinology 122:1933 (1988).
7. R.A.Roth, D.O. Morgan, J. Beaudoin, and V. Sara, Purification and characterization of the human brain insulin receptor, J. Biol. Chem. 261: 3753 (1986).
8. L. Bassas, F. DePablo, M.A. Lesniak, and J. Roth, Ontogeny of receptors for insulin-like peptides in chick embryo tissues: early dominence of insulin-like growth factor over insulin receptors in brain, Endocrinology 117:2321 (1985).
9. C.C. Yip, M.L. Moule, and W.T. Yeung, W.T., Characterization of insulin receptor subunits in brain and other tissues by photo affinity labeling, Biochem. Biophys. Res. Commun. 96:1671 (1980).
10. K.A. Heidenrich, P.R. Gilmore, D. Brandenburg,and E. Hatda, Peptide mapping on Northern blot analyses of insulin receptors in brain and adipocytes, Mol. Cell. Endocrinology 56:255 (1988).
11. S. Gammeltoft, R. Ballotti, A. Kowalski, B. Westermark, and E. Van Obberghen, Expression of two types of receptors for insulin-like growth factors in human malignant glioma, Cancer Research 48:1233 (1988).
12. M.K. Raizada, J. Shemer, J.J. Judkins, D.W. Clarke, B.A. Masters, and D. LeRoith, Insulin receptors in the brain: structural and physiological characterization, Neurochemical Research 13:297 (1988).
13. D.LeRoith, W.L. Lowe, J. Shemer, M.K. Raizada, and A. Ota, Development of brain insulin receptors, Int. J. Biochem. 20:225 (1988).

14. R.G. Rosenfeld, H. Pham, B.T. Keller, R.T. Borchardt and W.M. Pardridge, Demonstration and structural comparison of receptors for insulin-like growth factor-I and -II (IGF-I and -II) in brain and blood-brain barrier, Biochem Biophys Res Commun 149:159 (1987).

15 R.J. Waldbillig and G.J. Chader, G. Anomalous insulin binding activity in the bovine neural retina: a possible mechanism for regulation of receptor binding specificity, Biochem Biophys Res Commun 151:1105 (1988).

16. Y. Zick, A.M. Spiegel, and R. Eisenberg, Insulin-like growth factor I receptors in retinal rod outer segments, J Biol Chem 262:10259 (1987).

17. R.S. Bar, L.C. Harrison, R.C. Baxter, M Boes, B.L. Dake, B. Booth, and A. Cox, Production of IGF-binding proteins by vascular endothelial cells, Biochem Biophys Res Commun 148:734 (1987).

18. R.G. Elgin, W. H. Busby, and D.R. Clemmons, An insulin-like growth factor (IGF) binding protein enhances the biologic response to IGF-I, Proc Natl Acad Sci USA 84:3254 (1987).

19. B. Pfeifle, H. Hamann, R. Fussganger, and H. Ditschuneit H. Insulin as a growth regulator of arterial smooth muscle cells: effect of insulin of IGF I, Diabete Metab 13:326 (1987).

20. R.S. Bar, B.L. Dake, and S. Stueck, Stimulation of proteoglycans by IGF-I and II in microvessel and large vessel endothelial cells, Am J Physiol 253:E21 (1987).

21. M. Grant,J. Jerdan, T.J. Merimee, Insulin-like growth factor-I modulates endothelial cell chemotaxis, J Clin Endocrinol Metab 65:370 (1987).

22. H.L. Hachiya, J.L. Carpentier, and G.L. King, Comparative studies on insulin-like growth factor II and insulin processing by vascular endothelial cells, Diabetes 35:1065 (1986).

23. H.J. Frank, W.M. Pardridge, W.L. Morris, R.G. Rosenfeld, T.B. Choi, Binding and internalization of insulin and insulin-like growth factors by isolated brain microvessels, Diabetes 35:654 (1986).

24. R.S. Bar, M. Boes, M. Yorek, M., Processing of insulin-like growth factors I and II by capillary and large vessel endothelial cells, Endocrinology 118:1072 (1986).

25. K.F. Hanssen, K. Dahl-Jorgensen, T. Lauritzen, B. Feldt-Rasmussen, O. Brinchmann-Hansen, and T. Deckert, REVIEW: Diabetic control and microvascular complications: the near-normoglycaemic experience, Diabetologia 29:677 (1986).

26. W. Kiess, J.F. Haskell, L. Lee, L.A. Greenstein, B.E. Miller, A.L. Aarons, M.M. Reckler, and S.P. Nissley, An antibody that blocks insulin-like growth factor (IGF) binding to the Type II IGF receptor is neither an agonist nor an inhibitor of several IGF stimulated biologic responses in L6 myoblasts, J Biol Chem 262:12745 (1987).

27. D.O. Morgan, J.C. Edman, D.N. Standring, V.A. Fried, M.C. Smith, R.A. Roth, W.J. Rutter, Insulin-like growth factor II receptor as a multifunctional binding protein, Nature 329:301 (1987).

28. J. Haskell, D. J. Pillion, and E. Meezan, Specific high-affinity receptors for insulin-like growth factor II (IGF-II) in rat kidney glomerulus, Endocrinology 123:774 (1987).

29. D.J. Pillion, J.F. Haskell, and E. Meezan, Distinct receptors for insulin-like growth factor I in rat renal glomeruli and tubules, Amer J Physiol 255:E504 (1988)

30. R.C. Hunt, A.A. Davis, A. Dewey, J.F. Haskell, Transferrin and insulin-like growth factor-I receptors on retinal pigment epithelial cells, J. Cell Biology 105:236A (1987).

31. C. Polychronakos and R. Piscina, Endocytosis of receptor-bound insulin-like growth factor II is enhanced by mannose-6-phosphate in IM9 cells, Endocrinology 126:2943 (1988).

32. J. Haskell, L. Haws, and T. Lin, Insulin-like growth factor receptors in rat testicular vascular tissue, Endocrinology 70th Annual Meeting Abst. 519 (1988).

33. T. Braulke, C. Causin, A. Waheed, U. Junghans, A. Hasilik, P. Maly, R.E. Humbel, and K. von Figura, Mannose-6-phosphate/insulin-like growth factor II receptor: Distinct binding sites for mannose 6-phosphate and insulin-like growth factor II, Biochem Biophys Res Commun 150:1287 (1988).

34. R.G. Rosenfeld and R. Pham, Production of monoclonal antibodies to the rat insulin-like growth factor II (IGF-II receptor), Biochem Biophys Res Commun 146:717 (1987).

35. R.A. Roth, C. Stover, J. Hari, D.O. Morgan, M.C. Smith, V. Sara, V.A. Fried, Interaction of the receptor for insulin-like growth factor II with mannose-6-phosphate and antibodies to the mannose-6-phosphate receptor, Biochem Biophys Res Commun 149:600 (1987).

36. P.Y. Tong, S.E. Tollefsen, and S. Korafeld, The cation-independent mannose-6-phosphate receptor binds insulin-growth factor II, J Biol Chem 263:2585 (1988).

37. W. Kiess, G. Blickenstaff, M. Sklar, C. Thomas, S.P. Nissley, and G. Sahagian, Biochemical evidence that the Type II IGF receptor is identical to the cation-independent mannose-6-phosphate receptor, J Biol Chem 263:9339 (1988).

38. G. Todderud, and G. Carpenter, Presence of mannose phosphate on the epidermal growth factor receptor in A-431 cells. J Biol Chem 263:17893 (1988).

IGF I RECEPTOR PHOSPHORYLATION

Cary Moxham, Vincent Duronio, and Steven Jacobs

Burroughs Wellcome Co.
Research Triangle Park, NC 27709

Physiologically, cells are simultaneously exposed to a plethora of hormones. The way in which a cell responds to stimulation by a specific hormone depends not only upon the concentration of that hormone, but upon the presence of other hormones as well. Receptors for both insulin and IGF I are phosphorylated when cells are exposed to phorbol esters, presumably through activation of protein kinase-C (1-4). Since protein kinase C is physiologically activated by a number of hormones whose receptors are coexpressed with insulin and IGF I receptors, it seems possible that these hormones could regulate responsiveness of insulin and IGF I receptors by activating protein kinase C.

Phorbol esters have been shown to clearly effect two different insulin receptor functions. First, serine and threonine phosphorylation of the receptor mediated by protein kinase C inhibits its intrinsic tyrosine kinase activity (3). Second, incubation of cells with phorbol esters and insulin leads to an increase in the amount of insulin which accumulates intracellularly (5,6). This has been attributed to an increased rate of receptor mediated internalization of insulin (5) or to a decreased rate of retroendocytosis (6). Far less is known about the functional consequences of protein kinase C activation on the IGF I receptor. We have therefore attempted to examine some of these effects.

Effect of TPA on IGF I binding

Incubation of Hep G2 cells with 500 nM TPA for thirty minutes at 37 degrees, had no effect on IGF I receptor number or affinity if binding was subsequently measured at 15 degrees. However, if IGF I binding was measured at 37 degrees, TPA increased the amount of cell associated IGF I. At 37 degrees, IGF I is internalized by receptor mediated endocytosis. To determine which pool of IGF I, surface bound or internalized, was being increased by TPA, cells were incubated with IGF I for thirty minutes at 37 degrees, and then the total amount bound and the amount which remained bound after surface bound IGF I was removed by acid stripping was measured (Fig. 1). The TPA mediated increase in cell associated IGF I was due entirely to an increase in the intracellular pool. This increase was not due to a change in its rate of degradation since the TCA precipitable fraction of both cell associated IGF I and IGF I in the media was small and not significantly different in TPA treated cells.

To determine if the TPA mediated increase in internalized IGF I was accompanied by an increased rate of receptor internalization, HepG2 cell surface proteins were radioiodinated vectorially with lactoperoxidase. The trypsin sensitivity of the labeled receptors was then determined after incubation with 500 nM TPA, 100 nM IGF I or the combination for various time periods. Trypsin degraded the 135,000 Mr alpha subunit of the IGF I receptor to a 115,000 Mr fragment. At time zero, virtually all labeled receptors were susceptible to trypsin, confirming that only cell surface receptors were labeled. By 5 minutes, a fraction of the receptors had become resistant to trypsin and were presumably internalized, and it was clear that both IGF I and phorbol esters increased this fraction. By 15 minutes 33%, 45%, 39%, and 47% of the receptors were resistant to trypsin in cells that had been incubated without addition, in cells incubated with IGF I alone, in cells incubated with TPA alone, or in cells incubated with IGF I and TPA.

These results indicate that TPA increases the rate of IGF I receptor internalization. Since there was no net change in cell surface receptor number as determined by IGF I binding, there presumably is an equivalent increase in the rate of receptor exocytosis. It is tempting to attribute these changes in receptor flux to changes in its state of phosphorylation. However, it is equally likely that the cellular processes involved in receptor internalization are enhanced by phorbol ester treatment. This has been shown to be the case for EGF (7) and transferrin (8) receptors, since mutant receptors in which the major sites of protein kinase C phosphorylation have been altered are still internalized more rapidly following treatment with phorbol esters.

Phorbol esters inhibit IGF I receptor autosphophorylation

In order to determine the effects of phorbol esters on IGF I receptor autophosphorylation, which is a reflection of its tyrosine kinase activity, HepG2 cells were incubated with $H_3{}^{32}PO_4$ for two hours in phosphate free media to label their endogenous pool of ATP. The cells were then incubated with or without 500 nM TPA. After 10 minutes, 100 nM IGF I was added to half the cells in each group, and the incubation was continued for an additional 5 minutes. The reaction was then stopped, the cells solubilized, and IGF I receptor partially purified on a wheat germ agglutinin Sepharose column. Labeled IGF I receptors were then immunoprecipitated using one of two antibodies: alpha IR-3 which recognizes all IGF I receptors regardless of their state of phosphorylation, and antiphosphotyrosine antibody which recognizes only those receptors containing phosphotyrosine. As shown in Fig. 2, in the basal state, a phosphorylated band corresponding to the beta subunit was immunoprecipitated by alpha IR-3 but not by the antiphosphotyrosine antibody, indicating that in the basal state the IGF I receptor contained virtually no phosphotyrosine. TPA caused an approximately three fold increase in labeling of the band precipitated by alpha IR-3, but again there was no labeling of the beta subunit band in the antiphosphotyrosine immunoprecipitates, indicating that the sites of increased phosphorylation were on serine or threonine residues. IGF I also caused an approximate three fold increase in the total phosphorylation of the beta subunit band in alpha IR-3 immunoprecipitates and also resulted in the appearance of a prominent phosphorylated band in antiphosphotyrosine immunoprecipitates. When TPA was added in addition to IGF I, although there was further increase in phopshorylation of the receptor population immunoprecipitated by alpha IR-3, there was a 50% decrease in labeling of the beta subunit band immunoprecipitated with antiphophotyrosine, indicating that TPA inhibited autophosphorylation while stimulating serine-threonine phos- phorylation. This was confirmed by phosphopeptide maps.

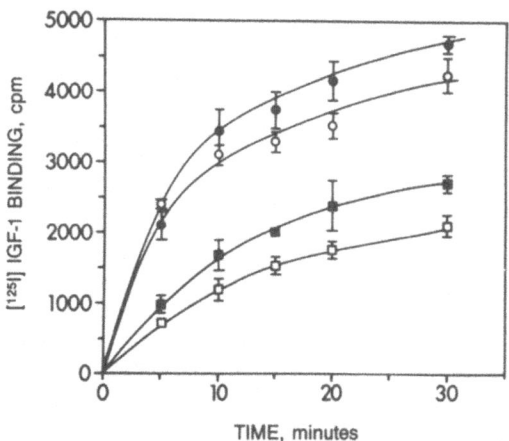

Fig. 1. Effect of TPA on total and intracellular pool
of IGF-I binding. ^{125}I-IGF-I bound to control
(open figures) and TPA treated (closed figures)
HEPG2 cells before (circles) or after (squares)
acid wash.

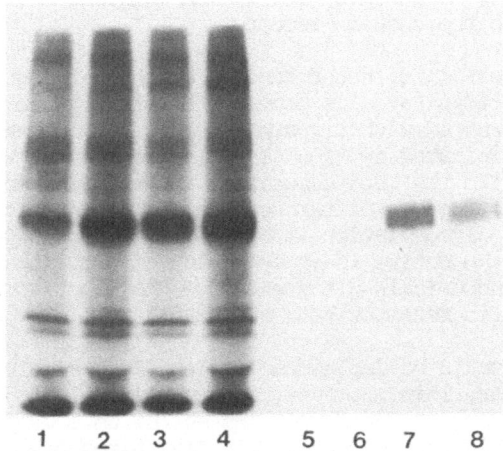

Fig. 2. Effect of TPA on IGF-I receptor autophosphorylation. HepG2 cells were treated with no
additions (1 & 5), TPA (2 & 6), IGF-I (3 & 7),
or TPA plus IGF-I. Autoradiogram of alpha IR-3
immunoprecipitates (1-4) or antiphosphotyrosine
immunoprecipitates (5-8).

Beta-subunit heterogeneity

In many cells, the beta subunit of the IGF I receptor appears as a doublet when analyzed on SDS polyacrylamide gels (9-11). The lower component of this doublet usually has a mobility similar to that of the beta subunit of the insulin receptor in the same cell. It has been suggested that this difference in mobility could be due to differences in glycosylation (12,13). We have recently obtained evidence, which although not entirely conclusive suggests that other factors might also contribute.

Fig. 3 shows phosphopeptide maps of receptors from HepG2 cells that had been treated with 500 nM TPA and immunoprecipitated with alpha IR-1, an antibody to the insulin receptor, or with alpha IR-3. A single phospho-threonine containing peptide (labeled T) was present in both immuno-precipitates. This was a major phosphopeptide in the basal state and was increased when cells were treated with TPA. The other major phosphopeptides contained phosphoserine. Several peptides, including those labeled 1, 5, and 6, were present in both immunoprecipitates, but phosphopeptides 2, 3, and 4 were present only in alpha IR-3 immunoprecipitates. Phosphopeptides 4 and 5 were present only in cells that had been treated with TPA. The other major phosphopeptides were present in the basal state and were stimulated somewhat when cells were treated with TPA.

When the beta subunit present in alpha IR-3 immunoprecipitates was separated into an upper and lower component and each individually analyzed by phosphopeptide maps, only phosphopeptides 1, 2, 3, and 4 were present in the upper component. The lower component contained these peptides, and in addition, the other phosphopeptides that were also present in the alpha IR-1 immunoprecipitates. A possible explanation for these results is that the lower component of the beta subunit immunoprecipitated by alpha IR-3 is contaminated with insulin receptor beta subunit. Since it is difficult to resolve these two components, it is even possible that the entire lower component is due to the insulin receptor.

A further indication of heterogeneity of the beta subunit contained in alpha IR-3 immunoprecipitates is provided by its trypsin sensitivity. When intact cells are treated with trypsin, the upper component of the beta subunit immunoprecipitated by alpha IR-3 is cleaved by trypsin, while the lower component, like the beta subunit of the insulin receptor, is resistant (Ref 13, Fig 7). Differences in trypsin sensitivity persist even when terminal glycosylation is blocked by mannodeoxynojirimycin. It thus appears that the entire lower component of the alpha IR-3 immuno-precipitated beta subunit is different from the upper component and resembles the insulin receptor beta subunit.

These results could be explained if alpha IR-3 simply crossreacts with the insulin receptor. This, however, seems unlikely. Alpha IR-3 completely inhibits IGF I binding at concentrations as low as 1 ug per ml, while concentrations as high as 100 ug per ml, five times higher than those used for immunoprecipitation, have no effect on insulin binding (14-16). Furthermore, 100 ng per ml insulin causes nearly maximal stimulation of the phosphorylation of the insulin receptor beta subunit but no detectable phosphorylation of the lower component of the receptor immunoprecipitated by alpha IR-3. Phosphopeptide maps of insulin receptor phosphorylated under these conditions reveal four phosphotyrosine containing peptides, two of which have greater intensity than the phosphothreonine containing peptide, while phosphopeptide maps of beta subunit immunoprecipitated by alpha IR-3 contained little if any phosphotyrosine but have the prominent phosphothreonine containing peptide.

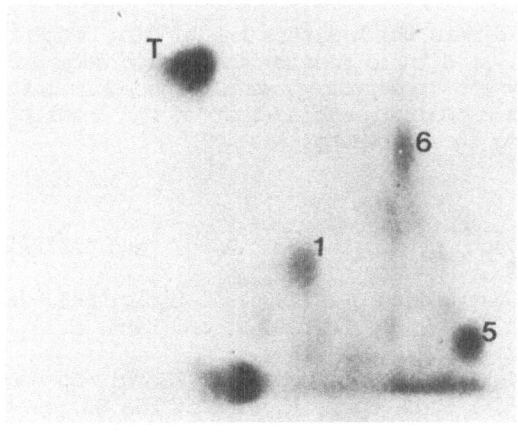

αIR 1

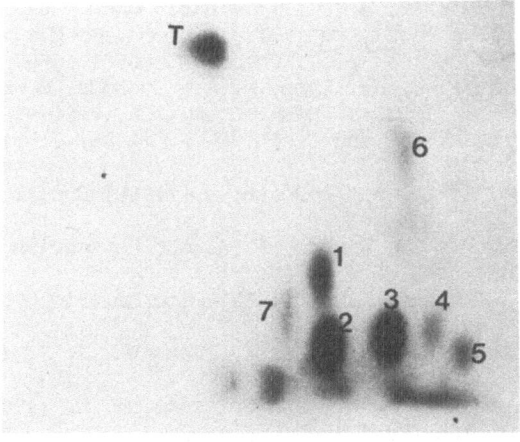

α IR 3

Fig. 3. Tryptic phosphopeptide maps of insulin receptor
beta subunit or IGF-I receptor beta subunit
doublet from TPA treated HepG2 cells.

Another possible explanation is based on recent evidence (Ken Siddle,
personal communication) indicating that disulfide linked hybrid tetramers
can form, containing one alpha-beta subunit contributed by the insulin
receptor and one contributed by the IGF I receptor. According to this
scenario, alpha IR-3 would immunoprecipitate insulin receptor not through
direct interaction but because of its covalent association with IGF I
receptor subunits. If this is correct, the properties of the insulin
receptor dimer contained in the hybrid must be different than that in the
symmetric tetramer, since its autophosphorylation is not stimulated by 100
ng per ml insulin, and it is not recognized by alpha IR-1. This is not
surprising since there is clear evidence for an allosteric interaction

between alpha-beta dimers which alters the binding properties of the insulin receptor (17). A third possibility, which seem unlikely, it that there is a third type of receptor whose alpha subunit interacts with alpha IR-3 and whose beta subunit resembles that of the insulin receptor. We are currently evaluating these possibilities.

References

1. Jacobs, S., Sahyoun, N. E., Saltiel, A. R., and Cuatrecasas, P. (1983) Proc. Natl. Acad. Sci. USA 80, 6211-6213
2. Jacobs, S. and Cuatrecasas, P. (1986) J. Biol. Chem. 261, 934-939
3. Takayama, S., White, M. F., Lauris, V. and Kahn, C. R. (1984) Proc. Natl. Acad. Sci. USA 81, 7797-77801
4. Bollag, G. E., Roth, R. A., Beaudoin, J., Mochly-Rosen, D. and Koshland, D. E. (1986) Proc. Natl. Acad. Sci. USA 83, 5822-5824
5. Hachiya, H. L., Takayama, S., White, M., and King, G. L. (1987) J. Biol. Chem. 262, 6417-6424
6. Blake, A. D. and Strater, C. D. (1986) Biochem J. 236, 227-234
7. Davis, R. J. (1988) J. Biol. Chem. 263, 9462-9469
8. Davis, R. J. and Meisner, H (1987) J. Biol. Chem. 262, 16041-16047
9. Kull, F. C. Jr, Jacobs, S., Su, Y-F., Svoboda, M. E., Van Wyk, J. J., and Cuatrecasas, P. (1983) J. Biol. Chem. 258, 6561-6566
10. Jacobs, S., Kull, F. C. Jr. Earp, H. S., Svoboda, M. E., Van Wyk, J. J. and Cuatrecasas, P. (1983) J. Biol. Chem. 258, 9581-9584
11. Yu, K-T., Peters, M. A., and Czech, M. P. (1986) J. Biol. Chem. 261, 11341-11349
12. Ota, A., Wilson, G. L., and LeRoith, D. (1988) Eur J. Biochem 174, 521-530
13. Duronio, V., Jacobs, S., Romero, P. A. and Herscovics, A. (1988) J. Biol. Chem. 263, 5436-5445
14. Flier, J. S., and Moses, A. C. (1985) Biochem. Biophys. Res. Commun. 127, 929-936
15. Jacobs, S., Cook, S., Svoboda, M. E., Van Wyk, J. J. (1986) Endocrinology 118, 223-226
16. Versohl, E. J., Madux, B. A., and Goldfine, I. D. (1988) J. Clin. Endocrin. Metab. 67, 169-174
17. Boni-Schnetzler, M., Kaligian, A, DelVecchio, R., and Pilch, P.F. (1988) J. Biol. Chem. 263, 6822-6828

REGULATION OF GROWTH FACTOR RECEPTORS BY PROTEIN KINASE C

Laura M. Mudd and Mohan K. Raizada

Department of Physiology
University of Florida
Gainesville, FL

Insulin, insulin-like growth factor-I (IGF-I) and epidermal growth factor (EGF) all act to stimulate growth in many tissues. All three growth factors must first bind to specific membrane receptors to exert their actions. Upon binding of the appropriate ligand, these growth factor receptors exhibit increased tyrosine kinase activity, which is thought to be the second messenger, or pathway by which they cause some, if not all, of their effects within the cell (Evered et al., 1985; Raizada et al., 1986). Activation of these receptors by ligand binding causes different physiological effects in different cell types. These receptors also differ with regard to regulation of the receptors and/or of events beyond the receptors. As protein kinase C (PKC) has been shown to mimic some of these receptor actions and inhibit others, much attention has been focused on the effects of PKC on the tyrosine kinase receptors in order to determine first whether, and then how, PKC regulates those receptors. The following is a brief review of those studies.

Protein kinase C (PKC) is a serine/threonine kinase which is present in many tissues but occurs at highest concentrations in brain (Nishizuka, 1986). The enzyme was first purified and characterized in 1977 (Takai et al., 1977). Later studies indicated that PKC was calcium-dependent and that 1,2-diacylglycerol, a product of membrane phospholipid metabolism, increases the affinity of the enzyme for calcium (Takai et al., 1979; Kaibuchi et al., 1981). Reports of the molecular weight of PKC vary. The different values may reflect the method by which the Mr is determined, as evidenced by a study in which values of 77,000 and 82,000 were obtained from sucrose density gradient and polyacrylamide gel electrophoresis, respectively (Kikkawa, 1982). Differences may also be attributable to subunit aggregation (Azhar et al., 1987) or to the existence of different isozymes of PKC (Kikkawa et al., 1987; Nishizuka, 1988). To date, seven highly homologous isozymes of PKC are known to exist. Four are single polypeptide chains with four constant and five variable regions. Three subspecies differ slightly. The isozyme distributions differ with respect to one another (Ono et al., 1987; Mochly-Rosen et al., 1987) and with respect to the development of the organism (Burgess et al., 1986; Kikkawa et al., 1987); however, the kinetic properties of the isozymes are very similar (Azhar et al., 1987; Kikkawa et al., 1987). For a review see Nishizuka (1988).

Tumor-promoting phorbol esters, such as 12-O-tetradecanoyl-phorbol-13-acetate (TPA), act as exogenous stimulators of PKC (Castagna et al., 1982). Upon activation by TPA or calcium, there is a rapid decrease in cytoplasmic PKC and a corresponding increase in membrane PKC (Kraft et al., 1983; Wolf et al., 1985). Following this translocation, the membrane-bound PKC catalyzes the phosphorylation of specific proteins (Jacobs, 1983; Cochet, 1984). Translocation appears to mediate other activities of PKC such as neuronal potentiation (Akers et al., 1986) and synaptic plasticity (Akers and Routtenberg, 1987) as well. Prevention of PKC-redistribution, as with concanavalin A, has been shown to block PKC activation (Patel et al., 1987). Once activated,

PKC acts in many cell types to block hormone-stimulated phosphoinositide hydrolysis and thus exerts negative feedback over its own activation (Orrelano et al., 1985; Drummond, 1985; Naccache et al., 1985; Vicentini et al., 1985).

PKC has different effects in different tissues (Nishizuka, 1986). In the brain, PKC is involved in the regulation of neuronal ion channels (Kaczmarek, 1986; Strong et al., 1987; Baraban et al., 1985), synaptic plasticity (Routtenberg, 1987; Baranyi et al., 1988), neurite outgrowth (Hall et al., 1988) and neurotransmitter release (Wang et al., 1987; Matthies et al., 1987; Tanaka et al., 1984) as well as astrocyte and oligodendrocyte differentiation (Honegger, 1986; Yong et al., 1988) and changes in astrocyte membrane conductance (MacVicar, 1987). Many of the varied effects of PKC appear to be the result of its interactions with tyrosine kinase growth factor receptors. PKC-induced receptor phosphorylations alter the affinities, activities and effects of some of these receptors. These growth factors, in turn, have been shown to regulate the level of PKC in certain cells.

PKC EFFECTS ON TYROSINE KINASE GROWTH FACTOR RECEPTORS

Ligand-Receptor Interactions

PKC appears to influence the receptor binding of insulin, insulin-like growth factor I (IGF I), and epidermal growth factor (EGF). Stimulation of PKC by phorbol esters inhibits insulin binding by increasing the K_M of the high-affinity receptor (Grunberger et al., 1982; Häring et al., 1985; Thomopoulos et al., 1982). The endogenous analogues of TPA, the diacylglycerols, also reversibly inhibit insulin binding to its receptor (Grunberger et al., 1986) by altering the receptor affinity. This effect is not universal because TPA has no effect on insulin binding in 3T3 cells or in hepatoma cells (Shoyab et al., 1979; Takayama et al., 1984). TPA also induces endocytosis of insulin receptors in HL60 cells (Iacopetta et al., 1986). While preincubation with TPA did not affect receptor binding, incubation in the presence of insulin led to receptor internalization, suggesting that TPA causes internalization of occupied receptors only. TPA and diacylglycerols also inhibit IGF-I receptor binding in lymphocytes, monocytes and adipocytes by altering the high-affinity binding site without altering receptor number (Grunberger et al., 1986). Thus, PKC-induced decreases in insulin and IGF-I binding may result from decreased receptor affinity and/or increased ligand-receptor complex internalization.

The effect of PKC on EGF receptors varies in different cell types. Diacylglycerol analogues decrease high-affinity binding in A431 cells (McCaffrey et al., 1984). This is a common pathway for EGF-receptor regulation as evidenced by the fact that insulin, IGF-I and platelet-derived growth factor (PDGF) all elicit similar effects in different cell types (Corps et al., 1988; Davis et al., 1985). TPA activation of PKC causes a decrease in high-affinity binding of EGF in many cell types (Davis et al., 1985; Shoyab et al., 1979), but it has also been shown to decrease the number of receptors in HeLa cells without decreasing the affinity of the receptor (Lee et al., 1978). TPA also induces a transient internalization of the EGF receptor, followed by apparent receptor recycling to the cell surface in KB epidermal carcinoma cells through PKC activation (Beguinot et al., 1985). Thus, as with the insulin receptor, PKC-induced changes in EGF-receptor binding may occur by more than one mechanism.

Receptor Phosphorylation

The mechanism for PKC-stimulated alterations in the insulin, IGF-I and EGF receptors appears to involve serine/threonine phosphorylation of those receptors. In 1983, TPA was shown to stimulate phosphorylation of both insulin and IGF-I receptors in IM-9 cells that had been incubated with $H_3{}^{32}PO_4$ (Jacobs et al., 1983). Subsequently, TPA was shown to enhance serine/threonine phosphorylation of the insulin receptor in rat hepatoma cells at nine sites (Takayama et al., 1984). In later studies on IM-9 and HepG2 cells, TPA was found to phosphorylate four major serine residues, which were not phosphorylated in untreated cells, and to increase the phosphorylation of one threonine residue on the insulin receptor (Jacobs et al., 1986). PKC was acting directly on the insulin receptor as it phosphorylated the insulin receptor in vitro (Bollag et al., 1986). Similar results were seen with the IGF-I receptor. Recent experiments indicate that

TPA enhances phosphorylation of predominantly one serine residue on the insulin receptor in hepatoma cells (Takayama et al., 1988).

In initial studies on the phosphorylation of the insulin receptor, insulin- and TPA-stimulated phosphorylation appeared to be additive, suggesting that there was no interaction between the sites (Jacobs et al., 1983). Later, insulin was shown to stimulate phosphorylation of tyrosine and serine residues at six sites, three of which were similar to the TPA-phosphorylated sites. In addition, the phorbol ester decreased insulin-stimulated phosphorylation, suggesting that there was an interaction between the sites (Takayama et al., 1984). Other investigators found that the TPA-phosphorylated serine residues were not phosphorylated by insulin, which, however, did phosphorylate three tyrosine residues (Jacobs et al., 1986). The different profiles of receptor phosphorylation induced by insulin and phorbol esters gave strong evidence that insulin and IGF I were not acting through PKC. TPA-treatment of cells inhibited insulin-stimulated receptor autophosphorylation of tyrosine as well as insulin-stimulated phosphorylation of exogenous substrates by 50%. These changes in the receptor were maintained when the receptors were isolated and were reversed by incubation with alkaline phosphatase, suggesting that PKC decreased insulin receptor kinase activity and that this decrease was due to the phosphorylative changes induced in the receptor. Studies on rat adipocytes show that TPA increased the K_M of the insulin receptor for ATP, suggesting a mechanism for adipocyte insulin resistance (Häring et al., 1985).

In 1984, several investigators reported that phorbol esters stimulate EGF receptor phosphorylation as well (Decker, 1984; Iwashita et al., 1984; Cochet et al., 1984; Davis et al., 1984). These reports indicate that, as with the insulin receptor, TPA-stimulated phosphorylation occurs on serine and/or threonine residues of the EGF receptor. Diacylglycerol (McCaffrey et al., 1984) and PDGF (Davis et al., 1985) alter EGF-receptor phosphorylation as well; both patterns are the same as that observed for TPA and there is evidence that the activities of both are mediated by PKC, suggesting that this is a normal method of EGF-receptor regulation. Further work has led to the characterization of specific amino acids in the EGF receptor which are phosphorylated (Heisermann et al., 1988). In contrast, EGF normally phosphorylates its receptor at tyrosine residues (Ushiro et al., 1980). TPA-treated cells exhibit decreases in both EGF-stimulated receptor autophosphorylation at tyrosine residues and tyrosine kinase activity (Cochet et al., 1984; Davis et al., 1985; Decker et al., 1984).

Receptor-Induced Effects

Insulin receptor tyrosine kinase activity is necessary for normal receptor function and downregulation (Maegawa et al., 1988; Russel et al., 1987). This has been demonstrated by studies in which kinase-defective mutant insulin receptors were used to transfect cells. The mutant receptors demonstrate normal insulin binding but do not possess tyrosine kinase activity, are not internalized and do not possess biological activity. Treatment of endogenous, biologically-active insulin receptors with monoclonal antibodies against the receptor kinase results in inhibition of insulin-stimulated effects as well (Morgan et al., 1987). TPA inhibition of receptor tyrosine kinase activity leads to the same types of defects. TPA-treatment has been associated with inhibition of insulin-mediated DNA synthesis (Takada et al., 1988), phosphorylation of metabolic enzymes (Takayama et al., 1984), glycogen synthesis (Caron et al., 1988) and glucose uptake (Kirsch et al., 1985). Thus, an impaired PKC pathway can have adverse consequences for the cell or organism. This is demonstrated by genetically obese (fa/fa) rats in whose hearts and hepatocytes both the basal distribution and the translocation of PKC are abnormal. The resultant insulin-insensitivity can be duplicated by treating lean rats with TPA to downregulate PKC (van de Werve et al., 1987). The same is not true for all tissues, however, as phorbol esters have only minimal effects on insulin sensitivity in rat skeletal muscle (Sowell et al., 1988).

The EGF receptor loses its biological activity in the absence of receptor tyrosine kinase activity as well. Acute actions of the receptor such as EGF-stimulated enzyme activation are lost in cells in which EGF-receptor mutations have eliminated the tyrosine kinase activity (Yarden et al., 1988). Phorbol ester inactivation of the tyrosine kinase activity also inhibits EGF-stimulated DNA synthesis and cell proliferation (Decker, 1984). The relationship between the EGF receptor and PKC is somewhat complicated by reports that synergism between the two has been reported with regard to DNA synthesis (Carpenter, 1987). The interaction of PKC and the insulin receptor is also complicated by reports of synergism in mitogenic stimulation (Shimizu et al., 1986). In addition, there are proteins which have phosphorylation sites for both PKC and receptor kinases (Zick et al., 1986) and are stimulated by both types of mitogens (Pelech et al., 1987). The latter mechanism would account for PKC's paradoxical inhibition of insulin receptor activity and mimicking of insulin-stimulated effects.

Growth Factor Effects on PKC

While the majority of studies on PKC/growth factor receptors in the literature focus on PKC regulation of growth factor receptors, these same receptors regulate PKC as well. Insulin treatment in the presence of glucose increases both PKC binding capacity and enzymatic activity of PKC in adipocytes (Draznin et al., 1988). As the insulin effect is eliminated in the presence of high glucose, the effect may be secondary to increased glucose uptake in the insulin-treated cells. Insulin also increases adipocyte levels of cytosolic Ca^{++} (Draznin et al., 1987), which could account for the increased PKC activity. Studies have also demonstrated insulin enhancement of PKC activity in myocytes (Cooper et al., 1987). The increase occurs in both the cytosolic and membrane fractions and is not inhibited by cycloheximide. This increase in PKC activity is reportedly mediated via increased diacylglycerol generated by phospholipid hydrolysis and phospholipid synthesis. There is some controversy on this point with other work suggesting that insulin does not increase myocyte PKC activity (Spach et al., 1986); it is suggested that the increases in phosphorylation observed after insulin administration are mediated by S6 kinase, which is activated by both insulin and TPA. Enhanced PKC activity has been demonstrated in mammary tumor cells in response to both insulin and EGF with the former being the most active (Gomez et al., 1988). EGF acts through its receptor to stimulate phosphatidylinositol turnover and calcium uptake, both of which activate PKC (Sawyer et al., 1981). Growth factors, then, may increase PKC activity indirectly by increasing cytosolic free Ca^{++} or diacylglycerol, or by acting at a point in the pathway beyond the PKC molecule itself. A direct effect of insulin or EGF on PKC must be demonstrated on the isolated enzyme.

PKC IN THE BRAIN

Distributions of PKC and Growth Factor Receptors in the Brain

Although PKC is present in many tissues, it occurs at highest concentration in the brain. The seven subspecies of PKC have different distributions in the brain (Nishizuka, 1988) and those change with brain development. The γ subspecies, which occurs only in the central nervous system in the rat and monkey, has a developmentally regulated distribution, with expression increasing from birth until it reaches a maximum at about three weeks of age. Total PKC is also developmentally regulated in brain in studies in cultured embryonic neurons and in the developing fetal brain *in vivo* (Burgess et al., 1986). Immunohistochemical analyses have shown different staining patterns with different antibodies. There appear to be subspecies which are present in neurons almost exclusively (Saito et al., 1988; Mochly-Rosen et al., 1987) in astrocyte-glial cells and in oligodendro-glia (Mochly-Rosen et al., 1987; Wood et al., 1986). PKC activity is also unevenly distributed, with the left cerebral hemisphere expressing more than the right in the rat (Ginobili de Martinez et al., 1988). Binding studies using phorbol esters in neuronal and astroglial cell cultures from the same rat brains show two to three times more PKC in neurons than in glial cells (Raizada et al., 1988). Although PDB binding is greater in neurons than in glial cells cultured from the same rat brains, PKC activity is higher in glial cells than in neurons (Mudd et al., 1989a). This discrepancy could be due to either an inactive PKC in neurons or a fraction of glial PKC which does not bind PDB. Either is possible. As described, there are differences in neurons and glial cells with regard to the isozymes of PKC which are expressed in the cells.

The subcellular distribution of PKC has been the subject of many studies. It is localized in dendrites, axons, perikarya, and nuclei (Saito et al., 1988) of neurons with particularly high concentrations in presynaptic terminals (Wood et al., 1986) and in growth cones (Girard et al., 1988). This is not unexpected, as PKC mediates both neurotransmission and neurite outgrowth (Recio-Pinto et al., 1988). Fractionation of glial cells demonstrated that the majority of the PKC was cytoplasmic (Mudd et al., 1989a; Neary et al., 1988). In addition, it was shown that the majority of PKC in whole brain is associated with the membrane (Neary et al., 1988). Consistent with this is the observation that the majority of PKC activity in neuronal cells in primary cultures is bound to membranes. This is true if immunocytochemical methods or assays of PKC activity are used to quantitate PKC (Mudd et al., 1989a). Membrane-bound PKC is latent in liver (Azhar et al., 1987, 1988). A latent membrane-bound PKC in neurons might explain why neurons bind more PDB than glial cells while exhibiting less PKC activity. The majority of both glial cell and neuronal cytoplasmic PKC can be translocated to the membrane within 30 to 60 min. of TPA treatment. This is similar to the situation seen in peripheral tissues.

Insulin-, IGF-I-, IGF-II- and EGF-receptors are distributed non-uniformly in brain tissue as well (Hill et al., 1986; Mendelsohn, L., 1986; Baskin, D., et al. 1986; Gomez-Pinilla et al., 1988). Insulin and IGF-I receptors in brain tissue are uniformly distributed between neurons and glia (Boyd et al., 1985; Clarke et al., 1984; Lowe et al., 1986; Shemer et al., 1987), while functional EGF receptors are localized primarily in glial cells (Wang et al., 1989). Insulin, IGF-I and EGF receptors are all developmentally regulated in the rat brain with increases in the first weeks of life, followed by a decline (Kappy et al., 1984; Bassas et al., 1985; Gomez-Pinilla et al., 1988). While the subunit structures of the insulin and IGF-I receptors are similar in brain and peripheral tissues, the α-subunit of each is somewhat smaller in the brain (Heidenreich, 1985; Heidenreich et al., 1986), due to decreased glycosylation of the brain receptor. In particular, with regard to insulin and IGF-I, the neuronal α- and β-subunits are smaller, while the glial cell subunits resemble the liver subunits in size (Masters et al., 1987; Shemer et al., 1987; Lowe et al., 1986). While the EGF receptor is the same size in glial cells from the brain as in peripheral tissues, neither $[^{125}I]$-EGF covalent crosslinking nor EGF-induced receptor phosphorylation or bioactivity could be demonstrated in neurons (Wang et al., 1989). That the glial EGF, IGF-I and insulin receptors are similar in the brain and liver is not surprising, as insulin and EGF appear to have traditional metabolic effects in glia (Wang et al., 1989; Clarke et al., 1985; Clarke et al., 1984), while insulin seems to have neuromodulatory/neurotransmitter effects in neurons (Boyd et al., 1985; Recio-Pinto et al., 1988).

PKC Interaction with Brain Insulin Receptors

In contrast to its effect in peripheral cells to decrease insulin receptor binding, TPA stimulates insulin binding to its receptor in cultured glial cells and has no effect on neuronal insulin binding (Mudd and Raizada, 1989). The stimulation of insulin binding in glia is not associated with an increase in insulin-stimulated glucose uptake, suggesting that the receptors are inactive. Inactivation of tyrosine-kinase activity has been shown to prevent internalization of insulin receptors in Chinese hamster ovary cells (Russell et al., 1987). A similar proposal could explain the increase in glial insulin receptors. The lack of effect of TPA on neuronal insulin binding is interesting. PKC does regulate neuronal receptors as demonstrated by an increase in angiotensin II binding in response to TPA (Sumners et al., 1988). The lack of effect of TPA on neuronal insulin binding could be due to structural differences in the neuronal receptor, to the different isotypes of PKC present in neurons, or to the smaller pool of PKC that is available to be translocated in neurons. We have found that only 31 and 6% of the basal and stimulated PKC activities, respectively, and 12% of the PKC immunoreactivity are present in the neuronal cytosol in contrast to much larger percentages in glial cells from the same rat brains.

Insulin also appears to regulate the level of immunoreactive PKC in glia, but not in neurons (Mudd et al., 1989b). Insulin stimulates PKC in a time-dependent manner in cultured glial cells but the same dose has no effect in neurons. Thus, in glial cells the decrease in insulin receptor activity in response to PKC acts as a form of negative feedback, while in neurons both arms of the circuit may be missing (Figure 1).

SUMMARY

Insulin, IGF-I and EGF receptors appear to be regulated by PKC in many cells. PKC-induced receptor regulation frequently takes the form of a decrease in receptor affinity for its ligand accompanied by a decrease in receptor tyrosine kinase activity and loss of both short- and long-term receptor-stimulated effects. This is not true in all cell types as the receptor affinity may be unaltered and receptor number may increase or decrease. The loss of receptor activity is a more general phenomenon. These changes in the receptor are due to increases in serine/threonine phosphorylation of the receptors subsequent to PKC activation and translocation from cytosol to membrane.

In several cell types, growth factor receptors stimulate PKC, although whether this occurs directly or as a consequence of increased intracellular free calcium and/or diacylglycerol has yet to be determined. In addition, growth factor receptors and PKC act on some of the same substrates. This could be of great advantage to the cell which is so regulated: activation of growth factor receptors activates PKC; PKC feeds back to inactivate the growth factor receptors without inhibiting essential responses to those receptors.

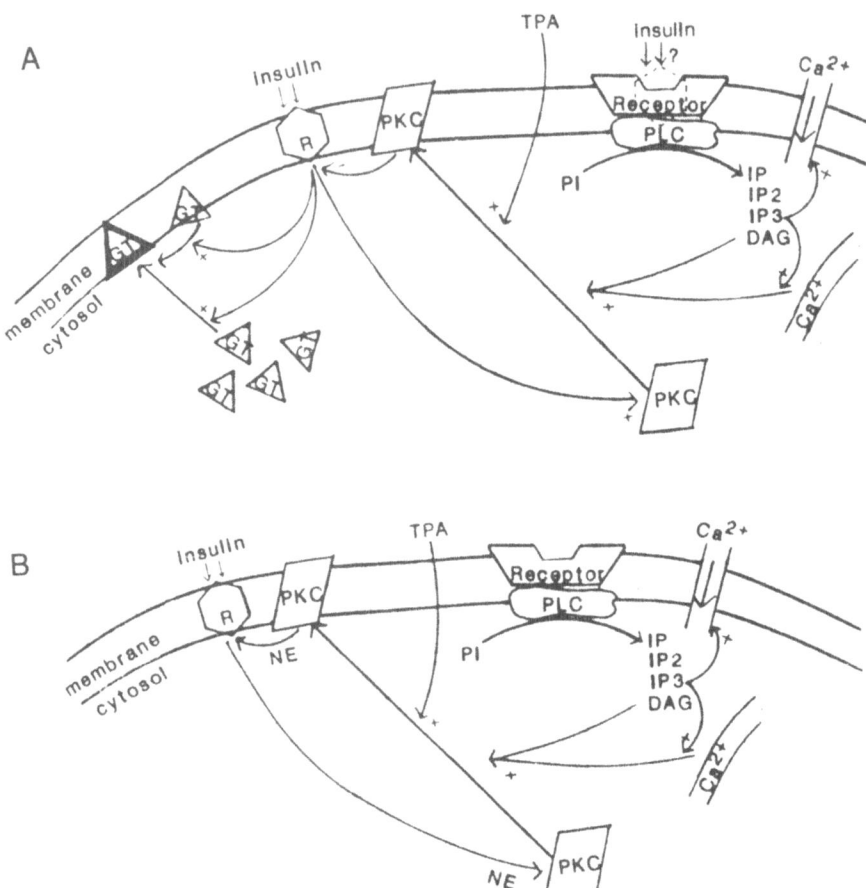

Figure 1. PKC interactions with insulin receptors in glia (A) and neurons (B) in primary culture from the brain. Receptor activation stimulates phospholipase C (PLC) activity. Phosphoinositides (PI) in the membrane are converted to inositol phosphates (IP, IP2 and IP3) and diacylglycerol. IP3 stimulates calcium mobilization. Diacylglycerol (DAG) increases the affinity of PKC for calcium. As PKC is activated it is translocated to the membrane. TPA acts in a manner which is analogous to that of DAG.
In glial cells, insulin binding to its receptor(s) stimulates glucose transporter (GT) activity as well as immunoreactive PKC. This may occur by way of a phosphorylated intermediate or insulin may activate phospholipase C. PKC feeds back to inhibit the receptor response to insulin. In neurons, insulin has no effect (NE) on immunoreactive PKC, nor does PKC alter insulin receptor binding.

REFERENCES

Akers, R. and Routtenberg, A. (1987). Calcium-promoted translocation of protein kinase C to synaptic membranes: Relation to the phosphorylation of an endogenous substrate (protein F1) involved in synaptic plasticity. J. Neurosci. 7:3976-83.

Akers, R., Lovinger, D., Colley, P., Linden, D. and Routtenberg, A. (1986). Translocation of protein kinase C activity may mediate hippocampal long-term potentiation. Science 231:587-9.

Azhar, S., Butte, J. and Reaven, E. (1987). Calcium-activated, phospholipid-dependent protein kinases from rat liver: Subcellular distribution, purification and characterization of multiple forms. Bchm. 26:7047-57.

Azhar, S., Butte, J. and Reaven, E. (1988). Identification of isoenzyme forms of hepatic calcium-activated phospholipid-dependent protein kinase in various animal models. Bchm. Bphys. Res. Comm. 155:1017.

Baraban, J., Snyder, S. and Alger, B. (1985). Protein kinase C-regulated ionic conductance in hippocampal pyramidal neurons: Electrophysiological effects of phorbol esters. Proc. Natl. Acad. Sci. USA 82:2538-42.

Baranyi, A., Szente, M. and Woody, C. (1988). Activation of protein kinase C induces long-term changes of postsynaptic currents in neocortical neurons. Brain Res. 440:341-7.

Baskin, D., Bohannon, N. and Figlewicz, D. (1986). Insulin-like growth factor receptors in the brain. In: *Insulin, Insulin-Like Growth Factors and Their Receptors in the Central Nervous System.* M. Raizada, M. Phillips and D. LeRoith, eds., Plenum Press, New York, pp. 245-60.

Bassas, L., De Pablo, F., Lesniak, M. and Roth, J. (1985). Ontogeny of receptors for insulin-like peptides in chick embryo tissues: early dominance of insulin-like growth factor over insulin receptor in brain. Endocr. 117:2321-9.

Beguinot, L., Hanover, J., Ito, S., Richert, N., Willingham, M. and Pastan, I. (1985). Phorbol esters induce transient internalization without degradation of unoccupied epidermal growth factor receptors. Proc. Natl. Acad. Sci. USA 82:2774-8.

Blackshear, P., Witters, L., Girard, P., Kuo, J. and Quamo, S. (1985). Growth factor-stimulated protein phosphorylation in 3T3-L1 cells: Evidence for protein kinase C-dependent and -independent pathways. J. Biol. Chem. 260:13304-15.

Bollag, G., Roth, R., Beaudoin, J., Mochly-Rosen, D. and Koshland, D. (1986). Protein kinase C directly phosphorylates the insulin receptor *in vitro* and reduces its protein tyrosine kinase activity. Proc. Natl. Acad. Sci. USA 83:5822-4.

Boyd, F., Clarke, D., Muther, T. and Raizada, M. (1985). Insulin receptors and insulin modulation of norepinephrine uptake in neuronal cultures from rat brain. J. Biol. Chem. 260:15880-4.

Burgess, S., Sahyoun, N., Blanchard, S., Le Vine, H., Chang, K. and Cuatrecasas, P. (1986). Phorbol ester receptors and protein kinase C in primary neuronal cultures: development and stimulation of endogenous phosphorylation. J. Cell Biol. 102:312-9.

Caron, M., Cherqui, G., Wicek, D., Capeau, J., Bertrand, J. and Picard, J. (1988). Effect of protein kinase C activation and depletion on insulin stimulation of glycogen synthesis in cultured hepatoma cells. Experientia 44:34-7.

Carpenter, G. (1987). Receptors for epidermal growth factor and other polypeptide mitogens. Ann. Rev. Bchm. 56:881-914.

Castagna, M., Takai, Y., Kaibuchi, K., Sano, K., Kikkawa, U. and Nishizuka, Y. (1982). Direct activation of calcium-activated, phospholipid-dependent protein kinase by tumor-promoting phorbol esters. J. Biol. Chem. 257:7847-51.

Clarke, D., Boyd, F., Kappy, M. and Raizada, M. (1984). Insulin binds to specific receptors and stimulates 2-deoxy-D-glucose uptake in cultured glial cells from rat brain. J. Biol. Chem. 259:11672-5.

Clarke, D., Boyd, F., Kappy, M. and Raizada, M. (1985). Insulin stimulates macromolecular synthesis in cultured glial cells from rat brain. Am. J. Physiol. 249:C484-9.

Cochet, C., Gill, G., Meisenhelder, J., Cooper, J. and Hunter, T. (1984). C-Kinase phosphorylates the epidermal growth factor receptor and reduces its epidermal growth factor-stimulated tyrosine protein kinase activity. J. Biol. Chem. 259:2553-8.

Cooper, D., Konda, T., Standaert, M., Davis, J., Pollet, R. and Farese, R. (1987). Insulin increases membrane and cytosolic protein kinase C activity in BC3H-1 myocytes. J. Biol. Chem. 262:3633-9.

Corps, A. and Brown, K. (1988). Insulin-like growth factor I and insulin reduce epidermal growth factor binding to Swiss 3T3 cells by an indirect mechanism that is apparently independent of protein kinase C. FEBS Lett. 233:303-6.

Davis, R. and Czech, M. (1984). Tumor-promoting phorbol diesters mediate phosphorylation of the epidermal growth factor receptor. J. Biol. Chem. 259:8545-9.

Davis, R. and Czech, M. (1985). Platelet-derived growth factor mimics phorbol diester action on epidermal growth factor receptor phosphorylation at threonine-654. Proc. Natl. Acad. Sci. USA 82:4080-4.

Decker, S. (1984). Effects of epidermal growth factor and 12-O-tetradecanoylphorbol-13-acetate on metabolism of the epidermal growth factor receptor in normal human fibroblasts. Mol. Cell. Biol. 4:1718-24.

Draznin, B., Kao, M. and Sussman, K. (1987). Insulin and glyburide increase cytosolic free Ca^{++} concentration in isolated rat adipocytes. Diabetes 36:174-8.

Draznin, B., Leitner, J., Sussman, K. and Sherman, N. (1988). Insulin and glucose modulate protein kinase C activity in rat adipocytes. Bchm. Bphys. Res. Comm. 156:570-5.

Drummond, A. (1985). Bidirectional control of cytosolic free calcium by thyrotropin-releasing hormone in pituitary cells. Nature 315:752-5.

Evered, D., Nugent, J. and Whelan, J., eds. (1985). Growth Factors in Biology and Medicine, Ciba Foundation Symp. 116, Pittman Press, London.

Ginobili de Martinez, M. and Barrantes, F. (1988). Ca^{2+} and phospholipid-dependent protein kinase activity in rat cerebral hemispheres. Brain Res. 440:386-90.

Girard, P., Wood, J., Freschi, J. and Kuo, J. (1988). Immunocytochemical localization of protein kinase C in developing brain tissue and in primary neuronal cultures. Dev. Biol. 126:98-107.

Gomez, M., Medrano, E., Cafferatta, E. and Tellez-Iñon, M. (1988). Protein kinase C is differentially regulated by thrombin, insulin and epidermal growth factor in human mammary tumor cells. Exp. Cell. Res. 175:74-80.

Gomez-Pinilla, F., Knauer, D. and Nieto-Sampedro, M. (1988). Epidermal growth factor receptor immunoreactivity in rat brain: Development and cellular location. Brain Res. 438:385-90.

Grunberger, G. and Gorden, P. (1982). Affinity alteration of insulin receptor induced by a phorbol ester. Amer. J. Physiol. 243:E319-24.

Grunberger, G., Iacopetta, B., Carpentier, J. and Gorden, P. (1986). Diacylglycerol modulation of insulin receptor from cultured human mononuclear cells. Diab. 35:1364-70.

Hall, F., Fernyhough, P., Ishii, D. and Vulliet, P. (1988). Suppression of nerve growth factor-directed neurite outgrowth in PC12 cells by sphingosine, an inhibitor of protein kinase C. J. Biol. Chem. 263:4460-6.

Häring, H., Kirsch, D., Obermaier, B., Ermel, B. and Machicao, F. (1986). Tumor-promoting phorbol esters increase the K_M of the ATP-binding site of the insulin receptor kinase from rat adipocytes. J. Biol. Chem. 261:3869-75.

Heidenreich, K. (1985). The Mr difference between insulin receptors in brain and peripheral tissues is due to variation in carbohydrate content. Diabetes (Suppl.) 34:56A.

Heidenreich, K., Freidenberg, G., Figlewicz, D. and Gilmore, P. (1986). Evidence for a subtype of insulin-like growth factor I receptor in brain. Reg. Peptides 15:301-10.

Heisermann, G. and Gill, G. (1988). Epidermal growth factor receptor threonine and serine residues phosphorylated in vivo. J. Biol. Chem. 263:13152-8.

Hill, J., Lesniak, M., Rojeski, M., Pert, C. and Roth, J. (1986). Receptors for insulin and insulin-related peptides in the CNS: Studies of localization in rat brain. In: Insulin, Insulin-Like Growth Factors and Their Receptors in the Central Nervous System. M. Raizada, M. Phillips and D. LeRoith, eds., Plenum Press, New York, pp. 261-8.

Honegger, P. (1986). Protein kinase C-activating tumor promoters enhance the differentiation of astrocytes in aggregating fetal brain cell cultures. J. Neurochem. 46:1561-6.

Iacopetta, B., Carpentier, J., Pozzan, T., Lew, D., Gorden, P. and Orci, L. (1986). Role of intracellular calcium and protein kinase C in the endocytosis of transferrin and insulin by HL60 cells. J. Cell. Biol. 103:851-6.

Iwashita, S. and Fox, C. (1984). Epidermal growth factor and potent phorbol tumor promoters induce epidermal growth factor receptor phosphorylation in a similar but distinctly different manner in human epidermoid carcinoma A431 cells. J. Biol. Chem. 259:2559-67.

Jacobs, S. and Cuatrecasas, P. (1986). Phosphorylation of receptors for insulin and insulin-like growth factor 1: Effects of hormones and phorbol esters. J. Biol. Chem. 261:934-7.

Jacobs, S., Sahyoun, N., Saltiel, A. and Cuatrecasas, P. (1983). Phorbol esters stimulate the phosphorylation of receptors for insulin and somatomedin C. Proc. Natl. Acad. Sci. USA 80:6211-13.

Kaczmarek, L. (1986). Phorbol esters, protein phosphorylation and the regulation of neuronal ion channels. J. Exp. Biol. 124:375-92.

Kaibuchi, K., Takai, Y. and Nishizuka, Y. (1981). Cooperative roles of various membrane phospholipids in the activation of calcium-activated, phospholipid-dependent protein kinase. J. Biol. Chem. 256:7146-9.

Kappy, M., Sellinger, S. and Raizada, M. (1984). Insulin binding in four regions of the developing rat brain. J. Neurochem. 42:198-203.

Kazlauskas, A. and Cooper, J. (1988). Protein kinase C mediates platelet-derived growth factor-induced tyrosine phosphorylation of p42. J. Cell Biol. 106:1395-1402.

Kikkawa, U., Takai, Y., Minakuchi, R., Inohara, S. and Nishizuka, Y. (1982). Calcium-activated, phospholipid-dependent protein kinase from rat brain: Subcellular distribution, purification and properties. J. Biol. Chem. 257:13341-8.

Kikkawa, U., Ogita, K., Ono, Y., Asaoka, Y., Shearman, M., Fujii, T., Ase, K., Sekiguchi, K., Igarashi, K. and Nishizuka, Y. (1987). The common structure and activities of four subspecies of rat protein kinase C family. FEBS Lett. 223:212-6.

Kirsch, D., Obermaier, B. and Häring, H. (1985). Phorbol esters enhance basal D-glucose transport but inhibit insulin stimulation of D-glucose transport and insulin binding in isolated rat adipocytes. Bchm. Bphys. Res. Comm. 128:824-32.

Kraft, A. and Anderson, W. (1983). Phorbol esters increase the amount of Ca^{++}, phospholipid-dependent protein kinase associated with plasma membrane. Nature 301:621-3.

Lee, L. and Weinstein, I. (1978). Tumor-promoting phorbol esters inhibit binding of epidermal growth factor to cellular receptors. Science 202:313-15.

Lowe, W., Boyd, F., Clarke, D., Raizada, M., Hart, C. and LeRoith, D. (1986). Development of brain insulin receptors: Structural and functional studies of insulin receptors from whole brain and primary cell cultures. Endocr. 119:25-35.

MacVicar, B., Crichton, S., Burnard, D. and Tse, F. (1987). Membrane conductance oscillations in astrocytes induced by phorbol ester. Nature 329:242-3.

Maegawa, H., Olefsky, J., Thies, S., Boyd, D., Ullrich, A. and McClain, D. (1988). Insulin receptors with defective tyrosine kinase inhibit normal receptor function at the level of substrate phosphorylation. J. Biol. Chem. 263:12629-37.

Masters, B., Shemer, J., Judkins, J., Clarke, D., LeRoith, D. and Raizada, M. (1987). Insulin receptors and insulin action in dissociated brain cells. Brain Res. 417:247-56.

Matthies, H., Palfrey, H., Hirning, L. and Miller, R. (1987). Down-regulation of protein kinase C in neuronal cells: Effects on neurotransmitter release. J. Neurosci. 7:1192-1206.

McCaffrey, P., Friedman, B. and Rosner, M. (1984). Diacylglycerol modulates binding and phosphorylation of the epidermal growth factor receptor. J. Biol. Chem. 259:12502-7.

Mendelsohn, L. (1986). Visualization of IGF-2 receptors in rat brain. In: *Insulin, Insulin-Like Growth Factors and Their Receptors in the Central Nervous System.* M. Raizada, M. Phillips and D. LeRoith, eds., Plenum Press, New York, pp. 269-76.

Mochly-Rosen, D., Basbaum, A. and Koshland, D. (1987). Distinct cellular and regional localization of immunoreactive protein kinase C in rat brain. Proc. Natl. Acad. Sci. USA 84:4660-4.

Morgan, D. and Roth, R. (1987). Acute insulin action requires insulin receptor kinase activity: Introduction of an inhibitory monoclonal antibody into mammalian cells blocks the rapid effects of insulin. Proc. Natl. Acad. Sci. USA 84:41-5.

Mudd, L. and Raizada, M. (1989). Phorbol esters stimulate insulin binding in glial but not neuronal cells from the brain. (In preparation)

Mudd, L., Azhar, S. and Raizada, M. (1989a). Differential effects of phorbol esters on protein kinase C translocation, insulin binding and glucose uptake in neurons and glial cells from the brain. (In preparation)

Mudd, L., Chung, H. and Raizada, M. (1989b). Differential effects of insulin and dexamethasone on protein kinase C in neuronal and glial cells from the brain. (In preparation)

Naccache, P., Molski, T., Borgeat, P., White, J. and Sha'afi, R. (1985). Phorbol esters inhibit the fMet-Leu-Phe- and Leukotriene B_4-stimulated calcium mobilization and enzyme secretion in rabbit neutrophils. J. Biol. Chem. 260:2125-31.

Neary, J., Norenberg, L. and Norenberg, M. (1988). Protein kinase C in primary astrocyte cultures: Cytoplasmic localization and translocation by a phorbol ester. J. Neurochem. 50:1179-84.

Nishizuka, Y. (1986). Studies and perspectives of protein kinase C. Science 233:305-12.

Nishizuka, Y. (1988). The molecular heterogeneity of protein kinase C and its implications for cellular recognition. Nature 334:661-4.

Ono, Y. and Kikkawa, U. (1987). Do multiple species of protein kinase C transduce different signals? Trends Biochm. Sci. 143:421-3.

Orrelano, S., Solski, P. and Heller-Brown, J. (1985). Phorbol ester inhibits phosphoinositide hydrolysis and calcium mobilization in cultured astrocytoma cells. J. Biol. Chem. 260:5236-9.

Patel, J. and Kassis, J. (1987). Concanavalin A prevents phorbol-mediated redistribution of protein kinase C and β-adrenergic receptors in rat glioma C6 cells. Bchm. Bphys. Res. Comm. 144:1265-72.

Pelech, S. and Krebs, E. (1987). Mitogen-activated S6 kinase is stimulated via protein kinase C-dependent and independent pathways in Swiss 3T3 cells. J. Biol. Chem. 262:11598-606.

Raizada, M., Phillips, M. and LeRoith, D., eds. (1986). *Insulin, Insulin-Like Growth Factors and Their Receptors in the Central Nervous System.* Plenum Press, New York.

Raizada, M., Morse, C., Gonzales, R., Crews, F. and Sumners, C. (1988). Receptors for phorbol esters are primarily localized in neurons: Comparison of neuronal and glial cultures. Neurochem. Res. 13:51-6.

Recio-Pinto, E. and Ishii, D. (1988). Insulin and related growth factors: Effects on the nervous system and mechanism for neurite growth and regeneration. Neurochem. Int. 12:397-414.

Rosenfeld, R., Hintz, R. and Dollar, L. (1982). Insulin-induced loss of an insulin-like growth factor-I receptor on IM-9 lymphocytes. Diabetes 31:375-81.

Rouis, M., Thomopoulos, P., Cherier, C. and Testa, U. (1985). Inhibition of insulin receptor binding by A23187: Synergy with phorbol esters. Bchm. Bphys. Res. Comm. 130:9-15.

Routtenberg, A. (1987). Phospholipid and fatty acid regulation of signal transduction at synapses: Potential role for protein kinase C in information storage. J. Neur. Transm. (Suppl.) 24:239-45.

Russell, D., Gherzi, R., Johnson, E., Chou, C. and Rosen, O. (1987). The protein-tyrosine kinase activity of the insulin receptor is necessary for insulin-mediated receptor down-regulation. J. Biol. Chem. 262:11833-40.

Saito, N., Kikkawa, U., Nishizuka, Y. and Tanaka, C. (1988). Distribution of protein kinase C-like immunoreactive neurons in rat brain. J. Neurosci. 8:369-82.

Sawyer, S. and Cohen, S. (1981). Enhancement of calcium uptake and phosphatidyl-inositol turnover by epidermal growth factor in A431 cells. Bchm. 20:6280-6.

Shemer, J., Raizada, M., Masters, B., Ota, A. and LeRoith, D. (1987). Insulin-like growth factor I receptors in neuronal and glial cells: characterization and biological effects in primary culture. J. Biol. Chem. 262:7693-9.

Shimizu, Y., Fujiki, H., Sugimura, T. and Shimizu, N. (1986). Mouse 3T3-L1 cell variants unable to respond to mitogenic stimulation of dihydroteleocidin B: Genetic evidence for the synergism of tumor promoters with growth factors. Canc. Res. 46:4027-31.

Shoyab, M., DeLarco, J. and Todaro, G. (1979). Biologically active phorbol esters specifically alter affinity of epidermal growth factor membrane receptors. Nature 279:387-91.

Sowell, M., Treutelaar, M., Burant, C. and Buse, M. (1988). Minimal effects of phorbol esters on glucose transport and insulin sensitivity in rat skeletal muscle. Diab. 37:499-506.

Spach, D., Nemenoff, R. and Blackshear, P. (1986). Protein phosphorylation and protein kinase activities in BC3H-1 myocytes. J. Biol. Chem. 261:12750-3.

Strong, J., Fox, A., Tsien, R. and Kaczmarek, L. (1987). Stimulation of protein kinase C recruits covert calcium channels in Aplysia bag cell neurons. Nature 325:714-7.

Sumners, C., Rueth, S., Myers, L., Kalberg, C., Crews, F. and Raizada, M. (1988). Phorbol ester-induced upregulation of angiotensin II receptors in neuronal cultures is potentiated by a calcium ionophore. J. Neurochem. 51:1-10.

Takada, K., Amino, N., Tetsumoto, T. and Miyai, K. (1988). Phorbol esters have a dual action through protein kinase C in regulation of proliferation of FRTL-5 cells. FEBS Lett. 234:13-16.

Takai, Y., Kishimoto, A., Inoue, M. and Nishizuka, Y. (1977). Studies on a cyclic nucleotide-independent protein kinase and its proenzyme in mammalian tissue. J. Biol. Chem. 252:7603-9.

Takai, Y., Kishimoto, A., Iwasa, Y., Kawahara, Y., Mori, T. and Nishizuka, Y. (1979). Calcium-dependent activation of a multifunctional protein kinase by membrane phospholipid. J. Biol. Chem. 254:3692-5.

Takayama, S., White, M., Lauris, V. and Kahn, R. (1984). Phorbol esters modulate insulin receptor phosphorylation and insulin action in cultured hepatoma cells. Proc. Natl. Acad. Sci. USA 81:7797-801.

Takayama, S., White, M. and Kahn, C. (1988). Phorbol ester-induced serine phosphorylation of the insulin receptor decreases its tyrosine kinase activity. J. Biol. Chem. 263:3440-7.

Tanaka, C., Taniyama, K. and Kusunoki, M. (1984). A phorbol ester and A23187 act synergistically to release acetylcholine from the guinea pig ileum. FEBS Lett. 175:165-9.

Thomopoulos, P., Testa, U., Gourdin, M., Hervy, C., Titeux, M. and Vainchenker, W. (1982). Inhibition of insulin receptor binding by phorbol esters. E. J. Bchm. 129:389-93.

Ushiro, H. and Cohen, S. (1980). Identification of phosphotyrosine as a product of epidermal growth factor-activated protein kinase in A-431 cell membranes. J. Biol. Chem. 255:8363-5.

van de Werve, G., Zaninetti, D., Lang, U., Vallotton, M. and Jeanrenaud, B. (1987). Identification of a major defect in insulin-resistant tissues of genetically obese (fa/fa) rats: Impaired protein kinase C. Diabetes 36:310-14.

Vicentini, L., Virgilio, F., Ambrosini, A., Pozzan, T. and Mendolesi, J. (1985). Tumor promoter phorbol 12-myristate, 13-acetate inhibits phosphoinositide hydrolysis and cytosolic Ca^{2+} rise induced by the activation of muscarinic receptors in PC12 cells. Bchm. Bphys. Res. Comm. 127:310-17.

Wang, H. and Friedman, E. (1987). Protein kinase C: regulation of serotonin release from rat brain cortical slices. Eur. J. Pharm. 141:15-21.

Wang, S., Shiverick, K., Ogilvie, S., Dunn, W. and Raizada, M. (1989). Characterization of epidermal growth factor receptors in astrocytic glial and neuronal cells in primary culture. Endocr. 124:240-7.

Wolf, M., Levine, H., May, W., Cuatrecasas, P. and Sahyoun, N. (1985). A model for intracellular translocation of protein kinase C involving synergism between calcium and phorbol ester. Nature 317:546-9.

Wood, J., Girard, P., Mazzei, G. and Kuo, J. (1986). Immunocytochemical localization of protein kinase C in identified neuronal compartments of rat brain. J. Neurosci. 6:2571-7.

Yarden, Y. and Ullrich, A. (1988). Molecular analysis of signal transduction by growth factors. Bchm. 27:3113-8.

Yong, V., Sekiguchi, M. and Kim, S. (1988). Phorbol ester enhances morphological differentiation of oligodendrocytes in culture. J. Neurosci. Res. 19:187-94.

Zick, Y., Sagi-Eisenberg, R., Pines, M., Gierschik, P. and Spiegel, A. (1986). Multisite phosphorylation of the $\propto$ subunit of transducin by the insulin receptor kinase and protein kinase C. Proc. Natl. Acad. Sci. USA 83:9294-7.

INSULIN AND INSULIN-LIKE GROWTH FACTOR I MEDIATED PHOSPHORYLATIONS

IN MOUSE NEUROBLASTOMA N18 CELLS: A MODEL FOR STUDYING INSULIN AND IGF-I

ACTION ON NEURAL TISSUE

Martin Adamo, Joshua Shemer and Derek LeRoith

Diabetes Branch, NIDDK, NIH, Bethesda, MD 20892

INTRODUCTION

Insulin and insulin-like growth factor I are traditionally believed to regulate metabolic and growth activities in peripheral target tissues (1). However, since both peptides are present in the brain, they probably also function as neuropeptides (2). Additional evidence for their ability to function on nervous tissue is the finding that specific insulin and IGF-I receptors are present throughout the brain (3,4). These receptors exhibit structures typical of insulin and IGF-I receptors, which are heterotetramers consisting of 2 alpha-subunits which bind the ligand and two beta-subunits which are autophosphorylated as a result of ligand binding. The subunits are covalently linked via disulfide bridges (5,6,7). An important difference between brain and peripheral insulin and IGF-I receptors is the size of the receptor alpha-subunit, which has been found to be approximately 10kDa smaller in receptors from the CNS (8,9). The basis for this difference lies primarily in the extent and pattern of alpha-subunit glycosylation (10,11). Upon autophosphorylation, insulin and IGF-I beta-subunits act as tyrosine kinases, toward endogenous and exogenous substrates, and considerable evidence has accumulated to suggest that this cascade of phosphorylations is critical to insulin action (12,13,14). Insulin and IGF-I affect CNS function in several ways including regulation of metabolic pathways, growth and differentiation, and regulation of behavioral processes such as food intake (2). Presumably, ligand stimulated receptor and substrate phosphorylation is also the signalling mechanism for these CNS actions.

Insulin and IGF-I receptor autophosphorylation and tyrosine kinase activity have been studied in cell free systems from whole brain tissue, primary neuronal and glial cell cultures and a variety of transformed neural-derived cells (4,15,16,17,18). We have extended these studies by examining insulin and IGF-I stimulated protein phosphorylations in intact N18 neuroblastoma cells (19). This approach is more physiologically relevant to insulin and IGF-I action in the CNS because it allows for determination of the pattern of receptor beta-subunit autophosphorylation in intact cells as well as offering the potential for discovering endogenous substrates for insulin and IGF-I receptor kinases in neural tissue.

METHODS

N18 neuroblastoma cells (kindly provided by Dr. M. Nirenberg, Bethesda, Maryland) were maintained in serum free DMEM for 12 hours, after which they were placed in serum- and phosphate-free Eagles MEM. Metabolic labeling of intracellular ATP pools was accomplished by incubating cells for two hours with ^{32}P orthophosphate (0.5 mCi/ml). Cells were then exposed to insulin or IGF-I at various concentrations for various time periods (19). Incubations were terminated by rapidly aspirating the medium and freezing the cells on liquid N_2. Cells were solubilized by addition of 1% Triton X-100 in 50mM Hepes pH 7.4 containing 10mM sodium pyrophosphate, 10mM NaF, 4 mM EDTA, 2 mM sodium orthovanadate and 2mM PMSF (19,20). Cell lysates were then treated in one of two ways: 1) lysates were centrifuged at 12,000xg for one hour at 4°C, and the supernatants were recovered, and phosphotyrosine containing proteins immunoprecipitated by incubation with anti-phosphotyrosine antibody (αPY, a kind gift of Dr. Y. Zick, Rehovot, Israel) at 1:50 dilution for 2 hrs at 4°C, followed by incubation with pansorbin for 1 hr at 4°C. The immunoprecipitates were collected by centrifugation and were washed five times with 50mM Hepes pH 7.4 containing 1% Triton X-100, 0.1% SDS, 150mM NaCl, 10mM NaF, 4mM EDTA and 2mM sodium orthovanadate. Phosphotyrosine containing phosphoproteins were then specifically eluted with 10mM p-nitrophenylphosphate (PNPP) (19,21), denatured, and were analyzed on 7.5% SDS/PAGE reducing gels.
2) Cell lysates were centrifuged at 40,000xg for 50 min at 4°C and supernatants were passed over 0.5ml wheat germ agglutinin agarose columns three times. The columns were washed with 90ml of 50mM Hepes pH 7.4 containing 0.1% Triton X-100, 10mM sodium pyrophosphate, 10mM NaF, 4mM EDTA, 2mM sodium orthovanadate and 2mM PMSF. Adsorbed glycoproteins were then eluted using 0.3M N-acetyl glucosamine in the above buffer. Fractions 1 and 2, containing the highest levels of radioactivity were immunoprecipitated with either αPY or anti-insulin receptor antibodies and pansorbin. Phosphoproteins were eluted from αPY with PNPP and from anti-receptor antibodies using Laemmli's sample buffer. Samples were denatured and analyzed on 7.5% SDS/PAGE gels (19). Phosphoamino acid composition of phosphoproteins was determined by modifications (19) of a previously described method (22).

RESULTS

Insulin and IGF-I stimulated phosphorylation of several proteins immunoprecipitated by anti-phosphotyrosine antibody in a time-dependent manner (Fig. 1). Insulin, at 10^{-7}M stimulated ^{32}P incorporation into phosphoproteins of 95 and 185 kDa while IGF-I at 10^{-7}M stimulated ^{32}P incorporation into phosphoproteins of 95, 105 and 185kDa. Ligand induced phosphorylation of all phosphoproteins was evident as early as 20 sec, achieved a maximum at one minute and declined thereafter. The amount of ^{32}P incorporated into the 185kDa protein decreased more rapidly than that in the 95 and 105kDa proteins for both ligands.

During one minute exposures of N18 intact cells to the ligands, insulin and IGF-I stimulated phosphorylations occurred in a dose dependent manner (Fig. 2). Insulin stimulated ^{32}P incorporation into 95kDa and 185kDa proteins while IGF-I stimulated ^{32}P incorporation into 95, 105 and 185kDa proteins. Both ligands effected increased ^{32}P incorporation at the lowest concentration tested, i.e., 1nM (Fig. 2).

The insulin and IGF-I sensitive phosphoproteins identified in Figs. 1 and 2 were extracted from whole cell lysates using anti-phosphotyro-

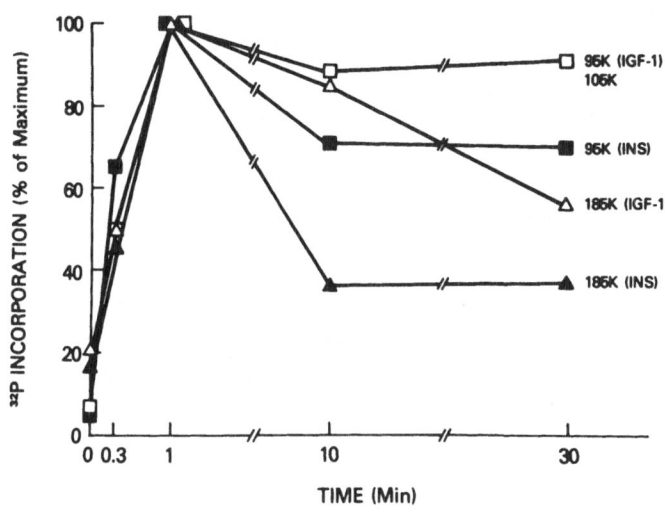

Figure 1

Time course of Insulin and IGF-I induced phosphorylation in intact
N18 neuroblastoma cells.

N18 cells were metabolically labeled with ^{32}P-orthophosphate and
stimulated with insulin or IGF-I (10^{-7}M) for 0, 0.3, 1, 10 and 30 min.
Whole cell extracts were immunoprecipitated with antiphosphotyrosine
antibody and PNPP eluates were electrophoresed on reducing 7.5%
SDS/PAGE gels (upper panel). Insulin stimulated ^{32}P incorporation
into 95kDa and 185kDa phosphoproteins (lanes B-E) while IGF-I
stimulated ^{32}P incorporation into 95, 105 and 185kDa phosphoproteins.
The gel bands representing the phosphoproteins were excised and
counted, and the ^{32}P incorporation has been plotted as a percentage
of the maximal in the bottom panel.

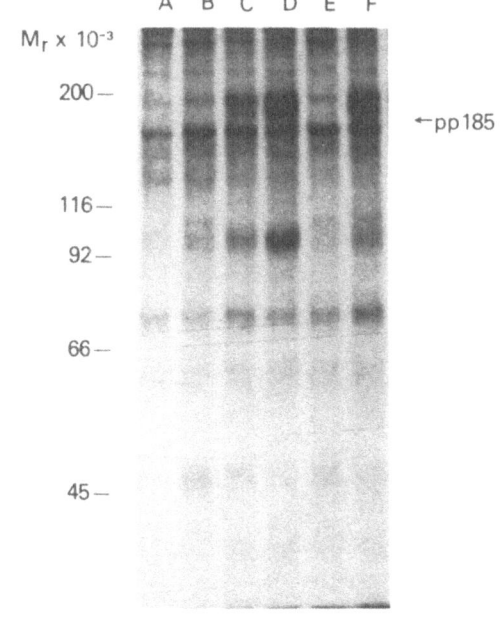

INSULIN (M) — 10⁻⁹ 10⁻⁸ 10⁻⁷ — —

IGF-1 (M) — — — — 10⁻⁹ 10⁻⁷

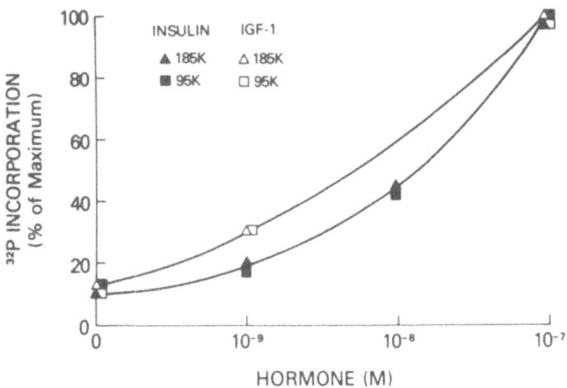

Figure 2

Dose response of Insulin and IGF-I stimulated phosphorylation in intact N-18 neuroblastoma cells.

^{32}P-labeled N-18 cells were incubated for one minute without (lane A) or with insulin at 10^{-9}M (lane B), 10^{-8}M (lane C) or 10^{-7}M (lane D), or with IGF-I at 10^{-9}M (lane E) or 10^{-7}M (lane F). Whole cell extracts were immunoprecipitated with antiphosphotyrosine antibody and PNPP eluates were run on reducing SDS/PAGE gels (upper panel). The bands representing the insulin stimulated 95 and 185 kDa phosphoproteins and the IGF-I stimulated 95 and 185kDa phosphoproteins were excised, counted, and ^{32}P incorporation has been plotted vs. ligand concentration in the bottom panel.

330

sine antibody. Based on other studies using cell free systems(23), it
appeared that the 95kDa phosphoproteins were the receptor β subunits
while the 105 kDa phosphoprotein was a subtype of the IGF-I receptor beta
subunit. The 185kDa phosphoprotein may represent an endogenous substrate. In order to determine if this were the case, anti-receptor
antibodies were employed to immunoprecipitate receptors from anti-phosphotyrosine antibody eluates. Two such antibodies were used, B10 and
B2. B10 specifically recognizes insulin receptors while B2 recognizes
both insulin and IGF-I receptors (24). As seen in Fig. 3, anti-receptor
antibodies, B10 and B2 immunoprecipitated the 95 kDa protein from insulin
stimulated cells, identifying it as the insulin receptor beta-subunit.
Antibody B10 failed to immunoprecipitate the IGF-I stimulated 95kDa protein, while antibody B2 immunoprecipitated both the 95 and 105 kDa phosphoproteins which resulted from IGF-I stimulation. Neither anti-receptor
antibody recognized the 185kDa phosphoprotein. These results strongly suggest that 1) insulin and IGF-I each stimulate phosphorylation of the 95kDa
beta-subunits of their respective receptors; 2) IGF-I beta-subunits exist
as 95kDa and 105kDa subunits; 3) The 185kDa phosphoprotein stimulated by
insulin and IGF-I is not immunologically related to their receptors and is
probably an endogenous substrate.

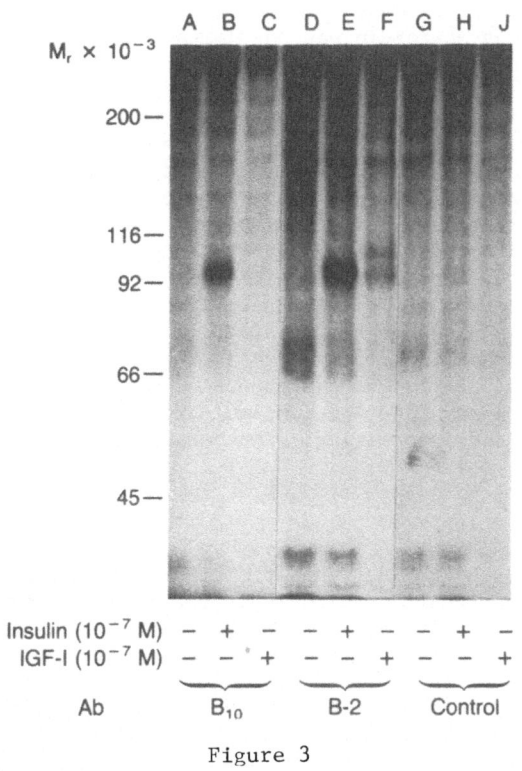

Figure 3

Anti-insulin and IGF-I receptor immunoprecipitable phosphoproteins
in intact N-18 neuroblastoma cells.

^{32}P labeled N-18 cells were stimulated with insulin or IGF-I
at 10^{-7}M for one minute and whole cell extracts were immunoprecipitated
with antiphosphotyrosine antibody. PNPP eluates were then re-immuno-
precipitated with anti-receptor antibodies B10 or B2 with or normal
serum (control). Immune complexes were collected using pansorbin and
phosphoproteins were eluted using Laemmli's sample buffer, boiled and
run on 7.5% SDS/PAGE gels.

In order to further confirm that the 105kDa protein, whose phosphorylation was stimulated by IGF-I, is indeed an IGF-I receptor beta-subunit subtype, receptors were partially purified from stimulated cells by WGA affinity chromatography prior to immunoprecipitation. IGF-I stimulated ^{32}P incorporation into both 95 and 105kDa phosphoglycoproteins which were immunoprecipitable with αPY antibody in a time and dose dependent manner (Figs 4 and 5). Insulin stimulated ^{32}P incorporation only into a 95kDa phosphoglycoprotein which was recognized both by αPY and B10 (Fig. 5). In contrast, B10 did not recognize any IGF-I stimulated phosphoproteins (Fig. 5). Furthermore, the 185kDa phosphoprotein was not evident in any immunoprecipitates of WGA-purified phosphoglycoproteins from control or insulin or IGF-I stimulated cells (Figs. 4 and 5).

In an attempt to determine the relationship of the 105kDa beta-sub-unit subtype to the oligomeric form of the IGF-I receptor, αPY immuno-precipitates from IGF-I stimulated N18 cells were run on non-reducing gels. As seen in Fig. 6, (left panel), two high molecular weight phosphoproteins were observed, one >300kDa, and one slightly <300 kDa. When these two bands were excised, reduced, and run on reducing SDS/PAGE gels the high molecular weight band yielded the 95 and 105kDa beta-subunit subtypes, while the smaller oligomer yielded only 105kDa sub-types (Fig.6, right panel; and data not shown).

The 185kDa phosphoprotein was not immunologically related to either the insulin or IGF-I receptors, and was assumed to represent an endogenous substrate of their receptor kinases. Both the EGF and PDGF receptors are tyrosine kinases, become autophosphorylated and are close in Mr to the 185kDa phosphophoprotein (pp185). In order to determine if pp185 was re= lated to these growth factor receptors, we utilized these ligands to stimulate ^{32}P incorporation into αPY-immunoprecipitable phosphoproteins. As seen in Fig. 7, (left panel) in contrast to insulin and IGF-I, neither of these growth factors stimulated phosphorylation of a 185kDa protein. Use of anti-EGF receptor antiserum revealed that EGF stimulated ^{32}P incorporation into its receptor of Mr 170 kDa in N18 cells; neither insulin nor IGF-I effected this phosphorylation (Fig. 7, right panel).

Finally, phosphoamino acid analysis of both the 95kDa beta-subunit (Fig. 8, upper) and the 185kDa endogenous substrate (Fig. 8, lower) revealed that insulin and IGF-I both stimulated ^{32}P incorporation into serine, threonine and interestingly, into tyrosine residues.

DISCUSSION

The present study was designed to better define the signal transduction pathway for insulin and IGF-I in the nervous system, so that it could be compared and contrasted with that in the periphery. Furthermore, we wished to utilize a more physiological system than the previously employed cell-free approach, one that would allow for the identification of possible endogenous substrates for the insulin and IGF-I receptor kinases in central nervous tissue. To accomplish these goals, we utilized a well characterized, neural-derived cell line, N18. N-18 cells possess substantial numbers of insulin and IGF-I receptors and in cell-free systems, these receptors undergo autophosphorylation and exhibit tyrosine kinase activity (23,25). In this study we determined, in intact cells, insulin and IGF-I stimulated phosphorylation of proteins immunoprecipitable with an anti-phosphotyrosine antibody as well as anti-receptor antibodies.

In these cells, insulin stimulates phosphorylation of its receptor

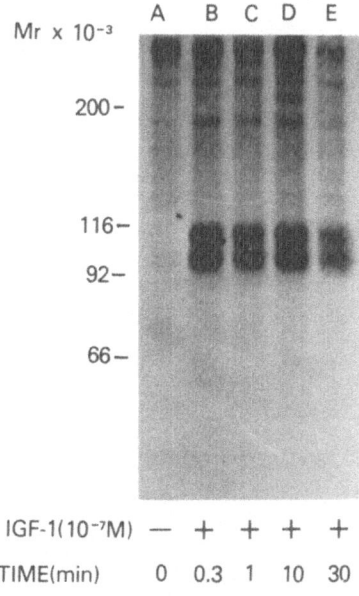

IGF-1(10^{-7}M) — + + + +

TIME(min) 0 0.3 1 10 30

Figure 4

Time course of IGF-I stimulated ^{32}P incorporation into wheat germ agglutinin adhering phosphoproteins in intact N-18 neuroblastoma cells.

^{32}P labeled N-18 cells were incubated with 10^{-7} M IGF-I for 0, 0.3, 1. 10 or 30 minutes. After solubiliztion and centrifugation, lysates were partially purified on wheat germ agglutinin agarose columns. Eluates were immunoprecipitated with antiphosphotyrosine antibody and PNPP eluates were run on 7.5% SDS/PAGE gels.

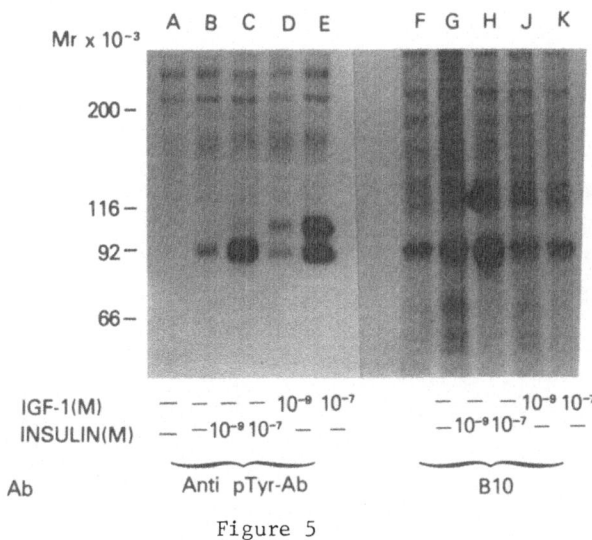

IGF-1(M)   — — — — 10^{-9} 10^{-7} — — — 10^{-9} 10^{-7}
INSULIN(M) — —$10^{-9}$$10^{-7}$ — —   —$10^{-9}$$10^{-7}$ — —

Ab   Anti pTyr-Ab   B10

Figure 5

Dose response of insulin and IGF-I stimulated phosphorylation of WGA adhering phosphoproteins in intact N-18 neuroblastoma cells.

^{32}P labeled N-18 cells were stimulated with insulin or IGF-I at 10^{-9} and 10^{-7} M and cell lysates were subjected to WGA affinity chromatography. Eluates were immunoprecipitated either with antiphosphotyrosine antibody, or anti-insulin receptor antibody B10 and run on SDS/PAGE.

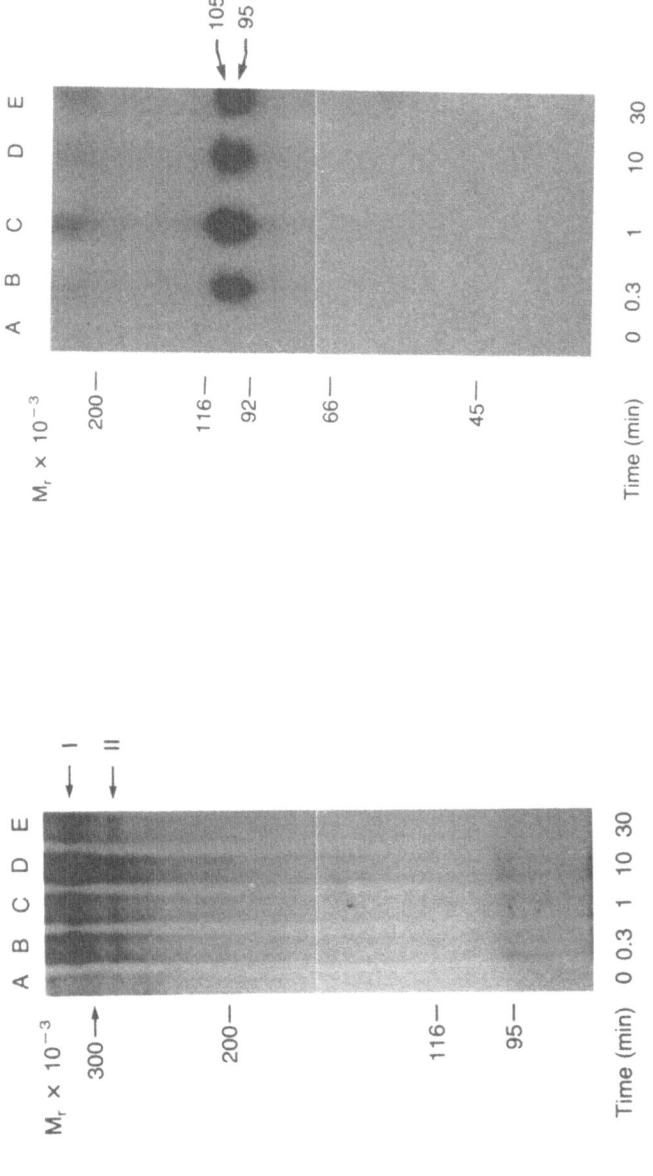

Figure 6

Analysis of oligomeric forms of IGF-I receptors in N18 neuroblastoma cells.

^{32}P labeled N-18 cells were incubated with 10^{-7}M IGF-I for 0, 0.3, 1, 10 and 30 minutes. After solubilization, WGA affinity chromatography and immunoprecipitation with antiphosphotyrosine antibody, PNPP eluates were run on non-reducing 5% SDS/PAGE gels. Autoradiography (left panel) revealed two labeled phosphoproteins. I (>300kDa) and II (<300kDa). Gel band I was excised and run on a 7.5% reducing SDS/PAGE gel yielding "spots" of 105 and 95kDa (right panel).

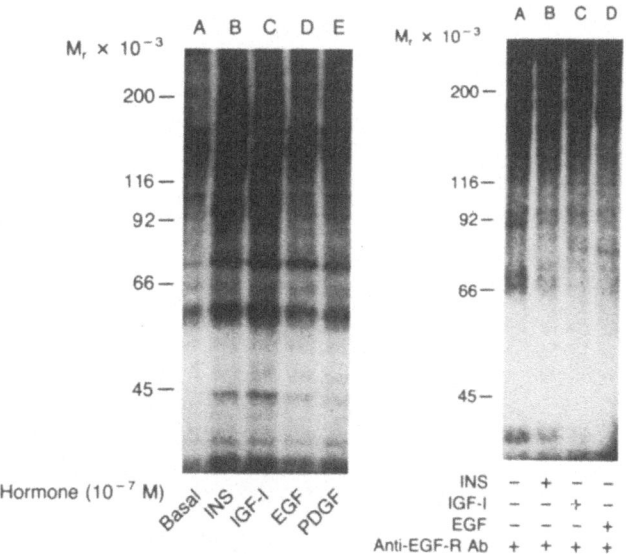

Figure 7

Substrate specificity of pp185

[32]P labeled N-18 neuroblastoma cells were incubated
without (lane A) or with insulin (lane B), IGF-I (lane C),
EGF (lane D) or PDGF (lane E), all at 10^{-7}M for one minute.
Cells were then solubilized and whole cell extracts immunoprecipitated with antiphosphotyrosine antibody (left panel).
Only insulin and IGF-I stimulated phosphorylation of the
185kDa phosphoprotein. In the experiment shown in the right
panel, ^{32}P labeled N-18 cells were incubated for one minute without (lane A) or with insulin (lane B), IGF-I (lane C)
or EGF (lane D) all at 10^{-7}M for one minute. Whole cell extracts were immunoprecipitated with an anti-EGF receptor
antibody kindly provided by Dr. J. Schlesinger. A 170 kDa
phosphoprotein stimulated by EGF was detected on SDS/PAGE gels.

β Subunit-95K

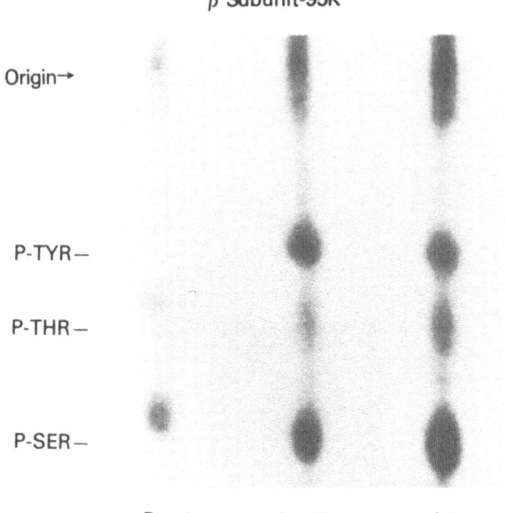

Origin→
P-TYR—
P-THR—
P-SER—

Basal   Insulin   IGF-1

PP-185

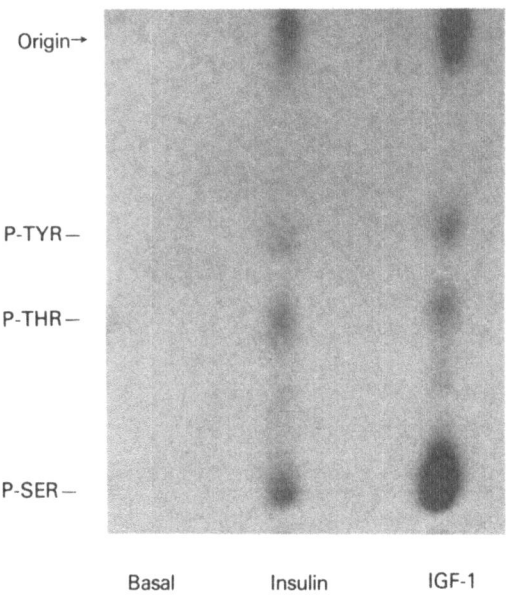

Origin→
P-TYR—
P-THR—
P-SER—

Basal   Insulin   IGF-1

Figure 8

Phosphoamino acid analysis of insulin and IGF-I stimulated
phosphoproteins in N18 cells.

^{32}P labeled N-18 cells were incubated without (basal) or with
insulin or IGF-I (10^{-7}M) for one minute, solubilized, and extracts
were immunoprecipitated with antiphosphotyrosine antibody and run
on 7.5% SDS/PAGE gels. Gel bands representing the 95kDa beta-subunits
(upper) and pp185 (lower panel) were excised and phosphoamino acids
were determined. P-Tyr (phosphotyrosine), P-Thr (phosphothreonine)
and P-Ser (phosphoserine) indicate migration of ninhydrin stained
phosphoamino acid standards on TLE plates.

beta subunit, which has an Mr of 95 kDa. IGF-I, on the other hand, stimulates ^{32}P incorporation into two beta-subunit subtypes, of 95 and 105kDa. In all cases, significant quantities of ^{32}P were incorporated into serine as well as tyrosine residues. These results should be contrasted with results obtained using partially purified detergent extracts from the N18 cells (23). Thus, in cell-free extracts, the 105kDa phosphoprotein is stimulated in response to insulin as well as IGF-I and beta-subunit phosphorylations are almost entirely on tyrosine residues. The difference between the intact cell and cell-free system results suggests that 1) in intact cells insulin and IGF-I do not cross-occupy the others receptors and/or the 105kDa IGF-I beta-subunit subtype is not a substrate for the insulin receptor kinase in the intact cell; and 2) possibly, a serine kinase activity, in addition to tyrosine kinase activity is increased after insulin and IGF-I bind to their receptors. A difference between the phosphorylated beta-subunit amino acids in intact cells versus cell-free systems has been observed in non-neural cell types, where considerable serine phosphorylation has been observed in intact cells (26,27). It is suggested that in terms of these two phenomena i.e., receptor "cross-talk", and amino acid phosphoacceptors, the results seen in intact cells may be more relevant to insulin and IGF-I signal transduction in vivo in the brain than those seen using cell-free systems.

The existence of IGF-I receptor beta-subunit subtypes as seen in this study in N18 cells is not unique, as a non-neural cell line, the human epidermoid carcinoma cell line KB, showed IGF-I stimulated beta-subunit subtypes of 92 and 98kDa (21). Our present results suggest that in N18 cells, IGF-I receptor heterotetramers may exist as two main types, one type containing 95 and 105kDa subunits and the other only the 105kDa subtype. Clearly, further studies will be needed to clarify this point. It should be noted, that in the N18 cells, using phosphopeptide mapping and glycosidase digestion of cell-free phosphorylated receptors, that the size difference between IGF-I receptor beta-subunit subtypes is probably due primarily to glycosylation differences (23).

Of considerable interest is the 185kDa protein whose phosphorylation is stimulated by insulin and IGF-I. The observations that anti-receptor antibodies do not recognize it, that it does not bind to WGA, and that ligand-induced phosphorylation is rapid and also on tyrosine residues, provide strong evidence that pp185 is a substrate common to and specific for the insulin and IGF-I receptor kinases. This is especially significant since a pp185 is an insulin and IGF-I receptor kinase substrate in several non-neural cell lines (20,21,28). The potential importance of pp185 in insulin receptor signal transduction has recently been further demonstrated by Kahn and coworkers (29). Transfected insulin receptors with Tyr-960 mutated demonstrated diminished ability to phosphorylate pp185 and diminished ability to stimulate glycogen synthase. Whether pp185 found in neural and non-neural tissues are the same protein or even if the insulin stimulated and IGF-I stimulated pp185 are identical is yet to be determined. The answers to these questions will await the molecular cloning of the gene or cDNA for pp185. As with the beta-subunits, pp185 was phosphorylated to a greater extent on serine residues than tyrosine residues suggesting that in intact cells, serine kinase activation may also be important in insulin and IGF-I signal transduction (27).

In summary, we have demonstrated that insulin and IGF-I stimulated receptor autophosphorylation and putative endogenous substrates are similar in intact cells of neural origin as compared to those of non-neural cells. These findings strongly suggest that the signal transduction pathway leading to insulin and IGF-I action is similar in the CNS and periphery. Future studies will be focused on more detailed analysis of the signal transduction pathway including the molecular identification of pp185 and determination of the potential significance of IGF-I beta-subunit subtypes in IGF-I action.

References

1. J.F. Perdue, Chemistry, structure and function of insulin-like growth factors and their receptors: a review. Can. J. Biochem. Cell Biol. 62:1237 (1984).
2. E. Recio-Pinto and D.N. Ishii, Insulin and related growth factors: effects on the nervous system and mechanism for neurite growth and regeneration. Neurochem. Int. 12:397 (1988)
3. J. Havrankova, J. Roth, and M. Brownstein, Insulin receptors are widely distributed in the central nervous system of the rat. Nature 272:827 (1978).
4. S. Gammeltoft, G. Hasselbacher, R. Humbel, M. Feldmann, E. Van Obber-ghen, Two types of receptor for insulin-like growth factors in mammalian brain. EMBO J. 4:3407 (1985).
5. A. Ullrich, J.R. Bell, E.Y. Chen, R. Herrera, L.M. Petruzzelli, T.J. Dull, A. Gray, L. Coussens, Y.-C. Liao, M. Tsubokawa, A. Mason, P.H. Seeburg, C. Grunfeld. O.M. Rosen and J. Ramachandran, Human insulin receptor and its relationship to the tyrosine kinase family of oncognes. Nature 313:756 (1985).
6. Y. Ebina, L. Ellis, K. Jarnagin, M. Edery, L. Graf, E. Clauser, J.H. Ou, F. Masiarz, Y.W. Kan, I.D. Goldfine, R. Roth and W. Rutter, The human insulin receptor cDNA: the structural basis for hormone-activated transmembrane signalling. Cell 46:747 (1985).
7. A. Ullrich, A. Gray, A.W. Tam, F. Yang-Feng, M. Tsubokawa, C. Collins, W. Henzel, T. LeBon, S. Dathuria, S. Chen, S. Jacobs, U. Francke, J. Ramachandran and Y. Fujita-Yamaguchi, Insulin-like growth factor I receptor primary structure: comparison with insulin receptor sug-gests structural determinants that define functional specificity. EMBO J. 5:2503 (1986).
8. C.C. Yip, M.L. Moule and C.W.T. Yeung, Characterization of insulin receptor subunits in brain and other tissues by photoaffinity label-ing. Biochem. Biophys. Res. Comm. 96:1671 (1980).
9. K.A. Heidenreich and D. Brandenberg, Oligosaccharide heterogeneity of insulin receptors. Comparison of N-linked glycosylation of in-sulin receptors in adipocytes and brain. Endocrinology 118:1835 (1986).
10. K.A. Heidenreich, N.R. Zahniser, P. Berhanu, D. Brandenberg and J.M. Olefsky, Structural differences between insulin receptors in the brain and peripheral target tissues. J. Biol. Chem. 258:8527 (1983).
11. K.A. Heidenreich, G.R. Freidenberg, D.P. Figlewicz and P.R. Gilmore, Evidence for a subtype of insulin-like growth factor I receptor in brain. Reg. Peptides 15:301 (1987).
12. C.K. Chou, T.J. Dull, D.S. Russell, R. Gherzi, D. Lebwohl, A. Ullrich, O.M. Rosen, Human insulin receptors mutated at the ATP-binding site lack protein tyrosine kinase activity and fail to mediate post-re-ceptor effects of insulin. J. Biol. Chem. 262:1842 (1987).
13. Y. Ebina, E. Araki, M. Taira, F. Shimada, M. Mori, C.S. Craik, K. Siddle, S.B. Pierce, R.A. Roth and W.J. Rutter, Replacement of lysine residue 1030 in the putative ATP-binding region of the insu-lin receptor abolishes insulin- and antibody stimulated glucose up-take and receptor kinase activity: introduction of an inhibitory monoclonal antibody into mammalian cells blocks the rapid effects of insulin. Proc. Natl. Acad. Sci. U.S.A. 84:41 (1987).
14. D.O. Morgan and R.A. Roth, Acute insulin action requires insulin receptor kinase activity: introduction of an inhibitory monoclonal antibody into mammalian cells blocks the rapid effects of insulin. Proc Natl. Acad. Sci. U.S.A. 84:41 (1987).

15. R.W. Rees-Jones, S.A. Hendricks, M. Quarum, and J. Roth, The insulin receptors of rat brain are coupled to tyrosine kinase activity. J. Biol. Chem. 260:4461 (1984).

16. W.L. Lowe, Jr., F.J. Boyd, D.W. Clarke, M.K. Raizada, C. Hart and D. LeRoith, Development of brain insulin receptors:structural and functional studies of insulin receptors from whole brain and primary cell cultures. Endocrinology 119:25 (1986).

17. J. Shemer, M.K. Raizada, B.A. Masters, A. Ota, and D. LeRoith, Insulin-like growth factor I receptors in neuronal and glial cells. Characterization and biological effects in primary culture. J. Biol. Chem. 262:7693 (1987).

18. A. Ota, G.L. Wilson, O. Spilberg, R. Pruss and D. LeRoith, Functional insulin-like growth factor I receptors are expressed by neural-derived continuous cell lines. Endocrinology 122:145 (1988).

19. J. Shemer, M. Adamo, G.L. Wilson, D. Heffez, Y. Zick and D. LeRoith, Insulin and insulin-like growth factor I stimulate a common endogeneous phosphoprotein substrate (pp185) in intact neuroblastoma cells. J. Biol. Chem. 262:15476 (1987).

20. M.F. White, R. Maron and C.R. Kahn, Insulin rapidly stimulates tyrosine phosphorylation of a Mr 185,000 protein in intact cells. Nature 318:183 (1985).

21. T. Kadowaki, S. Kayasu, E. Nishida, K. Tobe, T. Izumi, F. Takaku, H. Sakai, I. Yahara, and M. Kasuga, Tyrosine phosphorylation of common and specific sets of cellular proteins rapidly induced by insulin, insulin-like growth factor I, and epidermal growth factor in an intact cell. J. Biol. Chem. 262:7342 (1987).

22. J.A. Cooper, B.M. Sefton and T. Hunter, Detection and quantification of phosphotyrosine in proteins. Meth. Enzymol. 99:387 (1983).

23. A. Ota, G.L. Wilson and D. LeRoith, Insulin-like growth factor I receptors on mouse neuroblastoma cells. Two beta subunits are derived from differences in glycosylation. Eur. J. Biochem. 174:521 (1988).

24. M. Kasuga, N. Sasaki, C.R. Kahn, S.P. Nissley and M.M. Rechler. Antireceptor antibodies as probes of insulin-like growth factor receptor structure. J. Clin. Invest. 72:1459 (1983).

25. A. Ota, J. Shemer, R.M. Pruss, W.L. Lowe, Jr., and D. LeRoith, Characterization of the altered oligosaccharide composition of the insulin receptor on neural derived cells. Brain Res. 443:1 (1988).

26. M. Kasuga, F.A. Karlsson, and C.R. Kahn, Insulin stimulates the phosphorylation of the 95,000-dalton subunit of its own receptor. Science 215:185 (1982).

27. L.M. Petruzzelli, L. Stadtmauer, R. Herrera, M. Makowske, S. Ganguly, D. Tabarini, H. Lee, Y. Olowe, and O.M. Rosen, The insulin receptor as a tyrosine-specific protein kinase, in: "Mechanisms of Receptor Regulation" G. Poste and S.T. Crooke, eds. Plenum, New York and London (1985).

28. M. F. White, E.W. Stegmann, T.J. Dull, A. Ullrich and C.R. Kahn, Characterization of an endogenous substrate of the insulin receptor in cultured cells. J. Biol. Chem. 262:9769 (1987).

29. M.F. White, J.N. Livingston, J.N. Bacher, V. Lauris, T.J. Dull, A. Ullrich, and C.R. Kahn, Mutation of the insulin receptor at tyrosine 960 inhibits signal transmission but does not affect its tyrosine kinase activity. Cell 54:641 (1988).

INSULIN-LIKE GROWTH FACTOR RECEPTORS IN THE CENTRAL NERVOUS SYSTEM:

PHOSPHORYLATION EVENTS AND CELLULAR MEDIATORS OF BIOLOGICAL FUNCTION

Brian A. Masters[1], Joshua Shemer[2], Derek LeRoith[2] and Mohan K. Raizada[1]

[1]Department of Physiology, University of Florida
Gainesville, FL 32610
[2]Section of Molecular and Cellular Physiology
Diabetes Branch, NIADDK
National Institutes of Health
Bethesda, MD 20892

INTRODUCTION

A complex pattern of growth and development in the nervous system would seem inherent in consideration of its morphological design for control of physiological processes and psychological phenomena. Yet in spite of this tangled design it would appear reasonable that certain basics tenets must be preserved in development and growth in the cellular world. Using such reasoning, the complex mechanisms responsible for growth and development in the nervous system have begun to acquire a very rudimentary definition. It has become increasingly apparent that a functional family of peptides termed growth factors are basic to the regulation of growth and development in peripheral cell populations (Daughaday, 1989). Recent research addressing nervous system growth and development points at these same peptide factors as important determinants in the generation and maintenance of the distinct cell types that comprise the brain, peripheral nerves, and organs of neural ectodermal origin.

The somatomedins or insulin-like growth factors (IGF 1 and IGF 2), members of this growth-promoting peptide family, are important mediators of growth in peripheral tissues (Daughaday, 1989; Schoenle et al., 1982; Guler et al., 1988; Phillips et al., 1988; Van Wyk, 1984). Recently the IGFs have been postulated to act as differentiation and maturation promoting substances in the nervous system (McMorris et al., 1986; Mattsson et al., 1986; DiCicco-Bloom and Black, 1988; Aizenman and de Vellis, 1987; Recio-Pinto et al., 1986). Although only a few of these responses to IGF 1 or IGF 2 are well characterized, it would appear that these responses have profiles which are unique to the nervous system and are markedly different in each cell type that is represented in nervous tissue. Such differential responses to the IGFs suggest that the means by which the peptides' signal is translated into a functional response in these neural and glial cells may be distinct. While little of the mechanism(s) by which IGF 2 acts in the periphery or nervous system is understood at this time, investigations into the machinery of IGF 1's actions in the nervous system have elucidated basic differences and similarities between neural and glial cells, and peripheral mechanisms of IGF 1 responses.

BIOLOGICAL ACTIONS OF THE IGFs IN THE NERVOUS SYSTEM

The presence and production of the IGFs in the nervous system has been demonstrated numerous times (Sara et al., 1982; Noguchi et al., 1987; Lund et al., 1986; Murphy et al., 1987; Han et al., 1988; Rotwein et al., 1988; Ballotti et al., 1987; Adamo et al., 1988; Haselbacher et al., 1985; Shiu and Paterson, 1988; Brown et al., 1986; Hynes et al., 1988). Difficulties in delineating the effects of IGF 1 and IGF 2 in the nervous system arise from structural similarities that these peptides share and from the fact that, structurally, the IGFs are related to insulin (Perdue, 1984): Although production of insulin in the nervous system is somewhat more controversial (Perdue, 1984; Budd et al., 1986; Young, 1986; Clarke et al., 1987; Schechter et al., 1988; Baskin et al., 1983; Pardridge et al., 1985; Frank et al., 1986; Stein et al., 1983) than the production of the IGFs, insulin has a demonstrated range of functions in the central and peripheral nervous systems (Aizenman and de Vellis, 1987; Clarke et al., 1984; Rinaudo et al., 1987; Van der Pal et al., 1988; Coimbra and Migliorini, 1986; Sakaguchi and Bray, 1987; Boyd et al., 1985; Raizada et al., 1987; Kwok and Juorio, 1987; DiCicco-Bloom and Black, 1988; Puro and Agardh, 1984; Recio-Pinto and Ishii, 1984). Due to the structural similarities of IGF 1, IGF 2, and insulin, the biological responses they elicit overlap and depend greatly on the relative concentrations of each peptide and the presence of their appropriate receptors. Using assays which compare the effects of the IGFs and insulin with respect to their potency in producing the appropriate response, it has been possible to define responses that are shared and unique to each of these growth factors.

Insulin, IGF 1 and IGF 2 all share homology with mouse NGF (Ishii and Recio-Pinto, 1987), nerve growth factor, which has been shown to have diverse effects on neurons and glial tissues. IGF 2 has recently been shown to be a potent neurotrophic agent (Recio-Pinto et al., 1986). Expression of IGF 2 is developmentally regulated and follows a pattern similar to, but distinct, from that of NGF in nervous tissue (Ishii and Recio-Pinto, 1987). IGF 2 gene expression is tissue-specific and developmentally regulated (Soares et al., 1985). Transcript content is high in most fetal tissues, but only brain and spinal cord contain transcripts in adult animals (Soares et al., 1986). In studies in which the sciatic nerve is transected in young rats, IGF 2 mRNA transcripts in the innervated muscle are elevated as compared to the expression of transcripts in intact animals (Ishii and Recio-Pinto, 1987). In this way IGF 2 may be permissive for the reinnervation of muscle (Ishii and Recio-Pinto, 1987). NGF, IGF 2, and to a lesser extent insulin, increase the expression of mRNA for both alpha and beta tubulin in a neural derived cell line and increased tubulin expression can be correlated to the increased outgrowth of neurite processes in these cells. IGF 1 does not appear to mimic these effects (Mill et al., 1985). The presence of IGF 2 in growth medium supports the survival of neurons (Aizenman and de Vellis, 1987; Recio-Pinto et al., 1986) and oligodendroglia (McMorris, 1983) in culture in the absence of serum. Doses at which IGF 2 elicits such effects strongly suggest that these processes are mediated through the IGF 2 receptor protein.

IGF 1, which acts as the primary effector of growth hormone stimulated processes in peripheral organs, inhibits growth hormone synthesis and secretion at the hypothalamus and anterior pituitary (Berelowitz et al., 1981). In addition to this well documented action, IGF 1 appears to have proliferative and mitogenic functions in the nervous system similar to those which have been demonstrated in the periphery (McMorris et al., 1986; Mattsson et al., 1986; DiCicco-Bloom and Black, 1988; Aizenman and de Vellis, 1987; Recio-Pinto et al., 1986; Burgess et al., 1987; Shemer et al., 1987a; Ballotti et al., 1987; Van der Pal et al., 1988; Ota et al., 1988a). The incorporation of [^{3}H]-thymidine in relatively pure serum-deprived primary cultures of neurons and astroglia is stimulated by addition of 0.013 nM IGF 1. Although insulin is able to produce a stimulated [^{3}H]-thymidine content in the TCA-precipitable fraction it is approximately 100 times less potent than IGF 1 and appears more potent in stimulating glucose uptake into glia (Figure 1). Thus the proliferative response seems to be mediated through the IGF 1 receptor protein (Burgess et al., 1987). IGF 1 also stimulates the incorporation of [^{3}H]-uridine into RNA in neuroblastoma cell lines (Ota et al., 1988a). Stimulation of DNA and RNA synthesis can also be demonstrated in sympathetic neuroblasts (DiCicco-Bloom and Black, 1988). Although IGF 1 stimulates growth processes, it appears that it may play an even more basic role in neural and glial cells as is supported by studies which indicate that IGF 1 is sufficient for survival of some of these cell types in culture (Aizenman and de Vellis, 1987; Recio-Pinto et al., 1986). Also supportive of a proliferative role for IGF 1 are studies indicating that neural-ectodermal tumors produce and secrete high levels of IGF 1 (Yee, D., Favoni, R.E., Lebovic, G.S., Reynolds, C.P., Rosen, R., Abstract: Molecular and Cellular aspects of Insulin, IGFs and their Receptors: Implications for the CNS. Symposium at the University of Florida, Gainesville , FL. January, 1989).

IGF 1 is a potent inducer of oligodendrocyte development (McMorris et al., 1986). Thus, IGF 1 may an important factor regulating myelination in the CNS. Newborn rats that are made growth hormone (GH) deficient show decreased myelination (Pelton et al., 1977). Similarly, an inherited GH deficiency in the Snell dwarf mouse is known to be responsible for the hypomyelination that is

342

distinctive of this strain. Myelination returns to normal levels upon return to normal GH levels (Noguchi et al., 1982a,b). GH also increases the contents of a myelin-specific protein, myelin basic protein (MBP), in brain cell aggregates (Alzmann et al., 1984). Perhaps lack of GH reduces the level of IGF 1 in the CNS and thus oligodendrocyte proliferation and the resulting myelination are retarded. Direct evidence for IGF 1 mediated myelination in the CNS is the induction of the myelin-specific enzyme system CNPase activity by IGF 1 in fetal brain cell cultures (McMorris et al., 1986). Myelination is also decreased in newborn rats and humans in a reduced nutritional state (Wiggins, 1982). Plasma IGF 1 levels are also reduced in malnutrition (Daughaday, 1977; Phillips and Young, 1976; Phillips and Vassilopoulou-Sellin, 1979, 1980), but it is unclear if CNS IGF 1 levels are likewise reduced or whether such a reduction in IGF 1 would result in the hypomyelination that occurs during malnutrition.

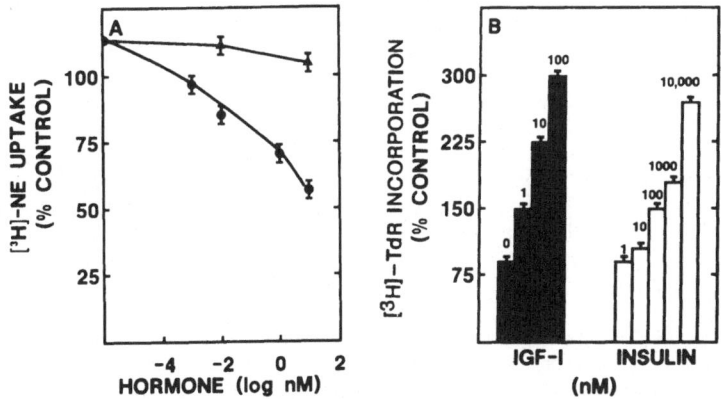

FIGURE 1. *Insulin and insulin-like growth factor (IGF 1) alterations in [^{3}H]-norepinephrine uptake and [^{3}H]-thymidine incorporation in intact primary neuronal and glial cultures from neonatal rat. **Panel a.** Dose-dependent effects of insulin and IGF 1 10-min pretreatment on specific [^{3}H]-norepinephrine uptake into cultured neurons after 5 min at 37°C (▲, IGF 1; •, insulin). **Panel b.** Incorporation of [^{3}H]-thymidine into a 10% TCA-precipitable fraction derived from serum-deprived neuronal and glial cells in culture that were incubated in the presence of either insulin or IGF 1 for 11 hr prior to addition of tracer.*

IGF 1 clearly elicits mitogenic effects in distinct nervous system cell types, but IGF 1 may also act as a neuromodulator for specific sub-populations on neurons in the CNS. Quite recently, evidence has accumulated that IGF 1 can regulate several aspects of cholinergic function, and in concert with NGF, may direct development, drive repair, and promote survival of cholinergic neurons. In cortical slices from adult rat, IGF 1 increases the potassium dependent release of [^{3}H]-acetylcholine in a dose-dependent manner, but fails to alter the basal release of the labeled neurotransmitter. IGF 2 and insulin are unable to elicit a similar increase in [^{3}H]-acetylcholine release, demonstrating that this effect is specific for IGF 1 and is mediated through the IGF 1 receptor (Nilsson et al., 1988). In mixed cell co-cultures the presence of IGF 1 increases ChAT activity (Franz Hefti, abstract, Miami Winter Symposium, Miami, FL, February, 1989). However in embryonic rat striatal neuron cultures IGF 1 actually decreases ChAT activity (B. Brass and J.N. Barrett, Abstract: Molecular and Cellular aspects of Insulin, IGFs and their Receptors: Implications for the CNS. Symposium at the University of Florida, Gainesville, FL, January, 1989). This specificity of IGF 1 on cholinergic neurons is not a unique phenomenon for growth factors. Insulin inhibits the reuptake of norepinephrine into cultured neurons and thus appears to interact specifically with catecholaminergic neurons in the brain (Boyd et al., 1985), whereas IGF 1 fails to alter reuptake (Shemer et al., 1987a). IGF 1 also increases the rate of GABA synthesis in striatal neurons (B. Brass and J.N. Barrett, abstract cited above). Further investigation will surely yield more specificity with respect to growth factor action in sub-populations in the nervous system.

INSULIN-LIKE GROWTH FACTOR RECEPTORS

Due to the variety of functions that IGF 1 and IGF 2 are able to induce in nervous tissues, it is necessary to understand the mechanism by which such diverse actions can occur. The first step in the mechanism of IGF action is interaction of the peptide with the receptor protein associated with the plasma membrane of the target cell. Biochemical studies in preparations from peripheral target organs and their cells in culture have contributed a lengthy profile of the receptor proteins for the IGFs. Structurally, the receptors for IGF 1 and IGF 2 are quite distinct. The IGF 1 receptor protein is structurally similar to the insulin receptor (Perdue, 1984; Morgan et al., 1986; Ullrich et al., 1986). It is a heterotetramer glycoprotein consisting of two extracellular associated alpha subunits (M.W. 136,000 daltons) and two transmembrane β subunits (M.W. 94,000 and 105,000 daltons) (Morgan et al., 1986; Ullrich et al., 1986; Ota et al., 1988b; Massagué and Czech, 1982). The β subunit of the IGF 1 receptor protein has intrinsic tyrosine kinase activity which is stimulated by IGF 1 binding to the alpha subunit. This kinase activity phosphorylates tyrosine residues of exogenous and endogenous proteins (Jacobs et al., 1983; Rubin et al., 1983; Zick et al., 1984). The IGF 1 receptor β subunit itself is an endogenous substrate for the tyrosine kinase activity (Jacobs et al., 1983; Rubin et al., 1983; Zick et al., 1984). The region of the β subunit which expresses the tyrosine kinase activity is highly homologous to sequences of amino acids producing the tyrosine kinase activity in receptors for epidermal growth factor (EGF) and insulin, and members of the src family of oncogene products (Ullrich et al., 1985, 1986; Ebina et al., 1985). Several studies indicate that the tyrosine kinase activity of these receptors is necessary to produce some biological responses, but others appear to occur independently of alterations in tyrosine kinase activity and are even elicited from receptors with mutated tyrosine kinase regions (Morgan and Roth, 1987; Ebina et al., 1987; Chou et al., 1987). It is thought that the tyrosine kinase activity mediates growth processes, and that hyperactivity of tyrosine kinase activity might promote tumorigenesis (Ullrich et al., 1985).

Radio-ligand binding assays have demonstrated a high degree of cross-reactivity of the IGFs and insulin at the receptor level. Thus, IGF 1 associates with the IGF 1, IGF 2, and insulin receptors, but preferentially binds to the IGF 1 receptor. IGF 2 binds with equal affinity to the IGF 2 receptor and the IGF 1 receptor, but has negligible affinity for the insulin receptor. Insulin associates predominately with the insulin receptor, but has approximately 100 times less affinity for the IGF 1 receptor and little affinity for the IGF 2 receptor protein. Competition binding with IGF 1, IGF 2, and insulin enables the identification of the different receptors (Morgan et al., 1986). Cross-reactivity in binding to the receptors suggests that perhaps the receptor protein determines function rather than the peptide which interacts with the binding subunit of the receptor.

The IGF 2 receptor protein is a monomer of M.W. 260-300 kilodaltons, and like IGF 1 and insulin receptors, is a glycoprotein (Morgan et al., 1986; Massagué and Czech, 1982; Rechler and Nissley, 1985). Unlike the IGF 1 and insulin receptors, the IGF 2 receptor does not have demonstrated tyrosine kinase activity (Massagué and Czech, 1982). Recent sequence analysis of the IGF 2 receptor confirms the lack of an integral tyrosine kinase activity and shows little homology in any region with either the IGF 1 or insulin receptors. Surprisingly, however, the IGF 2 receptor in rat has greater than 80% homology with the mannose-6-phosphate (M6P) receptor protein (Morgan et al., 1987; Roth et al., 1987; Roth, 1988; McDonald et al., 1988). The M6P receptor is hypothesized to target lysosomal enzymes to the lysosomal compartment following their production in the ER and processing within the golgi apparatus (Sly and Fischer, 1982). Recent evidence suggests that the IGF 2/M6P receptor protein contains two binding domains for lysosomal enzyme association and a single IGF 2 binding domain (Tang et al., 1988; Waheed et al., 1988). The functional consequences of such a protein are not understood at this time, but speculation is that the IGF 2/ M6P receptor may act at the plasma membrane to reduce the levels of circulating IGFs, in addition to targeting enzyme to intracellular compartments. The validity of this suggestion is diminished by demonstrated IGF 2 receptor-mediated responses to IGF 2.

INSULIN-LIKE GROWTH FACTOR RECEPTORS IN THE NERVOUS SYSTEM

Radio-ligand binding in membranes from whole brain homogenates has established the existence of IGF 1 and IGF 2 receptors in the central nervous system (Sara et al., 1983; Gammeltoft et al., 1985; Goodyer et al., 1984; Bassas et al., 1985). Likewise the presence of these receptors in the pituitary (Goodyer et al., 1984), isolated brain microvessels (Frank et al., 1986; Duffy et al., 1988), and retina (Zick et al., 1987) has been shown. Binding assays in membranes from various regions and quantitative autoradiography with labeled IGF 1 and IGF 2 localize IGF 1 and IGF 2 receptors to distinct regions of the brain and pituitary (Gammeltoft et al., 1985; Goodyer et al., 1984; Bassas et al., 1985; Bohannon et al., 1986, 1988a,b; Rosenfeld et al., 1987; Hill et al., 1987; Mendelsohn, 1987; Rosenfeld and Hoffman, 1987; Duffy et al., 1988; Valentino et al., 1988; Strum et al., 1989). IGF 1 receptors found in nervous tissue are distinct from the insulin receptor as demonstrated by inhibition of labeled ligand binding by unlabeled peptide using IGF 1, IGF 2, and insulin in both labeled and unlabeled form, and different autoradiographic patterns for these peptide receptors (Gammeltoft et al., 1985; Goodyer et al., 1984; Bassas et al., 1985; Frank et al., 1986; Zick et al., 1987; Rosenfeld et al., 1987). Such analysis of binding characteristics, however, fails to identify the nervous cell types which express these receptors. IGF 1 receptors are present in both neuronal cells and glial cells (Burgess et al., 1987; Shemer et al., 1987a; Ballotti et al., 1987). Membranes prepared from primary cultures of fetal neurons and astroglia that are 85% and 98% pure, respectively, as determined by immunohistochemical staining, contain specific receptors for both IGF 1 and insulin (Shemer et al., 1987a). Binding kinetics are similar in neurons and glia, and insulin is 50-100 times less potent than IGF 1 in displacing $[^{125}I]$-IGF 1 binding in membrane preparations of both cell types. Similar binding activity for IGF 1 can be isolated from WGA-purified membrane fractions; thus, IGF 1 binding is to a glycoprotein (Shemer et al., 1987a). In addition, specific $[^{125}I]$-IGF 1 and $[^{125}I]$-IGF 2 receptors can be demonstrated in transformed neural cell lines (Ota et al., 1988a,b; Strum et al., 1989). Due to cross-reactivity of these peptides at their receptors, it is apparent that the biological characteristics of the IGFs and insulin in the nervous system often overlap. However, as a further complication of correlating binding and response to these factors, both IGF 1 and insulin can elicit identical responses to binding to their appropriate receptors in HEP-G2 cells (Verspohl et al., 1988).

Immunohistological studies indicate that IGF 2 receptors in the CNS are found in all lobes of the pituitary, in the choroid plexus, and in vascular and ependymal structures (Strum et al., 1989). In addition, immunostaining is present around nerve sheaths and axon bundles (Strum et al., 1989). Specific immunostaining is also present in cultured adult hypothalamic neurons (Ocrant et al., 1988) and in cultured fetal neuronal and glial cells from rat (Figure 2). Immunostaining is of greatest intensity in the perinuclear region in both neurons and glia, but a diffuse stain occurs throughout the cell. The latter pattern may represent IGF-2 receptors associated with plasma membranes. PAB-II perinuclear immunostaining in glia co-localizes with endosomal and golgi compartment marker (not shown). A similar staining pattern occurs in neuroblastoma cell lines (Strum et al., 1989). Both immunological and radio-ligand binding data suggest a greater number of IGF 1 and IGF 2 receptors in fetal tissue than in adult tissue (Bassas et al., 1985; Ocrant et al., 1988). IGF 1 receptor levels were greater in the brain than in liver, while IGF 2 receptor levels in both brain and liver were reduced to the same extent (Ocrant et al., 1988).

Affinity cross-linking of labeled ligands to their receptors in coordination with antibodies which recognize the IGF 1 and IGF 2 receptor protein have enabled the identification and partial purification of receptors for these peptides in the central nervous system. Chemical cross-linking of radio-labeled IGF 1 and IGF 2 enables the determination of the molecular weight of the binding subunit of each receptor when labeled products are separated on reducing SDS-PAGE gels. In contrast to binding kinetics, which suggest that both the peripheral and CNS receptors for IGF 1 are similar, cross-linking studies indicate that the binding subunit of the receptor protein for IGF 1 in the CNS is structurally unique as compared to receptor protein isolated in peripheral tissues (Burgess et al., 1987; Shemer et al., 1987a; Ballotti et al., 1987; Gammeltoft et al., 1985; Heidenreich et al., 1986; McElduff et al., 1988). The IGF 1 receptor alpha subunit found in whole brain homogenate membranes from adult rat is of lower molecular weight than the alpha subunit which has been isolated from liver, placenta, and adipose membranes, differing by approximately 10 kDa (Gammeltoft et al., 1985; McElduff et al., 1988). Oddly, while fetal membranes prepared from primary neuronal cultures seem to express this smaller molecular weight version of the IGF 1 receptor, astrocytes in culture appear to express only the peripheral form of the receptor protein (Burgess et al., 1987; Shemer et al., 1987a; Ballotti et al., 1987). The lower molecular weight form of the receptor is also found in neuroblastoma cells (Ota et al.,

1988a,b); however, a peripheral form is isolated from retinal rod outer segments (Zick et al., 1987). A smaller receptor form of the insulin receptor is also found in fetal neuronal cultures from rat (Lowe et al., 1986). In both cases the difference in the molecular weight of the peripheral/glial and neuronal forms of the receptor appears to result from different glycosylation patterns (Ota et al., 1988b; Heidenreich et al., 1986; Lowe et al., 1986). The molecular weight of the β subunit of the IGF 1 receptor isolated by IGF 1 specific antibodies from brain membrane is identical to that of the peripheral receptor (as discussed below) (Burgess et al., 1987; Shemer et al., 1987a).

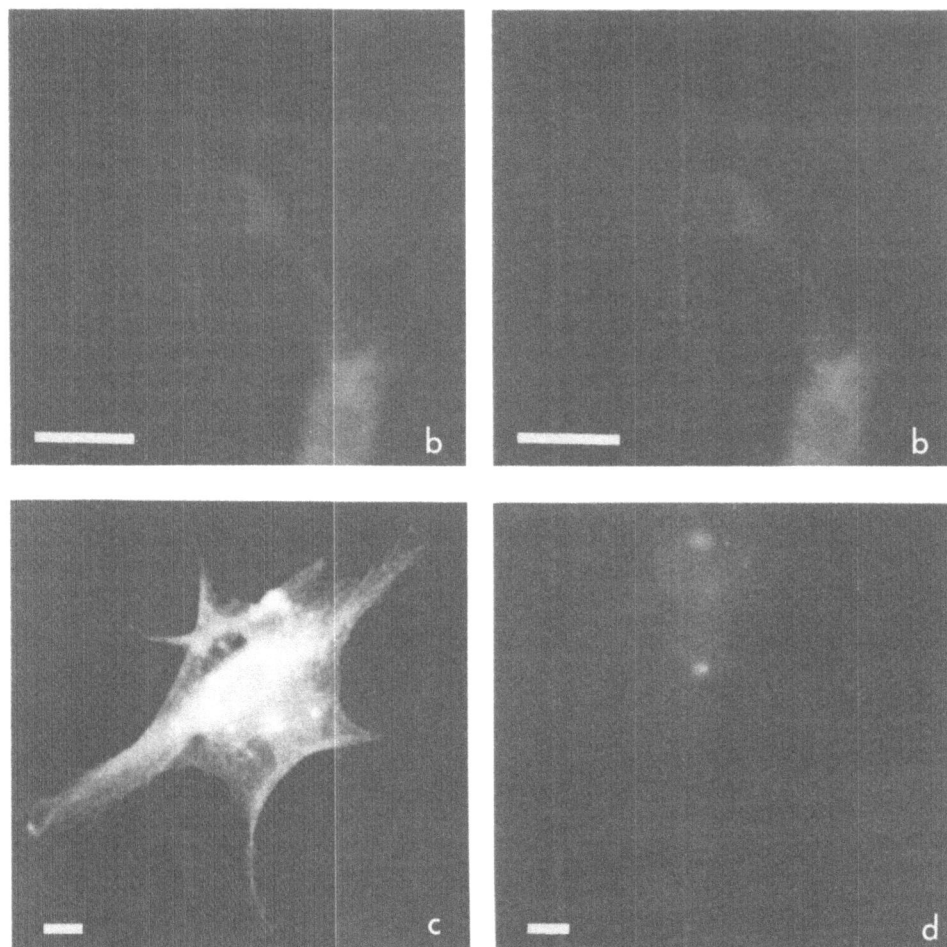

FIGURE 2. *Cellular localization of insulin-like growth factor 2 (IGF-2) receptor protein in neonatal rat neuronal and glial cells in culture, using PAB-II, a polyclonal antibody to the IGF-2 receptor. Neuronal and glial cultures were prepared as previously reported (Boyd et al., 1985), media was removed, cells on culture plates washed (3× in PBS pH 7.4), and were fixed in 4% paraformaldehyde/10% picric acid in PBS at 4°C for 30 minutes. Cells were washed again, made permeable with 0.1% Triton X-100 in PBS, and then incubated overnight at 4°C in 1:1000 dilution of ammonium persulfate purified PAB II in PBS pH 7.4 with 1% BSA and 0.1% NaN_3 or a 1:1000 of normal rabbit serum (NRS). Rhodamine-conjugated goat-anti-rabbit whole IgG (1:200 dilution, 30 min) was used with fluorescence-microscopy to detect primary antibody binding to cells in culture. Panel a. Neurons, PAB-II. Panel b. Neurons, NRS. Panel c. Glia, PAB-II. Panel d. Glia, NRS. Bar represents 20 μm. (PAB-II anti-IGF 2 receptor antibody was a gift from Dr. Ron Rosenfeld, Stanford University.)*

The functional significance of this smaller receptor form is not known, but for both the IGF 1 and insulin receptors, it appears that differential glycosylation of the receptor does not alter ligand binding or stimulated tyrosine kinase activity (discussed below). It is difficult to explain the presence of only the smaller molecular weight form of the IGF 1 receptor in whole brain homogenates, as both neurons and glia are present in this preparation. Further, this finding is strange when one considers the relative contents of glia and neurons in the adult brain; glia account for greater than two-thirds of the cells in the adult brain. Perhaps this dichotomy can be explained by differences in the size of the fetal and adult glial IGF 1 receptors, with a shift to the smaller form of the glial receptor occurring during development. Receptor ontogeny studies indicate the switch from the larger molecular weight form to a smaller form in embryonic development (Bassas et al., 1985). Another possibility is that a smaller molecular weight form would be found in oligodendroglia, thus tipping the balance of cell content to the smaller form of the IGF 1 receptor in whole brain membranes. Recent evidence suggests that smaller receptor forms may be an artifact of tissue preparation. When [^{125}I]-IGF 1 is crosslinked to its receptor in intact fetal rat neuron monolayers, only the peripheral form of the IGF 1 receptor is isolated on SDS-PAGE gels (Ocrant et al., 1988). Why preparation of glial membranes does not produce a similar change in the receptor has yet to be explained.

Only recently has information concerning the structure of the IGF 2 receptor begun to accumulate. Affinity-labeling studies as done with the IGF 1 receptor using [^{125}I]-IGF 2 have enabled the isolation on SDS-PAGE gels of the IGF 2 receptor protein from peripheral and brain membranes (Gammeltoft et al., 1985). Reported molecular weights for this receptor are 260 kDa in adult rat liver (Rechler and Nissley, 1985), while the adult rat brain IGF 2 receptor is approximately 250 kDa in membrane preparations (Gammeltoft et al., 1985; McElduff et al., 1987). A similar study found the lower molecular weight form of the IGF 2 receptor present only in adult brain as compared to fetal brain, fetal liver, or adult liver membranes (Ocrant et al., 1988). Affinity-labeling in intact fetal neuronal and glial cultures demonstrates the presence of a 250 kDa receptor in both neurons and glia (Ocrant et al., 1988). A 240 kDa form of the receptor is present in neuroblastoma cell lines (Strum et al., 1989). Decreased size of the receptor on SDS-PAGE gels appears to result from differences in glycosylation of the receptor protein (McElduff et al., 1987). The significance of the lower molecular weight form of the IGF 2 receptor in adult brain is as yet unestablished.

IGF-ASSOCIATED TYROSINE KINASE ACTIVITY AND PHOSPHORYLATION IN THE NERVOUS SYSTEM

IGFs elicit distinct responses in cultured neurons and glia, whereas the ability to bind IGF to its receptor is similar in both cell types (Shemer et al., 1987a). Structural differences appear to be due to differences in the carbohydrate moiety of the extracellular domain, but there is some suggestion that this alteration may occur as an artifact of the tissue preparation (see above). Whether receptor structural dissimilarities are responsible for biological differences in these cell types remains to be seen. However, it is difficult to reason a means by which alterations in glycosylation could alter function, if not by influences on ligand binding. It is therefore likely that distinct responses in neurons and glia are a consequence of differences in the IGF 1 β subunit and its inherent tyrosine kinase activity or in an intracellular post-receptor effector system.

As stated above the β subunit of the IGF 1 receptor exhibits tyrosine kinase activity which is stimulated by IGF binding to the alpha subunit of the receptor molecule (Jacobs et al., 1983; Rubin et al., 1983; Zick et al., 1984). The β subunit itself is a substrate for this kinase activity and IGF 1 can be used to label the β subunit with γ-[^{32}P]-ATP (autophosphorylation). Immunopurification using antibodies against the IGF 1 receptor results in the isolation of a single band from autoradiograms of SDS-PAGE gels. Using this strategy, it has been possible to isolate the IGF 1 receptor β subunit from the brain and study its tyrosine kinase activity. An IGF 1 receptor β subunit with M_r of approximately 94 kDa has been demonstrated in solubilized, lectin-purified brain cortex homogenates from human by immunoprecipitation with the IGF 1 receptor monoclonal antibody alpha-IR3 (Gammeltoft et al., 1985). IGF 1 at a concentration of 0.1 micromol/L markedly stimulates the phosphorylation of the immunoprecipitable protein band when analyzed by gel electrophoresis and autoradiography (Gammeltoft et al., 1985).

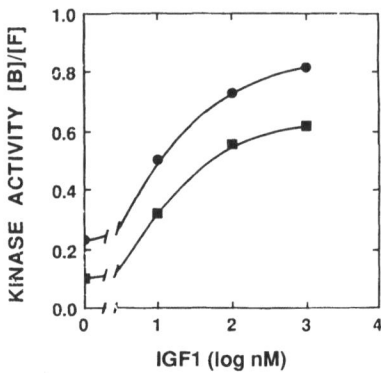

FIGURE 3. Dose-dependent IGF-1-stimulated autophosphorylation of IGF 1 receptor β subunit in WGA-purified membranes prepared from primary neuronal and glial cultures from neonatal rat brain. IGF 1 receptor autophosphorylation was performed as previously described for the insulin receptor (Lowe et al., 1986). Briefly, 20 μl of WGA-purified receptor preparation were incubated (30 min at 24°C) in a total volume of 70 μl of buffer (50 mM Hepes, 100 mM NaCl, 0.06% Triton X-100, 0.01% BSA, pH 7.6) in the presence or absence of IGF 1 (10^{-9} - 10^{-7} M). Phosphorylation was initiated by adding γ-[^{32}P]ATP (5 μM, specific activity ~60 μCi/nM), 1 mM CTP, and 3 mM manganese acetate, pH 7.6. The reaction was terminated after 10 min by adding 50 μl of SDS-PAGE (2.5 x) sample buffer to each 75 μl of assay samples. The samples were heated for 10 min at 95°C, and aliquots (55 μl) were analyzed by SDS gel electrophoresis and autoradiography. Following a 30-min incubation with IGF 1 at varying concentrations, autophosphorylation of the β subunit was initiated by the addition of reaction mix containing γ-[^{32}P]ATP. Aliquots were applied to SDS-PAGE and autoradiographed, and IGF 1 dose response was performed using concentrations of IGF 1 from 10^{-9} to 10^{-7} M (■, neurons; •, glia). The β-subunit bands were excised from duplicate gels and counted by liquid scintillation. The results are plotted as a mean of the duplicate estimations and represent the stimulated phosphorylation minus basal at each dose of IGF 1.

Autophosphorylation of the β subunit in response to IGF 1 can also be demonstrated in vitro in solubilized, WGA-purified membrane extracts prepared from fetal neurons and astroglia in primary culture (Shemer et al., 1987a). IGF 1 stimulates a dose-dependent increase in phosphorylation of a purified extract protein band from neurons with M_r 91 kDa after a 30 minute incubation with IGF 1 followed by incubation with γ-[^{32}P]-ATP. This stimulation also occurs in glial extracts, isolating a labeled band which with M_r 95 kDa. Maximal stimulation of [^{32}P]-incorporation is obtained in the presence of 100 nM IGF 1 in both neurons and glia. Insulin is much less potent in stimulating autophosphorylation of the IGF 1 receptor β subunit in both the neuronal and glial preparations, and co-incubation of extracts with 1 nM IGF and 1 nM insulin elicits a greater stimulation of β subunit phosphorylation than does 1 nM insulin alone (Shemer et al., 1987a). This is an additive effect in that stimulation is proportional to IGF 1's and insulin's individual effects. In addition, this study demonstrates a higher basal tyrosine kinase activity in glia than in neurons, and greater phosphorylation at maximal stimulation (Figure 3). The functional significance of this is unclear at the present time. Phosphoamino acid analysis of IGF 1-stimulated phosphorylation of the IGF 1 receptor β subunit from both neuronal and glial cultures reveals that IGF 1 stimulates [^{32}P]-incorporation primarily into tyrosine residues. In the basal state phosphoserine is the predominant phosphorylated amino acid, however neither the phosphoserine nor the phosphothreonine content of the extracts is stimulated in the presence of IGF 1 (Shemer et al., 1987a). Similar studies using astrocyte glial cultures confirm that IGF 1 induces autophosphorylation of the glial IGF 1 β subunit and acts specifically to increase phosphotyrosine content (Ballotti et al., 1987).

Receptor-specific antibodies against the IGF 1 receptor and the insulin receptor can be used to characterize the specificity of autophosphorylation of these receptors. In extracts from neuroblastoma cells maintained in culture, immunoprecipitation with the antibody alpha-IR3 isolates a M_r 96 kDa protein band on SDS-PAGE gels which is preferentially phosphorylated by IGF 1. Insulin stimulates phosphorylation of this band, but is 100 times less potent than IGF 1. When these same extracts are immunoprecipitated with the insulin-receptor-specific antibody, B 10, insulin is more potent than IGF 1 in stimulating the phosphorylation of a M_r 91 kDa protein band (Ota et al., 1988a). Thus, in neuroblastoma cultures, IGF 1 and insulin receptors are present and their stimulus for autophosphorylation is specific. Stimulation of the phosphorylation of the β subunit of the IGF 1 receptor can be demonstrated in intact cell cultures (Shemer et al., 1989). Incubation of primary neuronal and glial cultures from neonatal rat brains with [^{32}P]-orthophosphate and IGF 1 stimulates a time- and dose-dependent increase in the phosphorylation of 95 kDa phosphoprotein when immunoprecipitated by an anti-phosphotyrosine antibody. Insulin also stimulates phosphorylation of this phosphoprotein, or a phosphoprotein with similar molecular weight. IGF 1 however, stimulates the phosphorylation of a 102 kDa phosphoprotein which is not immunoprecipitated by phosphotyrosine antibodies in cultures exposed to insulin. The identity of this 102 kDa phosphoprotein has yet to be determined, but it is reasonable to conclude that it is a variant form of the IGF 1 receptor β subunit. Interestingly, stimulation of autophosphorylation in neuronal cultures occurs more rapidly than it does in glial cultures (Shemer et al., 1989). Whether this difference in the rate of kinase stimulation in neurons and glia has physiological significance is not known. Perhaps a faster rate of stimulation of the tyrosine kinase in neurons is suited to the more rapid responses to stimuli that are characteristic of excitable cells.

Activation of the IGF 1 receptor tyrosine kinase activity results not only in the autophosphorylation of the receptor β subunit, but also induces the phosphorylation of other tyrosine containing substances. Solubilized, WGA-purified rat brain homogenates demonstrate insulin stimulated tyrosine kinase activity by increased incorporation of ^{32}P into several artificial tyrosine-rich substrates. Tyrosine kinase activity in purified brain homogenates is approximately double that which is demonstrated in similarly prepared liver homogenates after standardization for receptor number. In addition, stimulation of brain kinase activity is more sensitive to insulin than is the stimulation of the liver kinase activity (Lowe et al., 1986). IGF 1-stimulated tyrosine kinase activity in the brain can also be demonstrated by immunodepleting purified brain homogenates with B-10 IgG. B-10 IgG recognizes the insulin receptor β subunit but fails to recognize the IGF 1 receptor β subunit. In immunodepleted samples, insulin-stimulated phosphorylation of exogenous, artificial substrate is markedly reduced while there is no reduction of IGF 1 induced phosphorylation of exogenous substrate. B-10 IgG decreases binding of [^{125}I]-insulin by 95% in these purified homogenates, so that the remaining insulin-stimulated, phosphorylated substrate after immunodepletion with B-10 IgG must represent cross-activation of the IGF 1 receptor by insulin (Lowe and LeRoith et al., 1986). Using immunodepletion of the insulin receptor with B-10, IGF 1 stimulated phosphorylation of exogenous substrates can be demonstrated in soluble extracts from neuronal and glial cultures from neonatal rat brains (Shemer et al., 1987a). IGF 1-induced phosphorylation of these exogenous substrates is dose-dependent and the immunodepletion with B-10 IgG reduces kinase activity by roughly 30% (Figure 4). Thus, it appears that the majority of the insulin-stimulated kinase activity in the brain is mediated through the IGF 1 receptor and not the insulin receptor. This is confirmed in retinal outer-segment membranes (Zick et al., 1987). In liver, on the other hand, insulin-stimulation of kinase activity occurs primarily as a result of insulin acting through the insulin receptor (Lowe and LeRoith et al., 1986).

IGF AND GROWTH FACTOR INDUCED PHOSPHORYLATION OF CELLULAR PROTEINS

IGF 1 induces the phosphorylation of several endogenous proteins in various tissues (Kadowaki et al., 1987; Shemer et al., 1987b; Madoff et al., 1988). Phosphorylation of these different phosphoproteins is tissue-specific in some cases, and may mediate tissue-specific responses to IGF 1 (Madoff et al., 1988). Likewise, common phosphoproteins in different tissues may direct the generalized proliferative responses that appear common to most IGF 1 responsive tissues (Kadowaki

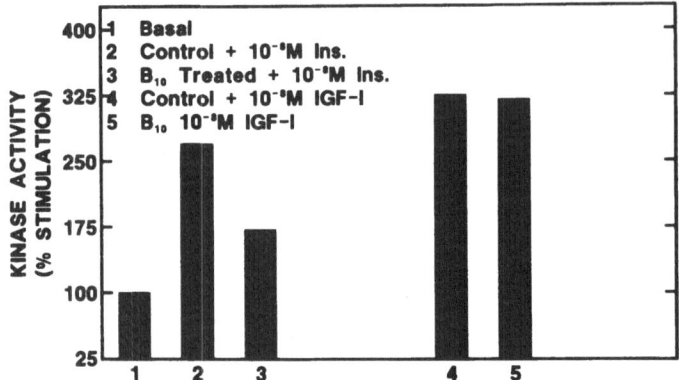

Figure 4. IGF-1-induced phosphorylation of the exogenous substrate poly(Glu,Tyr) 4:1. WGA-purified membranes of neuronal and glial cells were incubated with IGF 1 and insulin at varying concentrations, and the ^{32}P incorporation into the exogenous substrate poly(Glu,Tyr) 4:1 was measured. To exclude the contribution of the insulin receptor to these results, the insulin receptors were immunodepleted using anti-insulin receptor antiserum (B_{10}) prior to the phosphorylation experiment. Phosphorylation of artificial substrate by WGA-purified receptor was performed as previously described (Lowe et al., 1986). Briefly, 40 µl of WGA-purified receptors were incubated (30 min at 22-24°C) in a total volume of 140 µl of buffer (50 mM Hepes, 100 mM NaCl, 0.04% Triton X-100, 0.01% BSA, pH 7.6) with the artificial substrate poly(Glu,Tyr) 4:1 (400 µg) in the presence or absence of various concentrations of IGF 1 (10^{-10} - 10^{-6} M). Phosphorylation was initiated by addition of γ-[^{32}P]ATP (50 µM final concentration, specific activity ~3 µCi/nM), 1 mM CTP, and 20 mM $MgCl_2$. Aliquots (75 µl) were removed at various time intervals and spotted onto squares of Whatman No. 3 MM filter papers. The filters were treated with 10% trichloroacetic acid solution and 10 mM sodium pyrophosphate and were washed extensively. Filter-bound radioactivity was determined by counting in a liquid scintillation counter.

et al., 1987; Shemer et al., 1987b). The role which changes in phosphoprotein content has in intracellular processing of IGF receptor binding is clouded by the isolation of tissue-specific and common IGF 1-sensitive phosphoproteins from different cell types. In many cases insulin, either more or less potently, stimulates the phosphorylation of similar phosphoproteins in the same cell types as IGF 1 does (Kadowaki et al., 1987; Shemer et al., 1987b; Madoff et al., 1988; White et al., 1985, 1987; Machicao et al., 1987; Izumi et al., 1987; Gibbs et al., 1986; Beguinot and Smith, 1987). Phosphorylation of a 185 kDa phosphoprotein is stimulated by insulin or IGF 1 in an insulin-sensitive hepatoma cell line (White et al., 1985, 1987), human placental membrane preparations (Machicao et al., 1987), NRK cells (Izumi et al., 1987), and in intact human epidermoid carcinoma KB cells (Kadowaki et al., 1987). In KB cells EGF fails to stimulate the phosphorylation of pp185, but does stimulate the phosphorylation of a 190 kDa phosphoprotein; insulin, IGF 1 and EGF all stimulate the phosphorylation of a 240 kDa phosphoprotein (pp240) (Kadowaki et al., 1987). Insulin-induced phosphorylation of pp185 in peripheral preparations exhibits a similar dose-response curve to that for receptor autophosphorylation, suggesting that the phosphorylation of pp185 is closely associated with the receptor tyrosine kinase activation in response to insulin (Kadowaki et al., 1987). Interestingly, both IGF 1 and insulin in intact 3T3-L1 adipocyte cultures stimulate incorporation of ^{32}P into a 160 kDa protein (pp160) (Madoff et al., 1988). This phosphoprotein is also stimulated in BC3H-1 myocytes in response to these growth factors (Madoff et al., 1988). In these studies in both cell types no pp185 is labeled in response to either IGF 1 or insulin; however, in similar studies in these cell types, extraction isolates phosphorylated tyrosine-proteins of M_r 180 and 175 kDa, respectively (Gibbs et al.,

1986; Beguinot and Smith, 1987). Discontinuities in these studies are, as of yet, unexplained. Insulin also stimulates the phosphorylation of a membrane associated 46 kDa protein in 3T3-L1 adipocytes (Häring et al., 1987). It is difficult to speculate what function this phosphorylation might serve.

Three recent studies using nervous tissues indicate that IGF 1 stimulates the phosphorylation of endogenous substrates in the nervous system (Zick et al., 1987; Shemer et al., 1987b; Pierre et al., 1986). In intact mouse neuroblastoma N18 cells, which possess specific, high-affinity receptors for insulin and IGF 1, incubation with IGF 1 and insulin stimulates the phosphorylation of a 185 kDa phosphoprotein (pp185) in addition to the autophosphorylation of their respective receptors' β subunits (Shemer et al., 1987b). This phosphorylation is time and dose-dependent and labels both serine and tyrosine residues. pp185 is a cytosolic protein and stimulation of its phosphorylation is not induced by other growth factors (Shemer et al., 1987b). None of these growth factors induces phosphorylation of pp240 in neuroblastoma within the levels of detection. IGF 1- and insulin-induced phosphorylation of pp185 in the nervous system is also closely linked to the activation of the receptor kinase, since phosphorylation of pp185 is stimulated within 20 sec. after addition of insulin or IGF 1 (Shemer et al., 1987b). It is not known if this phosphoprotein is phosphorylated in response to IGF 1 and insulin in glial cells, but perhaps pp185 is a mediator of basic proliferative responses in IGF 1-responsive tissues. The identity and biochemical characteristics of pp185 are not known but their elucidation should clarify pp185's role in IGF 1 responses in these tissues.

GDP-bound transducin, an important mediator of phototransduction in the retina purified from retinal outer segment membranes, acts as a substrate for insulin- and IGF 1-stimulated tyrosine kinase activity present in liver homogenates (Zick et al., 1986). Recent studies using lectin-purified IGF 1 binding activity isolated from retinal outer segment partially purified membranes demonstrate that IGF 1 stimulates the incorporation of ^{32}P from γ-$[^{32}P]$-ATP into purified GDP-bound transducin added to receptor fractions. Insulin receptors in this preparation are not apparent by affinity labeling, and immunodepletion with the insulin receptor-specific antibody B-10 does not reduce phosphorylation of the GDP-transducin preparation (Zick et al., 1987). Therefore, it is likely that this stimulated phosphorylation of GDP-bound transducin occurs through the IGF 1 receptor. IGF 1-stimulation of GDP-transducin phosphorylation may represent all or a portion of a second messenger system for IGF 1 in the retina, and in contrast to the phosphorylation of pp185 in neuroblastoma and peripheral tissues, may define IGF responses specific to retinal function. It is not known whether phosphorylation of pp185 is stimulated by IGF 1 in the intact retina or in retinal cell cultures.

Cultured glial cells from cerebral hemispheres of the neonatal rat, when exposed to either IGF 1 or insulin, contain stimulated protein kinase activity which phosphorylates ribosomal protein S6 (Pierre et al., 1986). Pretreatment with cycloheximide does not inhibit S6 phosphorylation, so it appears stimulation is promoted through an existing factor in the cytosol. Ribosomal protein S6 acts as a regulatory factor for protein synthesis in cells. Although the identity of this cytosolic protein kinase is not known, stimulated phosphorylation of S6 in the presence of TPA (Pierre et al., 1986), a phorbol ester which binds to and activates protein kinase C in this assay system, suggests that protein kinase C may mediate IGF 1- and TPA-stimulated S6 phosphorylation. A role for extracellular electrolyte mediation of this process has been proposed (Novak-Hofer et al., 1988). Other growth factors which act through receptor tyrosine kinases and oncogenes stimulate phosphorylation of S6 in other cell types. Thus, IGF 1- and insulin-induced proliferative responses in the nervous system may involve the activation of protein kinase C and phosphorylation of S6, and perhaps other, as yet unknown, regulatory phosphoproteins.

PHOSPHORYLATION AND BIOLOGICAL RESPONSES TO THE IGFs

To what extent activation of tyrosine kinase activity or phosphoprotein induction is important for the translation of IGF binding into a biological response is not known. It is difficult to correlate the activation of kinase activity to physiological responses to IGF 1 in various cell types, due to the incompatibility of biochemical and physiological methods. Studies in which antibodies to the tyrosine kinase region of the insulin receptor are injected into intact cells suggests that tyrosine kinase activity may not be necessary for transduction of insulin's signal into all its biological responses (Morgan and Roth, 1987). Companion studies in which the tyrosine kinase portion of the insulin receptor was mutated clearly demonstrate that some function survives (Ebina et al., 1987; Chou et al., 1987). Although the IGF 1 receptor's inherent kinase activity may not be the mediator of all responses to IGF 1, the receptor kinase activity is an important part of the mechanism in eliciting many IGF 1-specific actions (Morgan and Roth, 1987; Ebina et al., 1987).

As noted previously, IGF receptors stimulate phosphorylation of specific phosphoproteins on tyrosine residues in a variety of cell types. The importance of phosphoprotein content in the response to growth factors is also suggested by responses to tumor promoting phorbol esters, such as TPA. Many of the biological responses to IGF 1 and insulin can be mimicked by incubation of intact cells with phorbol esters (Mudd and Raizada, 1989). Phorbol esters act to produce their effects by the activation of protein kinase C, which possesses high affinity receptor sites for the phorbol esters (Taylor et al., 1988). Both IGF 1 and phorbol ester 12,13-dibutyrate (PDBU) stimulate the incorporation of sulfate and thymidine into cultured chondrocytes (Novak-Hofer et al., 1988). Both IGF 1 and PDBU stimulation of these processes can be inhibited by specific PKC inhibitors (Novak-Hofer et al., 1988). This suggests that both IGF 1 and phorbol esters act through activation of protein kinase C, and increase phosphorylation of cell protein. Independent investigation in BC3H-1 myocytes also supports a mechanism by which IGF 1 acts through PKC. IGF 1, insulin and EGF all stimulate PKC activity and cause increases in diacylglycerol levels in a dose-dependent manner. These effects are non-additive and independent of phospholipid hydrolysis (M. Standaert, D. Cooper, G. Nair and R. Farese, Abstract: Molecular and Cellular Aspects of Insulin, IGFs and their Receptors: Implications for the CNS. Symposium at the University of Florida, Gainesville, FL, January, 1989). It is possible that tyrosine phosphorylation of the PKC molecule by the IGF 1 receptor could result in the activation of PKC, since the receptors for IGF 1, insulin, and EGF all have demonstrated tyrosine kinase activity. A more indirect route of PKC regulation may be involved in this activation of PKC by IGF 1. Further studies on the role of IGF 1 on protein kinase C are necessary to answer such questions, however IGF 1 does not appear to influence other characterized second messenger systems.

Although no direct evidence for IGF 2-induced phosphorylation of membrane-associated or cytosolic proteins exists, several studies suggest that phosphorylation is involved in a cellular mechanism by which IGF 2 acts. NGF stimulates neurite outgrowth in pheochromocytoma PC12 cells (Ishii and Recio-Pinto, 1987). An important event in the growth and stabilization of neurites is the formation of microtubule bundles that course the length of neurites (Black et al., 1986). Long-term exposure to NGF causes the stabilization of microtubule-associated proteins (MAPs), and stabilization of these MAPs is due to a post-translational phosphorylation (Black et al., 1986). NGF also increases the steady-state levels of certain MAPs (Aletta et al., 1988). It seems likely that these changes in MAP phosphorylation and stability and the stimulated increase in tubulin content in NGF-treated PC12 cells (Ishii and Recio-Pinto, 1987) act in concert to increase and stabilize microtubules in extending neurites. As noted earlier, IGF 2 also increases tubulin content and neurite outgrowth in PC12 cells (Ishii and Recio-Pinto, 1987). Perhaps by a similar mechanism, IGF 2 increases the stability of MAPs and therefore neurite outgrowth in cells of neural origin. Brain MAP can be phosphorylated in vitro by an insulin-stimulated protein kinase isolated from 3T3-L1 adipocyte cultures. This stimulated phosphorylation of protein kinase is specific for tyrosine residues (Ray and Sturgill, 1987). Brain membranes also contain a protein kinase activity which is capable of phosphorylating MAPs (H. Kim, S. Kim, D. Pillion, Abstract: Molecular and Cellular Aspects of Insulin, IGFs and their Receptors: Implications for the CNS. Symposium at the University of Florida, Gainesville, FL, January, 1989). Insulin is a less potent stimulus for neurite outgrowth than IGF 2 in PC12 cells (Ishii and Recio-Pinto, 1987), and thus the possibility that IGF 2 stabilizes MAPs through phosphorylation is reasonable. Further investigation is necessary to determine if indeed IGF 2 and NGF-stimulated neurite outgrowth share a common cytosolic pathway.

While a role for cellular phosphorylation in response to IGF 2 is a matter of speculation at this time, clearly IGF 1 receptor tyrosine kinase stimulates the phosphorylation of cytoplasmic elements in cells from the nervous system. In a mechanism analogous to PKC mediation of hormonal responses in peripheral tissues, it is possible that these IGF 1-sensitive cytoplasmic proteins are an early event in a biochemical cascade that transmits signal to function for IGF 1. In turn, it is possible that several steps may precede the phosphorylation of these identified phosphoproteins, as supported by phosphorylation of S6 in glial cultures. Whether pp185, GDP-transducin, S6, or a yet to be identified mediator fits into such a mechanism, and whether its phosphorylation is necessary to permit the diverse responses to IGF 1 in the nervous system remains to be determined.

REFERENCES

Adamo, M., Werner, H., Farnsworth, W., Roberts, C.T., Raizada, M.K, LeRoith, D. (1988). Dexamethasone reduced steady-state insulin-like growth factor messenger ribonucleic acid levels in rat neuronal and glial cells in culture. Endocrinology 123:2526.

Aizenman, Y., de Vellis, J. (1987). Brain neurons develop in a serum and glial free environment: Effects of transferrin, insulin, insulin-like growth factor 1 and thyroid hormone on neuronal survival, growth and differentiation. Brain Res. 406:32.

Aletta, J.M., Lewis, S.A., Cowan, N.J., Greene, L.A. (1988). Nerve growth factor regulates both the phosphorylation and steady-state levels of microtubule-associated protein 1.2 (MAP 1.2). J. Cell Biol. 106:1573.

Alzmann, G., Honegger, P., Guntert, B., Matthieu, J.M. (1984). Brain cell aggregates in culture: Effect of T_3 and BGH (bovine growth hormone) on myelination. Trans. Am. Soc. Neurochem. 15:258 (abstr. #354).

Ballotti, R., Nielsen, F.C., Pringle, N., Kowalski, A., Ricardson, W.D., Van Obberghen, E., Gammeltoft, S. (1987). Insulin-like growth factor 1 in cultured astrocytes: Expression of the gene, and receptor tyrosine kinase. EMBO J. 6:3633.

Baskin, D.G., Woods, S.C., West, D.B., van Houten, M., Posner, B.I., Dorsa, D.M., Porte, Jr., D. (1983). Immunocytochemical detection of insulin in rat hypothalamus and its possible uptake from cerebrospinal fluid. Endocrinology 113:1818.

Bassas, L., De Pablo, F., Lesniak, M., Roth, J. (1985). Ontogeny of receptors for insulin-like peptides in chick embryo tissues: Early dominance of insulin-like growth factor over insulin receptors in brain. Endocrinology 117:2321.

Beguinot, F., Smith, R.J. (1987). Insulin and IGFI stimulate phosphorylation of a differentiation-dependent 175,000 Mr protein in L6 skeletal muscle cells. Diabetes 36 (Suppl. 1):78A (abstr. #312).

Berelowitz, M., Szabo, M., Frohman, L., Firestone, S., Chu, L., Hintz, R. (1981). Somatomedin-C mediates growth hormone negative feedback by effects on both the hypothalamus and the pituitary. Science 212:1279.

Black, M.M., Aletta, J.M., Greene, L.A. (1986). Regulation of microtubule composition and stability during nerve growth factor-promoted neurite outgrowth. J. Cell Biol. 103:545.

Bohannon, N.J., Figlewicz, D.P., Corp, E.S., Wilcox, B.J., Porte, Jr., D., Baskin, D.G. (1986). Identification of binding sites for an insulin-like growth factor 1 (IGF-1) in the median eminence of the rat brain by quantitative autoradiography. Endocrinology 119:943.

Bohannon, N.J., Corp, E.S., Wilcox, B.J., Figlewicz, D.P., Dorsa, D.M., Baskin, D.G. (1988a). Characterization of insulin-like growth factor 1 receptors in the median eminence of the brain and their modulation by food restriction. Endocrinology 122:1940.

Bohannon, N.J., Corp, E.S., Wilcox, B.J., Figlewicz, D.P., Dorsa, D.M., Baskin, D.G. (1988b). Localization of binding sites for insulin-like growth factor 1 (IGF-1) in the rat brain by quantitative autoradiography. Brain Res. 444:205.

Boyd, F.T., Clarke, D.W., Muther, T.F., Raizada, M.K. (1985). Insulin receptors and insulin modulation of norepinephrine uptake in neuronal cultures from rat brain. J. Biol. Chem. 260:15880.

Brown, A.L., Graham, D.E., Nissley, S.P., Hill, D.J., Strain, A.J., Rechler, M.M. (1986). Developmental regulation of insulin-like growth factor 2 mRNA in different rat tissues. J. Biol. Chem. 261:13144.

Budd, G.C., Pansky, B., Cordell, B. (1986). Detection of insulin synthesis in mammalian anterior pituitary cells by immunohistochemistry and in situ RNA-DNA hybridization. J. Histochem. Cytochem. 34:673.

Burgess, S., Jacobs, S., Cuatrecasas, P., Sahyoun, N. (1987). Characterization of a neuronal subtype of insulin-like growth factor 1 receptor. J. Biol. Chem. 262:1618.

Chou, C.K., Dull, T.J., Russell, D.S., Gherzi, R., Lebwohl, D., Ullrich, A., Rosen, O.M. (1987). Human insulin receptors mutated at the ATP-binding site lack protein tyrosine kinase activity and fail to mediate postreceptor effects of insulin. J. Biol. Chem. 262:1842.

Clarke, D.W., Boyd, Jr., F.T., Kappy, M.S., Raizada, M.K. (1984). Insulin binds to specific receptors and stimulates 2-deoxyglucose uptake in cultured glial cells in rat brain. J. Biol. Chem. 259:11672.

Clarke, D.W., Poulakos, J.J., Mudd, L.M., Raizada, M.K., Cooper, D.L. (1987). Evidence for central nervous system insulin synthesis. In: Insulin, Insulin-like Growth Factors, and Their Receptors in the Central Nervous System, Raizada, M.K., Phillips, M.I., and LeRoith, D., eds., Plenum Press, New York, p. 121.

Coimbra, C., Migliorini, R.H. (1986). Insulin-sensitive glucoreceptors in rat preoptic area that regulate FFA mobilization. Am. J. Physiol. 251:E703.

Daughaday, W.H. (1977). Hormonal regulation of growth by somatomedin and other growth factors. Clin. Endocrinol. Metab. 6:117.

Daughaday, W.H. (1989). Somatomedins: A new look at old questions. In: *Molecular and Cellular Biology of Insulin-like Growth Factors and Their Receptors*, D. LeRoith and M. Raizada, eds., Plenum Press, New York.

DiCicco-Bloom, E., Black, I.B. (1988). Insulin growth factors regulate the mitotic cycle in cultured rat sympathetic neuroblasts. Proc. Natl. Acad. Sci. USA 85:4066.

Duffy, K.R., Pardridge, W.M., Rosenfeld, R.G. (1988). Human blood-brain barrier insulin-like growth factor receptor. Metab. 37:136.

Ebina, Y., Leland, E., Jarnagin, K., Edery, M., Graf, L., Clauser, E., Ou, J., Masiarz, F., Kan, Y.W., Goldfine, I.D., Roth, R.A., Rutter, W.J. (1985). The human insulin receptor cDNA: The structural basis for hormone-activated transmembrane signalling. Cell 40:747.

Ebina, Y., Araki, E., Taira, M., Shimada, F., Mori, M., Craik, C.S., Siddle, K., Pierce, S.B., Roth, R.A., Rutter, W.J. (1987). Replacement of lysine residue 1030 in putative ATP-binding region of the insulin receptor abolishes insulin- and antibody-stimulated glucose uptake and receptor kinase activity. Proc. Natl. Acad. Sci. USA 84:704.

Frank, H.J.L., Pardridge, W.M., Morris, W.L., Rosenfeld, R.G., Choi, T.B. (1986). Binding and internalization of insulin and insulin-like growth factors by isolated brain microvessels. Diabetes 35:654.

Gammeltoft, S., Haselbacher, G.K., Humbel, R.E., Fehlmann, M., Van Obberghen, E. (1985). Two types of receptor for insulin-like growth factors in mammalian brain. EMBO J. 4:3407.

Gibbs, M.E., Allard, W.J., Lienhard, G.E. (1986). The glucose transporter in 3T3-L1 adipocytes is phosphorylated in response to phorbol ester but not in response to insulin. J. Biol. Chem. 261:16597.

Goodyer, C.G., De Stéphano, L., Lai, W.H., Guyda, H.J., Posner, B.I. (1984). Characterization of insulin-like growth factor receptors in rat anterior pituitary, hypothalamus, and brain. Endocrinology 114:1187.

Guler, H.-P., Zapf, J., Scheiwiller, E., Froesch, E. (1988). Recombinant human insulin-like growth factor 1 stimulates growth and has distinct effects on organ size in hypophysectomized rats. Proc. Natl. Acad. Sci. USA 85:4889.

Han, V.K.M., Lund, P.K., Lee, D.C., D'Ercole, A.J. (1988). Expression of somatomedin/insulin-like growth factor messenger ribonucleic acids in human fetus: Identification, characterization, and tissue distribution. J. Clin. Endocrinol. Metab. 66:422.

Häring, H.U., White, M.F., Machicao, F., Ermel, B., Schleicher, E., Obermaier, B. (1987). Insulin rapidly stimulates phosphorylation of a 46-kDa membrane protein on tyrosine residues as well as phosphorylation of several soluble proteins in intact fat cells. Proc. Natl. Acad. Sci. USA 84:113.

Haselbacher, G.K., Schway, M.E, Pasi, A., Humbel, R.E. (1985). Insulin-like growth factor 2 (IGF2) in human brain: Regional distribution of IGF2 and of higher molecular mass forms. Proc. Natl. Acad. Sci. USA 82:2153.

Heidenreich, K.A., Freidenberg, G.R., Figlewicz, D.P., Gilmore, P.R. (1986). Evidence for a subtype of insulin-like growth factor 1 receptor in brain. Regl. Pep. 15:301.

Hill, J.M., Lesniak, M.A., Rojeski, M., Pert, C.B., Roth, J. (1987). Receptors for insulin and insulin-related peptides in the CNS: Studies of localization in rat brain. In: *Insulin, Insulin-like Growth Factors, and Their Receptors in the Central Nervous System*, Raizada, M.K., Phillips, M.I., and LeRoith, D., eds., Plenum Press, New York, p. 261.

Hynes, M.A., Brooks, P.J., Van Wyk, J.J., Lund, P.K. (1988). Insulin-like growth factor 2 messenger ribonucleic acids are synthesized in the choroid plexus of the rat brain. Mol. Endocrinol. 2:47.

Ishii, D.N., Recio-Pinto, E. (1987). Role of insulin, insulin-like growth factors and nerve growth factor in neurite formation. In: *Insulin, Insulin-like Growth Factors, and Their Receptors in the Central Nervous System*, Raizada, M.K., Phillips, M.I., and LeRoith, D., eds., Plenum Press, New York, p. 315.

Izumi, T., White, M.F., Kadowaki, T., Takaku, F., Akanuma, Y., Kasuga, M. (1987). Insulin-like growth factor 1 rapidly stimulates tyrosine phosphorylation of a M_r 185,000 protein in intact cells. J. Biol. Chem. 262:1282.

Jacobs, S., Kull, F.C., Earp, H.S., Svoboda, M.E., Van Wyk, J.J., Cuatrecasas, P. (1983). Somatomedin-C stimulates the phosphorylation of the β-subunit of its own receptor. J. Biol. Chem. 258:9581.

Kadowaki, T., Koyasu, S., Nishida, E., Tobe, K., Izumi, T., Takaku, F., Sakai, H., Yahara, I., Kasuga, M. (1987). Tyrosine phosphorylation of common and specific sets of cellular proteins rapidly induced by insulin, insulin-like growth factor 1, and epidermal growth factor in intact cells. J. Biol. Chem. 262:7342.

Kwok, R.P.S., Juorio, A.V. (1987). Facilitating effect of insulin on brain 5-hydroxytryptamine metabolism. Endocrinology 45:267.

Lenoir, D., Honegger, P. (1983). Insulin-like growth factor 1 (IGF1) stimulates DNA synthesis in fetal rat brain cell cultures. Dev. Brain Research 7:380.

Lowe, Jr., W., LeRoith, D. (1986). Tyrosine kinase activity of brain insulin and IGF-1 receptors. Biochem. Biophys. Res. Comm. 134:532.

Lowe, Jr., W.L., Boyd, F.T., Clarke, D.W., Raizada, M.K., Hart, C., LeRoith, D. (1986). Development of brain insulin receptors: Structural and functional studies of insulin receptors from whole brain and primary cell cultures. Endocrinology 119:25.

Lund, P.K., Moats-Staats, B.M., Hynes, M.A., Simmons, J.G., Jansen, M., D'Ercole, A.J., Van Wyk, J.J. (1986). Somatomedin C/insulin-like growth factor I and insulin-like growth factor II mRNAs in rat fetal and adult tissues. J. Biol. Chem. 261:14539.

Machicao, F., Häring, H., White, M.F., Carrascosa, J.M., Obermaier, B., Wieland, O.H. (1987). An M_r 180,000 protein is an endogenous substrate for the insulin-receptor-associated tyrosine kinase in human placenta. Biochem. J. 243:797.

Madoff, D.H., Martensen, T.M., Lane, M.D. (1988). Insulin and insulin-like growth factor 1 stimulate the phosphorylation on tyrosine of a 160 kDa cytosolic protein in 3T3-L1 adipocytes. Biochem. J. 252:7.

Massagué, J., Czech, M. (1982). The subunit structures of two distinct receptors for insulin-line growth factors 1 and 2 and their relationship to insulin receptor. J. Biol. Chem. 257:5038.

Mattsson, M.E.K., Enberg, G., Ruusala, A.-I., Hall, K., Pahlman, S. (1986). Mitogenic response of human SH-SY5Y neuroblastoma cells to insulin-like growth factor I and II is dependent on the stage of differentiation. J. Biol. Chem. 102:1949.

McDonald, R.G., Pfeffer, S.R., Coussens, L., Tepper, M.A., Brocklebank, C.M., Mole, J.E., Anderson, J.K., Chen, E., Czech, M.P., Ullrich, A. (1988). A single receptor binds both insulin-like growth factor 2 and mannose-6-phosphate. Science 239:1134.

McElduff, A., Poronnik, P., Baxter, R.C. (1987). The insulin-like growth factor 2 (IGF2) receptor from rat brain is of lower apparent molecular weight than the IGF2 receptor from rat liver. Endocrinology 121:1306.

McElduff, A., Poronnik, P., Baxter, R.C., Williams, P. (1988). A comparison of the insulin and insulin-like growth factor 1 receptors from rat brain and liver. Endocrinology 122:1933.

McMorris, F.A. (1983). Cyclic-AMP induction of the myelin enzyme 2',3'-cyclic nucleotide 3'-phosphohydrolase in rat oligodendrocytes. J. Neurochem. 41:508.

McMorris, F.A., Smith, T.M., DeSalvo, S., Furlanetto, R.W. (1986). Insulin-like growth factor 1/somatomedin C: A potent inducer of oligodendrocyte development. Proc. Natl. Acad. Sci. USA 83:822.

Mendelsohn, L. (1987). Visualization of IGF-2 receptors in rat brain. In: Insulin, Insulin-like Growth Factors, and Their Receptors in the Central Nervous System, Raizada, M.K., Phillips, M.I., and LeRoith, D., eds., Plenum Press, New York, p. 269.

Mill, J.F., Chao, M.V., Ishii, D.N. (1985). Insulin, insulin-like growth factor 2, and nerve growth factor effects on tubulin mRNA levels and neurite formation. Proc. Natl. Acad. Sci. USA 82:7126.

Morgan, D.O., Roth, R.A. (1987). Acute insulin action requires insulin receptor kinase activity: Introduction of an inhibitory monoclonal antibody into mammalian cells blocks the rapid effects of insulin. Proc. Natl. Acad. Sci. USA 84:41.

Morgan D.O., Jarnagin, K., Roth, R. (1986). Purification and characterization of the receptor for insulin-like growth factor 1. Biochemistry 25:5560.

Morgan, D.O., Edman, J.C., Standring, D.N., Fried, V.A., Smith, M.C., Roth, R.A., Rutter, W.J. (1987). Insulin-like growth factor 2 receptor as a multifunctional binding protein. Nature (London) 329:301.

Mudd, L.M., Raizada, M.K. (1989). Regulation of growth factor receptors by protein kinase C. In: Molecular and Cellular Biology of Insulin-like Growth Factors and Their Receptors, D. LeRoith and M. Raizada, eds., Plenum Press, New York.

Murphy, L.J., Bell, G.I., Friesen, H.G. (1987). Tissue distribution of insulin-like growth factor 1 and 2 messenger ribonucleic acid in the adult rat. Endocrinology 120:1279.

Nilsson, L., Sara, V.R., Nordberg, A. (1988). Insulin-like growth factor 1 stimulates the release of acetylcholine from rat cortical slices. Neurosci. Lett. 88:221.

Nishizuka, Y. (1986). Studies and perspectives of protein kinase C. Science 233:305.

Noguchi, T., Sugisaki, T., Tsukada, Y. (1982a). Postnatal actions of growth and thyroid hormones on the retarded cerebral myelinogenesis of Snell dwarf mice (dw). J. Neurochem. 38:257.

Noguchi, T., Sugisaki, T., Tsukada, Y. (1982b). Factors contributing to the poor myelination in the brain of the Snell dwarf mouse. J. Neurochem. 39:1693.

Noguchi, T., Kurata, L.M., Sugisaki, T. (1987). Presence of somatomedin-C-immunoreactive substances in the central nervous system: Immunochemical mapping studies. Neuroendocrinology 46:277.

Novak-Hofer, I., Kung, W., Eppenberger, U. (1988). Role of extracellular electrolytes in activation of ribosomal protein S6 kinase by epidermal growth factor, insulin-like growth factor, and insulin in ZR-75-1 cells. J. Cell Biol. 106:395.

Ocrant, I., Valentino, K.L., Eng, L.F., Hintz, R.L., Wilson, D.M., Rosenfeld, R.G. (1988). Structural and immunohistochemical characterization of insulin-like growth factor 1 and 2 receptors in the murine nervous system. Endocrinology 123:1023.

Ota, A., Wilson, G.L., Spilberg, O., Pruss, R., LeRoith, D. (1988a). Functional insulin-like growth factor receptors are expressed by neural-derived continuous cell lines. Endocrinology 122:145.

Ota, A., Wilson, G.L., LeRoith, D. (1988b). Insulin-like growth factor I receptors on mouse neuroblastoma cells. Two β-subunits are derived from differences in glycosylation. Eur. J. Biochem. 174:521.

Pardridge, W.F., Eisenberg, J., Young, J. (1985). Human blood brain barrier insulin receptor. J. Neurochem. 44:1771.

Pelton, E.W., Grindeland, R.E., Young, E. (1977). Effects of immunologically induced growth hormone deficiency on myelinogenesis in developing rat cerebrum. Neurology 27:282.

Perdue, J.F. (1984). Chemistry, structure, and function of insulin-like growth factors and their receptors. Can. J. Biochem. Cell Biol. 62:1237.

Phillips, A.F., Persson, B., Hall, K., Lake, M., Skottner, A., Sanengen, T., Sara, V.R. (1988). The effects of biosynthetic insulin-like growth factor 1 supplementation on somatic growth, maturation, and erythropoiesis on the neonatal rat. Pediatr. Res. 23:298.

Phillips, L.S., Vassilopoulou-Sellin, R. (1979). Nutritional regulation of somatomedin. Am. J. Clin. Nutr. 32:1082.

Phillips, L.S., Vassilopoulou-Sellin, R. (1980). Medical progress: Somatomedins (first of two parts). N. Engl. J. Med. 302:371.

Phillips, L.S., Young, H.S. (1976). Nutrition and somatomedin. I. Effect of fasting and refeeding on serum somatomedin activity and cartilage growth activity in rats. Endocrinology 99:304.

Pierre, M., Toru-Delbauffe, D., Gavaret, J.M., Pomerance, M., Jacquerin, C. (1986). Activation of S6 kinase activity in astrocytes by insulin, somatomedin C and TPA. FEBS Lett. 206:162.

Puro, D.G., Agardh, E. (1984). Insulin-mediated regulation of neuronal maturation. Science 225:1170.

Raizada, M.K., Boyd, F.T., Clarke, D.W., LeRoith, D. (1987). Physiologically unique insulin receptors on neuronal cells. In: Insulin, Insulin-like Growth Factors, and Their Receptors in the Central Nervous System, Raizada, M.K., Phillips, M.I., and LeRoith, D., eds., Plenum Press, New York, p. 191.

Ray, L.B., Sturgill, T.W. (1987). Rapid stimulation by insulin of a serine/threonine kinase in 3T3-L1 adipocytes that phosphorylates microtubule-associated protein 2 in vitro. Proc. Natl. Acad. Sci. USA 84:1502.

Rechler, M.M., Nissley, S.P. (1985). The nature and regulation of the receptors for insulin-like growth factors. Ann. Rev. Physiol. 47:425.

Recio-Pinto, E., Ishii, D.N. (1984). Effects of insulin, insulin-like growth factor 2 and nerve growth factor on neurite outgrowth in cultured human neuroblastoma cells. Brain Res. 302:323.

Recio-Pinto, E., Rechler, M.M., Ishii, D.N. (1986). Effects on insulin-insulin-like growth factor II, and nerve growth factor on neurite formation and survival in cultured sympathetic and sensory neurons. J. Neurosci. 6:1211.

Rinaudo, M.T., Curto, M., Bruno, R., Marino, C., Rossetti, V., Mostert, M. (1987). Evidence of an insulin generated pyruvate dehydrogenase stimulating factor in rat brain plasma membrane. Int. J. Biochem. 19:909.

Rosenfeld, R.G., Hoffman, A.R. (1987). Insulin-like growth factors and their receptors in the pituitary and hypothalamus. In: Insulin, Insulin-like Growth Factors, and Their Receptors in the Central Nervous System, Raizada, M.K., Phillips, M.I., and LeRoith, D., eds., Plenum Press, New York, p. 277.

Rosenfeld, R.G., Pham, H., Keller, B.T., Borchardt, R.T., Pardridge, W.M. (1987). Demonstration and structural comparison of receptors for insulin-like growth factor-1 and -2 (IGF-1 and -2) in brain and blood-brain barrier. Biochem. Biophys. Res. Comm. 149:159.

Roth, R.A. (1988). Structure of the receptor for insulin-like growth factor 2: The puzzle amplified. Science 239:1269.

Roth, R.A., Stover, C., Hari, J., Morgan, D.O., Smith, M.C., Sara, V., Fried, V.A. (1987). Interactions of the receptor for insulin-like growth factor 2 with mannose-6-phosphate and antibodies to the mannose-6-phosphate receptor. Biochem. Biophys. Res. Comm. 149:600.

Rotwein, P., Burgess, S.K., Milbrandt, J.D., Krause, J.E. (1988). Differential expression of insulin-like growth factor genes in rat central nervous system. Proc. Natl. Acad. Sci. USA 85:265.

Rubin, J.B., Shia, M.A., Pilch, P.F. (1983). Stimulation of tyrosine-specific phosphorylation in vitro by insulin-like growth factor 1. Nature (London) 305:438.

Sakaguchi, T., Bray, G.A. (1987). Intrahypothalamic injection of insulin decreases firing rate of sympathetic nerves. Proc. Natl. Acad. Sci. USA 84:2012.

Sara, V., Uvnas-Moberg, K., Uvnas, B., Hall, K., Wetterberg, L., Posslancec, B., Goiny, M. (1982). The distribution of somatomedins in the nervous system of the cat and their release following neural stimulation. Acta Physiol. Scand. 115:467.

Sara, V.R., Hall, K., Misake, M., Frylund, L., Christensen, N., Wetterberg, L. (1983). Ontogenesis of somatomedin and insulin receptors in human fetus. J. Clin. Invest. 71:1084.

Schechter, R., Holtzclaw, L., Sadiq, F., Kahn, A., Devaskar, S. (1988). Insulin synthesis by isolated rabbit neurons. Endocrinology 123:505.

Schoenle, E., Zapf, J., Humbel, R., Froesch, E. (1982). IGF-1 stimulates growth in hypophysectomized rats. Nature 296:252.

Shemer, J., Raizada, M., Masters, B., Ota, A., LeRoith, D. (1987a). Insulin-like growth factor I receptors in neuronal and glial cells. Characterization and biological effects in primary culture. J. Biol. Chem. 262:7693.

Shemer, J., Adamo, M., Wilson, G.L., Heffez, D., Zick, Y., LeRoith, D. (1987b). Insulin and insulin-like growth factor 1 stimulate a common endogenous phosphoprotein substrate (pp185) in intact neuroblastoma cells. J. Biol. Chem. 262:15476.

Shemer, J., Adamo, M., Raizada, M.K., Heffez, D., Zick, Y., LeRoith, D. (1989). Insulin and insulin-like growth factor 1 (IGF-1) stimulate phosphorylation of their respective receptors in intact neuronal and glial cells in primary culture. J. Mol. Neurosci., in press.

Shiu, R.P.C., Paterson, J.A. (1988). Characterization of insulin-like growth factor 2 peptides secreted by explants of neonatal brain and of adult pituitary of rats. Endocrinology 123:1456.

Sly, W.S., Fischer, H.D. (1982). The phosphomannosyl recognition system for intracellular and intercellular transport of lysosomal enzymes. J. Cell Biochem. 18:67.

Soares, M.B., Ishii, D.N., Efstratiadis, A. (1985). Developmental and tissue specific expression of a family of transcripts related to rat insulin-like growth factor II mRNA. Nuc. Acids Res. 13:1119.

Soares, M.B., Turken, A., Ishii, D.N., Mills, L., Episkopou, V., Cotter, S., Zeitlin, S., Efstratiadis, A. (1986). The rat insulin-like growth factor II gene: A single gene with two promoters expressing a multiscript family. J. Mol. Biol. 192:737.

Stein, L.J., Dorsa, D.M., Baskin, D.G., Figlewicz, D.P., Ikeda, H., Frankmann, S.P., Greenwood, M.R.C., Porte, Jr., D., Woods, S.P. (1983). Immunoreactive insulin levels are elevated in cerebrospinal fluid of genetically obese Zucker rats. Endocrinology 113:2299.

Strum, M.A., Conover, C.A., Pham, H., Rosenfeld, R.G. (1989). Insulin-like growth factor receptors and binding protein in rat neuroblastoma cells. Endocrinology 124:388.

Taylor, A.M., Dandona, P., Morrell, D.J., Preece, M.A.(1988). Insulin-like growth factor 1, protein kinase C, calcium and cyclic AMP: Partners in the regulation of chondrocyte mitogenesis and metabolism. FEBS Lett. 236:33.

Tong, P.Y., Tollefsen, S.E., Kornfeld, S. (1988). The cation-independent mannose-6-phosphate receptor binds insulin-like growth factor 2. J. Biol. Chem. 263:2585.

Ullrich, A., Bell, J.R., Chen, E.Y., Herrera, R., Petruzzelli, L.M., Dull, T.J., Gray, A., Coussens, L., Liao, Y.C., Tsubokawa, M., Mason, A., Seeburg, P.H., Grunfeld, C., Rosen, O.M., Ramachandran, J. (1985). Human insulin receptor and its relationship to the tyrosine kinase family of oncogenes. Nature (London) 313:756.

Ullrich, A., Gray, A., Tam, A.W., Yang-Feng, T., Tsubokawa, M., Collins, C., Henzel, W., LeBon, T., Kathuria, S., Chen, E., Jacobs, S., Francke, U., Ramachandran, J., Fujita-Yamaguchi, Y. (1986). Insulin-like growth factor 1 receptor primary structure: comparison with insulin receptor suggests structural determinants that define functional specificity. EMBO J. 5:2503.

Valentino, K.L., Pham, H., Ocrant, I., Rosenfeld, R.G. (1988). Distribution of insulin-like growth factor 2 receptor immunoreactivity in rat tissue. Endocrinology 122:2753.

Van der Pal, R.H.M., Koper, J.W., van Golde, L.M.G., Lopes-Cardozo, M. (1988). Effects of insulin and insulin-like growth factor 1 (IGF1) on oligodendrocyte-enriched glial cells. J. Neurosci. Res. 19:483.

Van Wyk, J.J. (1984). The somatomedins: Biological actions and physiological control mechanisms. In: *Hormonal Proteins and Peptides*, Li, C.H., ed., Academic Press, New York, p. 81.

Verspohl, R.J., Maddux, B.A., Goldfine, I.D. (1988). Insulin and insulin-like growth factor I regulate the same biological functions in HEP-G2 cells via their own receptors. J. Clin. Endocrinol. Metab. 67:169.

Waheed, A., Braulke, T., Junghans, U., von Figura, K. (1988). Mannose-6-phosphate/insulin-like growth factor 2 receptor: The 2 types of ligands bind simultaneously to one receptor at different sites. Biochem. Biophys. Res. Comm. 152:1248.

White, M.F., Marion, R., Kahn, C.R. (1985). Insulin rapidly stimulates tyrosine phosphorylation of a M_r-185,000 protein in intact cells. Nature (London) 318:183.

White, M.F., Stegmann, E.W., Dull, T.J., Ullrich, A., Kahn, C.R. (1987). Characterization of endogenous substrate of the insulin receptor in cultured cells. J. Biol. Chem. 262:9769.

Wiggins, R.C. (1982). Myelin development and nutritional insufficiency. Brain Res. Rev. 4:151.

Young, III, W.S., (1986). Periventricular hypothalamic cells in the rat brain contain insulin mRNA. Neuropeptides 8:93.

Zick, Y., Sasaki, N., Rees-Jones, R.W., Grunberger, G., Nissley, S.P., Rechler, M.M. (1984). Insulin-like growth factor 1 (IGF1) stimulates tyrosine kinase activity in purified receptors from a rat liver cell line. Biochem. Biophys. Res. Commun. 119:6.

Zick, Y., Sagi-Eisenberg, R., Pines, M., Gierschik, P., Spiegel, A.M. (1986). Multiple phosphorylation of $\propto$ subunit of transducin by the insulin receptor kinase and protein kinase C. Proc. Natl. Acad. Sci. USA 83:9294.

Zick, Y., Spiegel, A.M., Sagi-Eisenberg, R. (1987). Insulin-like growth factor 1 receptors in retinal rod outer segments. J. Biol. Chem. 262:10259.

THE INSULIN-LIKE GROWTH FACTOR-II/MANNOSE 6-PHOSPHATE RECEPTOR

Peter Nissley, Wieland Kiess, and Mark Sklar

Endocrinology Section, Metabolism Branch
National Cancer Institute, National Institutes of Health
Bethesda, MD

INTRODUCTION

Competitive binding experiments first demonstrated that insulin-like growth factor (IGF) receptors are distinct from insulin receptors and that there are two IGF receptors based on relative preference for IGF-I, IGF-II, and insulin[1]. The IGF-I receptor or type I IGF receptor binds IGF-I with higher affinity than IGF-II, and binds insulin with much lower affinity than IGF-II. The IGF-II receptor or type II IGF receptor binds IGF-II with considerably higher affinity than IGF-I and does not bind insulin at all. The structures of these two IGF receptors were defined by affinity crosslinking, biosynthetic labeling, purification, and finally, molecular cloning[2-5]. The IGF-I receptor is very similar to the insulin receptor, consisting of two alpha subunits of 130 kDa, and two beta subunits of 95 kDa. The beta subunit has intrinsic tyrosine kinase activity which is activated by autophosphorylation following ligand binding to the alpha subunit. The structure of the IGF-II receptor is quite different from the structure of the IGF-I receptor. The IGF-II receptor is a single 250 kDa glycoprotein which lacks tyrosine kinase activity.

Molecular cloning of the IGF-II receptor led to the startling discovery that the IGF-II receptor is identical to the cation independent mannose 6-phosphate receptor[2-5], which is one of two receptors which targets lysosomal enzymes to lysosomes. An important question is whether the binding of the growth factor, IGF-II, to the mannose 6-phosphate receptor results in modulation of the lysosomal enzyme targeting function of the receptor. A related question is whether extracellular lysosomal enzymes could modulate signaling by IGF-II through the IGF-II/Man-6-P receptor. We will discuss these questions in this brief review.

IDENTIFICATION OF THE IGF-II RECEPTOR AS THE CATION INDEPENDENT MANNOSE 6-PHOSPHATE RECEPTOR

Molecular Cloning

Morgan et al.[3] reported the primary structure of the human IGF-II receptor based on cloning and sequencing the receptor cDNA. The open reading frame encodes a protein with a predicted M_r of 274,353 which includes a 40 residue segment having the characteristics of a signal sequence. Ninety two percent of the receptor is extracellular. There are 19 N-linked glycosylation sites in the extracellular domain which consists of 15 conserved repeats of ~150 residues which are only 20% identical but share a highly conserved pattern of 8 cysteine residues. An insertion of 43 amino acids in repeat 13 is homologous to the type II region of fibronectin. There is one major hydrophobic segment of 23 residues which presumably represents the transmembrane domain of the receptor. The cytoplasmic domain is hydrophilic and includes several potential phosphorylation sites on tyrosine, threonine, and serine residues. There is no homology with known protein kinases within the cytoplasmic domain.

The human IGF-II receptor was found to be 80% homologous with the bovine cation independent mannose 6-phosphate receptor[4], consistent with a species difference for the same protein or two closely related proteins. Later, the human mannose 6-phosphate receptor was reported to be 99.4% homologous to the human IGF-II receptor[5] and the rat IGF-II receptor was found to be 79% homologous to the bovine mannose 6-phosphate receptor[6]. These results indicate that the two receptors are identical.

Biochemical and Immunochemical Evidence

Additional evidence for identity of the IGF-II receptor and the mannose 6-phosphate receptor was provided by biochemical and immunochemical experiments. IGF-II receptor purified to homogeneity by affinity chromatography on an IGF-II affinity column bound quantitatively to a lysosomal enzyme or phosphomannan affinity column and was eluted with mannose 6-phosphate[6-8]. Mannose 6-phosphate receptor purified by affinity chromatography on lysosomal enzyme or phosphomannan affinity columns bound IGF-II with high affinity and a binding stochiometry close to one[8-10]. IGF-II receptors and mannose 6-phosphate receptors purified in parallel from the same tissue by IGF-II affinity chromatography or ß-galactosidase affinity chromatography, were identical in size by SDS polyacrylamide gel electrophoresis

Experiments which demonstrated that each class of ligand (lysosomal enzymes and IGFs) influenced the binding of the other ligand provided evidence that the two ligands were binding to the same molecule. Thus ß-galactosidase inhibited the binding of ^{125}I-IGF-II to the receptor[8,11,12] and mannose 6-phosphate has been reported to increase the binding of ^{125}I-IGF-II to some preparations of receptor[6,7]. While the opposite behavior of a lysosomal enzyme and mannose 6-phosphate needs to be fully explained, these results lead to the conclusion that the two classes of ligands are binding to the same receptor molecule. The binding sites for the two classes of ligands are distinct since a monoclonal antibody for the mannose 6-phosphate receptor that blocked the binding of ^{125}I-IGF-II by 88%, had no effect on the binding of ^{125}I-pentamannosyl phosphate substituted bovine serum albumin to the receptor[10]. In addition, ^{125}I-pentamannosyl phosphate-lys-aprotinin and ^{125}I-IGF-II were crosslinked to different tryptic fragments of the mannose 6-phosphate receptor[13].

Immunochemical experiments which used antibodies that had been raised previously against either the mannose 6-phosphate receptor or the IGF-II receptor, provided additional evidence for identity of the two receptors[8]. Receptors prepared by IGF-II affinity chromatography or ß-galactosidase affinity chromatography from the same tissue, behaved identically in immunoprecipitation assays which utilized a panel of antibodies that had been raised against either the mannose 6-phosphate receptor or the IGF-II receptor. Also, binding of ^{125}I-IGF-II to the two receptor preparations was inhibited equally by the panel of antibodies.

TARGETING OF LYSOSOMAL ENZYMES BY THE IGF-II/MAN-6-P RECEPTOR

The role of the IGF-II/Man-6-P receptor in targeting acid hydrolases to lysosomes has been reviewed recently[14-16]. High mannose oligosaccharides are attached to some asparagine residues in lysosomal enzymes after synthesis of the hydrolases in the endoplasmic reticulum. In the cis Golgi, lysosomal enzymes are specifically recognized by N-acetyl glucosamine phosphotransferase which catalyses the attachment of P-glcNAc to the 6 hydroxyl group of mannose residues. In a second step, GlcNAc is removed, leaving mannose 6-phosphate residues in the lysosomal enzyme molecules. The mannose 6-phosphate molecules serve as a recognition marker for the IGF-II/Man-6-P receptor. In the trans Golgi network, proteins destined for secretion and plasma membrane proteins are segregated from lysosomal enzymes and membrane proteins destined for lysosomes. The binding of lysosomal enzymes to the IGF-II/Man-6-P receptor enables the hydrolases to move into an acidic, reticular-vesicular structure adjacent to the Golgi complex[17]. In the acid environment of this prelysosomal compartment, acid hydrolases dissociate from the IGF-II/Man-6-P receptors to form the contents of mature lysosomes while the receptors recycle to the trans Golgi to pick up more lysosomal enzymes. IGF-II/Man-6-P receptors are not found in mature lysosomes but are mainly found in the prelysosomal compartment; a minor population is found on the cell surface. During the segregation process in the trans Golgi network, approximately 10% of lysosomal enzymes are secreted rather than being targeted to lysosomes. Extracellular lysosomal enzymes which contain mannose 6-phosphate residues can bind to the cell surface IGF-II/Man-6-P receptors, be internalized with the receptors and travel to lysosomes by way of endosomes and the acidic prelysosomal compartment. Cell surface IGF-II/Man-6-P receptors are in equilibrium with intracellular receptors and cycle continuously from the cell surface, through the intracellular compartment, and back to the cell surface, at a rate that is independent of occupancy of the receptor by ligand.

A second Man-6-P receptor exhibits cation dependent binding of lysosomal enzymes and also targets lysosomal enzymes to lysosomes[14-16]. This receptor is considerably smaller than the IGF-II/Man-6-P receptor. The extracellular domain resembles the 150 amino acid cysteine rich segment that is repeated 15 times in the IGF-II/Man-6-P receptor[18,19]. This cation dependent Man-6-P receptor does not bind IGF-II[8,9].

SIGNALING BY IGF-II THROUGH THE MAN-6-P RECEPTOR

It seems clear that the IGF-II/Man-6-P receptor functions to target lysosomal enzymes to lysosomes. Does the IGF-II/Man-6-P receptor also signal IGF-II-stimulated growth responses, independent of its lysosomal enzyme targeting function?

For many cells in tissue culture, growth responses such as the synthesis of macromoles, nutrient transport, and ion flux, are stimulated by lower concentrations of IGF-I than IGF-II and insulin is active at high concentrations, suggesting mediation by the IGF-I receptor In addition, experiments using a monoclonal antibody (alpha IR-3) that blocks the binding of IGFs to the IGF-I receptor have provided strong evidence that signaling of growth responses to IGF-I and IGF-II in some cells is mediated by the IGF-I receptor. For example, in human fibroblasts in culture, alpha IR-3 has been shown to block IGF-I and IGF-II stimulated DNA synthesis[20-23]. Complementing these findings are experiments showing that antisera that blocked the binding of IGF-II to the IGF-II/Man-6-P receptor did not block growth responses to IGF-II in H-35 rat hepatoma cells[24] or L6 rat myoblasts[25] nor did these antisera mimick the effect of IGF-II when tested alone.

However, there are reports which suggest signaling of growth responses through the IGF-II/Man-6-P receptor. The evidence for mediation by the IGF-II/Man-6-P receptor include more sensitive dose response curves for IGF-II than for IGF-I and in some cases demonstration that an antibody to the IGF-II/Man-6-P receptor mimicks the biologic response to IGF-II. Hammerman and his colleagues[26-28] have reported stimulation of biologic responses by very low concentrations of IGF-II (<1 nM), using canine proximal tubular segments and brush border and basolateral membranes. IGF-II, but not IGF-I at 10 nM, stimulated Na^+/H^+ exchange across the brush border of proximal tubule segments, increased phosphorylation of proteins in basolateral membranes, and stimulated inositol trisphosphate and diacylglycerol generation in basolateral membranes.

The HepG2 human hepatoma cell line has receptors for IGF-I, IGF-II, and insulin and all three ligands stimulate glycogen synthesis in these cells[29]. Blocking antibodies for the insulin receptor (MC51) and the IGF-I receptor (alpha IR-3) only partially blocked the stimulation of glycogen synthesis by IGF-II in these cells[30]. An antibody directed against the human IGF-II/Man-6-P receptor stimulated glycogen synthesis 2.5 fold. In addition, the Fab fragment of this antibody was 20% less potent than the intact IgG in stimulating glycogen synthesis and was also 20% less potent in inhibiting the binding of ^{125}I-IGF-II to the cells.

Low concentrations of IGF-II (100 pM) were reported to stimulate Ca^{++} influx in 3T3 mouse fibroblasts that had been pretreated sequentially with platelet-derived growth factor and epidermal growth factor (primed, competent cells)[31,32]. IGF-I was reported to be less effective than IGF-II in stimulating Ca^{++} influx, although the relative potency of IGF-I and IGF-II was not clearly defined in these reports[31,33]. Signaling through a G protein was implicated because pertussis toxin blocked stimulation of Ca^{++} influx by IGF-II and GTP- S inhibited IGF-II binding to 3T3 cell membranes[31]. Recently, an IGF-II/Man-6-P receptor antibody has also been reported to stimulate Ca^{++} influx in 3T3 cells[32]. Low concentrations of IGF-II (1 nM) also stimulated [^{3}H]thymidine incorporation into DNA in primed, competent 3T3 cells and this stimulation was blocked by pertussis toxin. Moreover, the anti IGF-II/Man-6-P receptor antibody mimicked the effect of IGF-II. These results led to the proposal that the stimulation of Ca^{++} influx by IGF-II was an early signal leading to DNA synthesis.

Tally et al.[34] reported that an erythroleukemia cell line (K562) displayed IGF-II/Man-6-P receptors and insulin receptors but not IGF-I receptors.and a subclone (K562/cl 1) showed greater binding of ^{125}I-IGF-II and less binding of ^{125}I-insulin compared to the parental line. Using a clonal growth assay in semi-solid agar, they reported that in the parental cells, IGF-II and insulin were equipotent and IGF-I was 100 fold less potent. In the subclone, insulin was less active than IGF-II and IGF-I (100 ng/ml) did not stimulate. The relative potency of IGF-II versus IGF-I together with the apparent absence of IGF-I receptor suggested that stimulation of clonal growth by IGF-II was via the IGF-II/Man-6-P receptor. However, the K562 erythroleukemia cell lines carried in different laboratories appear not to exhibit identical properties. Blanchard et al.[35] reported that IGF-II was only slightly more potent than IGF-I in stimulating [^{3}H]thymidine incorporation into DNA In fact, recombinant IGF-I was more potent than IGF-II. Hizuka et al.[36] were able to demonstrate the presence of IGF-I receptors on K562 cells by affinity crosslinking and showed that IGF-I stimulated [^{3}H]thymidine incorporation into DNA and increase in cell number.

Thus these reports provide evidence that the IGF-II/Man-6-P receptor may be involved in signaling functions apart from the lysosomal enzyme targeting function. In most cases the argument for IGF-II/Man-6-P receptor involvement in signaling is based on relative potency data for IGF-II versus IGF-I; in two cases antibodies to the IGF-II/Man-6-P receptor were shown to mimic the growth responses, providing stronger evidence for mediation by the IGF-II/Man-6-P receptor. When considering the evidence that very low concentrations of IGF-II were capable of eliciting the biologic responses, potential signaling through a variant IGF-I receptor should be kept in mind. In IM-9 lymphocytes a high affinity binding site for ^{125}I-IGF-II was identified[37,38]. Crosslinking experiments demonstrated that the radiolabeled IGF-II was binding to a 130 kDa component consistent with binding to the alpha subunit of the insulin receptor or the IGF-I receptor. However, binding of ^{125}I-IGF-II was not inhibited by the monoclonal antibody, alpha IR-3, which blocks binding of ^{125}I-IGF-I to the IGF-I receptor. Similarily, Casella et al.[39] demonstrated high affinity binding of ^{125}I-IGF-II to a variant of the IGF-I receptor in human placental membranes. Again, alpha IR-3 did not inhibit binding of radiolabeled IGF-II to this site. Thus, possible signaling by IGF-II through this variant IGF-I receptor should be explored further before it can be definitely concluded that stimulation of biologic responses by very low concentrations of IGF-II means that the IGF-II/Man-6-P receptor is involved.

INTERACTION OF IGF-II AND LYSOSOMAL ENZYMES IN BINDING TO THE IGF-II/MAN-6-P RECEPTOR

Following the identification of the IGF-II receptor as the mannose 6-phosphate receptor, it was important to ask whether the two classes of ligands (lysosomal enzymes and IGF-II) bind independently to the receptor or modulate the binding of the other class of ligand.

Modulation of the Binding of ^{125}I-IGF-II to the IGF-II/Man-6-P Receptor by Lysosomal Enzymes and Mannose 6-phosphate

During the biochemical studies that compared the binding characteristics of IGF-II receptors and mannose 6-phosphate receptors it was observed that the lysosomal enzyme, ß-galactosidase, inhibited the formation of a ^{125}I-IGF-II: receptor complex as measured by affinity crosslinking[8]. The inhibition by ß-galactosidase was reversed by coincubation with mannose 6-phosphate, suggesting that the inhibition by ß-galactosidase depended upon binding to the mannose 6-phosphate recognition site. ß-galactosidase inhibited the binding of ^{125}I-IGF-II to crude preparations of IGF-II/Man-6-P receptor and pure receptors that had been prepared by affinity chromatography on IGF-II-Sepharose or ß-galactosidase-Sepharose[11]. Incubation of ^{125}I-IGF-II with the ß-galactosidase preparation did not result in degradation of the radioligand. Coincubation with an inhibitor of ß-galactosidase enzymatic activity had no influence on the ability of ß-galactosidase to inhibit the binding of ^{125}I-IGF-II to the receptor. The dose response curve for the reversal of ß-galactosidase inhibition by mannose 6-phosphate corresponded to the dose response curve for mannose 6-phosphate binding to the receptor. In addition, a subpopulation of ß-galactosidase that was isolated to exhibit enhanced cellular uptake also exhibited enhanced ability to inhibit the binding of ^{125}I-IGF-II to the receptor (P.N., W.K., C. Thomas, unpublished experiments). Thus it is clear that the lysosomal enzyme, ß-galactosidase, inhibits binding of ^{125}I-IGF-II to the IGF-II/Man-6-P receptor by binding to the mannose 6-phosphate recognition site. Scatchard analysis of IGF-II binding in the presence and absence of ß-galactosidase showed that ß-galactosidase decreased the affinity of IGF-II binding rather than decreasing the number of binding sites[11].

The earlier reports that mannose 6-phosphate actually increases the binding of ^{125}I-IGF-II to the IGF-II/Man-6-P receptor[6,7] seem paradoxical in view of the observations that the lysosomal enzyme, ß-galactosidase, inhibits the binding of ^{125}I-IGF-II to the same receptor. The positive effect of mannose 6-phosphate on ^{125}I-IGF-II binding is only seen with certain preparations of the IGF-II/Man-6-P receptor and in those receptor preparations which exhibit enhancement of ^{125}I-IGF-II binding by mannose 6-phosphate, the response is highly variable[11]. Thus microsomal membranes, solubilized microsomal membranes, and receptor purified on an IGF-II affinity column, show increased binding of ^{125}I-IGF-II when coincubated with mannose 6-phosphate, whereas receptors that have been purified on a lysosomal enzyme affinity column or a phosphomannan affinity column[9] do not exhibit the mannose 6-phosphate effect. Scatchard analysis of IGF-II binding data obtained in the presence and absence of mannose 6-phosphate shows that mannose 6-phosphate increases the binding affinity for IGF-II in those receptor preparations which exhibit the positive effect of mannose 6-phosphate[6,7]. The most straightforward explanation of the enhancement of ^{125}I-IGF-II binding by mannose 6-phosphate would be that the mannose 6-phosphate is causing the release of endogenous lysosomal enzymes from the receptor preparations, thus reversing the inhibitory effect of lysosomal enzymes on ^{125}I-IGF-II binding. Receptors that had been purified on a phosphomannan affinity column or a lysosomal enzyme affinity column would have exchanged their endogenous lysosomal enzymes for the phosphomannan or lysosomal enzyme coupled to the Sepharose. Having lost their endogenous lysosomal enzymes, such receptor preparations would no longer exhibit the stimulatory effect of mannose 6-phosphate on ^{125}I-IGF-II binding. This hypothesis has been tested experimentally with conflicting results. MacDonald et al.[6] incubated a plasma membrane preparation with mannose 6-phosphate, washed the membrane preparation repeatedly and then demonstrated that the washed membrane preparation still exhibited the positive effect of mannose 6-phosphate on ^{125}I-IGF-II binding. Polychronakos et al.[40] used the same experimental protocol but increased the volume of the membrane washes. They observed that the basal level of ^{125}I-IGF-II binding increased with washing and the positive effect of mannose 6-phosphate was lost, consistent with the hypothesis that the stimulation of ^{125}I-IGF-II binding by mannose 6-phosphate is explained by removal of endogenous lysosomal enzymes by mannose 6-phosphate.

Whatever the mechanism of the positive effect of mannose 6-phosphate on ^{125}I-IGF-II binding to some preparations of IGF-II/Man-6-P receptor, the more physiologically relevant phenomenon may be the inhibition of ^{125}I-IGF-II binding by ß-galactosidase, since lysosomal enzymes are found in extracellular fluids. The inhibition of ^{125}I-IGF-II binding is not limited to a single lysosomal enzyme since fucosidase and sphingomyelinase also caused inhibition[11];

additional lysosomal enzymes have not been tested. Thus if signaling of growth responses by IGF-II occurs via the IGF-II/Man-6-P receptor, extracellular lysosomal enzymes could modulate this response.

Insulin-like Growth Factor-II Inhibits the Cellular Uptake of ß-Galactosidase and the Binding of ß-Galactosidase to Purified IGF-II/Man-6-P Receptors

Not only do lysosomal enzymes inhibit the binding of ^{125}I-IGF-II to the IGF-II/Man-6-P receptor but there is evidence for reciprocal modulation of binding Thus IGF-II inhibits the uptake of ^{125}I-ß-galactosidase by cells in monolayer culture[41]. IGF-I is much less effective than IGF-II and insulin is inactive, suggesting that the inhibition of ^{125}I-ß-galactosidase uptake is via interaction with the IGF-II/Man-6-P receptor. This conclusion is strengthened by the observation that an antiserum to the IGF-II/Man-6-P receptor also inhibits the uptake of ^{125}I-ß-galactosidase. Although there may be other explanations for the inhibition of uptake of ^{125}I-ß-galactosidase by IGF-II such as down regulation of the cell surface receptor by IGF-II, inhibition of binding to the receptor would be the most straightforward explanation. This was confirmed by showing that IGF-II inhibits the binding of ^{125}I-ß-galactosidase to pure receptor. The inhibition of binding to the pure receptor was 60% whereas the inhibition of cellular uptake was approximately 80%. This result suggests that the inhibition of cellular uptake of ^{125}I-ß-galactosidase by IGF-II is only partly explained by the inhibition of binding to the receptor. Regardless of the mechanism, the modulation of cellular uptake of lysosomal enzymes by IGF-II suggests that in processes such as bone remodeling where secreted lysosomal enzymes digest matrix, extracellular IGF-II could affect this process by inhibiting the uptake of lysosomal enzymes. In a cell that is producing IGF-II it is possible that IGF-II, lysosomal enzymes, and receptor would be found in the same intracellular compartment within the Golgi. In that case, IGF-II could inhibit the binding of lysosomal enzymes to the IGF-II/Man-6-P receptor and thereby modulate the targeting of the lysosomal enzymes to lysosomes.

DEVELOPMENTAL EXPRESSION OF THE IGF-II/MAN-6-P RECEPTOR

Serum IGF-II is developmentally regulated in the rat, being high in fetal rat serum and then gradually declining to very low levels postnatally[42]. With the exception of brain, tissue IGF-II mRNA also shows a similar developmental pattern[43]. Because of the possibility that IGF-II could modulate the targeting of lysosomal enzymes via the IGF-II/Man-6-P receptor, it was of interest to examine the developmental expression of the receptor. The level of total IGF-II/Man-6-P receptor in various tissues of the fetal and postnatal rat was measured by quantitative Western blotting using highly purified receptor as standard[44]. In fetal tissues, the receptor concentration ranges from 0.16% to 1.7% of extracted protein in brain and heart, respectively. There is a striking developmental pattern characterized by a sharp decline occurring shortly after birth in most tissues but after day 5 in others. This developmental pattern has also been noted for a circulating form of the IGF-II/Man-6-P receptor. Thus the developmental pattern of the IGF-II/Man-6-P receptor is very similar to the developmental pattern of serum IGF-II and tissue IGF-II mRNA expression, suggesting important roles for IGF-II and the IGF-II/Man-6-P receptor in fetal growth and development. A key process in the developing embryo is tissue remodeling. We speculate that IGF-II could potentially play a regulatory role in this important function by modulating the routing of lysosomal enzymes via the IGF-II/Man-6-P receptor.

REFERENCES

1. M. M. Rechler and S. P. Nissley, The nature and regulation of the receptors for insulin-like growth factors, Ann. Rev. Physiol. 47:425 (1985).
2. A. Ullrich, A. Gray, A.W. Tam, T. Yang-Feng, M. Tsubokawa, C. Collins, W. Henzel, T. Le Bon, S. Kathuria, E. Chen, S. Jacobs, U. Frank, J. Ramachandran, and Y. Fujita-Yamaguchi, Insulin-like growth factor I receptor primary structure: comparison with insulin receptor suggests structural determinants that define functional specificity, EMBO J. 5:2503 (1986).

3. D. O. Morgan, J. C. Edman, D. N. Standring, V. A. Fried, M. C. Smith, R. A. Roth, and W. J. Rutter, Insulin-like growth factor II receptor as a multifunctional binding protein, Nature 329:301 (1987).

4. P. Lobel, N. M. Dahms, snd S. Kornfeld, Cloning and sequence analysis of the cation-independent mannose 6-phosphate receptor, J. Biol. Chem. 263: 2563 (1988).

5. A. Oshima, C. M. Nolan, J. W. Kyle, J. H. Grubb, and W. S. Sly, The human cation-independent mannose 6-phosphate receptor. Cloning and sequence of the full-length cDNA and expression of functional receptor in cos cells, J. Biol. Chem. 263: 2553 (1988).

6. R. G. MacDonald, S. R. Pfeffer, L. Coussens, M. A. Tepper, C. M. Brocklebank, J. E. Mole, J. K. Anderson, E. Chen, M. P. Czech, and A. Ullrich, A single receptor binds both insulin-like growth factor II and mannose 6-phosphate, Science 239:1134 (1988).

7. R. A. Roth, C. Stover, J. Hari, D. O. Morgan, M. C. Smith, V. Sara, and V. A. Fried, Interactions of the receptor for insulin-like growth factor II with mannose-6-phosphate and antibodies to the mannose-6-phosphate receptor, Biochem. Biophys. Res. Commun. 149: 600 (1987).

8. W. Kiess, G. D. Blickenstaff, M. M. Sklar, C. L. Thomas, S. P. Nissley, and G. G. Sahagian, Biochemical evidence that the type II insulin-like growth factor receptor is identical to the cation-independent mannose 6-phosphate receptor, J. Biol. Chem. 263:9399 (1988).

9. P. Y. Tong, S. E. Tollefsen, and S. Kornfeld, The cation-independent mannose 6-phosphate receptor binds insulin-like growth factor II, J. Biol. Chem. 263: 2585 (1988)

10. T. Braulke, C. Causin, A. Waheed, U. Junghans, A. Hasilik, P. Maly, R. E. Humbel, and K. von Figura, Mannose 6-phosphate/insulin-like growth factor II receptor: distinct binding sites for mannose 6-phosphate and insuin-like growth factor II, Biochem. Biophys. Res. Commun. 150: 1287 (1988).

11. S. P. Nissley, W. Kiess, C. Thomas, M. Sklar, and G. G. Sahagian, Lysosomal enzymes inhibit binding of insulin-like growth factor II (IGF-II) to the IGF-II/Mannose-6-phosphate receptor, J. Cell Biol. 107: 704a (1989).

12. G. D. Blickenstaff, G. Terres, and G. G. Sahagian, Insulin-like growth factor II (IGF-II) and lysosomal enzymes bind to distinct but interacting sites on the mannose 6-phosphate (Man-6-P)/IGF-II receptor, J. Cell Biol. 107:62a (1989).

13. A. Waheed, T. Braulke, U. Junghans, and K. von Figura, Mannose 6-phosphate/insulin like growth factor II receptor: the two types of ligands bind simultaneously to one receptor at different sites, Biochem. Biophys. Res. Commun. 152:1248 (1988).

14. K. von Figura and A. Hasilik, Lysosomal enzymes and their receptors, Ann. Rev. Biochem. 55:167 (1986).

15. S. Kornfeld, Trafficking of lysosomal enzymes, FASEB J. 1:462 (1987)

16. S. R. Pfeffer, Mannose 6-phosphate receptors and their role in targeting proteins to lysosomes, J. Membrane Biol. 103:7 (1988).

17. G. Griffiths, B. Hoflack, K. Simons, I. Mellman, and S. Kornfeld, The mannose 6-phosphate receptor and the biogenesis of lysosomes, Cell 52:329 (1988).

18. N. M. Dahms, P. Lobel, J. Breitmeyer, J.M. Chirgwin, and S. Kornfeld, 46 kd mannose 6-phosphate receptor: cloning, expression, and homology to the 215 kd mannose 6-phosphate receptor, Cell 50:181 (1987).

19. R. Pohlmann, G. Nagel, B. Schmidt, M. Stein, G. Lorkowski, C. Krentler, J. Cully, H. E. Meyer, K-H Grzeschik, G. Mersmann, A. Hasilik, and K. von Figura, Cloning of a cDNA encoding the human cation-dependent mannose 6-phosphate-specific receptor, Proc. Natl. Acad. Sci. USA. 84:5575 (1987).

20. J. S. Flier, P. Usher, and A. C. Moses, Monoclonal antibody to the type I insulin-like growth factor (IGF-I) receptor blocks IGF-I receptor-mediated DNA synthesis: clarification of the mitogenic mechanisms of IGF-I and insulin in human skin fibroblasts, Proc. Natl. Acad. Sci. USA 83:664 (1986).

21. J. J. Van Wyk, D. C. Graves, S. J. Casella, and S. Jacobs, Evidence from monoclonal antibody studies that insulin stimulates deoxyribonucleic acid synthesis through the type I somatomedin receptor, J. Clin. Endocrinol. Metab. 61:639 (1985).

22. C. A. Conover, P. Misra, R. L. Hintz, and R. G. Rosenfeld, Effect of an anti-insulin-like growth factor I receptor antibody on insulin-like growth factor II stimulation of DNA synthesis in human fibroblasts, Biochem. Biophys Res. Commun. 139:501 (1986).

23. R. W. Furlanetto, J. N. DiCarlo, and C. Wisehart, The type II insulin-like growth factor receptor does not mediate deoxyribonucleic acid synthesis in human fibroblasts, J. Clin. Endocrinol. Metab. 64:1142 (1987).

24. C. Mottola and M. P. Czech, The type II insulin-like growth factor receptor does not mediate increased DNA synthesis in H-35 hepatoma cells, J. Biol. Chem. 259:12705 (1984).
25. W. Kiess, J. F. Haskell, L. Lee, L. A. Greenstein, B. E. Miller, A. L. Aarons, M. M. Rechler, and S. P. Nissley, An antibody that blocks insulin-like growth factor (IGF) binding to the type II IGF receptor is neither an agonist nor an inhibitor of IGF-stimulated biologic responses in L6 myoblasts, J. Biol. Chem. 262:12745 (1987).
26. M. R. Hammerman and J. R. Gavin, Binding of insulin-like growth factor II and multiplication-stimulating activity-stimulated phosphorylation in basolateral membranes from dog kidney, J. Biol. Chem. 259:13511 (1984).
27. J. Mellas, J. R. Gavin, and M. R. Hammerman, Multiplication-stimulating activity-induced alkalinization of canine renal proximal tubular cells, J. Biol. Chem. 261:14437 (1986).
28. S. A. Rogers, and M. R. Hammerman, Insulin-like growth factor II stimulates production of inositol trisphosphate in proximal tubular basolateral membranes from dog kidney, Proc. Natl. Acad. Sci. USA 85:4037 (1988).
29. E. J. Verspohl, R. A. Roth, R. Vigneri, and I. D. Goldfine, Dual regulation of glycogen metabolism by insulin and insulin-like growth factors in human hepatoma cells (HEP-G2), J. Clin. Invest. 74:1436 (1984).
30. J. Hari, S. B. Pierce, D. O. Morgan, V. Sara, M. C. Smith, and R. A. Roth, The receptor for insulin-like growth factor II mediates an insulin-like response, EMBO J. 6:3367 (1987).
31. I. Nishimoto, Y. Hata, E. Ogata, and I. Kojima, Insulin-like growth factor II stimulates calcium influx in competent BALB/c 3T3 primed with epidermal growth factor, J. Biol. Chem. 262:12120 (1987).
32. I. Kojima, I. Nishimoto, T. Iiri, E. Ogata. and R. Rosenfeld, Evidence that type II insulin-like growth factor receptor is coupled to calcium gating system, Biochem. Biophys. Res. Commun. 154:9 (1988).
33. I. Nishimoto, E. Ogata, and I. Kojima, Pertussis toxin inhibits the action of insulin-like growth factor-I, Biochem Bipophys. Res. Commun. 148:403 (1987).
34. M. Tally, C. H. Li, and K. Hall, IGF-2 stimulated growth mediated by the somatomedin type 2 receptor, Biochem. Biophys. Res. Commun. 148:811 (1987).
35. M. M. Blanchard, B. Barenton, A. Sullivan, B. Foster, H. H. Guyda, and B. I. Posner, Characterization of the insulin-like growth factor (IGF) receptor in K562 erythroleukemia cells; evidence for a biological function for the type II IGF receptor, Molec. Cell. Endocrinol. 56:235 (1988).
36. N. Hizuka, I. Sukegawa, K. Takano, K. Asakawa, R. Horikawa, T. Tsushima, and K. Shizume, Characterization of insulin-like growth factor I receptors on human erythroleukemia cell line (K-562 cells), Endocrinol. Japon. 34:81 (1987).
37. R. L. Hintz, A. V. Thorsson, G. Enberg, and K. Hall, IGF-II binding on human lymphoid cells: demonstration of a common high affinity receptor for insulin like peptides, Biochem. Biophys. Res. Commun. 118:774 (1984).
38. P. Misra, R. L. Hintz, and R. G. Rosenfeld, Structural and immunological characterization of insulin-like growth factor II binding to IM-9 cells, J. Clin. Endocrinol. Metab. 63:1400 (1986).
39. S. J. Casella, V. K. Han, A. J. D'Ercole, M. E. Svoboda, and J. J. Van Wyk, Insulin-like growth factor II binding to the type I somatomedin receptor, J. Biol. Chem. 261:9268 (1986).
40. C. Polychronakos, H. J. Guyda, and B. I. Posner, Mannose-6-phosphate increases the affinity of its cation-independent receptor for insulin-like growth factor II by displacing inhibitory endogenous ligands, Biochem. Biophys. Res. Commun. 157:632 (1988).
41. W. Kiess, C. L. Thomas, L. A. Greenstein, L. Lee, M. M. Sklar, M. M. Rechler, G. G. Sahagian, and S. P. Nissley, Insulin-like growth factor-II (IGF-II) inhibits both the cellular uptake of ß-galactosidase and the binding of ß-galactosidase to purified IGF-II/mannose 6-phosphate receptor, J. Biol. Chem. 264:4710 (1989).
42. A. C. Moses, S. P. Nissley, P. A. Short, M. M. Rechler, R. M. White, A. B. Knight, and O. Z. Higa, Increased levels of multiplication-stimulating activity, an insulin-like growth factor, in fetal rat serum, Proc. Natl. Acad. Sci. USA 77:3649 (1980).
43. A. L. Brown, D. E. Graham, S. P. Nissley, D. J. Hill, A. J. Strain, and M. M. Rechler, Developmental regulation of insulin-like growth factor II mRNA in different rat tissues, J. Biol Chem. 261:13144 (1986).

44. M. M. Sklar, W. Kiess, C. L. Thomas, L. Lee, and S. P. Nissley, Developmental expression of tissue insulin-like growth factor-II (IGF-II)/mannose-6-phosphate (M6P) receptor in the rat. Measurement by quantitative immunoblotting, J. Biol. Chem. in press (1989).

THE INSULIN-LIKE GROWTH FACTOR-II/MANNOSE 6-PHOSPHATE RECEPTOR

Constantin Polychronakos

The Protein and Polypeptide Hormone Laboratory
Department of Pediatrics, McGill University
Montreal, Quebec, Canada

INTRODUCTION

Based on long-existing evidence (reviewed in Ref. 1), it has been known that the insulin-like growth factors (IGF-I and IGF-II) bind with high affinity to two different cell surface receptors. The type I receptor binds IGF-I with high affinity. It also binds IGF-II with somewhat lower affinity, and insulin very weakly. It has a heterotetrameric structure (Fig. 1) very similar to that of the insulin receptor, with which it shares a striking primary sequence homology. Like the insulin receptor, it consists of two identical extracellular alpha chains that contain the IGF binding domain and two identical beta chains that span the cell membrane and whose cytoplasmic domain, like that of insulin, displays ligand-activated tyrosine kinase activity, believed to be important in the transmission of the biologic effects of the IGFs[2].

The type II receptor binds only IGF-II with high affinity. It binds IGF-I very weakly if at all[3,4,5], and does not recognize insulin. It is a single-chain glycoprotein of a molecular size of approximately 250 kDa. This molecule was recently cloned and its precise sequence defined[6,7]. It was only then that it became evident that the type II IGF receptor is identical to the cation-independent mannose 6-phosphate receptor, a molecule that had been extensively studied and separately cloned as a receptor involved in the packaging and targeting of acid hydrolases to the lysosomes. These lysosomal enzymes are recognized by the receptor through a common carbohydrate group consisting of mannose residues with a terminal phosphorylation in the 6 position[8,9]. The identity of the two molecules was subsequently confirmed by extensive studies of ligand and immunologic cross-reactivity[10,11,12,13]. This totally unexpected observation led to the merging of two fields of research which, until then, were thought to be totally independent. This review aims at a synthesis of what is currently known about the receptor from each side, as well as of the small but growing bibliography on the interactions of the two ligands on their common receptor, henceforth abbreviated as M6P/IGF-II receptor. Emphasis is given to the possible implications for our understanding of the the actions of IGF-II as a stimulator of mitosis, metabolic effects and, possibly, as a regulator of other cellular processes.

IGF-II AND THE M6P/IGF-II RECEPTOR

Binding characteristics

The M6P/IGF-II receptor binds IGF-II with high affinity, the Kd being in the nano-

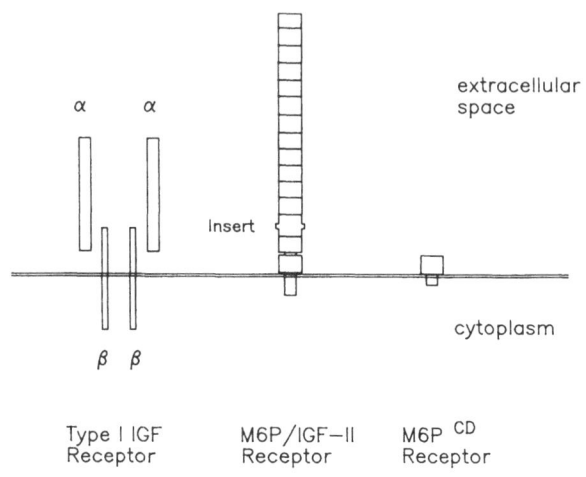

α   α

extracellular space

Insert

cytoplasm

β β

| Type I IGF Receptor | M6P/IGF—II Receptor | M6P CD Receptor |

Figure 1. Schematic representation of the three receptors that bind IGFs and M6P. The boxes represent the repeat sequences, distinguished from the rest of the chain by their width (adapted from references 6 and 8).

molar range[14]. The pH curve is rather broad, with an optimum near 8. This binding is highly specific. Thus, there is general agreement that the M6P/IGF-II receptor does not bind insulin[1]. Its affinity, however, for IGF-I remains a matter of some controversy. Using purified human IGF-I, many studies have found that this peptide displaces labeled IGF-II (or its rat equivalent, multiplication-stimulating activity) with a potency that is one to two orders of magnitude lower than that of IGF-II[1]. However, we[5] and others[3,4] have shown that unlabeled biosynthetic IGF-I is almost totally inactive in displacing labeled IGF-II from rodent tissues, where the type II receptor predominates. The small amount of displacement (15-20%), obtained at very high IGF-I concentrations, probably represents displacement of the labeled IGF-II from the IGF-I and the insulin receptor. This may be due to the substitution of a threonine for a methionine at position 58 in the biosynthetic IGF-I used, if this aminoacid is crucial for binding to the M6P/IGF-II receptor (it is not needed for binding to the type I receptor). Alternatively, the results may be explained by a small amount (in the order of 5%) of contamination of the purified IGF-I preparations with IGF-II. In any event, IGF-I does not appear to be a physiologically significant ligand for the M6P/IGF-II receptor.

Tissue distribution and developmental regulation.

The M6P/IGF-II receptor has been found and characterized in many tissues in the human, rat, and other species (bibliography in Reference 1). Very few tissues, if any, appear not to have M6P/IGF-II receptors.

The abundance of the M6P/IGF-II receptor appears to undergo a striking developmental regulation in early life, being much higher during the fetal and neonatal period. Using the model of the *in vitro* differentiating mouse embryonic limb buds, Bhaumick and Bala have shown that binding of IGF-II is higher than that of IGF-I as early as the eleventh day of gestation, with a progressive increase during maturation[15]. In the rat, the receptor is much more abundant in the neonate than in the adult, by as much as a factor of forty in some tissues[16,17,18]. Numbers rapidly decline to adult levels by the age of twenty days.

This pattern is in accordance with the generally held belief that at least in the rat, IGF-II is a growth factor specific for intrauterine and, perhaps, early extrauterine growth, as suggested by the higher levels of IGF-II in term fetal and neonatal plasma in the rat, and its regulation by placental lactogen that appears to be lost later in life[19]. In the mouse limb bud, IGF-II is mitogenic only during the first day of the *in vitro* culture of the embryonic limb buds. IGF-I appears to be ineffective at that time. In subsequent days, as the limb buds mature and differentiate, IGF-I begins to become active in promoting growth and differentiation and IGF-II loses activity[15].

An interesting observation, related to the distribution and developmental regulation of the M6P/IGF-II receptor, is that a soluble, truncated version of the molecule can be found in rat[20], monkey[21], and human[22] plasma. It consists essentially of the extracellular domain[23]. In the rat, concentrations are very high in fetal and early neonatal life, with a rapid decrease to much lower levels by 20 days of age, a fall that parallels the tissue abundance[20]. A similar developmental regulation seems to be occurring in the rhesus monkey[21]. As the M6P/IGF-II receptor is not degraded in the lysosomes, this surface shedding may be a mechanism of receptor deactivation. It is also possible that this molecule functions as yet another species of plasma binding protein for the IGFs.

Transmission of the biologic actions of IGF-II.

It would be reasonable to assume that, as a cell surface protein that binds a potent mitogen with high affinity, the M6P/IGF-II receptor is involved in the transmission of growth-regulating signals. Additional circumstantial evidence includes its abundance, in the rat, during fetal and neonatal life, at a time when *in vivo* cell proliferation is most rapid. Postnatally, we have reported that the total M6P/IGF-II receptor site number increases during the rapid phase of the propylthiouracil-induced hyperplasia in the rat thyroid[24] and the compensatory growth of the kidney following unilateral nephrectomy[25], in a manner that is specific for the organ involved in rapid growth.

Attempts, however, to directly demonstrate a role for this molecule have failed to clearly show that it is a classical mitogen receptor. Thus, Mottola and Czech, have shown that a polyclonal antibody that specifically inhibits IGF-II binding to the M6P/IGF-II receptor, without having biologic actions of its own, has absolutely no inhibitory effect on the stimulation of thymidine incorporation by IGF-II in the H-35 rat hepatoma cell line[26]. This observation, reproduced with a different antibody in L6 myoblasts[27], suggests that IGF-II may be acting as a mitogen through binding to the type I receptor or, conceivably, to the insulin receptor. This is concordant with parallel reports that a monoclonal antibody against the type I receptor blocks the mitogenic actions of IGF-II[28,29,30].

These observations, in conjunction with the very short cytoplasmic domain of the receptor, devoid of tyrosine kinase activity, or any other known intracellular transduction mechanism[6], have led to speculation that the M6P/IGF-II receptor may be only a membrane-anchored binding protein. However, other investigators have shown that antibodies specific for the M6P/IGF-II receptor seem to stimulate mitosis in competent mouse fibroblasts[31] and glycogen synthesis in liver cells[32], actions, suggesting an agonist activity that would be possible only if the receptor is, somehow, involved in transmembrane signaling. In fact, in the mouse fibroblast system, the mitogenic action of the antibody is associated with an increase in calcium influx[31], indicating that the receptor may be coupled to ion channels. The same authors have presented evidence that the receptor is coupled to a G protein[33].

Primary sequence

The determination of the cDNA sequences encoding for the rat[7], bovine[34], and human[6] M6P/IGF-II receptor has allowed a characterization of the molecule that considerably enhances previous knowledge obtained by conventional methods. The human receptor has a very large extracellular domain, consisting of 2,264 N-terminal aminoacid residues, or 92% of the entire sequence. The transmembrane and cytoplasmic domains are short, consisting of 23 and 164 residues, respectively (Fig 1.).

This number of aminoacids predicts a molecular weight, after removal of the signal peptide, of approximately 270 kDa which, with estimated N-linked glycosylation, would give a molecular mass approaching 300 kDa. This is a little higher than the 220 (unreduced molecule) or 270 (molecule unfolded by the reduction of disulfide bonds) estimated by electrophoretic methods. The discrepancy is probably within the limits of accuracy of these methods at this high molecular weight range.

The extracellular domain has a very unusual structure, consisting almost entirely of fifteen repeating sequences of approximately 150 aminoacids each. These repeat sequences have a 20% aminoacid identity to each other, but they all contain a highly conserved pattern of eight cystein residues and several conserved hydrophobic regions near the cysteins and the C terminals. These sequences are contiguous except for a small gap between repeats 14 and 15, and between 15 and the transmembrane domain[6]. Repeat 13 contains a 43-residue insertion that is highly homologous to a a disulfide-linked structure in the collagen-binding domain of fibronectin, that is also found in factor XII and in bovine seminal fluid protein CDC-109.

The cytoplasmic domain is highly hydrophilic and has no homology with a tyrosine kinase or any other known transduction mechanism. It contains two tyrosine phosphorylation sites, as well as several serine and threonine sites, including a consensus protein kinase A site (Arg-Arg-Ser-Ser)[35]. This is consistent with reports that the receptor can be phosphorylated on all three aminoacids[36,37].

The identity to the M6P receptor.

Perhaps the most unexpected and interesting result of the determination of the primary sequence of the human type II IGF receptor was the discovery of an 80% homology with the bovine cation-independent M6P receptor[6]. This homology was subsequently found to be practically 100% between the two human proteins, isolated and cloned separately[6,38]. Extensive ligand and immunologic cross-reactivity studies have confirmed that the mature products are also indistinguishable. Thus, the molecule obtained by expressing human IGF-II receptor cDNA in frog oocytes, cross-reacts with anti M6P-receptor antibodies[6]. Moreover, the receptors purified by either IGF-II or M6P-affinity chromatography bind both ligands equally well and are recognized by antibodies to either receptor in several species[10,11,12,13].

It is not clear what is the purpose of the same molecule binding two such disparate categories of ligands as a peptide growth factor and lysosomal enzymes. The merging of the two fields of research has raised obvious questions and added a large amount of information about the function of the molecule, new to most investigators who have been studying it as an IGF-II receptor.

The specific recognition of lysosomal enzymes

The lysosomes are acidified intracellular vesicles containing hydrolytic en-

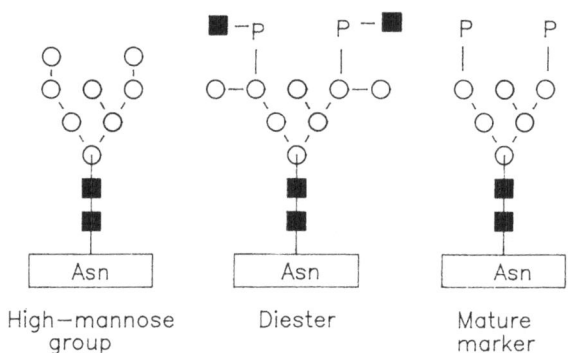

Figure 2. Schematic representation of the structure and biosynthesis of the M6P recognition marker. Open circles represent mannose and filled squares represent N-acetylglucosamine residues. Asn=asparagine. (Adapted from reference 8).

zymes capable of degrading most biologic macromolecules, either endogenous, or introduced into the cell by phagocytosis (bacteria and other particles) or receptor-mediated endocytosis (hormones, lipoproteins). As these enzymes are optimally active in an acid pH, they are also referred to as acid hydrolases. The existence of severe human disease states due to deficiencies in specific lysosomal enzymes or related functions underscores the importance of these organelles.

The posttranslational modifications of newly synthesized lysosomal enzymes, like that of other proteins, is completed in the Golgi apparatus. From there they are specifically concentrated in coated vesicles that transport them to the lysosome. The concentration process is achieved through specific recognition by two related receptors: The M6P/IGF-II receptor, and a smaller, 45 kDa membrane protein which binds lysosomal enzymes with similar binding characteristics except for a requirement for the presence of divalent cations and which is, therefore, called the cation-dependent M6P receptor. The extracellular part of this receptor has sequence homology with the repeating sequences of the M6P/IGF-II receptor. It does not bind IGF-II[12].

Both these receptors bind the same lysosomal enzyme recognition marker, an oligosaccharide determinant containing one or two mannose residues phosphorylated in position 6 (Fig. 2). Binding affinity is high, with a KD in the nanomolar range[9]. The receptor is saturable by free mannose 6-phosphate but the affinity for this interaction is much lower, the KD being in the milimolar range[9]. Glucose 6-phosphate and mannose 1-phosphate are not recognized.

Phosphorylation of the mannose residues is a process unique to the lysosomal enzymes. Lysosomal enzymes, along with secretory and plasma membrane proteins, acquire asparagine-linked carbohydrate groups of the high-mannose type in the rough endoplasmic reticulum. In the Golgi, the other categories of glycoprotein are converted to sialic-acid containing complex-type units. The mannose residues on the lysosomal enzymes undergo a completely different modification, consisting of phosphorylation in two steps[9,40]: First, a phosphodiester is formed with N-acetylglucosamine 1-phosphate by the action of a specific phosphotransferase (EC 2.7.8.17). Subsequently, the the N-acetylglucosamine is removed by N-acetyl glucosamine-1-phosphodiester alpha-N-acetylglucosaminidase (EC3.1.4.45).

The differential processing of the carbohydrate units of the lysosomal enzymes is determined through specific recognition by the phosphotransferase which probably involves the tertiary structure of the lysosomal enzyme[41,42]. The process is schematically depicted in Fig. 2.

Intracellular movement of the receptor

The lysosomal enzymes are delivered through fusion of the vesicle into which they are sorted, with a prelysosomal vesicle. The exact nature of that vesicle has been the subject of recent investigation. As the M6P/IGF-II receptor has never been found in mature lysosomes, the fusion must take place in a prelysosomal compartment. Evidence suggests that this compartment is part of the endocytic pathway[43].

Internalization of surface receptor-bound ligands occurs through accumulation of the receptor in clathrin-coated pits on the cell surface membrane. For those receptors whose endocytosis is ligand-mediated, the aggregation to the coated pits follows binding of the ligand at the surface (insulin, IGF-I, EGF)[44]. Other receptors such as the receptor for the low density lipoprotein (LDL), aggregate into coated pits constitutively[45] although, in the case of LDL there is an enhancement by ligand binding. The M6P/IGF-II receptor appears to aggregate into coated pits constitutively, as discussed below. Coated pits enter the cell to become coated vesicles which are the early component of the endocytic pathway. As they move further into the cell, they lose their clathrin coat, aggregate into larger structures, the endosomes, and the pH of their lumen becomes acidified. In the low pH, ligands dissociate and the free receptor recycles to the cell surface or the Golgi apparatus[44].

It is with this late, acidified, endosomal compartment that the vesicles containing lysosomal enzymes appear to fuse. Griffiths et al.[43] have shown the presence of vesicles enriched in both M6P/IGF-II receptor and the lysosomal marker lgp120, that are also labeled by endocytic tracers. Preventing acidification with weak bases (chloroquine or NH_4Cl) appears to block the fusion, as Brown et al. have reported that, under these conditions, the M6P/IGF-II receptors accumulate in vacuoles that are labeled by endocytic tracers (lucifer yellow and cationized ferritin) but are distinct from the vacuoles labeled with anti-lgp120 antibody[46]. Thus the lysosomal enzymes appear to be delivered· at the junction where the endocytic pathway converges with the lysosomes.

It is well known that most of the M6P/IGF-II receptor is found in intracellular organelles[47], cycling between the Golgi and the late endosomal-prelysosomal compartment. This cycling does not require passage through the cell surface pool, which contains 10-20% of the total. Cell surface receptors, however, appear to be in rapid equilibrium with the intracellular pools[48,49]. In addition to binding extracellular IGF-II, their function may be to recover the small proportion of lysosomal enzymes that reach the surface by escaping the initial sorting process. In fact, some of the phosphomannosylated proteins secreted by some cells and taken up again by receptor-mediated transport, appear to have actions not expected of lysosomal enzymes. Thus, MCF-7 cells, an estrogen-dependent breast carcinoma line, secrete, upon stimulation with estrogen, a 52 kDa cathepsin-like acid protease that is a potent mitogen for the same cells[50]. This protein is taken up by the cells through the M6P/IGF-II receptor. Proliferin, a protein with sequence homology to prolactin and growth hormone, found in various tissues of the mouse embryo, fetus and placenta, is also an M6P-dependent ligand, although it has no known enzymatic activity[18]. Uteroferrin, a protein secreted by the pregnant por-

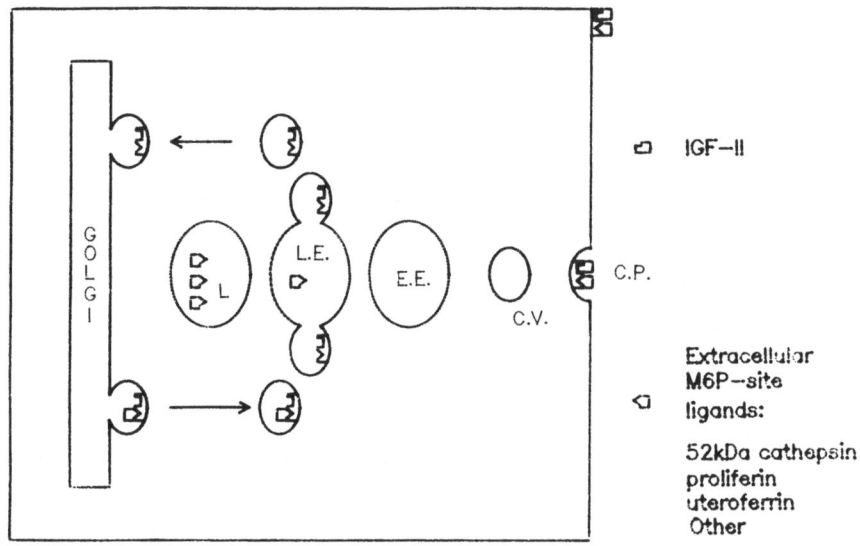

G O L G I

L.E.

E.E.

L

C.V.

C.P.

IGF—II

Extracellular
M6P—site
ligands:

52kDa cathepsin
proliferin
uteroferrin
Other

Figure 3. Schematic diagram of the delivery of lysosomal enzymes and the cycling of the receptor. C.P.:coated pit. C.V.:coated vesicle. E.E.:early encosome. L.E.: late endosome. L: lysosome. The surface receptor and ligands bound to it could equilibrate with the system through constitutive endocytosis. This is not depicted, as it has not been well studied.

cine uterus, probably involved in the transfer of iron to the fetus, also binds to this receptor through an M6P site[51].

Studies of the M6P/IGF-II receptor turnover[52] have shown that its half life is 16 hours, much longer than the length of a transport cycle between Golgi and lysosomes[53]. This, along with evidence that the receptor recycles to the Golgi indicates that the same receptor molecule can be used for a number of transport cycles.

Ligand regulation of the receptor movement

The movement of the M6P/IGF-II receptor from the Golgi to the lysosomal enzyme delivery site does not seem to require bound lysosomal enzymes. Thus, blocking the synthesis of new enzymes with cycloheximide in HepG2 cells does not alter the movement from Golgi towards the delivery site[54]. Return to the Golgi, however, requires dissociation of the enzymes. If this dissociation is blocked by preventing acidification with weak bases, the receptor becomes trapped and accumulates in endosomal vacuoles, while the Golgi apparatus is depleted[54]. An excess of free M6P, by competitively displacing the enzymes, provides an alternative dissociation mechanism and restores the normal distribution of the receptor in the Golgi[54].

Movement to and from the cell surface on the other hand, is totally independent of occupancy by M6P-related ligands. Thus, Braulke et al. have shown that neither blocking the synthesis of new enzymes, nor preventing their dissociation with weak bases has any effect on the rate of the M6P/IGF-II receptor movement to and from the cell surface of cultured fibroblasts[49]. In either situation, the cycling through the cell surface is rapid, with a $t_{1/2}$ of only a few minutes[48,49]. Using ligand-independent surface-labeling, Oka and Czech demonstrated that this constitutive rapid cycling is not altered by the presence of IGF-II in rat adipocytes[48]. Thus, movement of the M6P/IGF-II receptor to and from the cell surface appears to be constitutive, and independent of binding of either of the two known categories of ligands. The effect of IGF-II occupancy on the movement of the receptor between intracellular compartments has not been studied.

Insulin is one humoral stimulus known to affect the distribution of the M6P/IGF-II receptor between the cell surface and the intracellular compartments. In rat adipocytes[55] and H-35 hepatoma cells[37], it increases the proportion of receptors on the cell surface, without affecting the total number. At the same time, insulin decreases serine and threonine phosphorylation of the receptor in intact cells[37]. This has led to a suggestion that dephosphorylation of the receptor decreases its rate of endocytosis, resulting in an increased surface pool. This is consistent with a recent observation that phosphorylated receptor on the cell surface tends to aggregate in a clathrin-rich, detergent-insoluble fraction that probably represents the coated pits, thus indicating that phosphorylation on serine residues may be a signal to the receptor to internalize[56]. However, exposure to IGF-II stimulates phosphorylation of the M6P/IGF-II receptor in intact cells[36], without changing its rate of endocytosis.

Vanadate[57], also causes an increase in surface M6P/IGF-II receptors, in a manner that parallels its ability to stimulate the insulin receptor tyrosine kinase activity. There is no evidence, however, that the M6P/IGF-II receptor is directly a substrate for the insulin receptor kinase. Corvera et al. have shown that the M6P/IGF-II receptor from rat adipocytes can be tyrosine-phosphorylated *in vitro*, and that this phosphorylation is enhanced by IGF-II, presumably through its binding to the receptor. However, the same authors could demonstrate phosphorylation of only serine and a small amount of threonine residues in intact, [^{32}P] labeled cells, casting a doubt as to the physiologic significance of tyrosine phosphorylation of the M6P/IGF-II receptor.

INTERACTION OF IGF-II AND M6P ON THEIR COMMON RECEPTOR

Effects on binding

An obvious first step in exploring the possible interrelationships between the transport of lysosomal enzymes and the actions of IGF-II is to study the interaction of the two ligands on their common receptor. Early work demonstrated that free M6P does not displace IGF-II, establishing that the two ligands bind to distinct sites on the M6P/IGF-II receptor molecule[6].

In preliminary work, we found that free M6P actually increased IGF-II binding to particulate or solubilized rat liver microsomes by about 75%. Moreover, the increase was reproduced in IGF-II-affinity purified receptor and displacement studies demonstrated that it was due to increased affinity of the receptor for IGF-II, suggesting a positively cooperative interaction[58]. Essentially the same observations were published simultaneously and independently by both groups that purified and cloned the receptor by IGF-II-affinity chromatography[7,13].

However, Tong et al. were unable to see any effect of free M6P on IGF-II binding to receptor purified by M6P-affinity chromatography[10]. Kiess et al., also found the effect in receptor purified by IGF-II- but not by M6P- affinity chromatography (unpublished observations mentioned in Ref. 12). The same authors found that beta galactosidase, a M6P-linked lysosomal enzyme, inhibits [^{125}I] IGF-II affinity labeling of the purified receptor, an effect reversible by an excess of free M6P. These observations gave rise to the hypothesis that free M6P has no intrinsic effect on the M6P/IGF-II receptor molecule, but it acts by displacing endogenous phosphomannosylated ligands that are inhibitory to IGF-II binding. During homogenization and solubilization, preparations of the receptor are exposed to an excess of lysosomal enzymes. It is conceivable that some of these ligands copurify with the M6P/IGF-II receptor in an IGF-II-affinity column, attached to a site distinct from that of IGF-II. Recent observations in our laboratory provide strong support for this case[59]. We have found that the effect of M6P on IGF-II binding to the IGF-II-affinity purified receptor is considerably smaller than that seen with microsomes (40% vs 75% increase). Repurification of the receptor preparation by ultracentrifugation totally abolishes the effect.

It, therefore, appears that M6P by itself has no effect on IGF-II binding. However, at least some of the lysosomal enzymes appear to be inhibitory by reducing the affinity of the receptor for IGF-II, indicating negative cooperativity. Kiess et al. have also communicated data that suggest negative cooperativity in the oposite direction as well, as IGF-II inhibits the binding of beta galactosidase to the receptor[60].

Effects on receptor distribution between the cell surface and intracellular compartments

We have recently published observations showing that although an excess of M6P does not alter IGF-II binding to the surface of intact IM9 cells, it substantially (twofold to threefold) increases the amount of IGF-II taken up in the cells by receptor-mediated endocytosis[61]. This suggests that simultaneous occupancy of the receptor by IGF-II and free M6P may have effects on the distribution of the receptor between the cell surface and intracellular compartments that are not seen with occupancy by either ligand alone. It remains to be seen whether the phenomenon is due to the M6P itself or to displacement of endogenous phosphomannosylated ligands that modulate the movement of the receptor to or from the cell surface only when it is occupied by IGF-II.

CONCLUSIONS AND REMAINING QUESTIONS

The discovery that one of the two lysosomal enzyme transport proteins is also a high affinity receptor for IGF-II has forced a redefinition of the search for a role for IGF-II and its highly specific receptor. Although definitely a mitogen when acting through the type I IGF receptor and, under some circumstances, possibly through the binding to the M6P/IGF-II receptor, IGF-II has long been thought of as a molecule in search of a function. The recent developments reviewed here strengthen the speculation that this peptide may have actions that are not directly related to the stimulation of cell proliferation.

Lysosomal transport and sorting, on the other hand, has been extensively studied and defined without any knowledge of the existence of the IGF-II binding site on the receptor. Some of the established notions in the field may have to be reexamined under the light of a possible regulatory role for IGF-II in this function.

REFERENCES

1. Rechler M.M. and Nissley S.P. (1985) Receptors for insulinlike growth factors. In: Polypeptide Hormone Receptors, Posner B.I. (ed), Toronto, Marcel Dekker, Inc., p.227-297.
2. Zick, Y., Sasaki, N., Rees-Jones, R. N., Grumberger, G., Petri, S. P., Rechler, M., 1984, Insulin-like growth factor-I (IGF-I) stimulates tyrosine kinase activity in purified receptor from rat liver cell line, Biochem Biophys Res Comm, 119:6.
3. Ewton, D. Z., Falen, S. L., Florini, J. R., 1987, The type II insulin-like growth factor (IGF) receptor has low affinity for IGF-I analogs: pleiotypic actions of IGFs on myoblasts are apparently mediated by the type I repector, Endocrinology, 120:115.

4. Rosenfeld R.G., Conover C.A., Hodges D., Lee P.D.K., Misra P., Hintz R.L., Li C.H. 1987 Heterogeneity of IGF-I affinity for the IGF-II receptor: Comparison of natural, synthetic and recombinant DNA-derived IGF-I. Biochem Biophys Res Comm, 143:199

5. Barenton B., Guyda H.J., Goodyer C.G., Polycrhonakos C., Posner B.I. 1987 Specificity of IGF-II binding to type II IGF receptors in rabbit mammary gland and hypohysectomized rat liver Biochem Biophys Res Comm, 149:555

6. Morgan DO, Edman JC, Standring DN, Fried VA, Smith MC, Roth RA, Rutter WJ 1987 Insulin-like growth factor II receptor as a multifunctional binding protein. Nature 329:301

7. MacDonald RG, Pfeffer SR, Coussens L, Tepper MA, Brocklebank CM, Mole JE, Anderson JK, Chen E, Czech MP, Ullrich A 1988 A single receptor binds both insulin-like growth factor II and mannose-6-phosphate. Science 239:1134

8. Kornfeld, S., 1987, Trafficking of lysosomal enzymes, FASEB J, 1:462.

9. Von Figura K., Hasilik A. 1986 Lysosomal enzymes and their receptors. Ann Rev Biochem 55:167

10. Tong PY, Tollefsen SE, Kornfeld S 1988 The cation-independent mannose 6-phosphate receptor binds insulin-like growth factor II. J Biol Chem 263:2585

11. Braulke T., Causin C., Waheed A., Junghans U., Hasilik A., Maly P., Humbel R., von Figura K. 1987 Mannose 6-phosphate/IGF-II receptor: Distinct binding sites for mannose 6-phosphate and IGF-II. Biochem Biophys Res Comm, 150:1287

12. Kiess W., Blickenstaff G.D., Sklar M.M., Thomas C.L., Nissley S.P., Sahagian G.G. 1988 Biochemical evidence that the type II IGF receptor is identical to the mannose 6-phosphate receptor. J Biol Chem 263:9339

13. Roth R.A, Stover C, Hari J, Morgan D.O, Smith MC, Sara V, Fried V.A. 1987 Interactions of the receptor for insulin-like growth factor II with mannose-6-phosphate and antibodies to the mannose-6-phosphate receptor. Biochem Biophys Res Commun 149:600

14. Adams, S. O., Nissley, S. P., Kasuga, M., Foley, T. P., Rechler, M. M., 1983, Receptors for insulin-like growth factors and growth effects of multiplication-stimulating activity (rat insulin-like growth factor II) in rat embryo fibroblasts, Endocrinology, 112:971.

15. Bhaumick B., Bala R.M. 1987 Receptors for IGF-I and II in developing embryonic mouse limb bud Biochim Biophys Acta 927:117

16. Sklar, M. M., Kiess, W., Thomas, C. L., Lee, L., Nissley, S. P., 1988, The rat insulin-like growth factor-II (IGF-II)/mannose-6-phosphate (M6P) receptor is developmentally regulated in multiple tissues, 70th Annual Meeting of the Endocrine Society, New Orleans, Louisiana, Abstract #516.

17. Alexandrides T., Smith R.J. (1987) Distinct patterns of expression of receptors for insulin, IGF-I and IGF-II in skeletal muscle during development. Abstract #805, Endocrine Society Meeting, Indianapolis 1987.

18. Lee, S. J. Nathans, D., 1988, Proliferin secreted by cultured cells binds to mannose 6-phosphate receptors, J Biol Chem, 263:3521.

19. Adams, S. O., Nissley, S. P., Handwerger, S., Rechler M., 1983, Developmental patterns of of IGF-I and II synthesis and regulation in rat fibroblasts, Nature, 302:150.

20. Kiess, W., Greenstein, L. A., Lee, L., White, R. M., Rechler, M. M., Nissley, S. P., 1987, The type II insulin-like growth factor (IGF) receptor is present in rat serum, Proc Natl Acad Sci USA, 84:7720.

21. Gelato, M. C., Kiess, W., Lee, L., Malozowski, S., Rechler, M. M., Nissley, P., 1988, The insulin-like growth factor II/mannose-6-phosphate receptor is present in monkey serum, J Clin Endocrinol Metab, 67: 669.

22. Causin, C., Waheed, A., Braulke, T., Junghans, U., Maly, P., Humbel, R.E., von Figura, K., 1988, Mannose 6-phosphate/insulin-like growth factor II-binding proteins in human serum and urine. Their relation to the mannose-6-phosphate/insulin-like growth factor II receptor, Biochem J, 252:795.

23. MacDonald, R. G., Tepper, M. A., Clairmont, K. B., Perregaux, S. B., Brocklebank C. M., Czech, M. P., 1988, Antipeptide antibodies as domain-specific probes for cellular and serum forms of the insulin-like growth factor-II/mannose-6-phosphate receptor, 40th Annual Meeting ADA, New Orleans, Louisiana, Abstract #34.

24. Polychronakos C., Guyda H.J., Patel B., Posner B.I. 1986 Increase in the number

of type II IGF receptors during propylthiouracil-induced hyperplasia in the rat thyroid. Endocrinology 119:1204

25. Polychronakos C., Guyda H.J., Posner B.I., 1985 Increase in the type 2 IGF receptors in the rat kidney during compensatory growth. Biochem Biophys Res Commun 132:148

26. Mottola, C., Czech, M. P., 1984, The type II IGF recptor does not mediate increased DNA synthesis in H-35 hepatoma cells, J Biol Chem 259:12705.

27. Kiess, W., Haskell, J. F., Lee, L., Greenstein, L. A., Miller, B. E., Aarons, A. L., Rechler, M. M., Nissley, S. P., 1987, An antibody that blocks insulin-like growth factor (IGF) binding to the type II IGF receptor is neither an agonist nor an inhibitor of IGF-stimulated biologic responses in L6 myoblasts, J Biol Chem, 262:12745.

28. Furlanetto, R.W., DiCarlo, J. N., Wisehart, C., 1987, The type II insulin-like growth factor receptor does not mediate deoxyribonucleic acid synthesis in human fibroblasts, J Clin Endocrinol Metab, 64:1142.

29. Conover C.A. Misra P., HintZ R.L., Rosenfeld R.G. 1986 Effect of an anti-IGF-I receptor antibody on IGF-II stimulation of DNA synthesis in human fibroblasts Biochem Biophys Res Commun 139:501

30. Kadowaki, T., Koyasu, S., Nishida, E., Sakai, H., Takaku, F., Yahara, I., Kasuga, M., 1986, Insulin-like growth factors, insulin, and epidermal growth factor cause rapid cytoskeletal reorganization in KB cells, J Biol Chem, 261:16141.

31. Kojima, I., Nishimoto, I., Taroh, I., Ogata, E., Rosenfeld, R., 1988, Evidence that the type II IGF receptor is coupled to calcium gating system, Biochem Biophys Res Commun, 154:9.

32. Hari J, Pierce S.B, Morgan D.O, Sara V, Smith M.C, Roth R.A (1987) The receptor for IGF-II mediates an insulin-like response. EMBO J. 6:3367

33. Nishimoto, I., Yutaka, H., Etsuro, O., Kojima, I., 1987, Insulin-like growth factor II stimulates calcium influx in competent BALB/c 3T3 cells primed with epidermal growth factor, J Biol Chem, 262:12120.

34. Lobel P., Dahms N. M., Breitmeyer J., Chirgwin J. M., Kornfeld, S. 1987, Cloning of the bovine 215-kDa cation-independent mannose 6-phosphate receptor, Proc Natl Acad Sci USA, 84:2233.

35. Feramisco, J. R., Glass, D. B., Krebs, E. G., 1980, Optimal spatial requirements for the location of basic residues in peptide substrates for the cyclic AMP-dependent protein kinase, J Biol Chem, 255:4240.

36. Haskell J.F., Nissley S.P. Rechler M.M., Sasaki N., Greenstein L., Lee L. 1985 Biochem Biophys Res Commun 130:793

37. Corvera S., Roach, P.J., De Paoli-Roach A.A. Czech M.P. 1986 Insulin action inhibits IGF-II receptor phosphorylation in H-35 hepatoma cells. J Biol Chem 263:3116

38. Oshima, A., Nolan, C. M., Kyle, J. W., Grubb, J., Sly, W. S., 1988, The human cation-independent mannose 6-phosphate receptor. Cloning and sequence of the full-length cDNA and expression of functional receptor in COS cells, J Biol Chem, 263:2553.

39. Dahms, N. M., Lobel, P., Breitmeyer, J., Chirgwin, J.M., Kornfeld, S., 1987, 46 kD mannose 6-phosphate receptor: cloning, expression, and homology to the 215 kD mannose 6-phosphate receptor, Cell, 50:181.

40. Kornfeld, S. Trafficking of lysosomal enzymes in normal and disease states, 1986, J Clin Invest, 77:1.

41. Reitman M.L., Kornfeld S. 1981 Lysosomal enzyme targeting: N-acetyl-glucosaminylphosphotransferase selectively phosphorylates native lysosomal enzymes. J Biol Chem 256:11977

42. Lang, L., Reitman, M. L., Tang, J., Roberts, R. M., Kornfeld, S., 1984, Lysosomal enzyme phosphorylation. Recognition of a protein-dependent determinant allows specific phosphorylation of oligosaccharides present on lysosomal enzymes, J Biol Chem, 259:14663.

43. Griffiths, G., Hoflack, B., Simons, K., Mellman, I., Kornfeld, S., 1988, The mannose 6-phosphate receptor and the biogenesis of lysosomes, Cell, 52:329.

44. Posner, B.I., Khan M.N., and Bergeron J.M., Receptor-mediated uptake of peptide hormones and other ligands, in: "Polypeptide Hormone Receptors," Marcel Dekker, Inc., New York (1985).

45. Basu, S. K, Goldstein, J. L., Anderson, R. G. W., Brown, M. S., 1981, Monensin interrupts the recycling of low density lipoprotein receptors in human fibroblasts,,Cell 93:493
46. Brown, W.J., Goodhouse, J., Farquhar, M.G., 1986, Mannose-6 phosphate receptors for lysosomal enzymes cycle between the Golgi complex and endosomes, J Cell Biol, 103:1235.
47. Fischer HD, Gonzalez-Noriega A, Sly WS 1980 B-glucuronidase binding to phosphomannosyl-enzyme receptors in membranes from human and rat tissues. J Biol Chem 255:5069
48. Oka, Y., Czech, M.P., 1986, The type II insulin-like growth factor receptor is internalized and recycles in the absence of ligand. J Biol Chem, 261:909013.
49. Braulke, T., Gartung, C., Hasilik, A., Von Figura, K., 1987, Is movement of mannose 6-phosphate-specific receptor triggered by binding of lysosomal enzymes? J Cell Biol, 104:1735.
50. Vignon, F. et al., 1983, Autocrine stimulation of the MCF-7 breast cancer cells by the estrogen regulated 52K' protein, Endocrinology, 118:1537.
51. Saunders, P.T., Renegar, R. H., Raub, T.J., Baumbach, G. A., Atkinson, P. H., Bazer, F. W., Roberts, R. M., 1985, The carbohydrate structure of porcine uteroferrin and the role of the high mannose chains in promoting uptake by the reticuloendothelial cells of the fetal liver, J Biol Chem, 260:3658.
52. Sahagian, GG, Neufeld EF 1983 Biosynthesis and turnover of the mannose 6-phosphate receptor in cultured Chinese hamster ovary cells. J Biol Chem 258:7121
53. Sahagian G.G. The Mannose 6-Phosphate receptor. Action, biosynthesis and translocation. Biol Cell 51:207
54. Pfeffer, S. R., 1986, The endosomal concentration of a M6P receptor is unchanged in the absence of ligand synthesis, J Cell Biol, 105:229.
55. Wardzala, L. J., Simpson, I. A., Rechler, M. M., Cushman, S. W., 1984, Potential mechanism of the stimulatory action of insulin on IGF-II binding to the isolated rat adipose cell, J Biol Chem, 259:8378.
56. Corvera, S., Folander, K., Clairmont, K.B., Czech, M.P., 1988, A highly phosphorylated subpopulation of insulin-like growth factor II/mannose 6-phosphate receptors is concentrated in a clathrin-enriched plasma membrane fraction, Proc Natl Acad Sci USA, 85:7567.
57. Kadota S., Fantus I.G., Deragon G., Guyda H.J., Posner B.I. 1987 Stimulation of IGF-II receptor binding and insulin receptor kinase activity in rat adipocytes. J Biol Chem 262:8252
58. Polychronakos, C., Piscina, R., Guyda, H., Mannose-6-phosphate increases the affinity of the insulin-like growth factor II for its receptor on rat liver microsomes, Abstract #900, 70th Annual Meeting of the Endocrine Society, New Orleans 1988
59. Polychronakos C., Guyda H.J., Posner B.I. 1988 Mannose 6-phosphate increases IGF-II receptor binding by displacing endogenous phosphomannosylated ligands. Biochem Biophys Res Commun 157:632
60. Kiess, W., Sklar, M. M., Thomas, C. L., Lee, L., Nissley, S. P., Insulin-like growth factor II (IGF-II) inhibits binding of beta calactosidase to the IGF-II/mannose-6-phosphate (M6P) receptor, Abstract #899, 70th Annual Meeting of the Endocrine Society, New Orleans, 1988
61. Polychronakos C., Piscina R. 1988 Endocytosis of receptor-bound IGF-II is enhanced by mannose 6-phosphate in IM9 cells. Endocrinology 123:2943

THE ROLE OF INSULIN-LIKE GROWTH FACTOR BINDING PROTEINS IN CONTROLLING

THE EXPRESSION OF IGF ACTIONS

David R. Clemmons

University of North Carolina School of Medicine
Division of Endocrinology, CB #7170
Chapel Hill, NC 27599-7170

Although the insulin-like growth factors are secreted into blood and transported to peripheral tissues where they can act as traditional endocrine hormones, they are also synthesized and secreted by many types of cells and may exert their effects in the local microenvironment by autocrine or paracrine regulatory mechanisms. Specifically, IGF-I mRNA has been detected by in situ hybridization in connective tissue cells in several human tissues[1]. This mRNA appears to be translated and the peptide secreted since several types of cells in culture, such as fibroblasts, have been shown to secrete IGF-I[2,3]. Furthermore, blocking the binding of cell-secreted IGF-I to the type I IGF receptor will block mitogenic response in the fibroblasts that constitutively synthesize this peptide[4]. These data suggest that IGF-I is made by the connective tissue cell types within organs, and is secreted into interstitial fluids. Following secretion IGF-I not only binds to receptors on fibroblasts but it can bind also to receptors on cells of epithelial origin that contain receptors but do not contain IGF-I mRNA. Immunocytochemical localization studies have shown IGF-I to be present on the surfaces of many cell types that apparently do not synthesize and secrete this peptide[5]. Since following its secretion into extracellular fluids, IGF-I is capable of binding to multiple cell types, whether or not they are actually capable of synthesizing this peptide, there is a need for some mechanism to coordinate the effects of the connective tissue cells that are constitutively synthesizing this peptide and the epithelial and other cell types that are capable of responding to the peptide but are not actually engaged in synthesis.

Although variations in the rate of IGF-I synthesis, regulation of type I receptor number, or affinity would provide other mechanisms for modulating this response, these processes cannot directly control the quantity of IGF-I that is transported to individual cell types within a tissue and do not provide a selective mechanism for determining the specific cell types that are to be stimulated. One link between local secretion of IGF-I and expression of its activity on distinct cell types may be provided by high affinity, soluble binding proteins. These proteins are ubiquitously present in most types of extracellular and presumably interstitial fluids. Specifically, IGF binding proteins have been shown to be present in lymph, amniotic, spinal and ascitic fluids[6], as well as whole blood[7]. Likewise, these proteins are present in the supernatants of cultures of many types of cells suggesting that they are secreted into the interstitial microenvironment in vivo[8]. More recently, our laboratory, in collaboration with Dr. David Hill, has shown that one form of IGF-BP (Mr 26000) is present on the surfaces of multiple epithelial cell types in the human fetus[9]. Interestingly it is present on the surface of several epithelial cell types that do not contain IGF-I mRNA suggesting that the IGF-I that is also present on the surfaces of these cells is bound to this binding protein. This bound IGF-I may provide a reservoir for subsequent use by the epithelial cell types, even though they are not constitutively synthesizing this peptide. Therefore the binding proteins may function to control the amount of IGF that is available to bind to type I receptors and specific mechanisms may exist to regulate its availability if the growth requirements of a particular cell type are altered.

A more complete understanding of the role of these proteins in regulating expression of IGF-I action will require delineation of the specific types of binding proteins that are secreted by each cell type and definition of the variables that control the synthesis and secretion of these proteins. Likewise, the structural characteristics of each protein and the functional consequences of these differences should be analyzed. Finally the mechanisms that control the delivery of the IGF's from binding protein to receptors and whether or not this results in potentiation or inhibition of IGF actions will need to be determined.

STRUCTURAL ELUCIDATION

In the initial studies two classes of IGF binding proteins were characterized using gel filtration chromatography. The major binding component in plasma was shown to be a 150 kDa complex[10]. Its concentrations were a growth hormone dependent being elevated in acromegaly and low in hypopituitism[11]. This complex was subsequently discovered to be made up of an acid stable 53 kDa protein[12] and a 100 kDa acid labile component[13]. The acid stable subunit has recently been purified to homogeneity and shown to have a protein core of 291 amino acids[14]. It is a glycoprotein with three potential N-linked glycosylation sites. Differences in the extent glycosylation may account for the multiple molecular size estimates of this protein that have been reported[15,16]. When human plasma is electrophoresed under non-reducing conditions, both 42000 and 39000 Mr forms are detected, although no 54000 Mr form is seen[17]. These differences could be accounted by forms that contain sugar chains of variable length. In rat plasma the protein migrates by SDS gel electrophoresis as 47, 45 and 43000 Mr forms suggesting even further carbohydrate determined heterogeneity. The protein has a complicated disulfide bond structure with 18 cysteines and presumably no free sulfhydyl groups[16]. Presumably this characteristic accounts for anomolous migration patterns on SDS gels following reduction[18], and suggests a very compact structure due the extensive disulfide bond formation. cDNA sequence analysis has shown that the cDNA encoding in this protein contains 2587 base pairs with 873 have pairs in an open reading frame and a 1500 base pair 3' untranslated region. There appears to be only one gene and one species of mRNA of approximately 2.51 kB that is contained in human liver. The protein has a high affinity for IGF-I in range of approximately 10-10 M and binds IGF-I with slightly greater affinity than IGF-II.

A second class of IGF binding proteins also characterized originally by gel filtration chromatography have been shown to be inversely related to growth hormone concentrations[19]. The members of this group do not form a large molecular weight complex composed of subunits but migrate as a broad peak of binding activity between 25 to

40 kDa[20]. These proteins unlike the growth hormone dependent complex are not saturated with IGF-I or II and incubation of radioiodinated IGF-I with whole plasma followed by subsequent chromatography shows that a large portion of the radioactive material is associated with this peak[21]. Gel electrophoretic analysis of human and rat plasma has shown that in both species this peak of binding activity is composed of three proteins[17]. Specifically these have Mr estimates of 34, 30 and 25000. The putative 30000 Mr form is abundant in human amniotic fluid. It has been purified to homogeneity[22,23] and its N-terminal amino acid sequence determined[23]. The N-terminal sequence is identical to that reported for placental protein 12 and PP12 has been shown to bind IGF-I, therefore the two activities appear to be due to the same protein[24]. cDNA clones have been isolated from human decidual or human hepatoma cell (Hep G-2) libraries and contain 1550-1600 based pairs[25,26]. Since the mRNA's from these cell types are between 1.6-1.7 kB these cDNA's represent near full length clones. These clones have been sequenced and the complete amino acid sequence determined. The protein contains 259 amino acids and has a molecular weight of 25274 daltons. It contains 18 cysteine residues whose positions are identical to that contained in the 53 kDa growth hormone dependent IGF-BP therefore it probably has a very compact structure and this presumably accounts for its anomolous migration on SDS gels. Like the growth hormone dependent IGF-BP's there is only one gene encoding this protein within the human genome and only one species of mRNA has been detected in human decidua, liver, endometrium or Hep G2 cells[27]. This protein also has distinct structural differences. There are no N-linked glycosylation sites and multiple studies have shown that it does not contain carbohydrate. Furthermore, it contains an Arg-Gly-Asp (RGD) sequence near the C-terminus of the molecule[25]. This sequence is contained in extracellular matrix proteins and putatively can provide a cell attachment site for this protein to cell surfaces[28]. The general structure of the cDNA is similar to the growth hormone dependent binding protein. Specifically the clone has a short 5' untranslated sequence and a 610 base pair 3'untranslated region which contains three ATTTA sequences. These sequences may have an important regulatory function. In the GM-CSF gene these sequences have been shown to mediate mRNA degradation and are associated with rapid degradation of this labile mRNA species[29]. This is believed to be an important growth regulatory mechanism whereby the levels of this protein can be rapidly down-regulated. An additional interesting feature of this protein is its ability to form multimeric units in solution. The protein can form disulfide linked multimers under conditions that are achieved during the purification process. Likewise, it appears that these multimers may occur when the protein comes in contact with cell surfaces. Since the multimeric form of the protein is biologically active, this may confer a mechanism for switching from inhibitory to potentiating forms of IGF binding protein simply by forming multimeric subunits. The variables controlling this multimerization reaction are unknown at present.

Other IGF binding protein species have been clearly identified and are probably distinct gene products from those previously mentioned. Specifically, a 34000 Mr binding protein is abundant in cerebrospinal fluid and has a much greater affinity for IGF-II compared to IGF-I[19]. Likewise, Western blotting studies performed with specific antibodies against the forms of IGF binding protein purified from amniotic fluid or the growth hormone dependent forms do not react with the 34000 Mr protein. A bovine homologue of this protein has been partially sequenced and its amino terminus is similar to the GH dependent form. One other form of IGF binding protein that appears to be distinct, as determined by Western blotting studies, is the 25000 Mr form. This protein is the predominant form that is secreted by tumor cell types in culture (see below). Its structural characteristics have not been determined.

CONTROL OF SECRETION BY CELLS AND TISSUES

The recent development of radioimmunoassays for the 53 and 25 kDa forms of IGF binding proteins has enabled investigators to determine the major variables that control their secretion. Since antibodies to the human forms of IGF-BP's do not react with species other than primates, ligand blotting has been used to analyze the secretion of IGF-BP's by cells derived from other species. Secretion of the 25 kDa protein purified from human amniotic fluid has been extensively studied. Using polyclonal antisera prepared against placental protein 12, Rutanen et al. have shown that human decidual and endometrial explants synthesize this protein and synthesis by secretory endometrium is stimulated by the combination of estradiol and progesterone[30]. Likewise, decidual explants were shown to release PP-12 and the rate of release is significantly greater than explants of ammion or chorion[31]. In collaboration with Stuart Handwerger, our laboratory has shown that primary cultures of decidual cells synthesize the 25 kDa form of IGF-BP and cyclic AMP is a potent stimulant of secretion. In contrast, IGF-I and insulin are potent inhibitors, and these inhibitory effects are mediated through the type I receptor. After five days in culture, exposure to IGF-I results in approximately 8-fold reduction in the release of this protein from decidual cells[32]. Other cell types in culture have been shown to secrete this protein. Our laboratory in collaboration with Dr. David Hill has shown that human fibroblasts both fetal and postnatal, secrete small quantities of this protein[33]. Secretion appears to arise by de novo synthesis since exposure to cycloheximide is associated with inhibition of secretion. Interestingly IGF-I itself is a potent stimulus of both synthesis and secretion of this protein (figure 1).

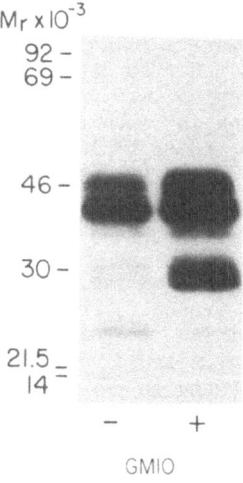

Fig. 1. Ligand blots of IGF binding protein secretion by human fibroblasts (GM-10 cells). The cells in the right hand panel were exposed to 50 ng/ml of IGF-I for 24 hours prior to media collection

384

Specifically, exposure of fibroblasts to IGF-I results in a 6-fold
increase in the concentration of the 25 kDa protein that is present in
tissue culture medium as determined by ligand blotting. In
contrast, the increases in other forms of IGF-BP were less than 2-fold.
Stimulation of IGF-BP secretion is mediated via the type I IGF receptor.
In addition to IGF-I, secretion of 25 kDa IGF-BP can be stimulated by
cyclic AMP and compounds that stimulate the synthesis of intercellular
cyclic AMP such as forskolin[34]. Likewise, beta adrenergic agonists such
as isoproternol, also increase its release. Dexamethasone and
hydrocortisone inhibit synthesis of this form of IGF binding protein by
fibroblasts.

Other human cell types that have been shown by radioimmunoassay to
secrete this protein include, breast carcinoma cells, uterine carcinoma
cells, ovarian granulosa cells and hepatoma cells. All have been shown
in culture to synthesize significant quantities of this protein and
release them into culture medium. In contrast to normal cell types,
these tumor cell types appear to constitutively synthesize higher levels
of 25000 Mr form of IGF-BP. Regulation of IGF-BP secretion may be
altered in tumor cells since IGF-I is not a potent stimulus in most
tumor cell lines whereas, the stimulatory effect of cAMP is usally
preserved. Muscle cells in culture do not appear to secrete the high
molecular weight growth dependent IGF-BP's, but they secrete various[35]
combinations of the 34000, 30000 and 25000 Mr forms[35]. Control of
secretion in muscle cells appears to correlate with the state of
cellular differentiation. When either BC3H-1 or L6 myoblasts are
examined in undifferentiated state they secrete the 31000 or 24000 Mr
forms of IGF-BP (figure 2). In undifferentiated L6 cells the ratio of

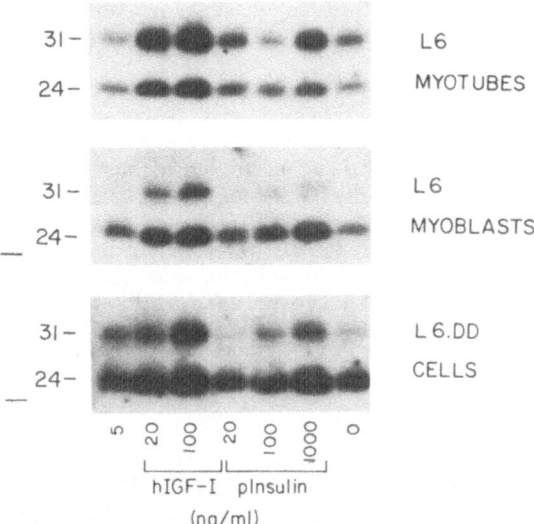

Fig. 2. Ligand blot of L6 cell conditioned media.
IGF-I or insulin were incubated with myotubes, myoblasts
or differentiation deficient cells for 24 hours and the
media collected.

24 kDa to 31 kDa protein is 5:1[36]. Following cellular differentiation
as assessed by myoblast fusion there is a shift in the ratio of the
forms of IGF-BP that are secreted to 1:1. Of further interest is the
fact that although IGF-I and insulin can stimulate the secretion of both

forms of these proteins when muscle cells differentiate the secretion of
BP-31 is preferentially induced[36]. Since insulin and IGF-I also
stimulate muscle cell differentiation secretion of this protein may be a
marker of insipient myoblast fusion and differentiation. IGF-BP
secretion by both L6 and BC3H-1 cells is very sensitive to inhibitory
effects of dexamethasone (figure 3). Since dexamethasone is a potent

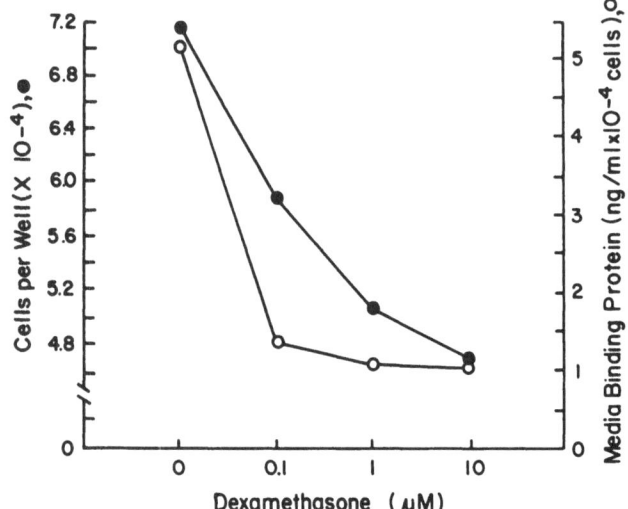

Fig. 3. Effect of dexamethasone on L6 cell growth
and IGF-BP secretion. L6 cultures were exposed to
increasing concentrations of dexamethasone and after
24 hours all numbers and IGF-BP concentrations were
determined.

inhibitor of smooth muscle cell growth and differentiation secretion of
the IGF-BP's appears to be related to these processes. The secretion of
IGF-BP, 53 kDa form, has been studied in human hepatocytes and
fibroblast cultures. Adams et al. demonstrated that cultured skin and
fibroblast released a 53 kDa form of IGF-BP and that release could be
inhibited by cycloheximide[37]. Martin and Baxter have reported that skin
fibroblasts release both 53 kDa IGF-BP and a 28 kDa form[38]. EGF
stimulated a 3-fold increase in the release of BP-53. Likewise,
cultured hepatocytes have been shown to release this form of IGF-BP and
its release is stimulated by GH[38].

CONTROL OF BLOOD CONCENTRATIONS

The levels of the 25 kDa forms of IGF-BP fluctuate widely in
normal human plasma. Specifically, nutritional deprivation is a major
stimulant to release this protein into the circulation. Following an
overnight fast the mean BP-25 concentration in 6 normal human subjects
showed a 10-12-fold increase from levels of 1-2 ng/ml at 1000 hours to
levels of 10-15 ng/ml at 0800 (figure 4)[39]. There appears to be no sex
difference in the degree of the response. The levels then decline 6-
fold following the ingestion of a meal and decline further after the
noon meal, reaching a nadir at approximately 1500 hours. A late
afternoon spike is then attenuated again by the evening meal, and the
levels remain depressed until approximately 1000. Baxter has shown that
this secretory pattern also occurs in short children with GH

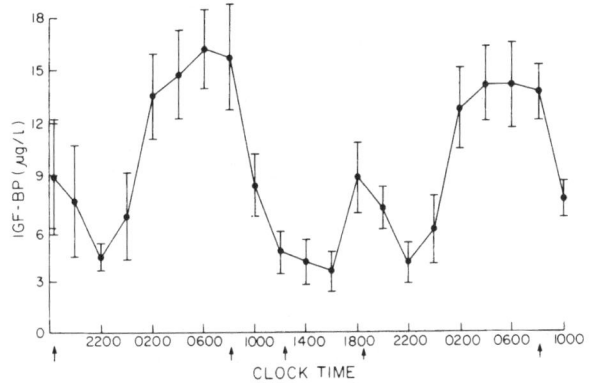

Fig. 4. Time dependent fluctuation in
25 kDa IGF-BP levels in six normal subjects.
Meals were ingested at (↑) the times indicated.

deficiency[40]. Prolonged fasting results in a further increase in the
fasting concentration of this protein. Specifically, when seven obese
subjects were fasted for nine days their levels rose from 8 ng/ml after
an overnight fast to 19 ng/ml by the fourth day of fasting and remained
elevated for the ensuing five days, then returned to normal control
levels after 48 hours of refeeding (figure 5). In contrast the plasma

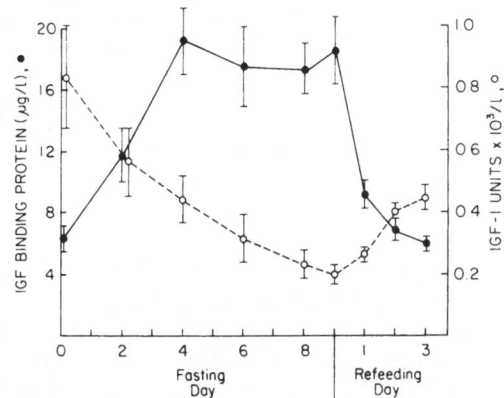

Fig. 5. Changes in 25 kDa IGF-BP and
IGF-I during fasting and refeeding.

IGF-I values fell 70% during the fasting interval. It appears then that
prolonged fasting is required to induce maximum secretion of this
protein at least in obese subjects. The specific food constituent that
is responsible for suppression of 25 kDa IGF-BP levels is carbohydrate.
In both type I and II diabeties, fasting BP-25 levels are elevated 2–4-
fold above control nondiabetic subjects. When glucose and insulin are
infused into those subjects the BP-25 concentrations fall 40–70%[41].
Whether this is due to provision of carbohydrate as a substrate, the
actual glucose transport event, or the ambient plasma insulin
concentration has not been definitely determined. Suikkari et al have
shown that physiologic concentrations of insulin can reduce 25 kDa BP
levels and that levels were reduced by 68% during a standard oral
glucose tolerance test[42]. A plasma insulin concentration of 1150 pmol/L
was required to induce maximal suppression. Therefore in these subjects

the plasma insulin concentration was a major determinant of 25 kDa IGF-BP levels. In addition to nutritional status multiple hormones appear to control plasma levels of this protein. Specifically administration of growth hormone to normal obese volunteers results in a 45% suppression in the ambient fasting levels of this protein[39]. Likewise, measurements in subjects with hypopituitarism are elevated approximately 3-fold and they are somewhat reduced in acromegaly[43]. Other factors that have been associated with significant changes in BP25 are age and pregnancy. Specifically, levels are elevated in newborn serum (mean fasting levels 6-fold > adults)[39] then fall progressively during childhood to reach the adult levels at adolescence. Likewise levels rise in maternal plasma during pregnancy and are further elevated by 50% in toxemia[44]. The protein may also serve as a tumor marker. Patients with hepatocellular carcinomas have been shown to have elevated levels as have some patients with colorectal cancer and ovarian cancer[45,46].

In contrast to BP-25, BP-53 levels do not vary even after 48 hours of fasting in humans[42]. This may be due to the marked GH dependency of this protein. The BP-53 levels are elevated 2.2-fold in acromegaly and reduced 50-80% in hypopituitarism. Following administration of GH there is a 4-fold increase in BP-53 levels within 6 hours. In contrast to BP-26 levels are low at birth and rise 3.5 fold at adolescence, then decline during adulthood. These changes parallel the age dependent changes in plasma IGF-I levels. In rats there are 3 proteins that represent the homologues of BP-53. If hypophysectomized rats receive an IGF-I infusion these three proteins are induced with IGF-I alone, although growth hormone can also achieve this induction, but it is only proportional to the degree to which growth hormone is able to increase the IGF-I concentration[5]. These data suggest that these proteins are IGF-I dependent and the capacity of growth hormone to induce their synthesis is dependent upon the ambient plasma IGF-I concentration. In summary, the plasma concentration of BP-53 and 25 appear to be controlled by distinct mechanisms and these findings suggest that although both proteins bind IGF-I their functional roles may be unique.

TARGET CELL ACTIONS

The purification of these two of forms of IGF-BP has made it possible to determine their target cell effects. When the 25 kDa form is coincubated with either cultured smooth muscle cells or fibroblasts there is no change in the rate of constituitive DNA synthesis. When IGF-I and this form of IGF-BP are coincubated IGF-I stimulated DNA synthesis responses are inhibited in both cell types[47]. However, if a low concentration of platelet poor plasma is added there is a 5-6 fold potentiation of the DNA synthesis response to IGF-I which is equal to or exceeds that achieved with 8.0% human serum (figure 6). This induction is specific for IGF-I since it cannot be achieved by coincubation of insulin plus binding protein suggesting that IGF-I must associate with the binding protein in order for this augmentation to occur and that activation of the type I receptor alone is insufficient. The factor in platelet poor plasma that is necessary for potentiation is also contained in cerebral spinal fluid but is not present in amniotic fluid. This suggests that the PPP factor may be present outside of the vascular space and therefore could be one variable in interstitial fluids that determines whether IGF binding proteins are positive or negative modulators of IGF action. However, this factor is not the sole determinant of whether the 25 kDa binding protein will act as a positive or negative modulator. In our initial studies we noted that binding protein must be associated with the cell surfaces or extracellular matrix in order to potentiate IGF-I actions[48]. The discovery that the

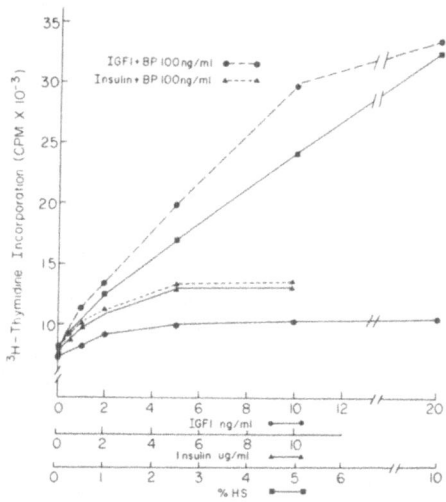

Fig. 6. Potentiation of smooth muscle
cell DNA synthesis response to IGF-I
by the 25 kDa IGF-BP.

binding protein contains an Arg-Gly-Asp sequence has led us to
reconsider this hypothesis and it is now apparent that this form of
binding protein is associated with the extracellular matrix of cells in
culture. The effect of this matrix associated binding protein on
potentiating IGF-I action can be blocked if a synthetic Arg-Gly-Asp
containing peptide is incubated with cells prior to exposure to IGF-I
and IGF-BP. This suggests that if association of the binding protein
that is present in the extracellular matrix with the cell surface is
blocked, potentiation of the DNA synthesis response to IGF-I is also
inhibited.

Another important variable in determining whether or not binding
protein is a positive or negative modulator appears to be its ability to
form multimers in solution. Specifically, BP-25 multimers have been
shown to have full potentiating biologic activity. It appears that
monomeric forms of BP-25 coincubated with cell cultures can multimerize
to form large disulfide linked multimers. If the multimerization
reaction is inhibited prior to initiation of the experiment, then
multimers do not form and the monomers will not potentiate the biologic
response suggesting that multimerization is an important for
potentiation of the IGF-I effect. Since matrix proteins such as
fibronectin must multimerize prior to exerting their biologic actions on
cells it is possible that the IGF binding protein specifically requires
multimerization before it can associate with cell surfaces though the
RGD site and activate DNA synthesis. Although this hypothesis will
require rigorous testing in multiple cell types, it appears that it
could provide molecular mechanism for the control of IGF-I mediated
growth stimulation.

A final interest of investigators in this area is determining the
mechanisms by which the multiple forms of IGF binding proteins might
interact to control cell growth. Specifically, most cells that we have
analyzed synthesize at least 2 to 4 separate forms of IGF binding
protein. Since many of these proteins can act as positive or negative
modulators of IGF action, these multiple combinations may interact to

389

influence cellular growth in either positive or negative manner. A recent paper by DeMellow and Baxter has shown that BP-53 can cause a 2-fold potentiation in IGF-I stimulated growth in fibroblasts if it is preincubated with fibroblast monolayer prior to addition of IGF-I[49]. If it is coincubated it is a negative modulator. This dual potential by at least 2 forms of IGF binding proteins could provide a mechanism for controlling the response of cells to IGF-I and II. Depending upon the needs of specific cell types for IGF dependent growth stimulation they could act as a positive or negative modulators.

In summary, understanding the structure and functional analysis of each form of binding protein will be required in order to form an integrated picture of how they function to regulate IGF-I mediated cellular replication. This family of proteins presents an interesting challenge to the ability of investigators to further dissect the molecular events that occur prior to growth factor binding to cell surface receptors before initiation of transmembrane signalling mechanisms. Elucidiation of the mechanisms by which these proteins effect these changes will be important for our further understanding of how the IGF's potentiate cellular growth.

ACKNOWLEDGEMENTS

The author gratefully acknowledges the technical assistance of Ms. Luanne Gardner and the editoral assistance of Ms. Anne Myers who prepared the manuscript. This work was supported by grants AG02331 and HL36313 from the National Institutes of Health.

REFERENCES

1. Han, V.KM.; D'Ercole, A.J.; Lund, P.K.: Cellular location of somatomedin (insulin-like growth factor) messenger RNA in the human fetus. Science 236: 193-197 (1987).

2. Clemmons, D.R.; Elgin, R.G.; Han, V.KM.; Casella, S.J.; D'Ercole, A.J.; Van Wyk, J.J.: Cultured fibroblast monolayers secrete a protein that alters the cellular binding of somatomedin-C/insulin-like growth factor I. J.Clin.Invest. 77: 1548-1558 (1986).

3. Hill, D.J.; Crace, C.J.; Fowler, L.; Holder A.T.; Milner D.G.: Cultured fetal rat myoblasts release peptide growth factors which are immunologically and biologically similar to somatomedin. J Cell Physiol 117: 349-358 (1984).

4. Clemmons, D.R.; Van Wyk, J.J.: Evidence for a functional role of endogenously produced somatomedin-like peptide in the regulation of DNA synthesis in cultured human fibroblasts and porcine smooth muscle cells. J Clin Invest 75:1914-1918 (1985).

5. Han, V.K.M.; Hill, D.J.; Strain, A.J.; Towle, A.C.; Lauder, J.M.; Underwood, L.E.; D'Ercole, A.J.: Identification of somatomedin/insulin-like growth factor immunoreactive cells in the human fetus. Ped Res 22:245-249 (1987).

6. Nissley, S.P.; Rechler, M.M.: Insulin-like growth factors: biosynthesis, receptors and IGF-carrier proteins; in Li, C.H., ed. Hormonal Proteins and Peptides XII, pp. 127-203 (Academic Press, New York 1985).

7. Hintz, R.L.; Liu, F.: Demonstration of specific plasma protein binding sites for somatomedin. J.Clin.Endocrinol.Metab. 45: 988-995 (1977).

8. Moses, A.C.; Freinkel, A.J.; Knowles, B.B.; Aden, D.P.: Demonstration that a human hepatoma cell line produces a specific insulin-like growth factor carrier protein. J.Clin.Endocrinol.Metab. 56: 1003-1008 (1983).

9. Hill, D.J.; Clemmons, D.R.; Wilson, S.; Hall, V.K.M.; Strain, A.J.; Milner, D.G.: Immunological distribution of one form of insulin-like growth factor (IGF) binding protein and IGF peptides in human fetal tissues. J Mol Endo 3: 154-159 (1988).

10. Drop, S.LS.; Kortleve, D.J.; Guyda, H.J.: Isolation of a somatomedin binding protein from preterm human amniotic fluid: development of a radioimmunoassay. J.Clin.Endocrinol.Metab. 59: 899-905 (1984).

11. Copeland, K.C.; Underwood, L.E.; Van Wyk, J.J.: Induction of immunoreactive somatomedin-C in human serum by growth hormone dose response relationships and effect on chromatograph profiles. J Clin Endocrinol Metab 50: 690-698 (1980).

12. Baxter, R.C.; Martin, J.L.: Radioimmunoassay of growth hormone dependent insulin-like growth factor binding protien in human plasma. J.Clin.Invest. 78: 1504-1512 (1986).

13. Baxter, R.C.: Characterization of the acid labile subunit of the growth hormone dependent insulin like growth factor binding protein complex. J.Clin.Endocrinol.Metab. 67: 265-272 (1988).

14. Wood, W.I.; Cathianes, G.; Henzel, W.J.; Winslow, G.A.; Spencer, S.A.; Hellmiss, R.; Martin, J.L.; Baxter, R.C.: Cloning and expression of the GH dependent insulin like growth factor binding protein. Mol.Endo. 2:1176-1185 (1988).

15. Martin, J.L.; Baxter, R.C.: Insulin like growth factor binding protein from human plasma. Purification and characterization. J.Biol.Chem. 261: 8754-8760 (1988).

16. Wilkins, J.R.; D'Ercole, A.J.: Affinity labelled somatomedin-C insulin-like growth factor binding proteins: Evidence of growth hormone dependent subunit structures. J Clin Invest 75: 1350-1358 (1985).

17. Hossenlopp, P.; Seurin, D.; Segovia, B.; Portolan, G.; Binoux, M.: Heterogeneity of insulin-like growth factor binding proteins and relationships between structure and affinity. 2 forms released by human and rat liver in culture. Eur.J.Biochem. 170: 133-142 (1987).

18. Baxter, R.C.; Martin, J.L.; Wood, M.H.: Two immunoreactive binding proteins for insulin-like growth factors in human amniotic fluid: relationship to fetal maturity. J.Clin.Endocrin.Metabol. 65: 423-431 (1987).

19. Drop, L.S.; Kortleve, D.J.; Guyda, H.J.; Posner, B.I.: Immunoassay of a somatomedin-binding protein from human amniotic fluid: levels in fetal, neonatal, and adult sera. J.Clin.Endocrinol.Metab. 59:908-915 (1984).

20. Zapf, J.; Waldvogel, M.; Froesch, E.R.: Binding of nonsuppressible insulin-like activity to human serum: Evidence for a carrier protein. Arch.Biochem.Biophys. 168: 638-645 (1975).

21. White, R.M.; Nissley, S.P.; Moses, A.C.; Rechler, M.M.; Johnsonbarvh, R.E.: The growth hormone dependence of somatomedin binding protein in human sera. J Clin Endocrinol Metab 53: 49–57 (1981).

22. Koistinen, R.; Huhtala, M.L.; Stenman, U.H.; Seppala, M.: Purification of placental protein PP12 from human amniotic fluid and its comparison with PP12 from placenta by immunological, physicochemical and somatomedin-binding properties. Clin Chim Acta 164: 293–303 (1987).

23. Povoa, G.; Enberg, G.; Jornvall, H.; Hall, K.: Isolation and characterization of a somatomedin binding protein from mid term human amniotic fluid. Eur.J.Biochem. 144: 199–204 (1984).

24. Koistinen, R.; Kalkkinen, N.; Huhtala, M.L.; Seppala, M.; Bohn, H.; Rutanen, E.M.: Placental protein 12 is a decidual protein and has an identical N-terminal amino acid sequence with somatomedin binding protein in human amniotic fluid. Endocrinology 118: 1375–1184 (1986).

25. Brewer, M.T.; Stetler, G.L.; Squires, C.H.; Thompson, R.C.; Busby, W.H.; Clemmons, D.R.: Cloning, characterization and expression of a human insulin- like growth factor binding protein. Biochem.Biophys.Res.Comm. 152: 1289–1297 (1988).

26. Lee, Y.L.; Hintz, R.L.; James, D.M.; Lee, P.DK.; Shively, J.E.; Powell, D.R.: Insulin like growth factor IGF binding protein complementary deoxyribonucleic acid form human Hep G2 hepatoma cells: predicted protein sequence suggests an IGF binding domain different from those of IGF-I and IGF-II receptors. Mol.Endo. 3: 404–411 (1988).

27. Julkunen, M.; Koistinen, R.; Setala, K.A.; Seppala, M.; Janne, O.; Kontula, K.: Primary structure of human insulin-like growth factor binding protein/placental protein 12 and tissue specific expression of its mRNA. FEBS Lett 236: 295–352 (1988).

28. Rouslahti, E.; Pierschbacher, M.D.: Arg–Gly–Asp: A versatile recognition signal. Cell 44: 517–518 (1988).

29. Shaw, G.; Kamen, R.P.: A conserved AU sequence from the 3' untranslated region of GM-CSF mRNA mediating selective mRNA degradation. Cell 46: 659–667 (1986).

30. Rutanen, E.-M.; Koistinen, R.; Wahlstrom, T.; Bohn, H.; Ranta, T.; Seppala, M.: Synthesis of placental protein 12 by human decidua. Endocrinol. 116: 1304–1309 (1985).

31. Rutanen, E.-M.; Koistinen, R.; Sjoberg, J.; Julkunen, M.; Wahlstrom, T.; Bohn, H.; Seppala, M.: Synthesis of placental protein 12 by human endometrium. Endocrinol. 118: 1067–1071 (1986).

32. Thrailkill, K.; Clemmons, D.R.; Handwerger, S.: Control of Secretion of IGF 25 kDa binding protein by human decidual cells. Annual Meeting of the American Society of Clinical Investigation. Washington, D.C.,: (1989).(Abstract)

33. Hill, D.J.; Camacho-Hubner, C.; Rashid, P.; Strain, A.J.; Clemmons, D.R.: Insulin like growth factor binding protein secretion by human fibroblasts: Dependency on cell density and IGF peptides. Journal of Endocrinology (1989).(in press)

34. Camacho-Hubner, C.; McCusker, R.H.; Clemmons, D.R.: Insulin like

growth factor binding protein secretion by human tumor cells is
hormonally regulated. 70th Annual Meeting of the Endocrine Society
(1988). Abstract #526.

35. McCusker, R.M.; Clemmons, D.R.: Insulin like growth factor binding
protein secretion by muscle cells: Effect of cellular differentiation
and proliferation. J.Cell Physiol. 137: 505-512 (1988).

36. McCusker, R.H.; Camacho-Hubner, C.; Clemmons, D.R.: Identification
of the types of insulin like growth factor binding proteins that
are secreted by muscle cells in vitro. J.Biol.Chem. in press: (1989).

37. Adams, S.O.; Kapadia, M.; Mills, B.; Daughaday, W.H.: Release
of insulin-like growth factors and binding proteins into serum-free
medium of cultured human fibroblasts. Endocrinol. 115: 520-526 (1984).

38. Martin, J.L.; Baxter, R.C.: Insulin like growth factor binding
proteins IGF-BP's produced by human skin fibroblasts immunological
relationship to other IGF-BP's. J Clin Endocrinol Metab 123: 1907-1915
(1989).

39. Busby, W.H.; Snyder, D.K.; Clemmons, D.R.: Radioimmunoassay of a
26000 dalton plasma insulin like growth factor binding protein: Control
by nutritional variables. J.Clin.Endocrinol.Metab. 67: 1231-1236 (1988).

40. Baxter, R.C.; Colwell, C.T.: Diurnal rhythm of growth hormone
dependent binding protein for insulin-like growth factors in human
plasma. J.Clin.Endocrinol.Metab. 65: 432-440 (1987).

41. Sukkari, A.M.; Koistinen, V.A.; Rutanen, E.M.; Jarvinen, H.;
Karonen, S.L.; Seppala, M.: Insulin regulates serum levels of low
molecular weight insulin like growth factor binding protein.
J.Clin.Endocrinol.Metab. 66: 266-273 (1988).

42. Sukkari, A.M.; Koivisto, V.A.; Koistinen, R.; Seppala, M.; Yki-
Jarvinen, H.: Dose response characteristics for suppression of low
molecular weight plasma insulin like growth factor binding proteins by
insulin. J Clin Endocrinol Metab 68: 135-140 (1989).

43. Povoa, G.; Roovete, A.; Hall, K.: Cross-reaction of serum
somatomedin-binding protein in a radioimmunoassay developed for
somatomedin-binding protein isolated from human amniotic fluid. Acta
Endocrinol. 107: 563-570 (1984).

44. Iino, K.; Sjoberg, J.; Seppala, M.: Elevated circulating levels
decidual protein placental protein 12 in preclampsia. Obst and Gyn 68:
58-60 (1986).

45. Iino, K.; Seppala, M.; Heinonen, P.K.; Seipponen, P.; Rutanen, E.M.:
Elevated levels of somatomedin binding protein PP12 in patients with
ovarian cancer. Cancer 58: 2294-3000 (1986).

46. Rutanen, E.M.; Wahlstrom, T.; Koistinen, R.; Seipponen, P.; Jalanko,
H.: Plancental protein 12 in primary liver cancer and cirrhosis. Tumor
Biology 5: 95-102 (1984).

47. Elgin, R.G.; Busby, W.H.; Clemmons, D.R.: An insulin-like growth
factor (IGF) binding protein enhances the biologic response to IGF-I.
Proc.Natl.Acad.Sci. 84: 3254-3258 (1987).

48. Busby, W.H.; Klapper, D.G.; Clemmons, D.R.: Purification of a

31000 dalton insulin like growth factor binding protein from human amniotic fluid. J.Biol.Chem. 263: 14203–14210 (1988).

49. De Mellow, J.SM.; Baxter, R.C.: Growth hormone dependent insulin-like growth factor binding protein both inhibits and potentiates IGF-I stimulated DNA synthesis in skin fibroblasts. Biochem. Biophys. Res. Commun. 156: 199–204 (1988).

CHARACTERIZATION AND CLONING OF A RAT INSULIN-LIKE GROWTH FACTOR BINDING PROTEIN

Matthew M. Rechler, Alexandra L. Brown, Craig C. Orlowski, Yvonne
W.-H. Yang, Joyce A. Romanus, Lorenzo Chiariotti*, and Carmelo B. Bruni*

Molecular, Cellular and Nutritional Endocrinology Branch, National
Institute of Diabetes and Digestive and Kidney Diseases, National Institutes
of Health, Bethesda, Maryland 20892

*Centro di Endocrinologica ed Oncologia Sperimentale del Consiglio
Nazionale delle Ricerche, Dipartimento di Biologia e Patologia Cellulare e
Molecolare, Universita degli Studi di Napoli, Italy 80122

INSULIN-LIKE GROWTH FACTOR (IGF) BINDING PROTEINS IN HUMAN SERUM

The insulin-like growth factors, IGF-I and IGF-II, occur complexed to specific
binding proteins in blood and other extracellular fluids.[1] IGF-binding protein complexes of
150 kDa predominate in adult human and rat serum[2,3], and also have been observed in human
and porcine milk [4,5], and human fibroblast conditioned media[6,7]. Acid pH irreversibly
dissociates the 150 kDa binding protein complex into an ~40 kDa acid-stable binding
subunit.[1] Recently, Baxter[8] provided evidence for the existence of a second subunit of ~100
kDa that is unstable at acid pH and does not bind IGFs.

IGF binding proteins of ~40 kDa have been observed in growth hormone-deficient
plasma, fetal plasma, extravascular fluids (such as amniotic fluid[9] and cerebrospinal fluid[10])
and cell culture media.[1] The ~40 kDa binding proteins and the binding subunit of the 150 kDa
complex bind IGF-I, IGF-II or both, with high affinity (10^9 to 10^{11} M^{-1}), but do not bind
insulin.[1,9,11] The IGF binding proteins are thought to play an important role in targeting
IGFs to the vascular compartment or tissues, and in modulating the mitogenic actions of IGFs
on their target tissues.[12]

The IGF binding components in human serum have been analyzed by ligand blotting,
a technique in which serum proteins size are fractionated by sodium dodecyl sulfate-gel
electrophoresis under nonreducing conditions, electroblotted to nitrocellulose, and the IGF
binding components identified by incubating the blots with ^{125}I-labeled IGFs and
autoradiography.[13] Electrophoresis under denaturing conditions dissociates endogenous
IGFs from the binding proteins so that both occupied and unoccupied binding proteins may be
examined. The observed molecular weight reflects only the IGF binding components. As
seen in Figure 1, five binding components were identified by this technique in human
serum.[13] The major proteins, 41.5 and 38.5 kDa, are thought to represent glycosylation
variants of the same protein, and occur predominantly in the 150 kDa region of a neutral
Sephadex G200 gel filtration column. Minor proteins of 34 and 30 kDa are present in the 40
kDa region of the column, and are not glycosylated. They correspond in size to the major
proteins in human amniotic fluid (30 kDa) and human cerebrospinal fluid (34 kDa). The
nature of the 24 kDa binding protein is presently unclear. The same five binding protein
components are present in all human cells and fluids, albeit in different proportions.

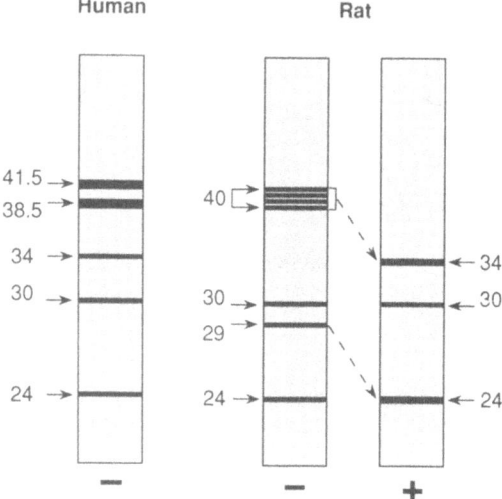

Figure 1. Schematic diagram illustrating the IGF binding components in adult human[13] and rat[14] serum as detected by ligand blotting. Serum proteins were fractionated by electrophoresis on sodium dodecyl sulfate-polyacrylamide gels under nonreducing conditions. Rat serum was examined before (-) or after (+) treatment with N-glycanase. Proteins were blotted to nitrocellulose, and the IGF binding components identified by incubation with ^{125}I-labeled IGFs and autoradiography. Apparent molecular weights (in kilodaltons) relative to protein standards are shown.

In rat serum, N-glycanase decreases the size of the 40 kDa cluster and the 29 kDa protein, indicating that they are N-glycoproteins.[14] The 30 kDa protein is unchanged, whereas the intensity of the 24 kDa protein is increased (presumably reflecting the contribution of the deglycosylated 29kDa glycoprotein fragment).

IGF BINDING PROTEINS IN RAT SERUM

Multiple IGF binding components also are observed in adult rat serum after ligand blotting.[14] A cluster of proteins at ~40 kDa appears to represent the binding components of the predominant 150 kDa complex. These proteins are growth hormone-dependent and contain N-linked oligosaccharides, since treatment with N-glycanase reduces their size to 34 kDa.[14] A 29 kDa growth hormone-dependent glycoprotein also is present in the 150 kDa region (after neutral gel filtration) whose size is decreased to 24 kDa after N-glycanase treatment.[14] Zapf et al.[15] have presented evidence that the 29 kDa protein represents an amino terminal fragment of the ~40 kDa glycoprotein cluster. The presence of the 29 kDa glycoprotein fragment in the 150 kDa column region suggests that, like the ~40 kDa glycoprotein cluster, it interacts with the rat counterpart of the acid-labile nonbinding subunit.

Nonglycosylated proteins of 30 kDa and 24 kDa are present in the 40 kDa region of the neutral gel filtration column. The 24 kDa protein may represent a deglycosylated 29 kDa binding protein fragment, a fragment of one of the other known binding proteins, or an as yet uncharacterized binding protein.

The 30 kDa binding protein in adult rat serum is similar in size to an IGF binding protein that is synthesized and secreted by the BRL-3A rat liver cell line[16-18], a cell line that has been extensively studied in our laboratory because it produces high concentrations of IGF-II. Polyclonal antibodies were raised in rabbits to purified BRL-3A binding protein and a radioimmunoassay established.[16] Radioimmunoassay[16], immunoblotting[14,17], and immunoprecipitation[14] studies indicate that the 30 kDa binding proteins in BRL-3A cells and in fetal and neonatal rat serum are immunologically related and possibly identical, and that they are distinct from the 30 kDa binding protein in adult rat serum.

BIOSYNTHESIS AND REGULATION OF THE BRL-3A BINDING PROTEIN (BP-3A)

We have studied the biosynthesis of the IGF binding protein in BRL-3A cells, BP-3A, using cell free translation in a reticulocyte lysate system, and biosynthetic labeling of intact cells.[19] Translation of BRL-3A RNA in the presence of microsomal membranes to provide a source of signal peptidase decreased the size of the biosynthetic precursor by ~2 kDa, indicating that the BP-3A precursor contains a prepeptide. In pulse-chase labeling experiments, BP-3A was labeled rapidly intracellularly, and secreted to the medium where it remained unchanged in size and concentration for up to 24 hours. The amino terminal amino acid sequence of biosynthetically labeled intracellular BP-3A[19] corresponded to that of BP-3A isolated from BRL-3A conditioned medium.[17,18,20] These results suggested that BP-3A does not contain a detectable propeptide, and that BP-3A is not degraded after secretion into BRL-3A medium.

Cell-free translation and immunoprecipitation of RNA from fetal and adult rat liver indicated that translatable RNA encoding BP-3A was present in fetal, but not in adult, rat liver.[19] This suggested that BP-3A RNA might have a similar developmental regulation to that of IGF-II RNA in rat tissues.[21]

CHARACTERIZATION OF cDNA CLONES FOR BP-3A

A λgt10 cDNA library was prepared with BRL-3A polyA+ RNA and screened with long oligonucleotide probes designed from the amino acid sequence.[22] Extensive regions of the 30 kDa BP-3A protein and of a ~14 kDa carboxyl terminal fragment of BP-3A[20] were purified from BRL-3A conditioned medium and sequenced by W. H. Burgess and T. Mehlman (American Red Cross, Rockville, MD). A cDNA clone was identified containing an ~2 kb insert that hybridized to three BP-3A oligonucleotide probes. The nucleotide sequence of this clone contained a long open reading frame that encoded the amino terminal sequence of BP-3A,[17-20] and the amino acid sequence that had been determined by protein sequencing (~70% of the protein). BP-3A contains a 34 residue hydrophobic prepeptide, and a mature protein of 270 residues (formula weight ~30 kDa) (Figure 2). Mature BP-3A contains no N-glycosylation sites, consistent with previous biochemical data.[16,18] It contains 18 cysteine residues clustered at the amino and carboxyl terminal regions. An Arg-Gly-Asp (RGD) sequence found in many extracellular matrix proteins[23] and thought to be recognized by cell adhesion receptors is present near the carboxyl terminus. The determined protein sequence extends to within five residues of the translation termination codon, indicating that if a carboxyl-terminal propeptide exists, it can not be longer than 5 residues. The protein and nucleic acid sequences of BP-3A have no significant homologies to those of other known proteins, including the IGF-I and the IGF-II/mannose-6-phosphate receptors[24,25], with the exception of other IGF binding proteins (see below).

A fragment of the BP-3A clone containing the coding region was inserted into a pGEM-3 expression plasmid, and a capped RNA transcript of the sense strand of the cDNA synthesized from the linearized plasmid using SP6 polymerase and m7GpppG.[22] The synthetic BP-3A RNA was translated in a reticulocyte lysate system and injected into Xenopus oocytes in the presence of [^{35}S]methionine. A labeled protein of the appropriate size (~30 kDa) was immunoprecipitated from both the translation incubation medium and from Xenopus medium by antibodies to mature BP-3A. The IGF binding proteins expressed in Xenopus had the same affinity and specificity for IGF-I and IGF-II as the native binding protein purified from BRL-3A cells: the Ka for IGF-I was ~10^9 M^{-1}, and IGF-II was 60% as potent as IGF-I.[20,22]

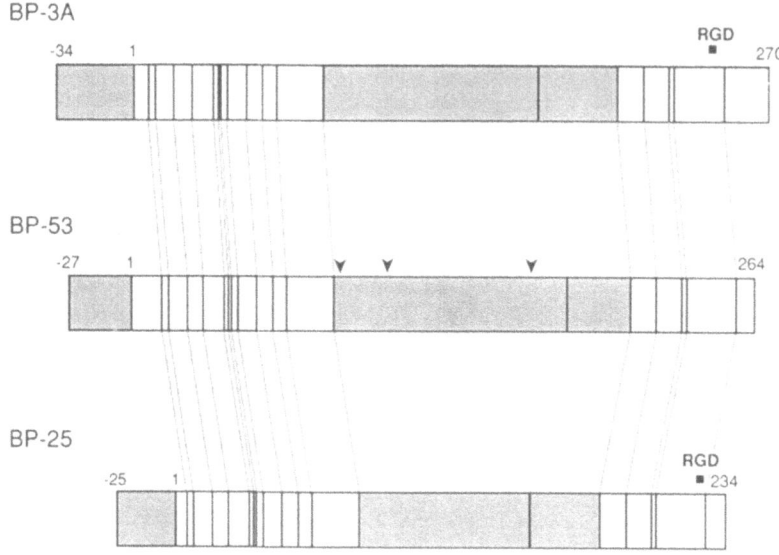

Figure 2. Structure of IGF-binding protein precursors. The structures are based on the nucleotide sequences of cDNA clones encoding BP-3A[22], BP-53[26], and BP-25[27-30]. The amino terminal residue of the mature binding proteins is designated residue 1. The carboxyl terminal residue is indicated. Other vertical lines represent cysteine residues, with the connecting lines illustrating the alignment of 17 of the 18 cysteines. The more highly conserved amino and carboxyl terminal regions are not shaded; the signal peptide and the middle non-conserved region are stippled. Carets indicate the positions of three potential N-glycosylation sites in BP-53. The position of the RGD sequence at the carboxyl terminal ends of BP-3A and BP-25 are shown.

Table 1. Comparison of Amino Acid Sequence Identities

	Amino Terminus	Middle	Carboxyl Terminus
BP-3A/BP-53	52% (45/86)	4% (4/92)	38% (21/56)
BP-3A/BP-25	45% (38/84)	16% (14/85)	50% (28/56)
BP-53/BP-25	48% (40/84)	19% (18/94)	32% (17/53)
BP-3A/BP-MDBK	82% (36/44)		96% (22/23)

Amino acid sequences of BP-3A[22], BP-53[26] and BP-25[27-30] were derived from the nucleotide sequences. The sequence of the binding protein from MDBK cells (BP-MDBK) is based on the partial amino acid sequence[31]. Alignments were chosen to maximize cysteine pairing and amino acid identities. The per cent of identical amino acids (neglecting gaps) is shown. The number of identical and total residues in each region are in parentheses.

Two human IGF binding proteins have recently been purified and cloned, BP-53[26] and BP-25[27-30]. BP-53 is the binding subunit of the 150 kDa complex in human serum.[11] It consists of a ~30 kDa protein moiety plus N-linked oligosaccharides to bring its weight to ~43 kDa on sodium dodecyl sulfate-gels under reducing conditions. BP-25 has been purified from human amniotic fluid, placental membranes, and the Hep G2 hepatocarcinoma cell line, and has been cloned from human uterine decidua[27-29] and Hep G2 cells.[30] It has a formula weight of 25 kDa and is not glycosylated. BP-25 is expressed in decidua and secretory endometrium, but not in placenta or proliferative endometrium.

The structures of the three cloned IGF binding proteins, BP-3A, BP-53 and BP-25 are compared in Figure 2. The mature proteins are similar in size (270, 264 and 234 residues, respectively), and are synthesized with similar size prepeptides (25-34 residues). Seventeen of the 18 cysteine residues are clustered in the amino and carboxyl terminal regions, and can be aligned in the 3 proteins. Despite these similarities, the three binding proteins differ in several important respects. Only BP-53 possesses N-glycosylation sites. BP-3A and BP-25, but not BP-53, have an RGD sequence at the carboxyl terminus.

Detailed comparison of the amino acid sequences deduced for the three proteins (Table 1) indicates that they are most closely related in the amino and carboxyl terminal regions (45-52% and 32-50% amino acid sequence identity, respectively). In the carboxyl terminal region, the homology of BP-25 and BP-3A is slightly greater than that of either protein to BP-53. The three proteins have no significant homology in the middle region. These results indicate that BP-3A is a distinct member of the IGF binding protein family, and not the rat homologue of BP-25 or BP-53.

A much closer relationship was observed between the BP-3A sequence and the partial amino acid sequence of a bovine IGF binding protein purified from MDBK kidney cells.[31] These proteins have 82% and 96% sequence identity in the amino and carboxyl terminal regions (Table 1), suggesting that the MDBK protein is the bovine homologue of BP-3A. Interestingly, the specificity of the MDBK binding protein differs from that of BP-3A in competitive binding experiments using IGF-II tracer, namely, it has 10-fold higher affinity for IGF-II than IGF-I.

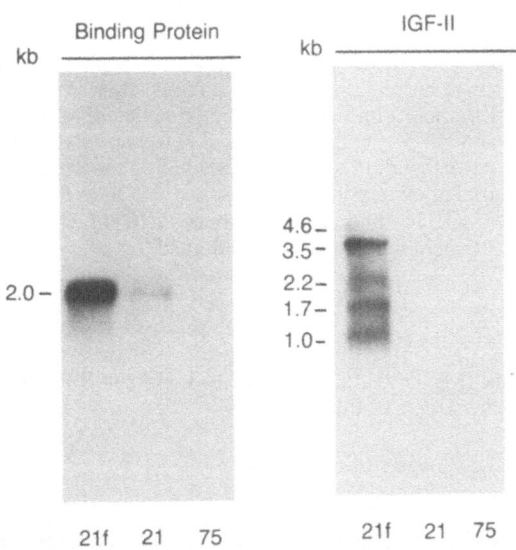

Figure 3. Hybridization of BP-3A[22] (left) and IGF-II[21] (right) cDNA probes to total RNA (15 µg/lane) prepared from livers of rats of different ages: 21 day gestation (21f), and 21 and 75 day postnatal. Approximate sizes (in kilobases) of the multiple IGF-II RNAs and the single BP-3A RNA are indicated. The abundance of both RNAs is greatly decreased after birth.

EXPRESSION OF BP-3A mRNA IN RAT TISSUES

BP-3A is expressed as a single ~2 kb RNA in BRL-3A cells (not shown) and in term fetal rat liver (Figure 3).[22] Lower amounts of BP-3A mRNA of the same size are expressed in term fetal rat kidney, lung, intestine, muscle, heart and stomach. The abundance of BP-3A RNA was 10-20 fold lower in livers from 21 and 75 day postnatal rats. A similar developmental regulation has been seen with IGF-II RNA. The relative abundance of BP-3A and IGF-II RNA differed in different fetal tissues, suggesting that they were not precisely coordinately regulated.

Unlike non-neural tissues, positive hybridization persisted with the BP-3A probe in brain stem, cerebral cortex and hypothalamus from 21 day postnatal rats and in brain stem from 75 day postnatal rats.[32,33] A similar persistence had been observed for IGF-II mRNA.[21] IGF-II mRNA has been predominantly localized to the choroid plexus by in situ hybridization.[34,35] Similar experiments using an oligonucleotide corresponding to the coding region of BP-3A also showed preferential localization of BP-3A mRNA to the choroid plexus.[32,33]

Localization of BP-3A RNA to the choroid plexus of adult rats suggested that BP-3A might be secreted into the cerebrospinal fluid. In fact, cerebrospinal fluid from adult rats contains an ~30 kDa protein that is recognized by antibodies to native BP-3A and binds IGF-I and IGF-II with a specificity and affinity indistinguishable from that of native BP-3A.[32,33] Preliminary results suggest that the 34 kDa IGF binding protein in human cerebrospinal fluid is the human homologue of BP-3A. The two proteins differ, however, in their specificity for IGF-I and IGF-II, since the human 34 kDa binding protein has higher affinity for IGF-II.[10,36]

The widespread expression of BP-3A in fetal rat tissues contrasts markedly with the limited expression of BP-25. Using a human BP-25 cDNA probe[29] (kindly provided by O. Janne, Rockefeller University, New York), we have identified a 1.5 kb mRNA in fetal rat liver, but not in other fetal or adult rat tissues (unpublished results).

CONCLUSIONS

At least three distinct IGF binding proteins appear to be present in both man and rat: BP-53, a glycosylated growth hormone-dependent binding subunit of the 150 kDa complex; BP-25, present in human amniotic fluid and human uterine decidua, and expressed to only a limited extent in the rat; and BP-3A, widely expressed in fetal rat tissues and in adult rat brain and choroid plexus, and corresponding to the 34 kDa binding protein in human cerebrospinal fluid. The different patterns of expression of BP-3A and BP-25 in the rat suggest that the biological roles of the two IGF binding proteins may be different. The role of IGF-II and BP-3A in the central nervous system remains to be determined.

ACKNOWLEDGMENTS

We wish to thank D.E. Graham, L. Tseng, N-J. Ge and S.P. Nissley for helpful discussions throughout the course of this work.

REFERENCES

1. Nissley, S.P. and Rechler, M.M. Insulin-like growth factors: biosynthesis, receptors, and carrier proteins, in: "Hormonal Proteins and Peptides", Vol. XII, C.H. Li, ed., Academic Press, New York and London (1984).
2. R.M. White, S.P. Nissley, A.C. Moses, M.M. Rechler, and R.E. Johnsonbaugh, The growth hormone dependence of a somatomedin-binding protein in human serum, J. Clin. Endocrinol. Metab. 53:49 (1981).

3. A.C. Moses, S.P. Nissley, K.L. Cohen, and M.M. Rechler, Specific binding of a somatomedin-like polypeptide in rat serum depends on growth hormone, Nature 263:137 (1976).
4. R.C. Baxter, Z. Zaltsman, and J.R. Turtle, Immunoreactive somatomedin-C/insulin-like growth factor I and its binding protein in human milk, J.Clin.Endocrinol.Metab. 58:955 (1984).
5. F.A. Simmen, R.C.M. Simmen, and G. Reinhart, Maternal and neonatal somatomedin C/insulin-like growth factor I (IGF-I) during early lactation in the pig, Dev. Biol. 130:16 (1988).
6. D.R. Clemmons, L.E. Underwood, and J.J. Van Wyk, Hormonal control of immunoreactive somatomedin production by cultured human fibroblasts, J. Clin. Invest. 67:10 (1981).
7. J.S.M. De Mellow, and R.C. Baxter, Growth hormone-dependent insulin-like growth factor (IGF) binding protein both inhibits and potentiates IGF-I-stimulated DNA synthesis in human skin fibroblasts, Biochem. Biophys. Res. Commun. 156:199 (1988).
8. R.C. Baxter, Characterization of the acid-labile subunit of the growth hormone-dependent insulin-like growth factor binding protein complex, J. Clin. Endocrinol. Metab. 67:265 (1988).
9. R.C. Baxter, J.L. Martin, and M.H. Wood, Two immunoreactive binding proteins for insulin-like growth factors in human amniotic fluid: relationship to fetal maturity, J.Clin.Endocrinol.Metab. 65:423 (1987).
10. P. Hossenlopp, D. Seurin, B. Segovia-Quinson, and M. Binoux, Identification of an insulin-like growth factor-binding protein in human cerebrospinal fluid with a selective affinity for IGF-II, FEBS Lett. 208:439 (1986).
11. J.L. Martin, and R.C. Baxter, Insulin-like growth factor-binding protein from human plasma: Purification and characterization, J.Biol.Chem. 261:8754 (1986).
12. G.R. Elgin, W.H.Jr. Busby, and D.R. Clemmons, An insulin-like growth factor (IGF) binding protein enhances the biologic response to IGF I, Proc.Natl.Acad.Sci.U.S.A. 84:3254 (1987).
13. S. Hardouin, P. Hossenlopp, B. Segovia, D. Seurin, G. Portolan, C. Lassarre, and M. Binoux, Heterogeneity of insulin-like growth factor binding proteins and relationships between structure and affinity: 1. circulating forms in man, Eur. J. Biochem. 170:121 (1987).
14. Y.W-H. Yang, J-F. Wang, J.A. Romanus, and M.M. Rechler, manuscript in preparation.
15. J. Zapf, W. Born, J-Y. Chang, P. James, E.R. Froesch, and J.A. Fischer, Isolation and NH2-terminal amino acid sequences of rat serum carrier proteins for insulin-like growth factors, Biochem. Biophys. Res. Commun. 156:1187 (1988).
16. J.A. Romanus, J.E. Terrell, Y.W. Yang, S.P. Nissley, and M.M. Rechler, Insulin-like growth factor carrier proteins in neonatal and adult rat serum are immunologically different: demonstration using a new radioimmunoassay for the carrier protein from BRL-3A rat liver cells, Endocrinology 118:1743 (1986).
17. R.M. Lyons, and G.L. Smith, Characterization of multiplication-stimulating activity (MSA) carrier protein, Mol. Cell. Endocrinol. 45:263 (1986).
18. C. Mottola, R.G. MacDonald, J.L. Brackett, J.E. Mole, J.K. Anderson, and M.P. Czech, Purification and amino-terminal sequence of an insulin-like growth factor-binding protein secreted by rat liver BRL-3A cells, J. Biol. Chem. 261:11180 (1986).
19 J.A. Romanus, Y.W-H. Yang, S.P. Nissley, and M.M. Rechler, Biosynthesis of the low molecular weight carrier protein for insulin-like growth factors in rat liver and fibroblasts, Endocrinology 121:1041 (1987).
20. J-F. Wang, B. Hampton, T. Mehlman, W.H. Burgess, and M.M. Rechler, Isolation of a biologically active fragment from the carboxy-terminus of the fetal rat binding protein for insulin-like growth factors, Biochem. Biophys. Res. Commun. 157:718 (1988).
21. A.L. Brown, D.E. Graham, S.P. Nissley, D.J. Hill, A.J. Strain, and M.M. Rechler, Developmental regulation of insulin-like growth factor II mRNA in different rat tissues, J.Biol.Chem. 261:13144 (1986).
22. A.L. Brown, L. Chiariotti, C. Orlowski, T. Mehlman, W.H. Burgess, E.J. Ackerman, C.B. Bruni, and M.M. Rechler, Nucleotide sequence and expression of a cDNA clone encoding a fetal rat binding protein for insulin-like growth factors, J. Biol. Chem. (1989). (In Press)

23. E. Ruoslahti, and M.D. Pierschbacher, New perspectives in cell adhesion: RGD and integrins, Science 238:491 (1987).
24. A. Ullrich, A. Gray, A.W. Tam, T. Yang-Feng, M. Tsubokawa, C. Collins, W. Henzel, T. Le Bon, S. Kathuria, E. Chen, S. Jacobs, U. Francke, J. Ramachandran, and Y. Fujita-Yamaguchi, Insulin-like growth factor I receptor primary structure: comparison with insulin receptor suggests structural determinants that define functional specificity, EMBO J. 5:2503 (1986).
25. D.O. Morgan, J.C. Edman, D.N. Standring, V.A. Fried, M.C. Smith, R.A. Roth, and W.J. Rutter, Insulin-like growth factor II receptor as a multifunctional binding protein, Nature 329:301 (1987).
26. W.I. Wood, G. Cachianes, W.J. Henzel, G.A.Winslow, S.A. Spencer, R. Hellmiss, J.L. Martin, and R.C. Baxter, Cloning and expression of the growth hormone-dependent insulin-like growth factor binding protein, Mol. Endocrinol. 2:1176.
27. M.T. Brewer, G.L. Stetler, C.H. Squires, R.C. Thompson, W.H. Busby, and D.R. Clemmons, Cloning, characterization, and expression of a human insulin-like growth factor binding protein, Biochem. Biophys. Res. Commun. 152:1289 (1988).
28. A. Brinkman, C. Groffen, D.J. Kortleve, A.G. van Kessel, and S.L.S. Drop, Isolation and characterization of a cDNA encoding the low molecular weight insulin-like growth factor binding protein (IBP-1), EMBO J. 7:2417 (1988).
29. M. Julkunen, R. Koistinen, K. Aalto-Setala, M. Seppala, O.A. Janne, and K. Kontula, Primary structure of human insulin-like growth factor-binding protein/placental protein 12 and tissue-specific expression of its mRNA, FEBS Lett. 236:295 (1988).
30. Y-L. Lee, R.L. Hintz, P.M. James, P.D.K. Lee, J.E. Shively, and D.R. Powell, Insulin-like growth factor (IGF) binding protein complementary deoxyribonucleic acid from human HEP G2 hepatoma cells: predicted protein sequence suggests an IGF binding domain different from those of the IGF-I and IGF-II receptors, Mol.Endocrinol. 2:404 (1988).
31. L. Szabo, D.G. Mottershead, F.J. Ballard, and J.C. Wallace, The bovine insulin-like growth factor (IGF) binding protein purified from conditioned medium requires the N-terminal tripeptide in IGF-1 for binding, Biochem. Biophys. Res. Commun. 151:207 (1988).
32. C.C. Orlowski, L.Y-H. Tseng, A.L. Brown, T. Taylor, Y.W-H. Yang, J.A. Romanus, and M.M. Rechler, Insulin-like growth factor II (IGF-II) and the BRL-3A IGF binding protein are expressed coordinately in adult rat choroid plexus and cerebrospinal fluid, Program of the Endocrine Society (1989).
33. L.Y-H. Tseng, A.L. Brown, T. Taylor, Y.W-H. Yang, J.A. Romanus, and M.M. Rechler, manuscript in preparation.
34. F. Stylianopoulou, J. Herbert, M.B. Soares, and A. Efstratiadis, Expression of the insulin-like growth factor II gene in the choroid plexus and the leptomeninges of the adult rat central nervous system, Proc. Natl. Acad. Sci. USA 85:141 (1988).
35. M.A. Hynes, P.J. Brooks, J.J. Van Wyk, and P.K. Lund, Insulin-like growth factor II messenger ribonucleic acids are synthesized in the choroid plexus of the rat brain, Mol. Endocrinol. 1:47 (1988).
36. M. Binoux, C. Lassarre, and M. Gourmelen, Specific assay for insulin-like growth factor (IGF-II) using the IGF binding proteins extracted from human cerebrospinal fluid, J. Clin. Endocrinol. Metab. 63:1151 (1986).

NEUROTROPHIC EFFECTS AND MECHANISM OF INSULIN, INSULIN-LIKE GROWTH FACTORS, AND NERVE GROWTH FACTOR IN SPINAL CORD AND PERIPHERAL NEURONS

Douglas N. Ishii[*], Gordon W. Glazner[*], Chiang Wang[*], and Paul Fernyhough[**]

[*]Physiology Department, Colorado State University
Fort Collins, CO, USA 80523

[**]Cell Biophysics Unit, Medical Research Council
Kings College, London WC2B 5RL, England

INTRODUCTION

Insulin and insulin-like growth factors (IGFs) are members of a gene family (Blundell and Humbel, 1980; Jansen et al., 1983; Dull et al., 1984; Bell et al., 1985; Soares et al., 1985). Their sites of synthesis, tissue distribution, and effects including those on feeding behavior, electrical activity, neuromodulation, and metabolism, and other data, lend strong support to the conviction that insulin homologs have important functions on the nervous system (reviewed in Recio-Pinto and Ishii, 1988a). Their neurotrophic properties, which were initially observed in cultured cells originating from the peripheral nervous system, support the hypothesis that these factors play roles in the development and maintenance of the circuitry and function of the nervous system (Recio-Pinto and Ishii, 1984; Recio-Pinto et al., 1986). Additional support for this hypothesis is obtained from studies which now show the effects of these factors on cells from the central nervous system, and the novel observation that the IGF-II gene is expressed in muscle during development and denervation in a manner correlated with synaptogenesis. The presence of multiple receptors for neurotrophic factors on the same population of neurons may permit flexibility in response.

In an attempt to elucidate the sequence of cellular events culminating in neurite formation, our previous work concentrated on identifying human neuroblastoma cell lines responsive and unresponsive to insulin, IGFs, and NGF; responses were then correlated with specific ligand receptors (reviewed in Ishii and Mill, 1987). The effects of insulin and IGFs on transcripts encoding proteins of the axonal cytoskeleton, beginning with tubulins, were studied. This work recently has been advanced to show the mechanism by which tubulin mRNA content is increased. In addition, effects on transcripts for neurofilament proteins have been studied. Other recent developments include a partial unraveling of the relationship of protein kinase C to the neurite growth pathway regulated by neuritogenic polypeptides. By comparing and contrasting the effects, mechanisms, and patterns of gene expression for insulin, IGF-I and IGF-II with those of the classic neurotrophic protein, nerve growth factor (NGF), a better understanding of the relative roles these factors play in the development and maintenance of the structure of the nervous system is beginning to emerge.

NEUROTROPHIC EFFECTS

Sympathetic and Sensory Cells

Subnanomolar concentrations of insulin, IGF-I and IGF-II increase both the proportion of cells with neurites and the average neurite length in cultured human neuroblastoma SH-SY5Y cells (Recio-Pinto and Ishii, 1984; Ishii and Recio-Pinto, 1987). SH-SY5Y cells are most likely of sympathetic origin, indicating insulin homologs might act on cells from peripheral ganglia. Subsequent investigation found that physiological concentrations of these factors enhance the survival of neurons as well as increase neurite formation in embryonic chick sensory (Bothwell, 1982; Recio-Pinto et al., 1986) and sympathetic cultures (Recio-Pinto et al., 1986). Insulin, but not IGF-II, is substantially less potent in sensory than in sympathetic neurons, showing that different populations of neurons may respond in a quantitatively different fashion to a particular concentration of ligand. Moreover, although all of these ligands can support survival, NGF has greater efficacy on survival than insulin or IGFs in both sensory and sympathetic neurons. Insulin and IGFs were shown to act directly on neurons because the presence of non-neuronal cells is not required for the responses. The effects of insulin are antagonized by anti-insulin but not anti-NGF antiserum. Insulin and IGFs' capacity to increase neurite outgrowth and support neuron survival are important neurotrophic properties shared with NGF.

Brain Neurons

The insulin homologs also have neurotrophic effects on cultured central nervous system neurons. Supraphysiological insulin concentrations support survival of fetal mouse cerebral (Bhat, 1983), postnatal mouse cerebellar (Huck, 1983) cells, and fetal rat cortical neurons (Aizenman and de Vellis, 1987). The latter authors showed however that when physiological concentrations were studied IGF-I, but not insulin is active.

NGF is a survival factor for lesioned cholinergic neurons. Transection of axons projecting to the hippocampus from the septum results in loss of cholinergic septal neurons; these cells can be rescued by administration of NGF (Hefti, 1986; Kromer, 1987). NGF, however, is not found to support survival of cultured fetal rat cortical neurons (Aizenman and de Vellis, 1987).

Indirect data suggest insulin homologs might influence neuronal proliferation. For example, thymidine uptake is stimulated by insulin (Raizada et al., 1980; Yang and Fellows, 1980) and IGF-I (Lenoir and Honegger, 1983) in brain cell cultures. This effect of IGF-I is also observed in cultures enriched in neurons (Shemer et al., 1987), but residual glial cells may potentially account for the thymidine incorporation. Thymidine incorporation and cell population density are increased in cloned neuroblastoma SH-SY5Y cells exposed to physiological concentrations of insulin, IGF-I and IGF-II (Recio-Pinto and Ishii, 1984; Mattsson et al., 1986; Ishii and Recio-Pinto, 1987). These data lead to the speculation that insulin homologs may regulate the proliferation of neuroblasts.

Spinal Cord Cells

Studies on the neurotrophic effects of insulin homologs have been extended to include spinal cord cells. Cells from embryonic rats were cultured with insulin, IGF-I, IGF-II, and NGF (Table I). All of these factors increased neurite outgrowth. The combination of insulin and NGF was not significantly better than insulin alone, indicating that insulin and NGF may have effects on the same or overlapping populations of spinal cord neurons. In other studies, the ED50 for insulin-stimulated neurite growth was approximately 2 nM (Glazner et al., 1988). It is of interest that NGF receptors are transiently expressed during development on alpha motor neurons (Raivich et al., 1985).

These studies show that physiological concentrations of insulin, IGF-I, IGF-II, and NGF can support neurite outgrowth in some spinal cord cells. About 25-30% of the plated cells appear to be responsive to insulin. Because serum contains these growth factors, and glia can produce IGFs and NGF, it was necessary to plate the cells at a low density and reduce the serum concentration in order to observe these effects.

Distribution of Brain Insulin and IGF Receptors

The widespread localization of insulin and IGF binding is consistent with a neurotrophic role, but does not exclude other functions. The distribution of radioactive insulin (Hill et al., 1986), IGF-I (Bohannon et al., 1988), and IGF-II (Smith et al., 1987) binding to rat brain regions has been studied by quantitative autoradiography. The type I IGF receptor binds to IGF-I with higher affinity than to IGF-II, is structurally related to insulin, and high experimental concentrations of insulin can cross-occupy this site. However, at low concentrations of radioactive insulin and IGFs the relative affinities greatly favor binding to their own receptors permitting fairly unambiguous localization. High densities of insulin and IGF binding are not found in the same brain regions, although it seems all areas contained at least low levels of ligand binding. Because IGF-I and IGF-II both bind to the functional type I IGF receptor, the discrimination of ligand effects may lie in the differential spatial and temporal pattern of production of each ligand in various tissues, and in localization of effects.

The type II IGF receptor binds more tightly to IGF-II than to IGF-I, is structurally dissimilar to the insulin receptor, and does not bind insulin even at high concentrations. A clear function has not been ascribed to this "receptor" and for the present it might be considered a binding site.

Table I. Effect of Insulin, IGF-I, IGF-II, and NGF on Neurite Outgrowth in Cultured Rat Spinal Cord Cells. Spinal cords were dissected from 17-19 day-old rat embryos, incubated with trypsin, and passed through a series of stainless steel wire gauzes of decreasing mesh size. The cells (100,000/ml) were plated on polylysine coated 35 mm dishes in F12/Dulbecco's modified Eagle's medium (1:1) supplemented with 2% horse serum, 2% fetal calf serum, and 100 ng/ml transferrin. After 2 days, viable neurons (based on trypan-blue dye exclusion) were scored for neurite outgrowth. Random fields were photographed and the proportion of neurons with neurites longer than 1 cell diameter was calculated from projected negatives using procedures similar to those used by Recio-Pinto et al. (1986). The non-neurons were flattened and spread out, whereas neurons were rounded and bright under modulation contrast microscopy. Generally, about 30-40% of the plated cells survived.

Experiment	Treatment	Neurite Outgrowth (%)
A	Untreated	29.3 $\pm$ 2.5 (3)
	Insulin, 1 uM	55.7 $\pm$ 5.5* (3)
B	Untreated	33.3 $\pm$ 1.8 (4)
	IGF-I, 10 nM	55.9 $\pm$ 2.8* (4)
C	Untreated	40.8 $\pm$ 4.5 (6)
	IGF-II, 10 nM	54.2 $\pm$ 2.2* (6)
D	Untreated	31.9 $\pm$ 3.1 (3)
	NGF, 0.4 nM	49.2 $\pm$ 5.6* (4)
E	Untreated	22.2 $\pm$ 3.3 (4)
	Insulin, 1 uM	57.1 $\pm$ 4.7* (4)
	NGF, 0.4 nM	47.8 $\pm$ 3.5* (4)
	Insulin, 1 uM plus NGF, 0.4 nM	61.8 $\pm$ 2.1* (4)

Values are means $\pm$ SEM, (N) shows number of replicate cultures.
*$P < 0.05$ between group and untreated control values.

Flexibility in Neurotrophic Responses in Different Neuronal Populations

Whole brain or brain regions have been utilized in most binding studies investigating the CNS. Due to the heterogeneity of cell types and the presence of receptors on glia as well as neurons, it has been understandably difficult to correlate specific responses with binding parameters which are averages from a complex cell population. To circumvent these objections, the relationship between binding and response has been studied in cell lines. Distinct receptors for NGF (Sonnenfeld and Ishii, 1982; 1985), insulin and IGFs (Recio-Pinto and Ishii, 1988b) are present on cloned SH-SY5Y cells. There is a strong correlation between concentrations of each ligand that occupies its own receptor and the neurite outgrowth response. Insulin and IGFs do not compete for binding to NGF receptors.

The binding data show that multiple independent receptors regulating neurotrophic responses may be present on the same cell. The responses in sensory, sympathetic, and spinal cord cells indicate that insulin, IGFs and NGF may have effects on the same or overlapping populations of neurons. The presence of multiple independent receptors may confer flexibility in neurotrophic response. For example, insulin and IGFs have similar potency for neurite outgrowth in sympathetic neurons, but insulin are substantially less potent in sensory neurons (Recio-Pinto et al., 1986). It is likely that insulin is acting through insulin receptors on sympathetic cells, but through IGF receptors on sensory cells. Certain populations of neurons may be responsive to some but not all of these neurotrophic factors. Moreover, although NGF is a more efficacious survival factor than insulin or IGFs in sympathetic neurons, nevertheless the same maximum neurite outgrowth response can be supported by all of these factors. This shows that although receptors for all three factors may be present on the same cell, some neurotrophic effects may be quantitatively different from others. Thus, the degree and kind of response may be dependent in part on the temporal and spatial regulation of both the relative concentrations of multiple neurotrophic factors in the microenvironment of the neuron and the selective expression of particular receptors. Summation of effects is probably also important.

SYNAPTOGENESIS

The IGF-II gene is expressed in a manner consistent with a role in the development of the nervous system. IGF-II transcripts are rare in adult rat tissues, except in spinal cord and brain (Soares et al., 1985; Soares et al., 1986; Rotwein et al., 1988) where abundance is particularly high in choroid plexus and leptomeninges (Stylianopoulou et al., 1988). Many peripheral tissues in embryos express high levels of mRNAs which subsequently are developmentally down regulated (Soares et al., 1985; Brown et al., 1986). The potential that this down regulation is related to the development of the nervous system was investigated.

IGF-II Gene Expression Correlates With Developmental Formation and Elimination of Polyneuronal Innervation

Coordinated movement and behavior are dependent on the formation of precise synaptic connections. Consistent with an effect on spinal cord neurons, the IGF-II gene is expressed in a manner closely correlated with the developmental formation of synapses at the neuromuscular junction (Ishii, 1989).

Polyneuronal innervation is a condition in which more than one neuron innervates a single muscle fiber. Using morphological criteria, motor endplates may first be distinguished at about the 16th day of gestation in rats (Kelly and Zacks, 1969). Polyneuronal innervation subsequently accumulates, for reasons not well understood, and reaches a maximum shortly before birth which occurs at about 21 days gestation (Redfern, 1970; Brown et al., 1976). The IGF-II gene is expressed in developing limb buds of 14-days-old embryos, prior to the formation of synapses, and IGF-II mRNAs accumulates and peaks in skeletal muscle concomitantly with polyneuronal innervation (Ishii, 1989).

The phenomenon of polyneuronal innervation may help ensure that all available muscle fibers become innervated. Superfluous synapses are eliminated in early postnatal life (Brown et al., 1976; O'Brien et al., 1978) leading by Day 15 to the adult condition in which each muscle fiber is innervated by only a single motor neuron. The amount of IGF-II mRNAs and extent of polyneuronal innervation are both observed to decline with the same developmental time course in littermates (Ishii, 1989). The developmental up and down regulation of IGF-II mRNA content is selective and are observed relative to total and poly(A)$^{+}$ RNA.

The Developmental Decline in IGF-II mRNAs and Polyneuronal Innervation Are Regulated Concomitantly by Nerve Activity

The elimination of polyneuronal innervation is governed by neuromuscular activity in a manner which is poorly understood. For example, the developmental loss of multiple synapses can be accelerated by stimulation (O'Brien et al., 1978) and retarded by inhibition (Thompson et al., 1979) of nerve activity. In light of this it is interesting, therefore, that nerve activity also modulates IGF-II gene expression in muscle. Nerve transection prevents the normal developmental down regulation of IGF-II mRNAs in ipsilateral denervated, but not contralateral intact muscles (Ishii, 1989). Nerve transection leads to loss of physical contact between nerve terminal and muscle basement membrane, release of neurotransmitter, and potentially the release of other substances from terminals.

Botulinum toxin does not affect the ultrastructure of nerve terminals but does block the release of acetylcholine and retard the elimination of polyneuronal innervation (Thompson et al., 1979; Brown et al., 1981). This toxin, like nerve transection, prevented the down regulation of IGF-II mRNAs (Fig. 1). IGF-II mRNA abundance was diminished in untreated 18-day-old muscles (C1, C2, C3) relative to 5 day muscle (Std), due to normal developmental down regulation. However, the levels were elevated in toxin-treated animals (B1, B2, B3).

These results suggest that neurotransmitter release may trigger down-regulation of IGF-II mRNA in muscle, but one may not entirely discount the alternative possibility that other factors released from nerve terminals are involved. To resolve this ambiguity, ongoing studies will examine the effects which inhibition of acetylcholine binding to its postsynaptic receptors may have on IGF-II gene expression in muscle.

IGF-II Gene Expression Correlates With Muscle Re-innervation

Adult muscle will generally not accept additional innervation. However, denervated mature muscle can become receptive to innervation (Elsberg, 1917; Hoffman, 1951). Nerve transection not only prevents the developmental down regulation of IGF-II transcripts in muscles of postnatal rats, it also causes up regulation of IGF-II mRNAs in mature animals. The up regulation of IGF-II transcripts in denervated adult muscle is correlated with the capacity for regeneration of synapses (Ishii, 1989). The unique IGF-II gene gives rise to multiple transcripts (Soares et al., 1985; 1986). These transcripts are expressed in the same ratio in both developing and denervated muscles.

In contrast, IGF-II gene expression and muscle growth are uncorrelated. Developmental down regulation of IGF-II transcripts occurs while rapid muscle growth continues for many additional weeks (Chiakulas and Pauly, 1965). Denervated muscles undergo atrophy even though IGF-II mRNA levels are increased many fold. A positive correlation has not been found between general somatic growth and circulating IGF-II concentrations. Also, growth hormone has little effect on IGF-II levels, and IGF-II levels decline in postnatal rats despite the continued growth spurt.

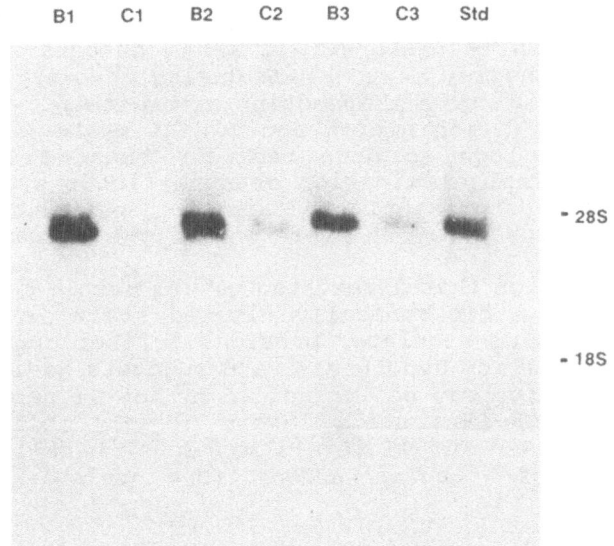

Fig. 1. Effect of nerve block with botulinum toxin on IGF-II mRNA abundance in rat muscle. Littermate rats (11-days-old) were randomly assorted into two groups. Groups were injected with 8.4 ul botulinum toxin (0.05 ng/ul), or vehicle alone, s.c. in the left hind leg. A booster of 4 ul was given when 15-days-old. RNA was isolated from left calf muscles of 18-day-olds, electrophoresed (40 ug/lane) in formaldehyde agarose gels, and transferred to nitrocellulose using methods previously described (Fernyhough and Ishii, 1987). The nitrocellulose blot was hybridized to nick-translated cDNA clone 27, which contains the coding region of rat IGF-II (Soares et al., 1985). The autoradiogram is shown: B, botulin toxin treated rats; C, untreated; Std, standard RNA from 4-day-old rat muscle. Ethidium bromide staining revealed the positions of 18 and 28S rRNA, and confirmed that equivalent amounts of undegraded RNA had been loaded.

A Provisional Model for The Relationship Between IGF-II Gene Expression and Synapse Formation

These observations are consistent with the hypothesis that the IGF-II mRNA is among factors which may regulate synaptogenesis (Ishii, 1989). High levels of IGF-II transcripts may predispose towards polyneuronal innervation. The great abundance of transcripts in the prenatal environment may not permit advancing axons to discriminate between uninnervated and innervated muscle fibers, resulting in accumulation of polyneuronal innervation.

The postnatal down regulation of IGF-II mRNAs may lead to an environment in which polyneuronal innervation is no longer supportable. After synaptic contacts are established, nerve terminals may be exposed to a microenvironment rich in IGF-II. IGF-II, like NGF, may become internalized and transported in a retrograde fashion back towards the neuronal soma. Alternatively, a transmembrane signal other than the ligand itself may be transported. A high level of IGF-II gene expression no longer may be required to retain communication between muscles and nerves once synapses have matured.

As synapses mature, a feed-back inhibitory signal may be produced to down regulate IGF-II mRNA content. In agreement with this hypothesis, nerve block during development, produced by either sciatic nerve transection or treatment with botulinum toxin, lifted the inhibition and IGF-II mRNAs were not down-regulated. Failure to down regulate these transcripts may prevent the normal elimination of superfluous synapses. This model, however, presently offers no explanation for how particular synapses may be preferentially eliminated.

The observation that denervated mature muscle can attract re-innervation in a biochemically altered state in which IGF-II mRNA content is up regulated provides further support for the feed-back inhibition hypothesis, and suggests additionally that nerve regeneration may be influenced by IGF-II gene expression. This assumes IGF-II protein levels closely follows its mRNA content. Increased IGF-II levels may contribute in part to the accumulation of tubulin mRNAs (see below) observed in regenerating nerves.

Following partial denervation, remaining intact nerve terminals are observed to sprout, but only within a distance of about 200 um from denervated muscle fibers (Pockett and Slack, 1982). This indicates that a diffusible neuritogenic substance is released from denervated muscle fibers. Extracts from muscle contain neuritogenic activity which is increased following denervation (Henderson et al., 1983). IGF-II is potentially responsible for at least part of this increase.

Some caution needs to be taken in the interpretation of the above data. While a strong positive correlation has been found between IGF-II gene expression and synaptogenesis, a causative relationship has not yet been established. However, it is worth noting that the IGF homolog, insulin, can accelerate the rate of development of evocable synaptic transmission between co-cultures of retinal ganglion and muscle cells (Puro and Agardh, 1984). Although the vertebrate neuromuscular junction was selected for study because it is well characterized, IGFs may be inferred to have effects on other types of synapses as well.

NGF Gene Expression During Development and Denervation

Consistent with expectations for a neurotrophic factor, the NGF gene is expressed in peripheral tissues receiving sensory and sympathetic innervation, and brain tissues receiving cholinergic innervation. There is a positive correlation between NGF mRNA concentration in target tissues and degree of sympathetic (Shelton and Reichardt, 1984; Goedert et al., 1986) and cholinergic (Korsching et al., 1985; Goedert et al., 1986) innervation.

Unlike the IGF-II gene, however, the developmental expression of the NGF gene in innervated tissues has not been found to correlate with synaptogenesis. For example, NGF mRNAs are not detected in whisker pads prior to the arrival of sensory axons from the trigeminal ganglion (Davies et al., 1987). Synaptogenesis may trigger the expression of the NGF gene, and it is believed that NGF may help neurons survive which otherwise are developmentally programmed to die.

Moreover, in contrast to the situation for the IGF-II gene, the expression of the NGF gene does not appear to be up regulated in target tissues of adult rats as a result of denervation. Although NGF content rises in denervated irises (Ebendal et al., 1983), the NGF mRNA levels do not (Shelton and Reichardt, 1986). Likewise, cholinergic deafferentiation of hippocampus and cerebral cortex does not cause a rise in NGF mRNA content in adult rats (Goedert et al., 1986; Korsching et al., 1986). However, fimbria transection does result in an increase in NGF levels in adult rats (Gasser et al., 1986; Korsching et al., 1986) and a transient increase in NGF mRNA in the hippocampus of neonatal rats (Whittemore et al., 1987). These data suggest that NGF content may become increased in denervated tissues as a result of increased translation, decreased degradation, or retention of protein which otherwise would be exported. The content of NGF and IGF in tissues appear to be regulated by denervation in fundamentally different ways.

MECHANISM OF NEURITE GROWTH

Receptors

The properties of insulin, IGF and NGF receptors which regulate neurite formation are discussed elsewhere (Ishii and Mill, 1987; Recio-Pinto and Ishii, 1988a; 1988b). It is possible that phosphorylation is the common transmembrane event activated by these factors. There is a strong correlation between occupancy of receptors, enhancement of neurite formation, and elevation of transcripts encoding structural proteins. The suggestion that there may be cooperative interactions between these factors is based on the observations that insulin and IGFs can modify the response to NGF by regulating the number of NGF receptors (Recio-Pinto et al., 1984) and can influence down-modulation of NGF receptors (Ishii and Recio-Pinto, 1987). Moreover, insulin and NGF can synergistically increase neurite formation and tubulin mRNA levels in PC12 cells (Recio-Pinto et al., 1984; Fernyhough and Ishii, 1987).

Axonal Transport

Target tissues release trophic substances which are picked up at nerve terminals. For example, NGF is internalized and transported in a retrograde fashion in axons back to the neuronal soma (Hendry et al., 1974; Seiler and Schwab, 1984). A substantial body of in vitro and in vivo evidence shows that the advancing growth cone can follow a concentration gradient of NGF (Gunderson and Barrett, 1980; Menesini-Chen et al., 1978); chemotactic guidance may be an important mechanism which helps to form the proper connections between nerve cells and their targets during development and regeneration. Although Schwann cells are not targets, they also can produce NGF. The role low affinity NGF receptors located on the surface of Schwann cells may play in chemotactic/contact guidance is discussed elsewhere (Ishii and Mill, 1987). Such a role may be particularly important during nerve regeneration when NGF content is increased in the distal nerve stump, which serves as a conduit for regrowth.

IGF-II, like NGF, potentially may be endocytosed at nerve terminals. Immunoreactive IGF I accumulates on both sides of ligated nerves, indicating IGFs are transported in axons (Hansson et al., 1987). It is not yet known, however, whether IGF-I transported in axons originates from target tissues. It is also unclear whether retrograde transport of the ligand is essential for activity, because insulin and type I IGF receptors are kinases and second messengers potentially could serve the same function.

Tubulin mRNAs

The stable growth of long axons and dendrites during development and regeneration requires synthesis of structural proteins and their assembly into the cytoskeleton of neurites. Of particular interest is whether the biochemical makeup of neurites formed in the presence of insulin and IGFs is the same as those formed in response to NGF, and whether these factors share a similar mechanism for the construction of neurites.

Alpha- and beta-tubulin heterodimers assemble into microtubules which are major structural elements of the cytoskeleton and which support axonal transport. Tubulin protein and mRNA levels are increased during regeneration of goldfish optic (Neumann et al., 1983) and rat sciatic (Hoffman et al., 1987) nerves. While the nature of the signals in these systems has not been identified, a potential link is suggested by the increase in IGF-II mRNA levels in denervated muscle. NGF increases tubulin (Drubin et al., 1985; Kolber et al., 1974) and microtubule (Levi-Montalcini, 1966; Luckenbille-Edds et al., 1979) content, as well as neurite growth in neurons, human neuroblastoma and rat pheochromocytoma PC12 cells. Microtubules become more resistant to depolymerization following prolonged exposure of cells to NGF (Black and Greene, 1982).

Alpha- and beta-tubulin mRNAs are increased by insulin and IGF-II in SH-SY5Y cells (Mill et al., 1985), and by NGF in sensory (Ishii and Mill, 1987) and PC12 (Fernyhough and Ishii,

1987) cells during neurite formation. Insulin by itself can neither increase tubulin mRNA levels nor enhance neurite formation in PC12 cells, but can synergistically potentiate both effects when administered together with NGF (Fernyhough and Ishii, 1987). The concentrations of these ligands which elevate tubulin transcripts are similar to those which occupy their respective receptors and enhance neurite elongation.

The increase in tubulin mRNA content appears to result largely from transcript stabilization. Tubulin mRNAs are stabilized by NGF in PC12 cells (Fernyhough and Ishii, 1987), and by insulin and IGF-II in SH-SY5Y cells (manuscript submitted). Alterations in nuclear run-off rates are not observed in response to insulin (unpublished). A transient 5-60 min increase in nuclear run-off rates is observed in NGF-treated PC12 cells (Greenberg et al., 1985). However, this transient increase can not explain the slow accumulation of tubulin transcripts which occurs over 2-4 days in PC12 (Fernyhough and Ishii, 1987) and 12-24 h in SH-SY5Y (Mill et al., 1985) cells. Tubulin synthesis is increased in NGF-treated PC12 (Drubin et al., 1988) and insulin-treated SH-SY5Y (unpublished) cells. Taken together, these data suggest that tubulin mRNA content is increased primarily by stabilization in cells treated with neuritogenic polypeptides. Because tubulin transcripts comprise a large fraction of the total message pool, stabilization may be an energy efficient mechanism to modulate neurite formation.

Multiple alpha- and beta-tubulin genes (Lemischka et al., 1981; Hall et al., 1983; Bond and Farmer, 1983) give rise to isotypes of alpha- and beta-tubulins, some of which are neuron specific (Bond et al., 1984). Arai and Matsumoto (1988) have recently found an asymmetry in the distribution of beta-tubulin isotypes in squid neurons which may be related to microtubules with separate functions. Segregation of function is observed in microtubules; some are involved in vesicular transport whereas others contribute to the axonal architecture (Miller et al., 1987).

Microtubule associated proteins (MAPs) are also regulated by neuritogenic polypeptides. For example, NGF can increase the content of MAPs such as tau, MAP1 (Drubin et al., 1988) and MAP1.2 (Aletta et al., 1988) in PC12 cells.

Neurofilament mRNA

The formation of the axonal cytoskeleton during neurite extension involves other structural elements as well. Prominent among these are neurofilaments (NF), a class of intermediate filaments restricted to neurons. NFs are comprised of subunit proteins with apparent masses of approximately 68 kD (light), 160 kD (medium), and 200 kD (heavy) (Hoffman and Lasek, 1975). Their cDNAs have been isolated and sequenced (Julien et al., 1985; Julien et al., 1986; Lewis and Cowan, 1986; Robinson et al., 1986). The 200 kD subunit is observed during postnatal brain development, whereas the other two subunits are observed earlier in embryos in some (Shaw and Weber, 1982; Willard and Simon, 1983), but not all (Cochard and Paulin, 1984) studies. Neurofilament proteins determine, in part, the caliber of axons (Hoffman et al., 1985; 1987). Their regulation appears to be at the level of gene expression.

NGF increases the content of 68 and 170 kD NF proteins within 1 day in PC12 cells, but not that of the 200 kD subunit until after 14 days (Lindenbaum et al., 1988). Accumulation of 68 and 170 kD proteins is associated with an increase in the abundance of their specific transcripts due to increased transcription (Dickson et al., 1986; Lindenbaum et al., 1988). An increase in 200 kD NF mRNAs is not seen. The heavy NF protein appears to form cross-bridges between filaments (Hirokawa et al., 1984). Its delayed expression during neurite elongation may be desireable so that early formation of cross-bridges do not inhibit the assembly of other proteins.

Do insulin and IGFs, like NGF, increase the expression of neurofilament genes? To study this question, SH-SY5Y cells were exposed to insulin for various times and the relative content of 68 and 170 kD NF mRNAs was measured (Fig. 2). Both 68 and 170 kD mRNAs were transiently increased to a maximum at about 24 h. In other studies it was found that these transcripts were clearly elevated by physiological concentrations of insulin and IGFs (Wang et al., 1988).

It is curious that these transcripts are not transiently expressed, but remain elevated in NGF-treated PC12 cells (Lindenbaum et al., 1988). The kinetics of NF mRNA expression

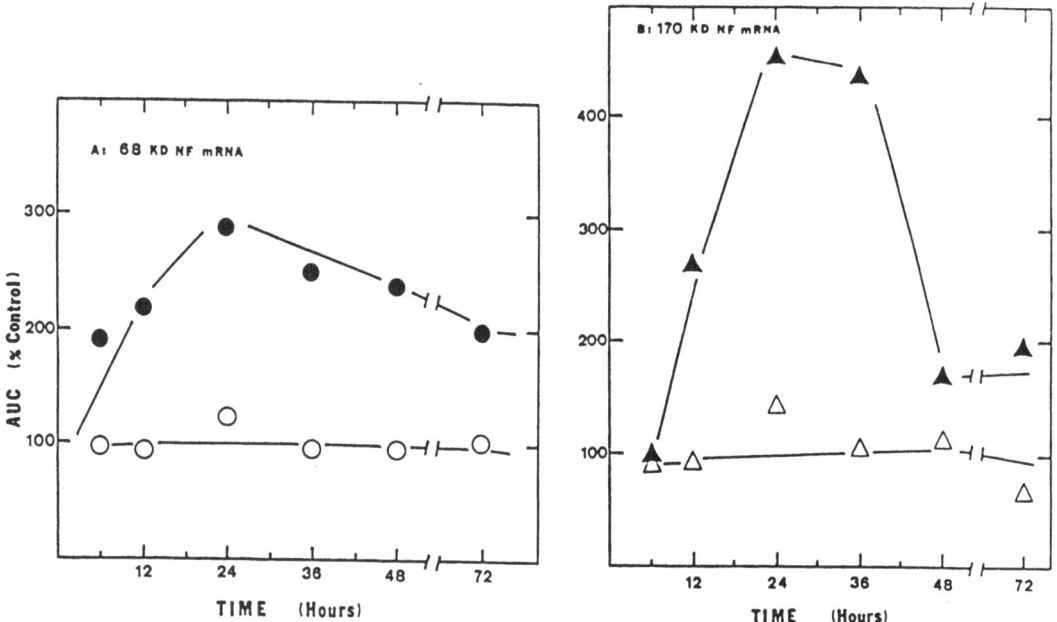

Fig. 2. Effect of insulin on neurofilament mRNA abundance in SH-SY5Y cells. Cloned human neuroblastoma SH-SY5Y cells were incubated without (open symbols) or with 1 uM insulin (closed symbols) for various times as shown up to 3 days under previously established culture conditions (Mill et al., 1985). RNA was purified and analyzed on Northern blots (40 ug total RNA/lane) by hybridization to nick-translated cDNAs containing the coding regions of 68 kD (Part A) and 170 kD (Part B) neurofilament proteins. The autoradiograms were scanned on a densitometer coupled to a computer to calculate areas under curves (AUC). Values are expressed in percentages relative to 6 h untreated cultures.

in that cell line possibly is slower, which may be related to the much slower rate of neurite growth in PC12 than in SH-SY5Y cells. Tubulin mRNAs likewise are transiently elevated for about a day in insulin-stimulated SH-SY5Y (Mill et al., 1985) and NGF-stimulated PC12 (Fernyhough et al., 1987) cells. Neurites, on the other hand, do not retract following the spontaneous decline in transcripts. Some reasons for transient expression of tubulin mRNAs have been discussed (Mill et al., 1985). High levels of transcripts may be required only during the elongation phase of neurites. This view is consistent with the dynamic instability model for microtubule growth in which the acquisition of a larger GTP cap predisposes towards a stable, growing microtubule. In addition, once structural proteins are assembled, they may become more stable, due to formation of cross-bridges and other structural modifications, and high transcript levels may no longer be required.

These results show that NGF, insulin, and IGFs share the common capacity to increase the transcripts for neurofilament as well as tubulin proteins preceding neurite growth. These observations provide additional support for the hypothesis that insulin and IGFs play a role in the development and maintenance of the structure of the nervous system. The increases in tubulin and neurofilament mRNAs are selective, as shown by their accumulation relative to total, poly(A)$^+$, actin, and histone 3.3 RNAs in insulin and IGF-treated cells. The selective accumulation of these transcripts is consistent with the observation that polymerase II activity is not increased indiscriminately. Insulin and IGFs do not increase neurite growth by increasing the size of the pool of total transcripts, but rather increase the relative abundance of select transcripts which code for proteins contributing to the axonal cytoskeleton.

The kinetics of increase and decline of transcripts is revealing. Alpha-tubulin, beta-tubulin, 68 kD neurofilament and 170 kD neurofilament mRNAs all appeared to increase at approximately the same time in SH-SY5Y cells. This suggests up regulation among these transcripts is coordinated and potentially through the same or a similar transmembrane mechanism. It is interesting that stabilization of tubulin transcripts is regulated in common by the neuritogenic proteins, but their effects on stabilization of neurofilament transcripts are still under examination. The time course to peak and rate at which the cellular content is decreased is somewhat different for individual transcripts, indicating that complex independent feed-back regulatory mechanisms may be present. Such mechanisms might also explain the variable rates for down regulation of transcripts between SH-SY5Y and PC12 cells. Further study is needed to understand the transcriptional and post-transcriptional events regulating the abundance of transcripts behind the growth of neurites.

Protein Kinase C

It has been nearly a decade since protein kinase C was implicated in the process of neurite outgrowth (Ishii, 1978). NGF-directed neurite formation can be reversibly modulated in cultured sensory ganglia by tumor promoting, but not inactive, phorbol ester congeners. Tumor promoting phorbol esters, such as 12-O-tetradecanoylphorbol-13-acetate (TPA), are persistent activators of protein kinase C. Neurite outgrowth can be speci-

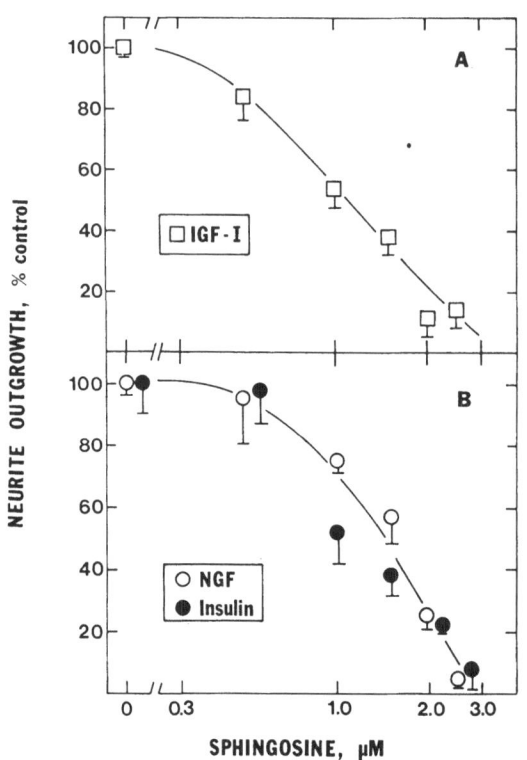

Fig. 3. Effect of various concentrations of sphingosine on IGF-I, NGF, and insulin-directed neurite outgrowth in cultured sympathetic neurons. Sympathetic ganglia from embryonic 12-day-old chicks were dissociated and cells cultured as previously described (Recio-Pinto et al., 1986). Neurite outgrowth was induced with 0.1 uM insulin, 10 nM IGF-I, or 10 ng/ml NGF. Various concentrations of sphingosine were included as shown. Neurite outgrowth was scored relative to untreated cultures (100%) after 2 days. Values are means $\pm$ S.E. (N = 4 replicate cultures.

fically and reversibly enhanced in cultured SH-SY5Y cells by tumor promoting phorbol esters (Spinelli et al., 1982) in a manner closely correlated with the capacity of phorbol esters to bind to protein kinase C (Spinelli and Ishii, 1983; Ishii et al., 1985). Physiological importance is suggested from the observations that brain has the highest concentration among tissues of protein kinase C (Blumberg et al., 1981), and concentrations are developmentally regulated (Nagle et al., 1981; Murphy et al., 1983).

However, it was not known until recently whether protein kinase C is situated in the pathway of NGF, insulin, and IGF-directed neurite growth, or whether phorbol esters could provoke neurite formation through an entirely different mechanism. The development of reversible inhibitors of protein kinase C, such as sphingosine, has rekindled interest in this problem. This compound competitively inhibits the binding of phorbol esters to protein kinase C. NGF-directed neurite growth in PC12 cells can be reversibly inhibited by sphingosine, and TPA can antagonize this inhibition (Hall et al., 1988). Similar concentrations of sphingosine inhibit NGF-directed neurite growth, binding of phorbol esters to protein kinase C, and phorbol ester activated protein phosphorylation. K-252a, another inhibitor of protein kinase C, also antagonizes various NGF dependent events in PC-12 cells, including enhanced neurite outgrowth, enhanced phosphatidylinositol breakdown, and decreased phosphorylation of Nsp100 protein (Koizumi et al., 1988). These results suggest that NGF-directed neurite growth is mediated in part through protein kinase C.

An important assumption underlying our early analysis was that insulin, IGFs, and NGF share activation of a common pathway leading to neurite elongation (Ishii et al., 1985). This assumption has now been tested (Fig. 3). Neurite outgrowth in cultured sympathetic neurons, whether directed by insulin, IGF-I, or NGF, was inhibited with similar sensitivity by sphingosine. These results indicate that protein kinase C mediates neurite growth and is situated in a pathway commonly activated by each of these factors.

Protein kinase C inhibitors can antagonize some but not all of the actions of the neuritogenic polypeptides. For example, they cannot inhibit NGF-dependent events such as cell flattening, increased tubulin mRNA abundance (Hall et al., 1988), or internalization of NGF (Koizumi et al., 1988). Enhanced phosphatidylinositol breakdown and decreased phosphorylation of Nsp100 are associated events which seem unlikely to be in a pathway preceding accumulation of tubulin mRNA, because these effects, but not the elevation of tubulin mRNA levels, are antagonized by protein kinase C inhibitors. These data are consistent with the interpretation that a cyclic AMP-dependent kinase may additionally be involved in the NGF pathway (Cremins et al., 1986).

ACKNOWLEDGMENTS

We are grateful to Diane Guertin and Yi Li for fine technical assistance. This work was supported in part by grant 5 R01 NS24327 from the National Institutes of Neurological and Communicative Disorders and Stroke.

REFERENCES

Aizenman, Y. and de Vellis, J., 1987, Brain neurons develop in a serum and glial free environment: Effects of transferrin, insulin, insulin-like growth factor-I and thyroid hormone on neuronal survival, growth and differentiation, Brain Res., 406: 32-42.

Aletta, J.M., Lewis, S.A., Cowan, N.J. and Greene, L.A., 1988, Nerve growth factor regulates both the phosphorylation and steady-state levels of microtubule-associated protein 1.2 (MAP1.2), J. Cell Biol., 106: 1573-1581.

Arai, T. and Matsumoto, G., 1988, Subcellular localization of functionally differentiated microtubules in squid neurons: Regional distribution of microtubule-associated proteins and beta-tubulin isotypes, J. Neurochem., 51: 1825-1838.

Bell, G.I., Gerhard, D.S., Fong, N.M., Sanchez-Pescador, R. and Rall, L.B., 1985, Isolation of the human insulin-like growth factor genes: Insulin-like growth factor II and insulin genes are contiguous, Proc. Natl. Acad. Sci. USA, 82: 6450-6454.

Bhat, N.R., 1983, Insulin dependent neurite outgrowth in cultured embryonic mouse brain cells, Dev. Brain Res., 11: 315-318.

Black, M.M., and Greene, L.A., 1982, Changes in the colchicine susceptibility of microtubules associated with neurite outgrowth: Studies with nerve growth factor-responsive PC12 pheochromocytoma cells, J. Cell Biol., 95: 379-386.

Blumberg, P.M., Declos, K.B. and Jaken, S., 1981, Tissue and species specificity for phorbol ester receptors, in: "Organ and species specificity in chemical carcino-genesis", Langenbach, R., Nesnow, S., and Rice, J.M. (eds.), Plenum Pub. Corp., New York, pp. 201-227.

Blundell, T.L., and Humbel, R.E., 1980, Hormone families: pancreatic hormones and homologous growth factors, Nature, 287: 781-787.

Bohannon, N.J., Corp, E.S., Wilcox, B.J., Figlewicz, D.P., Dorsa, D.M. and Baskin, D.G., 1988, Localization of binding sites for insulin-like growth factor-I (IGF-I) in the rat brain by quantitative autoradiography, Brain Res., 444: 205-213.

Bond, J.F. and Farmer, S.R., 1983, Regulation of tubulin and actin mRNA production in rat brain: expression of a new beta-tubulin mRNA with development, Molec. Cell. Biol., 3: 1333-1342.

Bond, J.F., Robinson, G.S. and Farmer, S.R., 1984, Differential expression of two neural cell-specific beta-tubulin mRNAs during rat brain development, Molec. Cell. Biol., 4: 1313-1319.

Bothwell, M., 1982, Insulin and somatomedin MSA promote nerve growth factor-independent neurite formation by cultured chick dorsal root ganglionic sensory neurons, J. Neurosci. Res., 8: 225-231.

Brown, A.L., Graham, D.E., Nissley, S.P., Hill, D.J., Strain, A.J. and Rechler, M.M., 1986, Developmental regulation of insulin-like growth factor II mRNA in different rat tissues, J. Biol. Chem., 261: 13144-13150.

Brown, M.C., Jansen, J.K.S. and Van Essen, D., 1976, Polyneuronal innervation of skeletal muscle in new-born rats and its elimination during maturation, J. Physiol., 261: 387-422.

Brown, M.C., Holland, R.L. and Hopkins, W.G., 1981, Restoration of focal multiple innervation in rat muscles by transmission block during a critical stage of development, J. Physiol., 318: 355-364.

Chiakulas, J.J. and Pauly, J.E., 1965, A study of postnatal growth of skeletal muscle in the rat, Anat. Rec., 152: 55-62.

Cochard, P. and Paulin, D., 1984, Initial expression of neurofilaments and vimentin in the central and peripheral nervous system of the mouse embryo in vivo, J. Neurosci., 4: 2080-2094.

Cremins, J., Wagner, J.A. and Halegoua, S., 1986, Nerve growth factor action is mediated by cyclic AMP- and Ca^{+2}/phospholipid-dependent protein kinases, J. Cell Biol., 103: 887-893.

Davies, A.M., Brandtlow, C., Heumann, R., Korsching, S., Rohrer, H. and Thoenen, H., 1987, Timing and site of nerve growth factor synthesis in developing skin in relation to innervation and expression of the receptor, Nature, 326: 353-358.

Dickson, G., Prentice, H., Julien, J.-P., Ferrori, G., Leon, A., and Walsh, F.S., 1986, Nerve growth factor activates thy-1 and neurofilament gene transcription in rat PC12 cells, EMBO J., 5: 3449-3453.

Drubin, D.G., Feinstein, S.C., Shooter, E.M., and Kirschner, M.W., 1985, Nerve growth factor-induced neurite outgrowth in PC12 cells involves the coordinate induction of microtubule assembly and assembly-promoting factors, J. Cell Biol., 101: 1799-1807.

Drubin, D., Kobayashi, S., Kellogg, D. and Kirschner, M., 1988, Regulation of microtubule protein levels during cellular morphogenesis in nerve growth factor-treated PC12 cells, J. Cell Biol., 106: 1583-1591.

Dull, T.J., Gray, A., Hayflick, F.S., and Ullrich, A., 1984, Insulin-like growth factor II precursor gene organization in relation to the insulin gene family, Nature, 310: 777-781.

Ebendal, T., Olson, L., and Seiger, A., 1983, The level of nerve growth factor (NGF) as a function of innervation, Exptl Cell Res., 148: 311-317.

Elsberg, C.A., 1917, Experiments on motor nerve regeneration and the direct neurotization of paralyzed muscles by their own and by foreign nerves, Science, 45: 318-320.

Fernyhough, P. and Ishii, D.N., 1987, Nerve growth factor modulates tubulin transcript levels in pheochromocytoma PC12 cells, Neurochem. Res., 12: 891-899.

Gasser, U.E., Weskamp, G., Otten, U., Dravid, A.R., 1986, Time course of the elevation of nerve growth factor (NGF) content in the hippocampus and septum following lesions of the septohippocampal pathway in rats, Brain Res., 376: 351-356.

Glazner, G.W. and Ishii, D.N., 1988, Insulin, insulinlike growth factor-I and nerve growth factor stimulate neurite formation in rat spinal cord cultures, Soc. Neurosci. Abs., 14: 1040.

Goedert, M., Fine, A., Hunt, S.P. and Ullrich, A., 1986, Nerve growth factor mRNA in peripheral and central rat tissues and in the human central nervous system: Lesion effects in the rat brain and levels in Alzheimer's Disease, Molec. Brain Res., 1: 85-92.

Greenberg, M.E., Greene, L.A., and Ziff, E.F., 1985, Nerve growth factor and epidermal growth factor induce rapid transient changes in protooncogene transcription in PC12 cells, J. Biol. Chem., 260: 14101-14110.

Gunderson, R.W. and Barrett, J.N., 1980, Characterization of the turning responses of dorsal root neurites toward nerve growth factor, J. Cell Biol., 87: 546-554.

Hall, F.L., Fernyhough, P., Ishii, D.N., Vulliet, P.R., 1988, Suppression of nerve growth factor-directed neurite outgrowth in PC12 cells by sphingosine, an inhibitor of protein kinase C, J. Biol. Chem., 263: 4460-4466.

Hall, J.L., Dudley, L., Dobner, P.L., Lewis, S.A. and Cowan, N.J., 1983, Identification of two human beta-tubulin isotypes, Molec. Cell. Biol., 3: 854-862.

Hansson, H.A., Rozell, B. and Skottner, A., 1987, Rapid axoplasmic transport of insulin-like growth factor I in the sciatic nerve of adult rats, Cell Tissue Res., 247: 241-247.

Hefti, F., 1986, Nerve growth factor (NGF) promotes survival of septal cholinergic neurons after fimbrial transection, J. Neurosci., 6: 2155-2162.

Henderson, C.E., Huchet, M., Changeux, J.P., 1983, Denervation increases a neurite-promoting activity in extracts of skeletal muscle, Nature, 302: 609-611.

Hendry, I.A., Stockel, K., Thoenen, H. and Iversen, L.L., 1974, The retrograde axonal transport of nerve growth factor, Brain Res., 68: 103-121.

Hill, J.M., Lesniak, M.A., Pert, C.B. and Roth, J., 1986, Autoradiographic localization of insulin receptors in rat brain: Prominence in olfactory and limbic areas, Neuroscience, 17: 1127-1138.

Hirokawa, N.M., Glicksman, M.A., and Willard, M.B., 1984,

Organization of mammalian neurofilament polypeptides within neuronal cytoskeleton, J. Cell Biol., 98: 1523-1536.

Hoffman, H., 1951, Study of factors influencing innervation of muscles by implanted nerves, Aust. J. Exp. Biol. Med. Sci., 29: 289-308.

Hoffman, P.N. and Lasek, R.J., 1975, The slow component of axonal transport. Identification of major structural polypeptides of the axon and their generality among mammalian neurons, J. Cell Biol., 66: 351-366.

Hoffman, P.N., Thompson, G.W., Griffin, J.W. and Price, D.L., 1985, Changes in neurofilament transport coincide temporally with alterations in the caliber of axons in regenerating motor fibers, J. Cell Biol., 101: 1332-1340.

Hoffman, P.N., Cleveland, D.W., Griffin, J.W., Landes, P.W., Cowan, N.J. and Price, D.L., 1987, Neurofilament gene expression: A major determinant of axonal caliber, Proc. Natl. Acad. Sci. USA, 84: 3472-3476.

Huck, S., 1983, Serum-free medium for cultures of the postnatal mouse cerebellum: only insulin is essential, Brain Res. Bull., 10: 667-674.

Ishii, D.N., 1978, Effect of tumor promoters on the response of cultured embryonic chick ganglia to nerve growth factor, Cancer Res., 38: 3886-3893.

Ishii, D.N., 1989, Relationship of insulinlike growth factor-II gene expression in muscle to synaptogenesis, Proc. Natl. Acad. Sci. USA, in press.

Ishii, D.N. and Mill, J.F., 1987, Molecular mechanisms of neurite formation stimulated by insulin-like factors and nerve growth factor, Curr. Topics Membranes Transport, 31: 31-78.

Ishii, D.N. and Recio-Pinto, E., 1987, Role of insulin, insulinlike growth factors, and nerve growth factor in neurite formation, in: "Insulin, IGFs, and Their Receptors in the Central Nervous System", M.K. Raizada, M.I. Phillips, and D. Le Roith, eds., Plenum Publ. Corp., New York, NY, pp. 315-348.

Ishii, D.N., Recio-Pinto, E., Spinelli, W., Mill, J.F., and Sonnenfeld, K.H., 1985, Neurite formation modulated by nerve growth factor, insulin, and tumor promoter receptors, Intern. J. Neurosci., 26: 109-127.

Jansen, M., van Schaik, F.M., Ricker, A.T., Bullock, B., Woods, D.E., Gabbay, K.H., Nussbaum, A.L., Sussenbach, J.S., and Van den Brande, J.L., 1983, Sequence of cDNA encoding human insulin-like growth factor I precursor, Nature, 306: 609-611.

Julien, J.P., Ramachandran, K. and Grosveld, F., 1985, Cloning of a cDNA encoding the smallest neurofilament protein from the rat, Biochim. Biophys. Acta, 825: 398-404.

Julien, J.-P., Meyer, D., Flavell, D., Hurst, J. and Grosveld, F., 1986, Cloning and developmental expression of the murine neurofilament gene family, Mol. Brain Res., 1: 243-250.

Kelly, A.M. and Zacks, S.I., 1969, The fine structure of motor endplate morphogenesis, J. Cell Biol., 42: 154-169.

Koizumi, S., Contreras, M.L., Matsuda, Y., Hama, T., Lazarovici, P. and Guroff, G., 1988, K-252a: A specific inhibitor of the action of nerve growth factor on PC12 cells, J. Neurosci., 8: 715-721.

Kolber, A.R., Goldstein, M.N., and Moore, B.W., 1974, Effect of nerve growth factor on the expression of colchicine-binding activity and 14-3-2 protein in an established line of human neuroblastoma, Proc. Natl. Acad. Sci. USA, 71: 4203-4207.

Korsching, S., Auburger, G., Heumann, R., Scott, J. and Thoenen, H., 1985, Levels of nerve growth factor and its mRNA in the central nervous system of the rat correlate with cholinergic innervation, EMBO J., 4: 1389-1393.

Korsching, S., Heumann, R., Thoenen, H., and Hefti, F., 1986, Cholinergic denervation of the rat hippocampus by fimbrial transection leads to a transient accumulation of nerve growth factor (NGF) without change in mRNA NGF content, Neurosci. Lett., 66: 175-180.

Kromer, L.F., 1987, Nerve growth factor treatment after brain injury prevents neuronal death, Science, 235: 214-217.

Lenoir, D. and Honegger, P., 1983, Insulin-like growth factor I (IGF I) stimulates DNA synthesis in fetal rat brain cell cultures, Dev. Brain Res., 7: 205-213.

Lemischka, I.R., Farmer, S., Racaniello, V.R. and Sharp, P.A., 1981, Nucleotide sequence and evolution of a mammalian alpha-tubulin messenger RNA, J. Molec. Biol., 151: 101-120.

Levi-Montalcini, R., 1966, The nerve growth factor: Its mode of action on sensory and sympathetic nerve cells, Harvey Lect., 60: 217-259.

Lewis, S.A., and Cowan, N.J., 1986, Anomalous placement of introns in a member of the intermediate filament multigene family: An evolutionary conundrum, Mol. Cell Biol., 6: 1529-1534.

Lindenbaum, M.H., Carbonetto, S., Grosveld, F., Flavell, D. and Mushynski, W.E., 1988, Transcriptional and post-transcriptional effects of nerve growth factor on expression of the three neurofilament subunits in PC-12 cells, J. Biol. Chem., 263: 5662-5667.

Luckenbille-Edds, L., Van Horn, C., and Greene, L.A., 1979, Fine structure of initial outgrowth processes induced in a pheochromocytoma cell line (PC12) by nerve growth factor, J. Neurocytol., 8: 493-511.

Mattsson, M.E.K., Enberg, G., Ruusala, A-I., hall, K. and

Pahlman, S., 1986, Mitogenic response of human SH-SY5Y neuroblastoma cells to insulin-like growth factor I and II is dependent on the stage of differentiation, J. Cell Biol., 102: 1949-1954.

Menesini-Chen, M.G.M., Chen, J.S. and Levi-Montalcini, R., 1978, Sympathetic nerve fibers ingrowth in the central nervous system of neonatal rodent upon intracerebral NGF injections, Arch. Ital. Biol., 116: 53-84.

Mill, J.F., Chao, M.V., and Ishii, D.N., 1985, Insulin, insulin-like growth factor II, and nerve growth factor effects on tubulin mRNA levels and neurite formation, Proc. Natl. Acad. Sci. USA, 82: 7126-7130.

Miller, R.H., Lasek, R.J., and Katz, M.J., 1987, Preferred microtubules for vesicle transport in lobster axons, Science, 235: 220-222.

Murphy, K.M.M., Gould, R.J., Oster-Granite, M.L., Gearhart, J.D. and Snyder, S.H., 1983, Phorbol ester receptors: Autoradiographic identification in the developing rat, Science, 222: 1036-1038.

Nagle, D.S., Jaken, S., Castagna, M. and Blumberg, P.M., 1981, Variation with embryonic development and regional localization of specific (^{3}H)phorbol-12,13-dibutyrate binding to brain, Cancer Res., 41: 89-93.

Neumann, D., Scherson, T., Ginzburg, I., Littauer, U.Z., and Schwartz, M., 1983, Regulation of mRNA levels for microtubule proteins during nerve regeneration, FEBS Lett., 162: 270-276.

O'Brien, R.A.D., Ostberg, A.J.C. and Vrbova, G., 1978, Observations on the elimination of polyneuronal innervation in developing mammalian skeletal muscle, J. Physiol., 282: 571-582.

Pockett, S., and Slack, J.R., 1982, Source of the stimulus for nerve terminal sprouting in partially denervated muscle, Neuroscience, 7: 3173-3176.

Puro, D.G. and Agardh, E., 1984, Insulin-mediated regulation of neuronal maturation, Science, 225: 1170-1172.

Raivich, G., Zimmermann, A. and Sutter, A., 1985, The spatial and temporal pattern of beta NGF receptor expression in the developing chick embryo, EMBO J., 4: 637-644.

Raizada, M.K., Yang, J.W. and Fellows, R.E., 1980, Binding of ^{125}I-insulin to specific receptors and stimulation of nucleotide incorporation in cells cultured from rat brain, Brain Res., 200: 389-400.

Recio-Pinto, E. and Ishii, D.N., 1984, Effects of insulin, insulin-like growth factor-II and nerve growth factor on neurite outgrowth in cultured human neuroblastoma cells, Brain Res., 302: 323-334.

Recio-Pinto, E. and Ishii, D.N., 1988a, Insulin and related growth factors: Effects on the nervous system and mechanism for neurite growth and regeneration, Neurochem. Int., 12: 397-414.

Recio-Pinto, E. and Ishii, D.N., 1988b, Insulin and insulinlike growth factor receptors regulating neurite formation in cultured human neuroblastoma cells, J. Neurosci. Res., 19: 312-320.

Recio-Pinto, E., Lang, F.F. and Ishii, D.N., 1984, Insulin and insulinlike growth factor-II permit nerve growth factor binding and the neurite formation response in cultured human neuroblastoma cells, Proc. Natl. Acad. Sci. USA, 81: 2562-2566.

Recio-Pinto, E., Rechler, M.M. and Ishii, D.N., 1986, Effects of insulin, insulin-like growth factor-II, and nerve growth factor on neurite formation and survival in cultured sympathetic and sensory neurons, J. Neurosci., 6: 1211-1219.

Redfern, P.A., 1970, Neuromuscular transmission in newborn rats, J. Physiol., 209: 701-709.

Robinson, P.A., Wion, D., Anderton, B.H., 1986, Isolation of a cDNA for the rat heavy neurofilament polypeptide (NF-H), FEBS Lett., 209: 203-205.

Rotwein, P., Burgess, S.K., Milbrandt, J.D. and Krause, J.E., 1988, Differential expression of insulin-like growth factor genes in rat central nervous system, Proc. Natl. Acad. Sci. USA, 85: 265-269.

Seiler, M. and Schwab, M.E., 1984, Specific retrograde axonal transport of nerve growth factor (NGF) from neocortex to nucleus Basalis in the rat, Brain Res., 300: 33-39.

Shaw, G., and Weber, K., 1982, Differential expression of neurofilament triplet proteins in brain development, Nature, 298: 277-279.

Shelton, D.L. and Reichardt, L., 1984, Expression of the beta-nerve growth factor gene correlates with the density of sympathetic innervation in the effector organs, Proc. Natl. Acad. Sci. USA, 81: 7951-7955.

Shelton, D,L., and Reichardt, L.F., 1986, Studies on the regulation of beta-nerve growth factor gene expression in the rat iris: the level of mRNA-encoding nerve growth factor is increased in irises placed in explant cultures in vitro, but not in irises deprived of sensory or sympathetic innervation in vivo, J. Cell Biol., 102: 1940-1948.

Shemer, J., Raizada, M.K., Masters, B.A., Ota, A. and LeRoith, D., 1987, Insulin-like growth factor I receptors in neuronal and glial cells, J. Biol. Chem., 262: 7693-7699.

Smith, M., Clemens, J., Kerchner, G.A. and Mendelsohn, L.G., 1988, The insulin-like growth factor-II (IGF-II) receptor of rat brain: regional distribution visualized by autoradiography, Brain Res., 445: 241-246.

Soares, M.B., Ishii, D.N., and Efstratiadis, A., 1985,

Developmental and tissue-specific expression of a family of transcripts related to rat insulin-like growth factor II mRNA, Nucleic Acids Res., 13: 1119-1134.

Soares, M.B., Turken, A., Ishii, D., Mills, L., Episkopou, V., Cotter, S., Zeitlin, S. and Efstratiadis, A., 1986, Rat insulin-like growth factor II gene: A single gene with two promoters expressing a multitranscript family, J. Mol. Biol., 192: 737-752.

Sonnenfeld, K.H. and Ishii, D.N., 1982, Nerve growth factor effects and receptors in cultured human neuroblastoma cell lines, J. Neurosci. Res., 8: 375-391.

Sonnenfeld, K.H. and Ishii, D.N., 1985, Fast and slow nerve growth factor binding sites in human neuroblastoma and rat pheochromocytoma cell lines: Relationship of sites to each other and to neurite formation, J. Neurosci., 5: 1717-1728.

Spinelli, W., Sonnenfeld, K.H., and Ishii, D.N., 1982, Effects of phorbol ester tumor promoters and nerve growth factor on neurite outgrowth in cultured human neuroblastoma cells, Cancer Res., 42: 5067-5073.

Spinelli, W. and Ishii, D.N., 1983, Tumor promoter receptors regulating neurite formation in cultured human neuroblastoma cells, Cancer Res., 43: 4119-4125.

Stylianopoulou, F., Herbert, J., Soares, M.B. and Efstratiadis, A., 1988, Expression of the insulin-like growth factor II gene in the choroid plexus and the leptomeninges of the adult rat central nervous system, Proc. Natl. Acad. Sci. USA, 85: 141-145.

Thompson, W., Kuffler, D.P. and Jansen, J.K.S., 1979, The effect of prolonged, reversible block of nerve impulses on the elimination of polyneuronal innervation of new-born rat skeletal muscle fibres, Neuroscience, 4: 271-281.

Wang, C., Wible, B., Angelides, K., and Ishii, D.N., 1988, Insulin and insulinlike growth factor-I increase neurofilament mRNA levels and neurite formation, Soc. Neurosci. Abs., 14: 1169.

Whittemore, S.R., Larkfors, L., Ebendal, T., Holets, V.R., Ericsson, A. and Persson, H., 1987, Increased beta-nerve growth factor messenger RNA and protein levels in neonatal rat hippocampus following specific cholinergic lesions, J. Neurosci., 7: 244-251.

Willard, M., and Simon, C., 1983, Modulations of neurofilament axonal transport during the development of rabbit retinal ganglion cells, Cell, 35: 551-559.

Yang, J.W. and Fellows, R.E., 1980, Characterization of insulin stimulation of the incorporation of radioactive precursors into macromolecules in cultured rat brain cells, Endocrinology, 107: 1717-1724.

INSULIN-LIKE GROWTH FACTORS AS REGULATORY PEPTIDES IN THE

ADULT RAT BRAIN

Dianne Figlewicz Lattemann, Michael G. King,
Patricia Szot, and Denis G. Baskin

Depts. of Psychology, Biological Structure,
Pharmacology, and Medicine, University of
Washington and the V.A. Med. Center, Seattle WA

Much of the present understanding of the actions of IGF-1 and IGF-2 in nervous tissue has come from *in vitro* studies using either primary brain cultures derived from fetal or neonatal material, or derived CNS cell lines such as neuroblastomas or gliomas. This line of research has established a potential role for the IGFs as growth factors in the developing nervous system, where it has been proposed that they stimulate neuronal sprouting and synaptogenesis (1-4), and induce glial proliferation (5).

Although this *in vitro* approach has yielded exciting hypotheses for IGF action in the developing brain, a major challenge remains in elucidating the functions of the IGFs in the adult brain. It is clear that the IGFs are abundant both within adult CNS tissue and cerebrospinal fluid (6). This implies that these growth factors have an important action within the mature nervous system. IGF-1 appears to play a role in the feedback inhibition of growth hormone secretion through the stimulation of somatostatin release from the hypothalamus (7). IGF-2 has been implicated in the regulation of feeding behavior. Lauterio *et al.* demonstrated a dose-related suppression of 24-hr food intake following intracerebroventricular injection of IGF-2 (8). In the present article we discuss some of the approaches that we are using in our laboratory to understand the physiologic role of the IGFs as regulatory peptides in the adult brain.

Our understanding of the physiological roles of IGFs in the intact CNS is hindered by several major unsolved problems. First, the origin of the IGFs within nervous tissue, and the regulation of CNS IGF levels, have not been fully elucidated. IGF-1 has been localized within numerous brain regions, however, IGF-1 mRNA has only been demonstrated within the fetal brain but not in adult brain (9). This suggests that IGF-1 may be sequestered by various brain cells but is not synthesized in adult brain. Both the expression of message for, and the secretion of, an IGF-2 molecule within the adult nervous system have been reported. Specific hybridization of an IGF-2 cDNA probe to choroid plexus and meningeal membranes has been shown, with no detectable hybridization to other brain regions (10,11). Expression of IGF-2 mRNA in adult human hypothalamus has been observed (12). Shiu and Paterson reported secretion of IGF-2 peptides from neonatal rat cortex and hypothalamus (13). These studies support the concept that IGF-2 is synthesized locally in the adult CNS.

The mechanism for access of IGF-I to the adult CNS is not known. For example, it is unclear if circulating IGF-1 is transported directly into the CSF or into the brain interstitial fluid compartment. One potential mechanism of transport may be transcytosis across capillary endothelial cells. High affinity receptors for

IGF-1 and IGF-2 have been localized to blood vessel endothelial cells from brain and other tissues and these receptors are structurally similar to those of other cell types (14-16). The IGFs appear to be retained intracellularly within endothelial cells for long periods of time; IGF-1 is subsequently released predominantly in an intact form, whereas IGF-2 is extensively degraded (17,18). It has been hypothesized that the capillary endothelial cells may thus serve as a physiologic reservoir for the IGFs (18). These studies further support the concept that IGF-1 is predominantly transported into the adult CNS but IGF-2 is synthesized there.

A second major unresolved issue is the cellular location of IGF action in nervous tissue. It is not known whether IGFs have effects on both neuronal and glial cell types in the intact brain. One approach to this problem has been to localize IGF binding sites throughout the CNS. Quantitative autoradiography (QAR) of IGF binding sites in brain slices with iodinated IGF has been reported by three laboratories and the results from the different groups are generally comparable (19-22). IGF receptor localization in the brain as detected by immunocytochemistry with anti-receptor antibodies (21,23,24) is similar to that obtained by QAR, although some discrepancies have been noted. These studies taken together clearly show that binding sites for IGF-1 and IGF-2 are concentrated in anatomically specific regions of the brain. Furthermore, the binding sites for IGF-1 and IGF-2 are for the most part located in different brain regions. For example, in the olfactory bulb, IGF-1 binding is highest in the glomerular and olfactory nerve fiber layer, whereas IGF-2 binding sites are concentrated in the mitral cell body layer. It is important to note that the anatomical location of IGF binding sites is distinctly different from that of insulin binding sites (25). The distinctive anatomical patterns of binding exhibited by IGF-1, IGF-2, and insulin support the conclusion that these peptides have different sites of action within the brain. We have also used QAR to characterize regional coefficients of IGF-I receptor number (Bmax) and affinity (Kd) which can provide an approach for studying the receptor regulation in specific neuronal populations (Table I).

One brain region we have focused on is the hypothalamus, where IGF-1 binding sites are concentrated in the dorsomedial nucleus and the median eminence, whereas IGF-2 binding sites are densest in the ventromedial nucleus (Figure 1). It is tempting to speculate that these binding sites may mediate the reported satiety effects of the IGFs in the CNS (7,8) because the dorsomedial and ventromedial hypothalamic nuclei are known to have important roles in the neural control of food intake and appetitive behavior. However, it must be remembered that IGF-1 and -2 binding sites are found in many other brain regions, some of which influence motivation, motor behavior, sensory processing, and energy metabolism. Thus, it is possible that IGF's central effects on food intake may result indirectly from IGF action in regions outside of the hypothalamus. Nevertheless, the striking concordance of IGF binding sites with hypothalamic regions involved with food intake raises the hypothesis that IGFs may affect feeding behavior by acting on specific neural pathways in the hypothalamus.

While compelling evidence supports the presence of IGF receptors on both neurons and glia in culture, the identity of specific cells bearing IGF receptors in the intact CNS is still unknown. The autoradiographic methods that have been used to localize binding sites in tissue slices have inadequate resolution to answer this question. New morphological approaches and probes for these receptors will have to be developed to localize CNS IGF receptors at the cellular level. Neurons and glia have been isolated with limited success from mature brain (23), and future studies measuring binding to membranes prepared from such isolated cells may provide insight into the pharmacology of IGF binding in specific CNS cell types.

There is very little information about how or if CNS IGF receptors are physiologically regulated, either in the developing or the mature brain. Anatomical localization of IGF binding sites in the fetal brain, and studies of how the location or concentration of these receptors may change during postnatal development and aging, remain to be carried out. Although IGF-1 and -2 are present in brain and CSF, it is not known if brain or CSF IGF levels regulate the brain receptors or whether plasma IGF levels may be involved. We have recently demonstrated that the concentrations of CNS IGF-1 receptors may be altered by nutritional status (26) or obesity (27). We observed that IGF-1 binding sites in the median eminence upregulated in response to low plasma IGF-1 levels associated with prolonged food restriction, but IGF receptors in the thalamus and choroid plexus were unchanged (26). Presumably, these results can be explained by the lack of a blood-brain barrier in the median eminence. The absence of regulation in the thalamus and choroid plexus suggests that most IGF-1 receptors in these regions are on the brain side of the blood-brain barrier, and not regulated by plasma IGF-1 levels. In another study, we measured IGF-1 binding to membranes prepared from the olfactory bulb of lean (Fa/Fa), heterozygous (Fa/fa), and obese (fa/fa) Zucker rats (27). As observed previously, percent specific binding of 0.1 nM insulin was decreased in Fa/fa and fa/fa olfactory bulb membranes as compared with Fa/Fa olfactory bulb (Fa/Fa=100%; Fa/fa=56%; and fa/fa=33%). Binding of IGF-1 was decreased also, although not to as great an extent as insulin binding (% specific binding of trace IGF-1=100%, 104%, and 64% for Fa/Fa, Fa/fa, and fa/fa). No genotype-related change of IGF-1 binding to hypothalamic or cortex membranes was observed, suggesting that IGF-1 receptor concentrations were not decreased uniformly throughout the brain. Together, our studies suggest that IGF-1 receptor concentration may be regulated in a site-specific fashion within the CNS.

In attempting to interpret binding and physiological studies on IGFs in the intact CNS, we are faced with the rapidly emerging literature on IGF binding proteins. Since it is now known that some of these binding proteins are of high affinity and are located on cell surfaces (and would presumably be present in the brain slices used for autoradiographic studies of IGF binding) the status of CNS IGF binding sites as "receptors" can be validly questioned. In defense of the conclusion that the IGF-1 binding sites localized by autoradiography are principally type 1 IGF receptors, we can point to the evidence that these binding sites have the specificity expected of a type 1 receptor. In particular, these IGF-1 binding sites recognize insulin (19,21) which does not bind to IGF binding proteins. Further, Lesniak et al (21) have shown that an antibody to the rat IGF-2 receptor localizes to the same regions as iodinated IGF-2. We have found that an antibody to the rat IGF-2 receptor which blocks binding of IGF-2 to the type 2 receptor but not to IGF-2 binding protein (28) blocks a portion of IGF-2 binding to brain regions possessing high concentrations of IGF-2 binding sites. Thus, it appears that a portion of IGF-2 binding visualized by autoradiography may represent a binding protein. It is not yet clear if the regions which have IGF-1 receptors also have IGF-1 binding proteins. Future anatomical studies on CNS IGF receptors with autoradiography will need to consider the possible contribution of binding proteins to the "specific" binding of IGF-1 and IGF-2.

Table I. Regions of high specific binding for IGF-I in adult rat brain by QAR. (mean ± s.e.m., n=5)

Region	Specific Binding (DPM/sq mm)	Bmax (pmol/sq mm)	Kd (nM)
Median Eminence	275 ± 33	1.8 ± 0.1	1.7 ± 0.3
Choroid Plexus	211 ± 16	2.1 ± 0.4	3.5 ± 0.8
Gelatinosus Nuc. Thalamus	170 ± 12	1.9 ± 0.3	2.3 ± 0.5
Ventral-Medial Thalamus	147 ± 7	1.9 ± 0.3	2.8 ± 0.4
Dentate Gyrus	143 ± 8	2.1 ± 0.4	2.9 ± 0.6
Hippocampus (CA3)	149 ± 8	1.6 ± 0.4	2.2 ± 0.5
Piriform Cortex	160 ± 9	2.2 ± 0.3	3.1 ± 0.4
Somatosensory Cortex			
laminae 1,2	114 ± 7	1.5 ± 0.2	2.9 ± 0.4
laminae 3,4	100 ± 5	1.3 ± 0.2	2.7 ± 0.4
laminae 5,6	148 ± 7	1.5 ± 0.2	2.3 ± 0.3

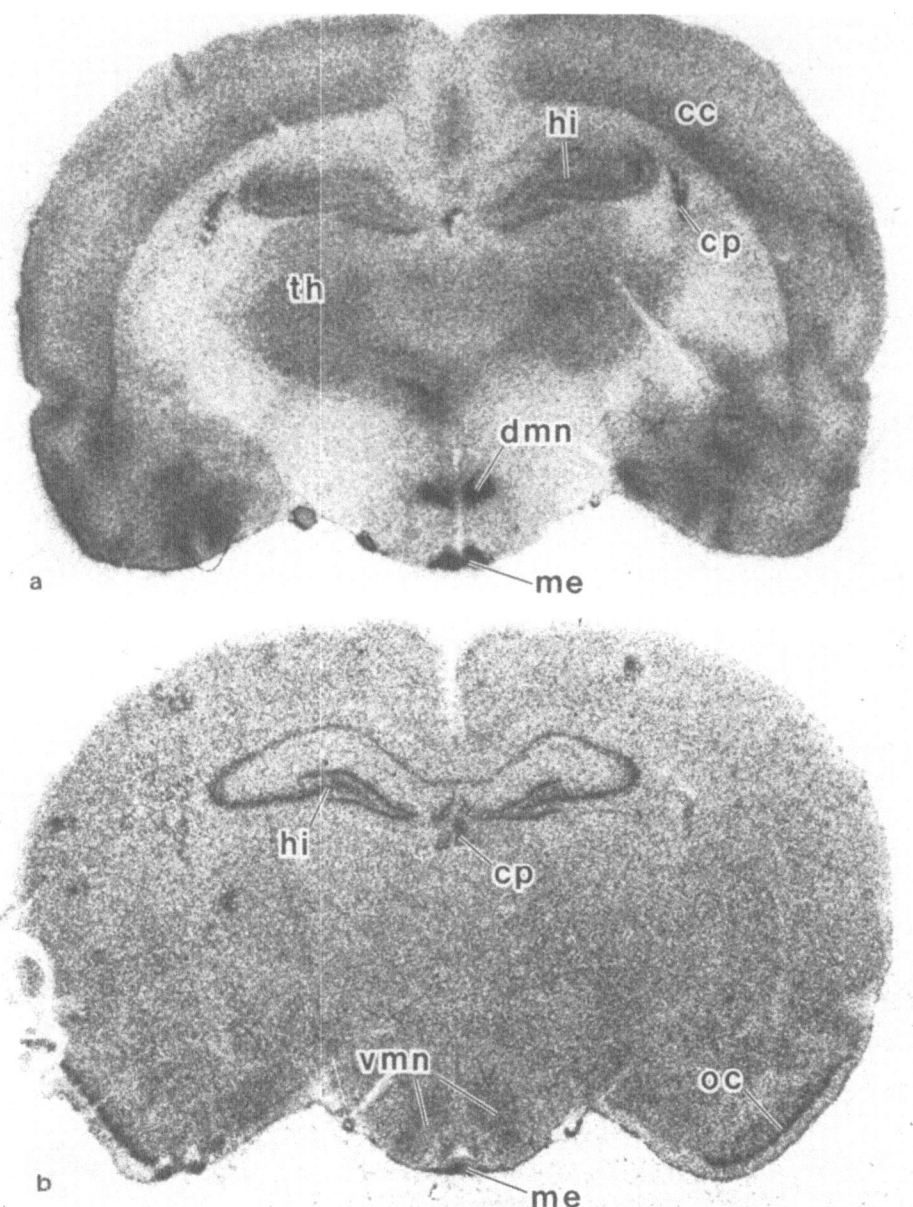

Figure 1. Autoradiographs of brain slices labeled with [^{125}I]-IGF-1 (a) and [^{125}I]-IGF-2 (b). IGF-1 binding sites are concentrated in the dorsomedial nucleus (*dmn*) of the hypothalamus, whereas moderate densities of IGF-2 binding sites are present in the ventromedial nucleus (*vmn*). Both types of IGF binding sites are present in the median eminence (*me*), hippocampus (*hi*), and choroid plexus (*cp*). IGF-1 binding sites are also present in the thalamus (*th*) and cerebral cortex (*cc*), and high densities of IGF-2 binding sites are located in the olfactory cortex (*oc*).

A third unresolved issue is the cellular level action of the IGFs once they have bound to target cells within the CNS. Post-receptor mechanisms of action of IGF-1 have been explored in neural-derived cell lines. The receptor appears to function as a tyrosine kinase, as does the insulin receptor (see (6) for reviews). However, little is known regarding the cellular actions of the IGFs in the mature or intact brain. We recently tested the ability of the IGFs to alter phospholipid

430

metabolism within slices prepared from the olfactory bulb and hippocampus of 17-22 day old rats. Our findings are summarized in Figures 2 and 3. IGF-1, tested over a range of doses of 10-100 nM, stimulated the incorporation of ^{3}H-myoinositol into IPI and membrane lipids within the olfactory bulb but not the hippocampus. The IGF-1 effect within the olfactory bulb was significant (p<0.05 for combined IGF-1 doses vs. paired controls), and IGF-1 was significantly more effective within the olfactory bulb vs. the hippocampus.

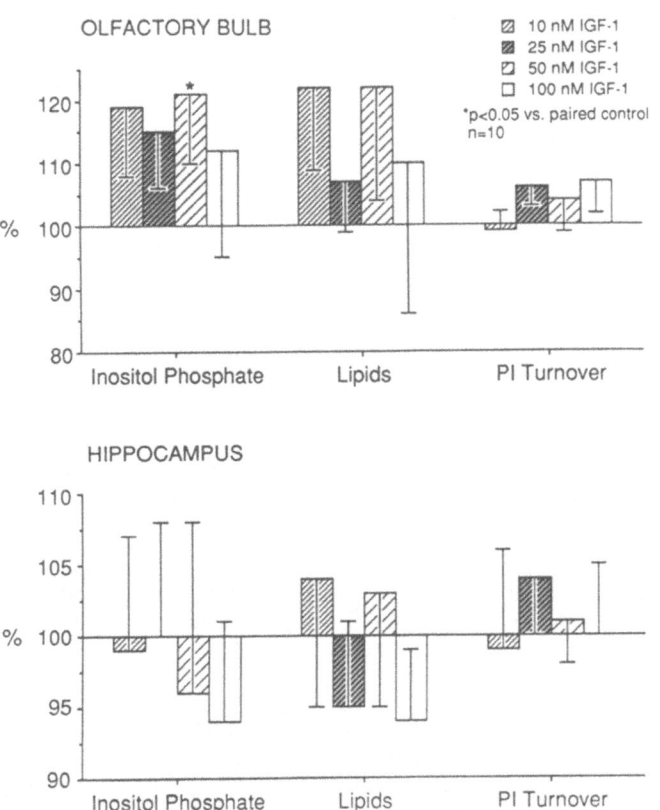

Figure 2. Effect of IGF-1 on ^{3}H-myoinositol incorporation into inositol phosphate and membrane lipids in the olfactory bulb and hippocampus of 17-22 day old rats. Data are expressed as percent of paired control values.

IGF-2 over the same concentration range (10-100 nM) inhibited IP1 formation within the hippocampus, and the effect was significant (p<0.05 for combined IGF-2 doses vs. paired controls). Within the olfactory bulb, IGF-2 had no significant or consistent effects. Both IGF-1 and IGF-2 binding sites were localized in the olfactory bulb and hippocampus by QAR. Thus, these data suggest that the IGFs may have cellular-level actions which differ both between brain regions and between growth factors. Additionally, these data together with the behavioral findings of Lauterio et al (that IGF-2 but not IGF-1 infused into the CSF decreases 24-hr food intake in rats) argue against the idea that both IGF-1 and IGF-2 act solely via the IGF-1 receptor in the adult rat brain (29). The physiological data suggest that some of the actions of IGF-2 in the CNS probably act through the type 2 IGF receptor.

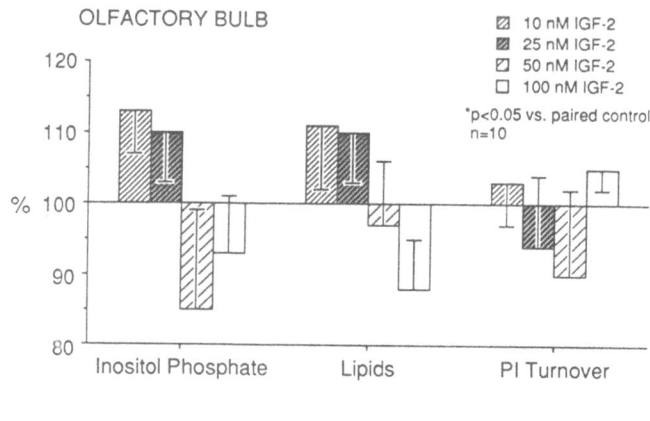

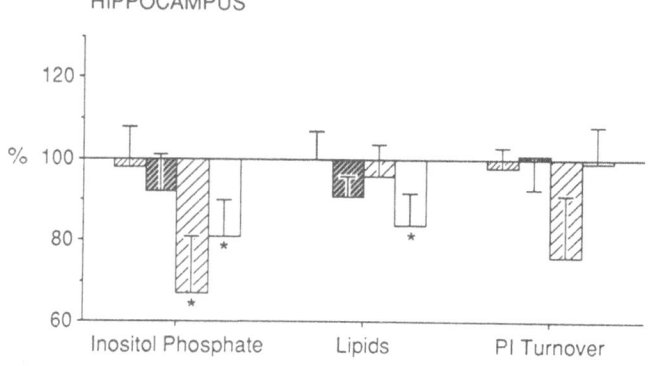

Figure 3. Effect of IGF-2 on ^{3}H-myoinositol incorporation into inositol phosphate and membrane lipids in the olfactory bulb and hippocampus of 17-22 day old rats. Data are expressed as percent of paired control values.

It is possible that the IGFs act by altering the release of other neurotransmitters within the brain slices; Nilsson *et al* have recently reported that IGF-1 can enhance the stimulated release of acetylcholine from rat cerebral cortex slices (30). These data suggest that the IGFs may behave as neuromodulators within the mature brain. An additional cellular action of the IGFs in the adult brain could be the maintenance of synaptic integrity, *i.e.*, a trophic action. At present there is no direct *in vivo* evidence to support this hypothesis.

In summary, much work remains to be done to elucidate the actions of the IGFs within the mature brain. This remains an open area for investigation, particularly since one may hypothesize that regulation could occur at a variety of levels: (a) synthesis of the IGFs and transport to their sites of action; (b) regulation of receptor number at target tissues; (c) regulation of local binding protein concentration which may compete with receptor sites; (d) regulation of postreceptor actions of IGFs; and (e) regulation of other neurotransmitter systems to which the IGFs may be coupled or modulatory. Future research should include the identification of specific cells and cell types on which receptors are localized, as well as specific neurochemical or functional anatomic pathways with which IGFs or their receptors are associated. Further functional and morphological studies aimed at the cellular level within the intact brain should provide new information regarding IGF function in the intact CNS.

ACKNOWLEDGMENTS

The authors would like to thank Dr. D. Dorsa for collaborative assistance with the measurement of inositol phosphate turnover, Dr. M.R.C. Greenwood for collaborative arrangements for Zucker rats, Dr. Thomas Wimpy for assistance with the quantitative autoradiography, and Lylian Merkley for technical assistance with the measurement of brain membrane IGF binding studies. Chare Vathanaprida assisted with the tissue preparation for autoradiography. ^{125}I-IGF-1 for Zucker brain membrane binding studies was a generous gift from Drs. J.J. VanWyk and L. Underwood. Unlabelled insulin was a gift of Dr. Ronald Chance, Lilly Corp. Labelled IGF-II and antibodies to the rat IGF-II receptor were graciously provided by Dr. Ronald Rosenfeld. These studies were supported by the American Diabetes Association, NIH grants DK40963-01, NS24809, and DK17047, and the Veterans Administration.

REFERENCES

1. E. Recio-Pinto and D.N. Ishii, Effects of insulin, insulin-like growth factor-II and nerve growth factor on neurite outgrowth in cultured human neuroblastoma cells, Brain Res. 302:323 (1984).
2. Y. Aizenman and J. de Vellis, Brain neurons develop in a serum and glial free environment: effects of transferrin, insulin, insulin-like growth factor-I and thyroid hormone on neuronal survival, growth and differentiation, Brain Res. 406:32 (1987).
3. E. Recio-Pinto and D.N. Ishii, Insulin and insulinlike growth factor receptors regulating neurite formation in cultured human neuroblastoma cells, J. Neurosci. Res. 19:312 (1988).
4. E. DiCicco-Bloom and I.B. Black, Insulin growth factors regulate the mitotic cycle in cultured rat sympathetic neuroblasts, Proc. Natl. Acad. Sci. 85:4066 (1988).
5. F.A. McMorris T.M. Smith, S. DeSalvo, and R.W. Furlanetto, Insulin-like growth factor I/somatomedin C: a potent inducer of oligodendrocyte development, Proc. Natl. Acad. Sci. 83:822 (1986).
6. M.K. Raizada, M.I. Phillips, and D. LeRoith, "Insulin, Insulin-like Growth Factors, and Their Receptors in the Central Nervous System," Plenum Press, New York (1987).
7. G.S. Tannenbaum, H.J. Guyda, and B.I. Posner, Insulin-like growth factors: a role in growth hormone negative feedback and body weight regulation via brain, Science 220:77 (1983).
8. T.J. Lauterio, L. Marson, W.H. Daughaday, and C.A. Baile, Evidence for the role of insulin-like growth factor II (IGF-II) in the control of food intake, Physiol. Behav. 40:755 (1987).
9. P.K. Lund, B.M. Moats-Staats, M.A. Hynes, J.G. Simmons, M. Jansen, A.J. D'Ercole, and J.J. VanWyk, Somatomedin-C/insulin-like growth factor I and insulin-like growth factor-II mRNAs in rat fetal and adult tissues, J. Biol. Chem. 261:14539 (1986).
10. M.A. Hynes, P.J. Brooks, J.J. Van Wyk, and P.K. Lund, Insulin-like growth factor II messenger ribonucleic acids are synthesized in the choroid plexus of the rat brain, Mol. Endocrinology 2:47 (1988).
11. F. Stylianopoulou, J. Herbert, M.B. Soares, and A. Efstratiadis, Expression of the insulin-like growth factor II gene in the choroid plexus and leptomeninges of the rat central nervous system, Proc. Natl. Acad. Sci. 85:141 (1988).
12. J.-C. Irminger, K.M. Rosen, R.E. Humbel, and L. Villa-Komaroff, Tissue-specific expression of insulin-like growth factor II mRNAs with distinct 5' untranslated regions, Proc. Natl. Acad. Sci. 84:6330 (1987).
13. R.P.C. Shiu, and J.A. Paterson, Characterization of insulin-like growth factor II peptides secreted by explants of neonatal brain and of adult pituitary from rats, Endocrinology 123:1456 (1988).

14. I. Jialal, M. Crettaz, H.L. Hachiya, C. R. Kahn, A. C. Moses S.M. Buzney, and G. L. King, Characterization of the receptors for insulin and the insulin-like growth factors on micro- and macrovascular tissues, Endocrinology 117:1222 (1985).

15. R.S. Bar, M. Boes, and M. Yorek, Processing of insulin-like growth factors I and II by capillary and large vessel endothelial cells, Endocrinology 118:1072 (1986).

16. R.G. Rosenfeld, H. Pham, B.T. Keller, R.T. Borchardt and W.M. Pardridge, Demonstration and structural comparison of receptors for insulin-like growth factor-I and -II (IGF-I and -II) in brain and blood-brain barrier, Biochem. Biophys. Res. Comm. 149:159 (1987).

17. H.L. Hachiya, J.-L. Carpentier, and G.L. King, Comparative studies on insulin-like growth factor II and insulin processing by vascular endothelial cells, Diabetes 35:1065 (1986).

18. N.K. Banskota, J.-L. Carpentier, and G.L. King, Processing and release of insulin and insulin-like growth factor I by macro- and microvascular endothelial cells, Endocrinology 119:1904 (1986).

19. N.J. Bohannon, E.S. Corp, B.J. Wilcox, D.P. Figlewicz, D.M. Dorsa, and D.G. Baskin, Localization of binding sites for insulin-like growth factor-I (IGF-I) in the rat brain by quantitative autoradiography, Brain Res. 444:205 (1988).

20. M. Smith, J. Clemens, G.A. Kerchner and L.G. Mendelsohn, The insulin-like growth factor-II (IGF-II) receptor of rat brain: regional distribution visualized by autoradiography, Brain Res. 445:241 (1988).

21. M.A. Lesniak, J.M. Hill, W. Kiess, M. Rojeski, C.B. Pert, and J. Roth, Receptors for insulin-like growth factors I and II: autoradiographic localization in rat brain and comparison to receptors for insulin, Endocrinology 123:2089 (1988).

22. D.G. Baskin, N.J. Bohannon, M.G. King, and R.G. Rosenfeld, IGF-II binding sites in the rat brain: Localization by quantitative autoradiography, Soc. Neurosci. Abstr. 13:1021 (1987).

23. I.Ocrant, K.L. Valentino, L.F. Eng, R.L. Hintz, D.M. Wilson, and R.G. Rosenfeld, Structural and immunohistochemical characterization of insulin-like growth factor I and II receptors in the murine central nervous system, Endocrinology 123: 1023-1034, (1988).

24. K.L. Valentino, H. Pham, I. Ocrant, and R.G. Rosenfeld, Distribution of insulin-like growth factor II receptor immunoreactivity in rat tissues, Endocrinology 122:2753 (1988).

25. D.G. Baskin, B.J. Wilcox, D.P. Figlewicz, and D.M. Dorsa, Insulin and insulin-like growth factors in the CNS, Trends Neurosci. 11:107 (1988)

26. N.J. Bohannon, E.S. Corp, B.J. Wilcox, D.P. Figlewicz, D.M. Dorsa, and D.G. Baskin, Characterization of insulin-like growth factor-I receptors in the median eminence of the rat brain and their modulation by food restriction, Endocrinology 122:1940 (1988).

27. D.P. Figlewicz, D.G. Baskin, S.C. Woods, and D. Porte, Jr., Insulin and insulin-like growth factor-I (IGF-1) binding is decreased in the olfactory bulb (OB) but not the cerebral cortex (C) of the obese Zucker (fa/fa) rat, Int. J. Obesity 9:A31 (1985).

28. R.G. Rosenfeld, D. Hodges, H. Pham, P.D.K. Lee, and D.P. Powell, Purification of insulin-like growth factor-II (IGF-II) receptor from an IGF-II producing cell line and generation of an antibody which both immunoprecipitates and blocks the type 2 IGF receptor, Biochem. Biophys. Res. Comm. 138:304 (1986).

29. S.J. Casella, V.K. Han, A.J. D'Ercole, M.E. Svoboda, and J.J. VanWyk, Insulin-like growth factor II binding to the type I somatomedin receptor. Evidence for two high affinity binding sites, J. Biol. Chem. 261:9268 (1986).

30. L. Nilsson, V.R. Sara, and A. Nordberg, Insulin-like growth factor I stimulates the release of acetylcholine from rat cortical slices, Neurosci. Letters 88:221 (1988).

RECEPTORS FOR INSULIN AND INSULIN-LIKE GROWTH FACTORS

I AND II IN PCD MICE AND DIABETIC RATS ARE UNALTERED:

AN AUTORADIOGRAPHIC STUDY

Maria Rojeski and Jesse Roth

Diabetes Branch, NIDDK
National Institutes of Health
9000 Rockville Pike, Building 10
Room 8N244, Bethesda, MD

Specific receptors for insulin and the insulin-like growth factors, IGF-I and IGF-II, are present in the central nervous system. Insulin binding sites were initially identified and localized to discrete areas of rat brain using membrane preparations (Havrankova et al., 1978) with the highest concentration in the olfactory bulb, limbic system and hypothalamus. Later studies, which combined autoradiography and radioreceptors assays using fresh frozen rat tissue slices (Hill et al, 1986), showed that specific binding of insulin is distributed throughout the brain in well defined neuroanatomic areas, and confirmed the earlier study by showing greatest density in the olfactory bulb and limbic system. Specific binding was demonstrated with competition studies which were like those of typical insulin receptors. We have also used autoradiography to study the binding sites of $[^{125}\text{-I}]$ IGF-I and $[^{125}\text{I}]$IGF-II in rat brain and compared the pattern of binding to that of insulin (Lesniak et al, 1988). This study demonstrated that both types of IGF receptors are also widely distributed in brain and localize to specific anatomic areas. The three receptors are found in many of the same regions but usually localize to different layers or groups of cells.

Although the anatomic distribution of the receptors for insulin and insulin-like growth factors is well documented, their function remains unknown. Finding an experimental condition which changes the amount or affinity of these binding sites might provide a clue to elucidate their function. Previous studies of insulin binding in membrane preparations (Havrankova et al, 1979) demonstrated that insulin receptors in brain, unlike classic peripheral insulin receptors, do not upregulate in response to hypoinsulinemia (the streptozotocin diabetic rat), or

435

downregulate in response to hyperinsulinemia (the genetically obese ob/ob mouse). This is probably not due solely to the protective effect of the blood brain barrier since insulin delivered directly to the CNS does not downregulate brain receptors except when massive amounts are given (Devaskar et al., 1986).

In the present study we extend our previous work by using autoradiography to measure insulin, IGF-I and IGF-II receptors in two models: the streptozotocin-induced diabetic rat as a model of hypoinsulinemia and metabolic derangement, and the pcd mouse as a model of specific neuronal degeneration. The advantage of autoradiography is that it is a very sensitive method, and small localized changes can be detected even when total binding is similar. In addition, the architecture is preserved so changes in distribution and cytopathology can be observed.

Methods

Brains and other tissues were dissected from freshly decapitated animals and frozen in isopentane immersed in a dry ice bath. The fresh frozen tissues were sliced (36 microns) on a cryostat at -20ºC and then mounted on gelatin coated slides, and dehydrated overnight at 4ºC under vacuum before use. The dehydrated sections were preincubated with buffer, Trizma pH 7.4 (Sigma), with 30 mM KCl, twice for 10 minutes each followd by incubation with radiolabeled ligand in HEPES buffer at 15º for 3 hours. Nonspecific binding was measured by incubating adjacent sections with an excess of cold ligand (1000 ng/ml for insulin, 1000 ng/ml for IGF and 1000 ng/ml for IGF-2). Sections were washed with ice cold phosphate buffered saline, air dried, and exposed to LKB Ultrofilm for 7 days. Adjacent sections were exposed to Kodak NTB-2 nuclear emulsion and then stained to establish the relationship of binding to cytoarchitecture.

To demonstrate the specificity of the binding of the labeled ligand to its receptor, we did competition studies with consecutive slices through a receptor rich area; each labeled ligand was coincubated with each of the three unlabeled ligands over a wide range of concentrations. An image analysis system was used to measure densities of the film representing binding in specific neuroanatomic regions.

Streptozotocin, 65 mg/kg IV was administered into the tail vein of 6-week-old male Sprague Dawley rats. Urine glucose was monitored to document that they were hyperglycemic within 36 hours of drug administration. Control rats were age, sex and weight matched, received at the same time, and housed under identical conditions during the experimental period. Animals were killed in pairs, experimental and control, 2 hours post prandially. Two pairs were studied 1 week after diabetes was induced; three pairs were studied after 1 month.

Pcd mice were obtained 3 months of age and 8 months of age. Heterozygote litter mates were used as controls.

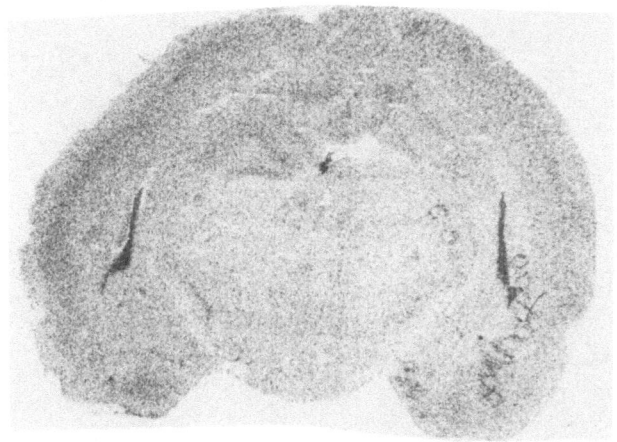

Figure 1A. Autoradiograph of dorsal hippocampus of diabetic rat brain labelled with [^{125}-I] insulin.

Results

Streptozotocin induced diabetes is a model of hypoinsulinemic diabetes. All of our treated animals became diabetic and none died. Binding studies showed no difference in the density or location of binding for [^{125}-I] insulin, [^{125}-I] IGF-I or [^{125}-I] IGF-II, in all pairs of animals studied (Fig. 1).

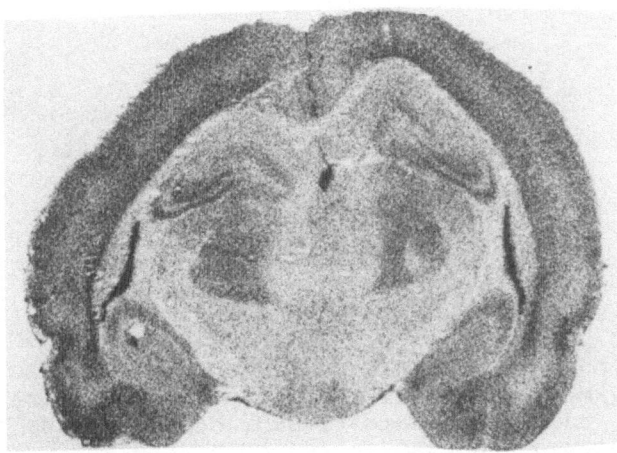

Figure 1B. Autoradiograph of dorsal hippocampus of diabetic rat brain labeled with [^{125}-I]IGF-I.

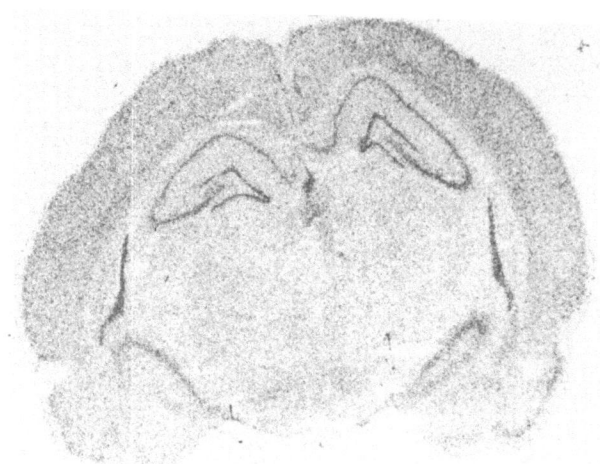

Figure 1C. Autoradiograph of dorsal hippocampus of
diabetic rat brain labeled with
[^{125}-I]IGF-II.

Radiolabeled insulin was displaced by unlabeled ligand as
expected insulin > IGF-I > IGF-II with a half maximal
displacement by insulin of approximately 10-20 ng/ml
(Fig. 2).

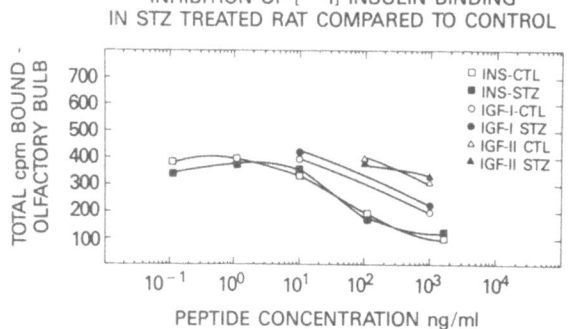

Figure 2A. Competition for [^{125}-I]insulin binding in the
olfactory bulb of streptozotocin-treated diabetic rat brain
and control. Serially obtained coronal sections (36 um)
coincubated with [^{125}-I]insulin and increasing
concentrations of cold insulin (.2-5000 ng/ml), 1GF-1
(20-1000 ng/ml) and IGF-II (100-1000 ng/nl). After
incubation, the slides were rinsed, dried and the tissue
sections scraped and transferred to a tube where
radioactivity was measured with a gamma counter. Binding
of I-125 insulin measured as total cpm scraped is plotted
against the peptide concentration.

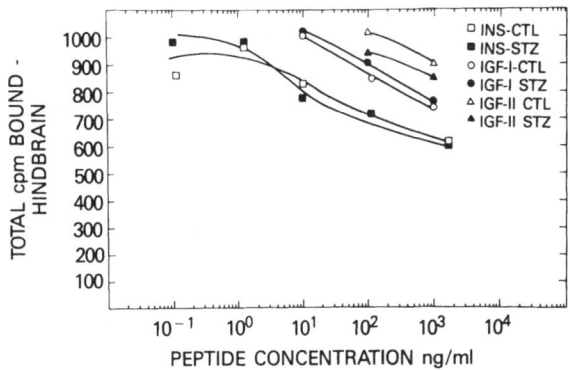

INHIBITION OF [^{125}-I] INSULIN BINDING
IN STZ TREATED RAT COMPARED TO CONTROL

Figure 2B. Competition for [^{125}-I]insulin binding in the
hindbrain of streptozotocin-treated diabetic rat brain and
control.

   In both the diabetic and control animal IGF-I or IGF-II,
stained sections of NTB-2 developed slides showed no change
in the cytoarchitecture.
      The pcd (Purkinje cell deficient) mouse is a well
characeterized mouse which loses its cerebellar Purkinje
cells during post natal development (Landis and Mullen,
1977). Purkinje cells are present in these animals in
normal number and location until postnatal day 18 after
which these cells rapidly degenerate and disappear by day
30. How the mutant allele effects this neuronal
degeneration is unknown. Heterozygotes are
indistinguishable from normal. In our study we compared
[^{125}-I]insulin, [^{125}-I]IGF-I and [^{125}-I]IGF-II
binding in an 8 month old homozygous pcd mouse compared to
an 8 month old heterozygote control. Serial sections were
cut through brain as well as sections through liver and
testicle. Specificity of binding was confirmed by
coincubating adjacent sections with an excess of cold
ligand. The insulin, IGF-I, and IGF-II binding were
identical in location and amount for homozygotes and
**heterozygotes (Fig. 3, 4, 5). Competition studies showed
similar curves between the pairs (Fig. 6).**

Discussion
      As an extension of our previous autoradiographic
studies, this study shows that in the streptozotocin-
induced diabetic rat there is no change in insulin receptor
binding, even in local areas of high insulin binding. In
addition, IGF-I and IGF-II binding are unchanged despite
the metabolic derangement. The pcd mouse also shows no
change in insulin, IGF-I or IGF-II receptors compared to
control and this study suggests these growth factors are
apparently not related to or affected by the neuronal
degeneration seen.

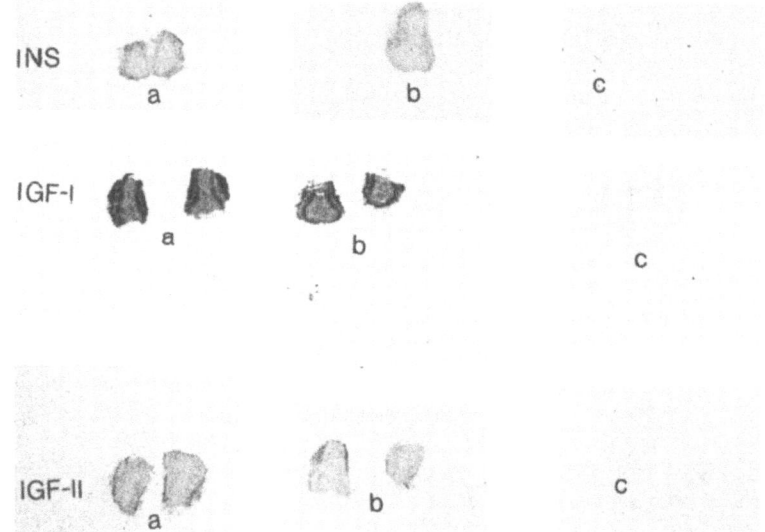

Figure 3.   Autoradiograph of olfactory bulb of mouse brain
labeled with [^{125}-I]insulin, [^{125}-I]IGF-1
and [^{125}-I]IGF-II.
a) pcd homozygote mouse, b) pcd heterozygote;
c) pcd homozygote coincubated with 1000 ng/ml
unlabeled peptide (nonspecific binding).

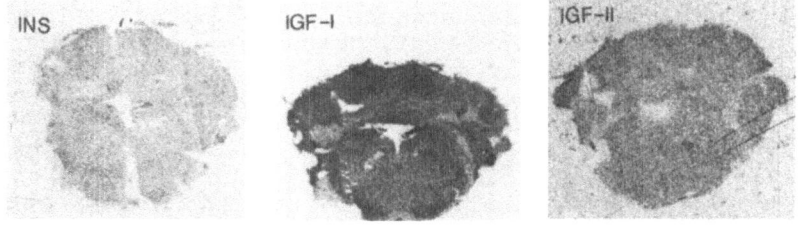

Figure 4.   Autoradiograph of cerebellum of pcd homozyogote
with [^{125}-I]insulin, [^{125}-I]IGF-I and
[^{125}-I]IGF-II.

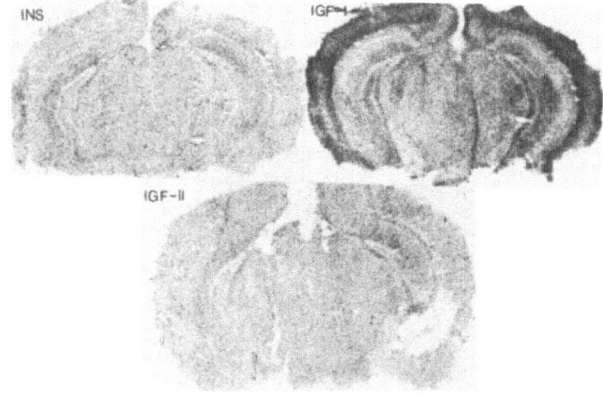

Figure 5.   Autoradiograph of hindbrain of pcd homozygote
labeled with [^{125}-I]insulin, [^{125}-I], IGF-I
and [^{125}-I]IGF-II.

a)

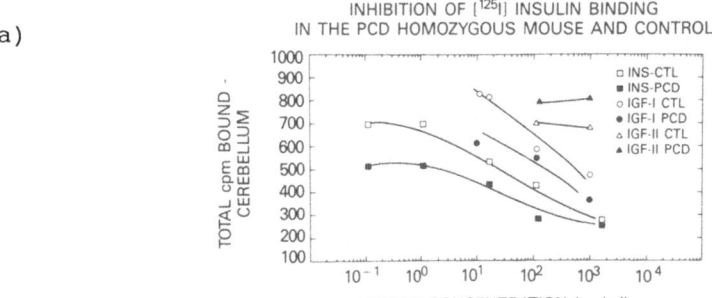

b)

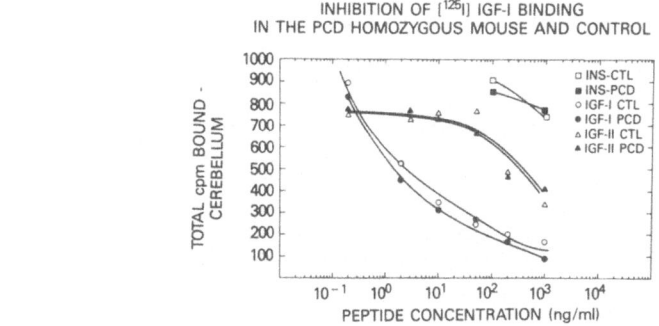

c)

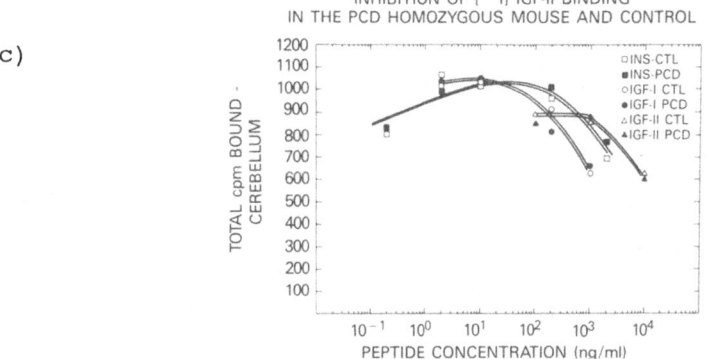

Figure 6.   Competition in the cerebellum of pcd hemozygote
             and pcd heterozygote mouse brain for a)
             [^{125}-I]insulin, b) [^{125}-I]IGF-I and c)
             [^{125}-I]IGF-II.   Coronal sections (36um) were
             incubated with labeled peptide and varying
             concentrations of each unlabeled peptide in both
             afflicted and control animals.   Total binding
             measured as the total cpm from a scraped section
             is plotted as a function of peptide
             concentration as in Fig. 2.   Each represents a
             single experiment.

Acknowledgments

We are grateful to Charles A. Greer of Yale University for providing the PCD mice and to Robert Waldbillig, Joanna Hill and Maxine Lesniak for their suggestions and expert opinions.

References

1. Havrankova J and Roth J, 1978. Insulin receptors are widely distributed in the central nervous system of the rat, Nature 272:827-829.
2. Havrankova J, Roth J, Brownstein, MJ, 1979. Concentration of insulin and insulin receptors in brain are independent of peripheral insulin levels, J Clin. Invest. 64:636-642.
3. Hill JM, Lesniak MA, Pert CB, Roth J, 1986. Autoradiographic localization of insulin receptors in rat brain: prominence in olfactory and limbic areas. Neuroscience 17:1127-1138.
4. Lesniak MA, Hill JM, Kiess W, Rojeski M, Pert CB and Roth, 1988. Receptors for Insulin-like Growth Factors I and II: Autoradiographic localization in rat brain and comparison to receptors for insulin. Endocrinology 123:2089-2099.
5. Davaskar SV, Karycki L, Devaskar VP, 1986. Varying brain insulin concentrations differentially regulate the fetal brain insulin receptor. Biophys. Res. Comm. 136:208-219.
6. Landis S and Mullen RJ, 1978. The development and degeneration of Purkinje cells in pcd mutant mice. J Comp. Neur. 177:125-144.

INSULIN-LIKE GROWTH FACTORS AND HEART DEVELOPMENT

Gary L. Engelmann [A], Joyce F. Haskell [B], and Keith D. Boehm [C]

Research Institute, Cleveland Clinic Foundation [A]
Department of Obstetrics and Gynecology, University of
Alabama at Birmingham [B], and Department of
Reproductive Biology, Case Western Reserve University [C]

INTRODUCTION

Growth of the adult mammalian ventricular myocyte (cardiomyocyte) is the topic of considerable interest since heart muscle cells rapidly respond to increases in functional demands by an increase in cellular mass(1). Although the cardiomyocyte of the adult ventricle occupies up to 90% of the area of the left ventricle (LV) free wall, these cells comprise a unique population of a permanently post-mitotic, structurally and functionally delimited cell type that can only expand in width (primarily) and length to increasing functional demand on this vital organ. Therefore, the adult cardiomyocyte population is a static number of muscle cells that define the life-long structure and function of the ventricular portion of the heart. It is the function of the cardiomyocytes that can then be modulated by adjacent cell types (i.e., fibroblasts and endothelium), blood vessels (capillaries and arterioles) and neural/humoral stimuli in the mature organ.

Nevertheless, the fetal and neonatal periods of ventricular development represent the only time periods wherein cardiomyocyte replication occurs (1,2). It is during this period of development that cardiomyocytes proliferate, migrate, undergo a morphogenic transformation into a mature morphology, and begin to assume the functional characteristics necessary to maintain the structure and function of the heart for the duration of the life-span of the organism (1-3). It has only recently been hypothesized that the developing heart may also initiate some of its own proliferative and development signals during fetal and neonatal stages of growth. It is this latter aspect of heart development that represents the topic of this article. The focus of the article will be upon the role the insulin-like growth factor (IGF) family of peptides, associated IGF receptors and IGF binding proteins may play in fetal and neonatal heart development. The working hypothesis that we have proposed (4) (Figure 1) is that the growing heart responds to growth factor stimuli that originate within the developing organ during fetal and neonatal development.

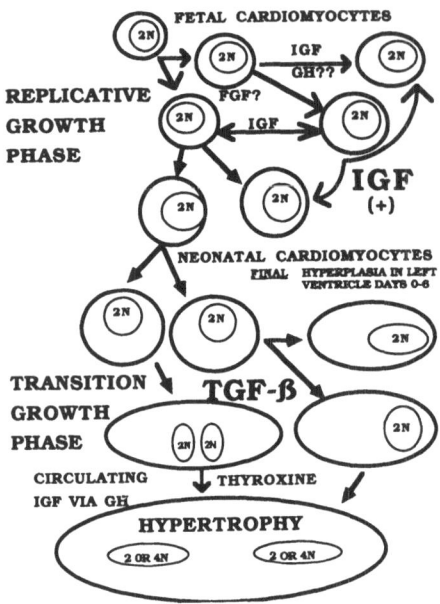

Figure 1. Potential Autocrine or Paracrine Influence of Growth
Factors on Fetal and Neonatal Cardiomyocyte Development

GROWTH FACTORS AND FETAL/NEONATAL HEART DEVELOPMENT

All vertebrate hearts are derived from splanchnic mesoderm and the
bilateral heart primordium completely differentiates during early fetal
development such that the heart is one of the first morphologically
recognizable organs (1,2,5). Because of the essential nature of the heart to
the developing circulatory system that is necessary for any subsequent fetal
development, structure/function relationships are rapidly established and
the muscle cells that comprise the ventricle of the heart quickly become
terminally differentiated. Hemodynamic effects on fetal heart development
and cardiomyocyte growth remain speculative, yet both ventricles pump
blood into the systemic circulation of the fetus. A proliferation-binucleation-
hypertrophy growth sequele for fetal/neonatal cardiomyocytes has been
proposed (6,7). Although cardiomyocytes become terminally differentiated
and express muscle specific proteins, such as myosin heavy chain, early in
fetal development, a major portion of them retain their proliferative
capabilities throughout fetal development and into the early phases of
neonatal development. In addition, intraventricular (8) and possibly
interventricular proliferative differences may exist in the developing heart
such that global growth characteristics may not be appropriate.

Growth of the heart can arise from an increase in one or more factors
that increase: 1) cardiomyocyte cell number (hyperplasia); 2) cardiomyocyte
cell size (hypertrophy); 3) number of non-myocyte cell types; or 4) a
combination of some or all of the above. Little is known of the mechanism(s)
that control cardiomyocyte replication since few cell types restrict their
proliferation to as great an extent as the cardiomyocyte (1,2,9). Developing
cardiomyocytes cease cell division when thrust into a more aerobic
environment, such as is found in the neonate (10). A potentially important

component of this shift in oxidative potential is the change in NAD^+ and associated poly-ADP ribosylation of nuclear proteins that may be essential for proliferation. Another nuclear characteristic of the cardiomyocyte is the process of polyploidation. We have shown that mature cardiomyocytes in normal and hypertrophy-prone rats are polyploid, and a unique intraventricular variation in nuclear ploidy was found between the left ventricle (LV) and the right ventricle (RV) (11). The widespread distribution of polyploid cells in mammalian tissues and the age-related intervals at which polyploidation occurs, indicates that cell polyploidy is a programmed part of development (12) and its relationship to cardiomyocyte development remains to be determined. Nevertheless, the effects of well characterized growth factors (GFs) on cardiomyocyte growth of the fetal and neonatal heart remain to be established.

Data on the expression of growth factors, growth factor-related proto-oncogenes, and proto-oncogenes in fetal and neonatal hearts, both human and rodent, is limited. Claycomb and Lanson (13) have recently reported the expression of 13 proto-oncogenes in neonatal and adult rodent ventricular tissue and isolated cardiomyocytes. They showed that cellular and tissue RNA levels and corresponding proto-oncogene production can be disparate. Schneider et al. (14) and Zimmerman et al. (15) have reported that c-*myc* and r-*fos* (*fos* -related oncogene) are expressed during fetal and neonatal mouse heart development. The expression of r-*fos* was most tightly associated with the proposed cessation of cardiomyocyte hyperplasia that occurs during the first neonatal week. Schneider et al. (14) showed that c-*myc* expression was greatest in embryonic heart and declined to one-third of this value by one-week of age. Mugrauer et al. (16) have shown that N-*myc* expression is limited to day 12-15 of fetal heart development in the mouse. Kasik et al. (17) have reported that c-*fos* expression is highest at birth in mice and drops to basal levels, yet expression is still evident in the 5-day old neonatal heart. Muller et al. (18) have shown that c-*ras* [Ki] is expressed in 8-day old mouse heart. The *ras* homology to the G-protein of the adrenergic receptor-coupling system may be related to its elevated expression at this time when adrenergic innervation is occurring within the ventricle (19). Certain proto-oncogenes have been shown to be expressed after growth factor stimulation. This expression is considered a "post-receptor" mechanism that is a transient nuclear event possibly involved with modulation of the expression of other genes (20). Since IGF-I has been shown to increase c-fos RNA abundance in L6 muscle cells (21), the expression of c-fos in the fetal/neonatal heart may represent a post-receptor IGF-mediated event. Expression of ventricular proto-oncogenes may also be a consequence of stimulation, at least in part, by IGF-receptor interactions in the developing rat heart.

IGFs AND HEART DEVELOPMENT

There is little evidence that "classical" maternal peptide hormones (i.e, growth hormone, thyroid hormone, insulin, etc.) regulate somatic cell growth in the mammalian fetus (22-24). This does not include those peptide hormones secreted by the placenta or fetal endocrine glands or steroids passed through or synthesized by the placenta. It is well known that insulin is essential for normal fetal growth, yet insulin may be permissive for fetal growth by mediating its anabolic effects rather than through direct stimulation of cellular proliferation (25). There is strong evidence that neither maternal or fetal insulin or IGFs pass transplacentally and it is therefore assumed that any insulin-like effects in utero are the result of endogenous fetal insulin or somatomedins (insulin-like growth factors, IGFs). Hill et al. (25) have recently reviewed the role of GFs, particularly the IGFs, in embryogenesis, although no direct evidence of IGF involvement in embryonic heart development was presented.

Somatomedins, IGFs, are a family of polypeptides found in fetal and neonatal tissues as well as circulating in the serum bound to specific carrier proteins (26,27). The potential regulatory role of the IGF binding proteins in fetal and neonatal development will not be discussed, but their biological and structural characteristics have recently been reviewed (28). IGFs have insulin-like anabolic activities, yet they are more potent stimulators of cellular proliferation than insulin (25-27). Many of the proliferative effects of insulin may actually be mediated indirectly through IGF synthesis or directly by interaction with the IGF receptors. Both IGF-I and IGF-II (multiplication stimulating activity, MSA) have been found to be associated with fetal development (25). IGF-I appears to be more important to early embryogenesis, while IGF-II may be more directly involved in later fetal and early neonatal development. In the adult, IGF-I is growth hormone (GH) dependent and a proposed mediator of GH effects via hepatic synthesis and secretion of IGF-I (26). In contrast, IGF-II appears to be GH-independent, binds to specific carrier proteins, yet remains enigmatic with regard to defined growth promoting effects.

Two specific membrane receptors for IGFs have been identified and well characterized in multiple cell types (29). We have recently identified Type 1 and Type 2 IGF receptors to be present in neonatal rat hearts and determined that the Type 1 receptor concentration decreases with increasing age, while receptor affinity increases (4). Receptor density for IGFs in several tissues change during fetal-to-adult transitions (30) and many tissues express both IGF receptors and insulin receptors. Multiple post-receptor events mediated by IGF stimulation have been identified (26,27,31), which include increases in DNA, RNA and protein synthesis often preceding cell replication. Recent evidence has shown that the Type 2 IGF receptor is genetically, structurally and possibly functionally analogous to the cation independent mannose-6-phosphate receptor (32,33). Since the physiological role for IGF-II is unknown, particularly in the fetus and neonate, the relationship of IGF-II to mannose-6-phosphate remains unresolved but may involve intracellular events associated with inhibition of protein synthesis (33). Whether elevated cellular IGF-II levels in fetal rat development are involved in protein half-life modulation and/or cellular phenotypic stability and maturation remains to be determined.

Studies of specific growth factor gene transcript expression in fetal and neonatal rat hearts is limited. Several developmental studies have performed survey-type analysis of IGF expression in multiple tissues which also included the heart (34-37). In all cases, abundant and multiple IGF-II gene transcripts were detected in the developing heart. We have recently concluded a detailed study of IGF gene expression during neonatal rat heart development (4). During the first neonatal week of age, the abundance of IGF-II gene transcripts declines rapidly, and coincides with the reported cessation of proliferation of the cardiomyocyte (1,2). We also localized, by *in situ* hybridization, the IGF-II gene transcripts to the cardiomyocytes of the developing ventricle. We also detected, although at a much lower level, ventricular IGF-I transcripts. The IGF-I transcripts may have been produced by non-myocytes, such as endothelial or vascular smooth muscle cells, which are beginning to populate the developing ventricle via capillary and arteriole growth in such numbers that they eventually represent 50-75% of the total cell number in a mature heart (1,38).

To further develop our working hypothesis (Figure 1), with particular reference to the IGFs, it remains to be shown that fetal and neonatal hearts fulfil three fundamental criteria that provide evidence of an autocrine or paracrine regulation of cardiomyocyte growth. The first requirement is that the developing heart contains growth factor peptides and their corresponding gene transcripts. Fetal heart tissue has been shown to contain both IGF-I and -II by radioimmunoassay (39) or tissue immunohistochemistry (40). D'Ercole et al. (39) demonstrated that the amount of IGF-I obtained in the conditioned medium of a fetal heart organ culture exceeded that which could be obtained from a tissue homogenate, suggesting active IGF biosynthesis. This criteria has been strengthened for IGF-II by Romanus et al. (41) who have shown that IGF-II gene transcripts from fetal rat hearts can be translated into pre-pro-IGF-II peptides. What remains to be shown is that biologically active IGFs are synthesized, secreted, and function within the local mileu of the developing rat heart.

A second component of the GF paradigm of the developing heart is that specific GF receptors are detected for the GFs generated by the gene transcripts described above. We have recently documented that the neonatal rat heart contains receptors for both IGFs (see Figure 2) (4). The presence of IGF receptors in the developing fetal human heart has also been confirmed, yet their expression, as well as the insulin receptor, are found on numerous human fetal tissues (25). While human fetal tissues are sensitive to the metabolic actions of insulin, insulin does not apparently act as a mitogen (25). Adamson and Meek (42) have reported that EGF receptors are first measurable at day 11 of mouse gestation and increase in all tissues until term. Because Wilcox and Derynck (43) have reported that mRNA for TGF-alpha, an EGF-like molecule, is not localized to the developing mouse heart by *in situ* hybridization, this suggests that EGF may not function in an autocrine or paracrine manner within the developing heart. There are no other published reports, to the authors knowledge, that identify or quantitate any other defined GF receptor specifically to the developing heart.

Tissue and/or cellular responsiveness to the above GFs represents the third criteria that must be established to support the autocrine or paracrine GF-mediated hypothesis of the developing heart. Philips et al. (44) have just reported that twice daily injections of IGF-I between neonatal day 3-15 resulted in an increase in heart weight. Direct evidence of the *in vivo* effects of either IGF or indirect IGF effects via growth hormone (GH) induction remains to be conclusively demonstrated using transgenic animals. Nevertheless, suggestive evidence has been reported by Turner et al. (45) that GH-secreting GH_3 cells induce a three- to four-fold increase in IGF-I and IGF-II transcript levels, respectively, in hypertrophied adult rat hearts. Extrapolation of these results to the fetal and neonatal heart can not be readily made since the adult cardiomyocyte is postmitotic and limited to cellular hypertrophic responses. Nevertheless, these data suggest that IGFs may influence both proliferation and differentiation in the same target tissue with the divergence of results dependent upon specific maturational events unique to that target tissue (25). The focus of the data presented in this manuscript are directed toward in vitro studies utilizing primary cultures of neonatal cardiomyocytes. This was undertaken since these studies represent the most direct method of assessment of the role IGFs may play on the cardiomyocyte in an environment generally free of other cellular, humoral and/or neural influences.

447

MATERIALS and METHODS

Neonatal animals were housed, bred and maintained in our AAALAC approved facilities. Nucleic acid extraction and isolation materials were from BRL or Sigma Chemical Company. ^{125}I-IGF-I, ^{32}P-dCTP, and ^{3}H-deoxy nucleotide triphosphates were from Amersham. ^{32}P-UTP was from NEN. ^{3}H-leucine and ^{3}H-thymidine were from ICN. Ilford K.5D liquid emulsion was from Polysciences. Electrophoresis supplies, high molecular weight standards and protein assay reagent were from Bio-Rad.

IGF Receptor Crosslinking: Frozen rat heart ventricles were processed as described (4,46). Using a 20,000 x g membrane fraction, membrane proteins were extracted with 1% Triton X-100 for 30 min on ice with vortexing every 10 min (Keiss et al., 1987). Samples were centrifuged at 14,000 x g for 2 min , supernatant collected, frozen in liquid nitrogen and stored at -75°C. This extract was used directly for receptor crosslinking using ^{125}I-IGF-I in the presence or absence of 1 mM mannose-6-phosphate or antibody (Ab) 3637 (46) that specifically binds to the Type 2 IGF receptor. Receptor peptide interactions were crosslinked after 18 hr incubation at 4°C with 250 µM DSS, separated under reducing conditions in a 6% PAGE gel, and detected by autoradiography after 5 days. Molecular weight determinations were made using stained protein standards.

RNA Isolation and Analysis: Rapidly frozen ventricular tissues were processed under liquid nitrogen and RNA extracted, purified and quantitated as previously described (4,47). Poly-A^{+} RNA fractions were obtained by oligo (dT)-cellulose chromatography. Slot blot hybridization analysis was performed with serially diluted (1-0.125 µg/slot), denatured Poly A^{+} RNA using a cDNA probe to human Type 1 IGF receptor (48) or human acidic-FGF (aFGF) (49). ^{32}P-labelled cDNA inserts were nick translated to 1×10^{8} cpm/µg DNA and hybridization performed as previously described (4,47). Nitrocellulose filters were washed with 2 X SSC at 55°C three times for 30 min. Equivalence of RNA immobilization was determined by oligo-dT hybridization of 25 mers as described (4).

Cardiomyocyte Cell Culture: Neonatal (2-day) American Wistar ventricles were dissociated with collagenase and an enriched population of cardiomyocytes cultured under serum-free conditions (50). Myocytes were maintained for 36-48 hrs prior to stimulation by exogenous recombinant IGF-I (Amgen) or rat IGF-II (MSA) (Sigma). After 10-12 hrs of incubation with the IGF peptide, labeled precursors (either leucine or thymidine) were added for the remaining 4 hrs of incubation. Protein synthesis was assessed by leucine incorporation into acid precipitable material (dpm's per µg protein) by liquid scintillation spectroscopy. DNA synthesis was assessed by both thymidine incorporation into acid precipitable material (expressed as dpm's per 10^{6} nuclei) or cellular autoradiography (50). Numerical data are presented as the percentage change relative to untreated controls with increasing concentration of IGFs added.

In Situ Hybridization Studies: Cardiomyocytes cultured on glass coverslips were fixed with 4% paraformaldehyde for 30 min, dehydrated with alcohol, adhered to a glass slide with permount, and processed as previously described (4,47). Hybridizations were performed using mouse IGF-I or IGF-II cDNA probes (51,52). The cDNA inserts were nick translated using all four ^{3}H-dNTPs to a specific activity of 1 x 10^7 cpm/µg DNA. Coverslips were washed in 0.1 x SSC or 1 x SSC for 30 min three times for IGF-I or IGF-II hybridizations, respectively. Isolated adult rat hepatocytes and chicken red blood cells were used as positive and negative control cells, respectively. Slides were coated with emulsion, exposed for 14-21 days at -20°C and grains developed with Kodak D-19. Photographic images were produced using standard bright field or dark field microscopy.

Analysis of Transcription from Isolated Nuclei: Cardiomyocytes were cultured as described above and nuclei isolated after a 40 hr incubation with 20ng/ml of IGF-I. Nuclei were isolated and labelled RNA generated by the method of Introna et al. (53). Nuclei were stored frozen in liquid nitrogen for 2 weeks prior to use. Plasmids containing inserts of cDNAs for specific gene products were applied to a nitrocellulose filter with a slot blotter and baked for 2 hrs at 80°C under vacuum. Ornathine decarboxylase (ODC) cDNA was from Kahana and Nathens (54). The JE and KC cDNAs were from Cochran et al. (55). The myosin cDNA insert was from Mahdavi et al. (56). 2 x 10^6 cpm of RNA from each nuclei preparation was added for hybridization to the immobilized cDNAs. Hybridization was for 72 hrs at 65°C and the filters were washed at RT with 2 X SSC for 30 min twice and at 65°C for 10 min once prior to film exposure for 36-48 hrs.

RESULTS

Crosslinking studies using membrane preparations from spontaneously hypertensive (SHR) and normotensive (Wistar Kyoto, WKY) rats confirmed our previous studies (4) showing that both IGF receptor subtypes (Type 1 alpha subunit molecular weight of 130 kilodaltons, kd) are present in the developing neonatal rat heart (Figure 2). An age and strain variation in Type 1 IGF receptor levels can be seen in Lanes 1-4 (SHR, 1,3,7 and 14 days of age) and Lanes 5-8 (WKY, 1,3,7 and 14 days of age). In addition, the crosslinking of IGF-I to the Type 2 receptor (250 kd band) illustrates that similar age and strain variations appear to also exist for this IGF receptor subtype. The apparent decrease in the levels of both IGF receptors coincides with the natural decrease in proliferative potential of the cardiomyocytes in the developing rat heart (1,2,50). The more rapid decrease in the amount of Type 2 IGF receptor detected in the 7-day old SHR may be related to the diminished incorporation of thymidine into ventricular DNA that also occurs in this general time frame (50).

Modulation of the affinity of the Type 2 receptor for IGF-I by mannose-6-phosphate is shown in Lanes 9 and 10. Using membranes from 1 day old SHR, increased binding of IGF-I to the Type 2 IGF receptor band is evident (Lane 9, Figure 2). Confirmation of this effect is shown in Lane 10 where the binding of IGF-I to the Type 2 receptor is inhibited by Ab 3637 (46). Nevertheless, some residual IGF-I binding (270 kd band) is detected which has been proposed to represent crosslinked Type 1 IGF alpha subunits (57).

Analysis of changes in the gene expression levels for the Type 1 IGF receptor in the ventricles of the two strains during neonatal development are shown in Figure 3. Slot blot hybridization analysis of rat ventricular Poly A+ RNA from animals at 1, 3, 7, and 14 days of age supported the cross linking studies which demonstrated that the concentration of Type 1 receptor declines with age in both strain. Interestingly, the slot blot transcript abundance data at 1 and 3 days of age does not reflect the much greater difference in the estimated membrane concentration of Type 1 receptor seen between the two strains at these ages (4). This suggests that translational or membrane turn over times may also be different for the Type 1 receptor between the SHR and WKY strains. Because additional nuclear run off experiments (Figure 7) suggest that IGF-I stimulation may increase the transcription of the Type 1 receptor gene, this may also have influenced the level of the Type 1 receptor in vivo since SHR ventricular expression of IGF-I is elevated relative to the WKY at these ages (4).

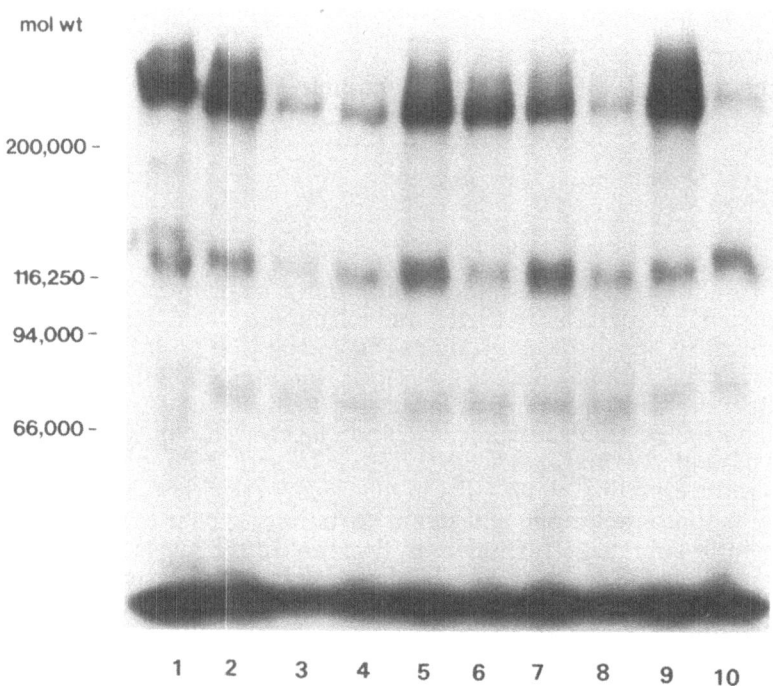

Figure 2. Crosslinking of IGF-I to Type 1 and Type 2 IGF Receptors

Covalent crosslinking of ^{125}I-IGF-I to detergent soluabilized neonatal rat cardiac membranes from SHR (Lanes 1-4, 9,10) and WKY (Lanes 5-8). Approximately equal amounts of membrane protein were loaded in each lane, incubated with IGF-I for 18 hrs. and crosslinked with DSS. Samples in lanes 9 and 10 were incubated with 1 mM mannose-6-phosphate in the presence (lane 10) or absence (lane 9) of Ab 3637 (1:1000 dilution). Autoradiography was for 5 days.

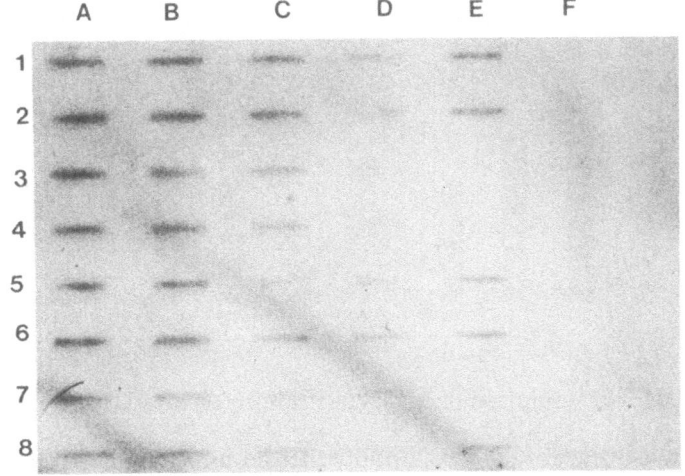

Figure 3. Analysis of Type 1 IGF Receptor Gene Expression

Poly A+ RNA from ventricles of SHR and WKY was serially diluted
(1.0-0.125 µg/lane, Columns A-D) and hybridized to a radiolabelled
cDNA insert to human Type 1 IGF receptor. Filter was exposed for
4 days. Rows 1, 3, 5 and 7 are WKY RNA of 1, 3, 7 and 14 day old
animals. Rows 2, 4, 6 and 8 are SHR RNA from 1, 3 7 and 14 day old
animals. Column E has two serial dilutions of human placenta
Poly A+ RNA as a positive control (rows 1-4 and 5-8 are separate
dilutions). Column F contained rat liver Poly A+ RNA.

Biochemical (Figure 4) and morphological (Figure 5) evidence of the
effects IGFs have on cardiomyocyte protein and DNA synthesis have been
obtained using primary cultures of neonatal myocytes. Increasing
concentrations of IGF-I or rat IGF-II (MSA) stimulated both protein (Figure
4A) and DNA synthesis (Figure 4B), with a maximal response seen at
approximately 10 ng/ml.

Documentation that the DNA synthetic stimulation was found in the
cardiomyocyte population, autoradiography of the cultured cells was
performed (Figure 5). Under control conditions (Figure 5A) only 5-7% of the
cells were labeled (6.85 ± 0.64) while IGF-I treatment for 12-16 hrs increased
the percentage of labeled cells to 10-12% (10.72 ± 0.61) (Figure 5B) (p<0.001).
Concomitant immunofluoescent microscopy of the cultured cells with a
desmin antibody (Chemicon Corporation) , a muscle cell specific marker,
showed that the majority of the cells labeled with thymidine contained the
desmin antigenic determinant (data not shown).

An important characteristic of the cardiomyocytes in culture is that
many of the cells become binuclear, a characteristic of the cells in vivo.
Further analysis is required to ascertain whether IGFs stimulate the
process of binucleation or nuclear polyploidation since thymidine
incorporation into cardiomyocyte nuclei is required for both developmental
processes to occur. The process of polyploidation will require flow cytometric
analysis of cellular and nuclear ploidy levels since the process of
binucleation renders the cardiomyocytes polyploid.

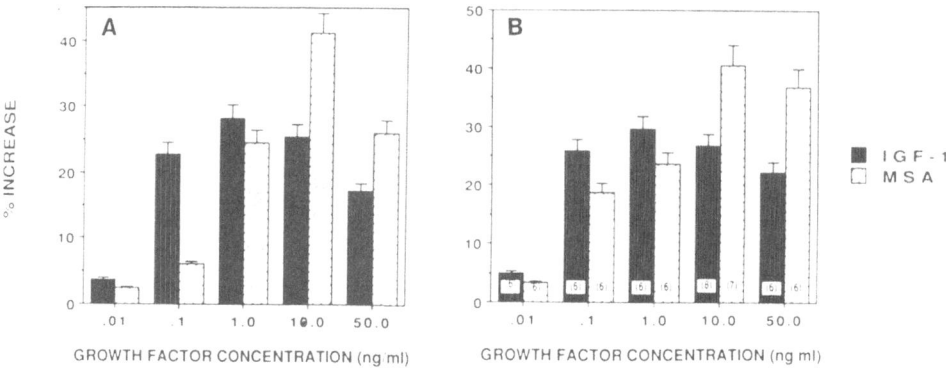

Figure 4. IGF Stimulation of Myocyte Protein (A) and DNA Synthesis (B)

Cardiomyocytes were isolated and cultured under serum-free conditions. After 36-48 in culture, cultures were treated with IGF-I or IGF-II (MSA) at the concentrations indicated for 10-12 hrs prior to the addition of radioactive precursor for the last 4 hrs.
Figure 4A: Protein Synthesis; Figure 4B: DNA synthesis.

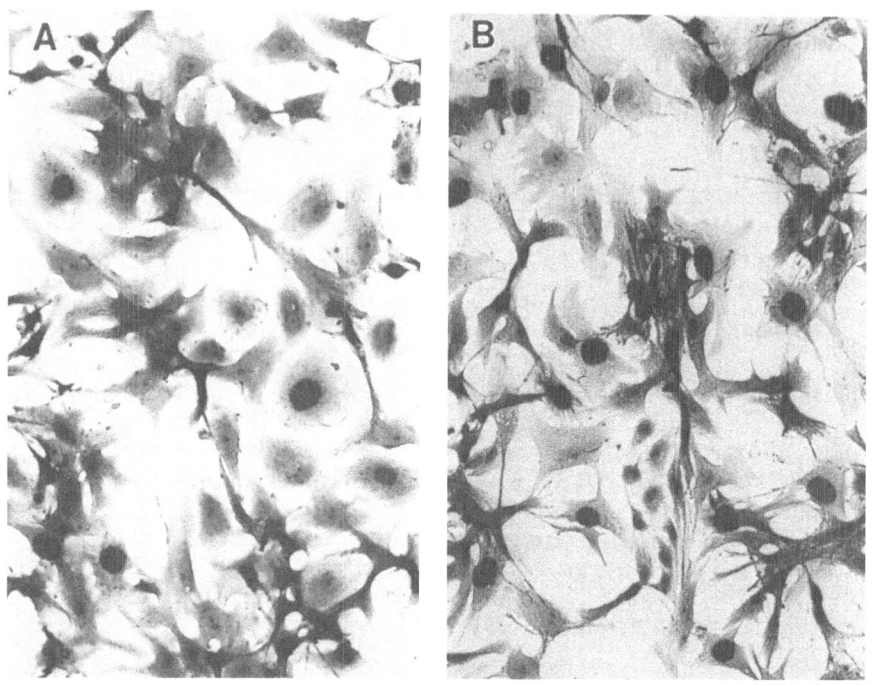

Figure 5. Autoradiography of IGF-I Stimulation of Cardiomyocyte DNA Synthesis.

Cardiomyocytes were cultured as described and maintained under basal conditions (Figure 5A) or stimulated with 10 ng/ml IGF-I for 12 hrs (Figure 5B) prior to the addition of radioactive thymidine. Cultures were fixed with 4% paraformaldehyde and thymidine incorporation detected by autoradiography.

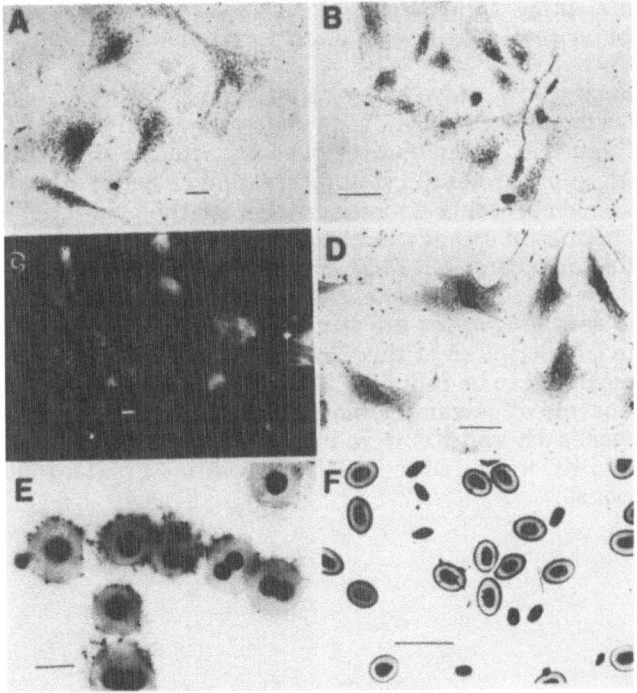

Figure 6. IGF Gene Expression in Cultured Cardiomyocytes: In Situ
Hybridization Analysis

Cardiomyocytes were isolated and cultured as described. Cells were
fixed with 4% paraformaldehyde and assessed for IGF gene expression
by in situ hybridization. IGF-II (Figure 6A-6C) and IGF-I (Figure 6D)
gene transcripts were detected in cultured cardiomyocytes. Figure 6C
is the dark field image of IGF-II hybridization. IGF-I positive control,
isolated rat hepatocytes (Figure 6E) and IGF-I negative control,
chicken red blood cell (Figure 6F) are also shown. Bar is 20 μm.

Localization of IGF gene transcripts to isolated, cultured
cardiomyocytes is shown in Figure 6. In situ hybridization studies using
cultured cardiomyocytes has substantiated our previous data on ventricular
IGF gene expression (4). Total ventricular RNA and tissue in situ
hybridization have shown that IGF-II gene transcripts are the most
abundant in the developing rat heart. Tissue in situ hybridization localized
these transcripts to the ventricular myocyte, yet the ability of the cultured
cardiomyocyte to mimic the in vivo IGF gene expression characteristics has
not been shown. Using primary cultures of neonatal rat cardiomyocytes, we
have shown that IGF-II gene transcripts are abundantly expressed (Figure
6A-6C) and IGF-I gene transcripts minimally expressed (Figure 6D). This
data is in agreement with our Northern blot and tissue in situ hybridization
analysis and strengthens our working hypothesis that developing
cardiomyocytes may regulate their own growth by autocrine or paracrine
methods. Further analysis will investigate the ability of cultured
cardiomyocytes to synthesize and secrete biologically active IGFs into the
culture medium. Culture studies will also be undertaken with

cardiomyocytes from both the SHR and WKY strains to assess whether the IGF gene transcription and IGF receptor variations detected on whole hearts can also be analyzed in vitro under controlled conditions.

Preliminary studies have begun to focus on the direct effects IGF stimulation has on cardiomyocyte gene expression by nuclear run off experiments (Figure 7). This analysis was undertaken since direct evidence of IGF stimulation of cardiomyocyte proliferation or maturation has yet to be demonstrated and represents a major criteria for the evaluation of our hypothesis. These studies have indicated that IGF-I stimulation increases the transcription of several cell cycle specific genes as well as a cardiomyocyte muscle specific gene, the myosin heavy chain (MHC) genes. When cultured cardiomyocytes are stimulated by exogenous IGF-I, an increase in the transcription of three genes associated with cellular proliferation appeared to be induced. The hybridization signals for ODC , rate limiting enzyme of polyamine biosynthesis, and the PDGF-associated "competence" genes JE and KC were increased by IGF-I stimulation. The effects on JE and KC appeared to be selective, with a greater induction seen in the KC response.

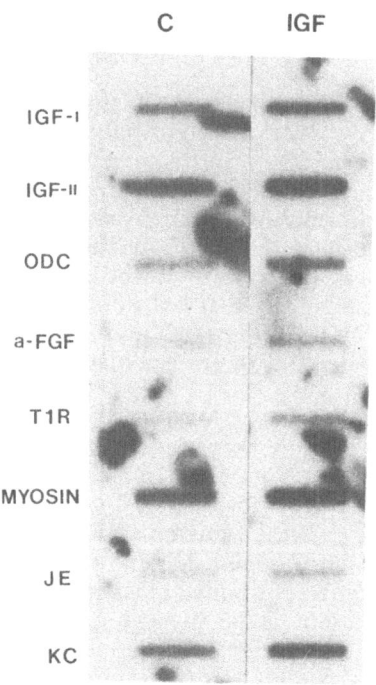

Figure 7. IGF-I Stimulation of Cardiomyocyte Gene Expression

Primary cultures of neonatal cardiomyocytes were treated with 20 ng/ml of IGF-I for 40 hrs prior to nuclei isolation. Isolated nuclei were used in a nuclear run-off protocol and the labeled RNA hybridized to specific cDNAs. Mouse IGF-I and IGF-II, ornithine decarboxylase (ODC), acidic FGF (aFGF), Type 1 IGF receptor (T1R), myosin heavy chain (MYOSIN), and PDGF-inducible genes (JE and KC).

In addition, the transcription of acidic-FGF (aFGF), another "competence" growth factor, and the Type 1 IGF receptor (T1R) genes were also induced. This suggests that IGF stimulation of cardiomyocytes may modulate cell surface GF receptor levels as well as other potential GF peptide genes which may all function within the mileu of the developing neonatal heart in an autocrine or paracrine manner. IGF treatment also appeared to increase the transcription of a muscle specific, differentiated gene, the MHC, which is characteristic of the cardiomyocyte. Although the increase in MHC gene expression was not profound, it suggests that IGFs may influence both proliferation and differentiation of the cardiomyocytes.

FUTURE PROSPECTS IN HEART DEVELOPMENT

Molecular and cellular aspects of heart development have "arrived" at the forefront of cardiovascular studies. This area of research is beginning to be explored, defined and catalogued and the key players identified. We feel that the IGFs represent one of the major growth factor families important in the proliferation and development of the fetal and neonatal cardiomyocyte. As illustrated in our original hypothesis (Figure 1), we must not forget the potential role other autocrine/paracrine or endocrine GF stimuli may have on heart development. In concert with the expression and response to autocrine or paracrine GF stimuli, the cardiomyocytes of the developing heart are also subject to growth regulation by the extracellular matrix, neural/humoral, and functional influences. We must not be content with mere phenomenologic identification of GF genes and receptors in the developing heart, but must strive for useful models and testable hypothesis as to their biological significance.

The role of the extracellular matrix in heart development is beginning to be evaluated (58). The extracellular matrix has taken on additional significance since it can provide both an extracellular scaffolding necessary for migration and cell-to-cell interactions as well as a "sink" for growth factors such as the FGFs (59). The modulation of extracellular matrix formation by specific growth factors, such as transforming growth factor-beta (TGF-ß), represents an intriguing focus of future studies. We have begun to examine the possible role TGF-ß may have in cardiomyocyte development as an autocrine/paracrine regulator of proliferation since readily detectable TGF-ß gene expression occurs during the first neonatal week of age (data not shown). The pleotrophic response of multiple cell types to TGF-ß (60) makes this specific GF family of particular significance to heart development.

A major shortcoming to the analysis of heart development at the cellular level is the lack of a defined cardiomyocyte cell line. Ongoing studies in several laboratories are attempting to establish a cardiomyocyte cell line utilizing cellular transformation by retroviral or SV-40 temperature-sensitive mutants. Other possible methods of analysis of cardiomyocyte cell biology could make use of cellular heterokaryons in a manner analogous to the studies of skeletal muscle myoblast-to-myotube transitions (61). Irregardless of the method of development, the cardiomyocyte cell line must express differentiated cellular functions yet retain its proliferative capabilities.

The rapid developments in the ability to establish transgenic animals will eventually provide one of the best methods of analysis of heart development. Since embryonic and fetal heart growth and development can not be directly manipulated, specific variations in gene expression during these periods of time in transgenic animals and subsequent effects on the developing heart would provide a more direct method of analysis. Recent studies using GH or IGF-I modified transgenic animals represents an important step in this direction (62). Detailed analysis of heart growth in these and other transgenic animals during fetal and neonatal development will provide essential temporal information on the specific role a defined GF family and/or its receptor may play in cardiac morphogenesis and development.

ACKNOWLEDGEMENTS: The authors thank Mr. J Yun, Mr. M. Kelley and Ms. L. Haws for their technical assistance. The authors also thank those who generously provided access to their cDNA probes: Drs. G. Bell, A. Ulrich, M. Jaye, D. Nathans, V. Mahdavi and C. Stiles. Portions of this project were supported by: NIH HL 42218, Diabetes Assoc. Greater Cleveland, and Edison Foundation of Ohio (GLE), NIH HL 38442 and VA Medical Research Funds (JFH) and Kidney Foundation of Ohio Grant-in-Aid (KDB).

REFERENCES

1. R Zak. "Growth of the Heart in Health and Disease", Raven Press, NY, 1984.
2. VJ Ferrans, G Rosenquist and C Weinstein. "Cardiac Morphogenesis", Elseiver, NY, 1985.
3. JM Pfeffer, MA Pfeffer, MC Fishbein and ED Frolich. Am. J. Physiol. 237: H461-H468, 1979.
4. GL Engelmann, KD Boehm, JF Haskell, PA Khairallah and J Ilan. Mol. Cell. Endocr. 61: in press, 1989.
5. JM Icardo. IN: "Growth of the Heart in Health and Disease", Raven Press, NY, pp. 41-79, 1984.
6. FJ Clubb and SP Bishop. Lab. Invest. 50: 571-579, 1984.
7. S Oparil, SP Bishop and FJ Clubb, Jr. Hypertension (Suppl III) 38-43, 1984.
8. P Anversa, G Olivetti and AV Loud. Circ. Res. 46: 495-502, 1980.
9. LB Bugaski, M Rabinowitz and R Zak. Cardio. Res 19: 89-95, 1985.
10. I Harary. IN: "The Heart Cell in Culture", CRC Press, Florida, pp. 33-52, 1987.
11. GL Engelmann, JC Vitullo & RG Gerrity. Circ. Res. 58: 137-147, 1986.
12. WY Brodsky & IV Uryvaeva. "Genomic Multiplication in Growth and Development", Cambridge Univ. Press, London, 1985.
13. WC Claycomb and NA Lanson. Biochem. J. 247: 701-706, 1987.
14. MD Schneider, PA Payne, H Uneo, MB Perryman and R Roberts. Mol. Cell. Biol. 6: 4140-4143.
15. KA Zimmerman, GD Yancopoulos, RG Collum, RK Smith, NE Khol, KA Denis, MM Nau, ON White, D Toran-Allerand, CE Gee, JD Minna and FW Alt. Nature 319: 780-783, 1986.
16. G Mugrauer, FW Alt and P Ekblom. J. Cell Biol. 107: 1325-1335, 1988.
17. JW Kasik, YY Wang and K Ozato. Mol. Cell. Biol. 7: 3349-3352, 1987.
18. R Muller, DJ Slamon, ED Anderson, JM Tremblay, D Muller, MJ Cline and IM Verma. Mol. Cell. Biol. 3: 1061-1069.
19. W Schaffer & RS Williams. Biochem. Biophys. Res. Comm. 138: 387-391, 1987.
20. JM Bishop. Ann. Rev. Biochem. 52: 301-354.

21. J Ong, S Yamashita and S Melmed. Endocr. 120: 353-357, 1987.
22. LE Underwood and AJ D'Ercole. Clin. Endocr. Metab. 13: 69-89, 1984.
23. RDG Milner and DJ Hill. Clin. Endocr. 21: 415-433, 1984.
24. AJ D'Ercole and LE Underwood. IN: "Human Growth: A Comprehensive Treatise", Plenum Press, NY, pp. 327-338, 1986.
25. DJ Hill, AJ Strain & RDG Milner. Oxf. Rev. Repr. Biol. 9: 398-455, 1987.
26. ER Froesch, C Schmid, J Schwander and J Zapf. Ann. Rev. Physiol. 47: 443-467, 1985.
27. JJ Van Wyk. IN: "Hormonal Proteins and Peptides", Academis Press, NY, pp. 81-125, 1984.
28. GT Ooi and AC Herington. J. Endocr. 118: 7-18, 1988.
29. MM Rechler and SP Nissley. IN: "Polypeptide Hormone Receptors", Dekker, NY, pp. 227-297, 1985.
30. AJ D'Ercole, DB Foushee and LE Underwood. J. Clin. Endocr. Metab. 43: 1069-1077, 1976.
31. JR Florini. Muscle & Nerve 10: 577-598, 1987.
32. DO Morgan, JC Edman, DN Strandring, VA Fried, MC Smith, RA Roth and WR Rutter. Nature 329: 301-307, 1987.
33. RG MacDonald, SR Pfeffer, L Coussens, MA Tepper, CM Brocklebank, JE Mole, JK Anderson, E Chen, MP Czech and A Ullrich. Science 239: 1134-1137, 1988.
34. PK Lund, M Moats-Staats, MA Hynes, JG Simmons, M Jansen, AJ D'Ercole and JJ Van Wyk. J. Biol. Chem. 261: 14539-14544, 1986.
35. MB Soares, DH Ishii and A Efstratiadis. Nucleic Acid Res. 13: 1119-1134, 1985.
36. AL Brown, DE Graham, SP Nissley, DJ Hill, AJ Strain and MM Rechler. J. Biol. Chem. 261: 13144-13150, 1986.
37. VKM Han, PK Lund, DC Lee and AJ D'Ercole. J. Clin. Endocr. Metab. 66: 422-429, 1988.
38. K Rakusan, PW Hrdina, Z Turek, EG Lakatta, HA Spurgeon and GD Walford. Basic Res. Cardiol. 79: 389-395, 1984.
39. AJ D'Ercole, G Applewhite & L Underwood. Dev. Biol. 75: 315-328, 1980.
40. VKM Han, DJ Hill, AJ Strain, AC Towle, JM Lauder, LE Underwood and AJ D'Ercole. Ped. Res. 22: 245-249, 1987.
41. JA Romanus, YWH Yang, SO Adams, AN Sofair, LYH Tseng, SP Nissley and MM Rechler. Endocr. 122: 709-716, 1988.
42. ED Adamson and J Meek. Dev. Biol. 103: 62-72, 1984.
43. JN Wilcox and R Derynck. Mol. Cell. Biol. 8: 3415-3422, 1988.
44. AF Philipps, B Persson, K Hall, M Lake, A Skottner, T Sanengen and VR Sara. Ped. Res. 23: 298-305, 1988.
45. JD Turner, P Rotwein, J Novakofski & PJ Bechtel. Am. J. Physiol. 255: E513-E517, 1988.
46. W Keiss, JF Haskell, L Lee, LA Greenstein, BE Miller, AL Aarons, MM Rechler & SP Nissley. J. Biol. Chem. 262: 12745-12751, 1987.
47. CY Wang, M Daimon, S Shen, GL Engelmann and J Ilan. Mol. Endocr. 2: 217-229, 1988.
48. A Ullrich, A Gray, AW Tam, T Yang- Feng, M Tsubokawa, C Collins, W Henzel, T LeBon, S Kathuria, E Chen, S Jacobs, U Francke, J Ramachandran & Y Fujita-Yamaguchi. EMBO J. 5:2503-2512, 1986.
49. M Jaye, R Howk, W Burgess, GA Ricca, IM Chiu, MW Ravera, SJ O'Brian, WS Modi, T Maciag and WN Drohan. Science 233: 541-545, 1986.
50. GL Engelmann & RG Gerrity. J. Mol. Cell. Cardiol. 20: 169-177, 1988.
51. GI Bell, MM Stempien, NM Fong and LB Rall. Nucleic Acid Res. 14: 7873-7882, 1986.

52. MM Stempien, NM Fong, LB Rall and GI Bell. DNA 5: 357-361, 1986.
53. M Introna, RC Bast, PA Johnston, DO Adams and TA Hamilton. J. Cell. Physiol. 131: 36-42, 1987.
54. C Kahana and D Nathans. Proc. Natl. Acad. Sci. 82: 1673-1677, 1985.
55. BH Cochran, AC Reffel and CD Stiles. Cell 33: 939-947, 1983.
56. V Mahdavi, M Periasamy and B Nadal-Ginard. Nature 297: 659-665.
57. DZ Ewton, SL Falen and JR Florini. Endocr. 120: 115-123, 1987.
58. TK Borg, RE Gay and LD Johnson. Collagen Rel. Res. 2: 211-218, 1982.
59. DB Rifkin and M Klagsbrun. "Angiogenesis: Mechanisms and Pathobiology", CSH Laboratory, NY, 1987.
60. MB Sporn, AB Roberts, LM Wakefield and B de Crombrugghe. J. Cell Biol. 105: 1039-1045, 1987.
61. CH Clegg and SD Hauschka. J. Cell Biol. 105: 937-947, 1987.
62. LS Mathews, RE Hammer, RL Brinster and RD Palmiter. Endocr. 123: 433-437, 1988

ROLE OF INSULIN-LIKE GROWTH FACTOR I AND II ON SKELETAL REMODELING

Ernesto Canalis, Michael Centrella, and Thomas L. McCarthy

St. Francis Hospital and Medical Center, Hartford,
Connecticut and the University of Connecticut Health
Center, Farmington, Connecticut

INTRODUCTION

The two major processes involved in bone remodeling are bone formation and bone resorption, and these two functions are regulated by systemic hormones and by local growth factors (1,2). Hormones seem to act on skeletal cells either directly or indirectly by modulating the synthesis or effects of local factors, which in turn stimulate or inhibit bone formation or bone resorption. Growth factors can affect bone remodeling either by increasing the number of cells or by affecting the differentiated phenotype of cells of the osteoblast or osteoclast lineage.

Bone matrix has been shown to contain a number of growth factors (3-8, Table 1). The cellular origin of some of these factors has not been determined, but both insulin-like growth factor I (IGF I) and transforming growth factor β (TGFβ) have been shown to be synthesized by normal skeletal cells (9,10). Studies on synthesis of the bone matrix associated factors are important since systemic factors could be trapped by the matrix and act as putative local agents: It is also necessary to consider that factors such as the IGF's, which are present in the systemic circulation, are also synthesized by skeletal cells. Although recent work has focused primarily on the role of IGF's as local regulators of bone function, it is likely that both the systemic and local IGF's have an important and possibly different role in bone remodeling.

Table 1. Growth factors isolated from
skeletal tissue

---

1. Platelet-derived growth factor
2. Acidic fibroblast growth factor
3. Basic fibroblast growth factor
4. Transforming growth factor Beta 1 and 2
5. Insulin-like growth factor I and II
6. Osteoinductive factors or bone
   morphogenetic proteins

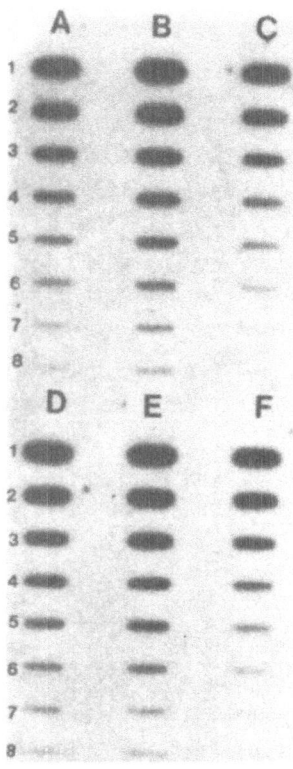

Figure 1. IGF I-mRNA from rat
calvarial cells detected by
Slot blot analysis. RNA was
extracted, blotted onto charged
nylon, immobilized and hybridized with a [32P]labeled rat
IGF I-cDNA clone and detected
autoradiography. Lanes A
through E represent individual
parietal bone cell populations
obtained by sequential collagenase digestion. Lane A
represents mostly fibroblasts,
whereas Lanes C, D, and E
represent mostly osteoblasts.
Lane F is a pool of these osteoblast-enriched cells.

Synthesis of insulin-like growth factors (IGF's) by skeletal cells

Work from our laboratory has shown that bone cells secrete and
synthesize various growth factors, including IGF I (9). In addition,
other investigators have shown that bone matrix contains IGF II, and what
was initially named skeletal growth factor is, in fact, IGF II (6).
Whereas definitive work on the synthesis of IGF II by bone cells has not
been reported, IGF II is present in bone cell cultures and is likely to
be synthesized by skeletal cells. On the other hand, Slot blot and

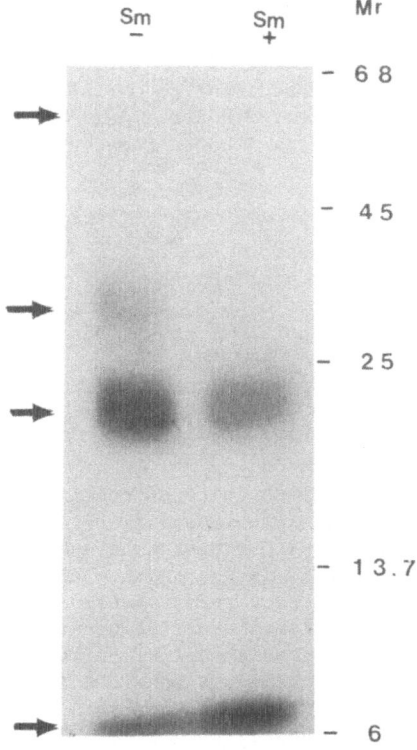

Figure 2. Polyacrylamide gel analysis of IGF-binding proteins from cultures of 21-day fetal rat calvariae. Conditioned medium was acidified and fractionated by gel filtration chromatography using 1 M acetic acid. The fractions eluting with Mr of 25,000 and higher were lyophilized, incubated with 125I-IGF I in the presence (+Sm) and absence (-Sm) of IGF I, cross-linked with disuccinimidyl suberate and fractionated on a 7 to 15% polyacrylamide gel in the presence of 2-mercaptoethanol.

Northern blot analysis, using a rat IGF I-cDNA probe (kindly provided by Dr. L. Murphy) revealed that a variety of cells present in fetal rat calvariae express IGF I transcripts (Figure 1, 11). These cells include bone fibroblasts and osteoblasts, suggesting that IGF I may act as a paracrine or autocrine regulator of bone function. These studies also showed that skeletal fibroblasts and osteoblasts secrete IGF I to their culture medium, and the endogenous concentration of IGF I is about 1 nM. Recently, we started to examine the regulation of IGF I synthesis by systemic and local factors, and found that growth hormone (GH) as well as parathyroid hormone (PTH) stimulates the synthesis of IGF I by osteoblas-

461

tic cells (11). Work by other investigators confirmed the effect of GH and revealed that 17 Beta estradiol (17βE2) increases the concentrations of IGF I and its binding proteins in osteoblast cultures (12,13). It is, therefore apparent that, in addition to GH, two hormones known to have major effects on the skeleton, PTH and 17βE2, regulate IGF I synthesis. This suggests that IGF I is an important mediator of hormonal action in the skeleton.

Hormones may act by affecting IGF I synthesis or by regulating the binding of IGF I to its receptor. While studies on the IGF receptor in osteoblasts are underway in our laboratory, information about the regulation of IGF-binding by systemic hormones is still not available. However, work by other investigators has shown that glucocorticoids enhance IGF I-binding to its receptor and this may explain the transient stimulatory effect of glucocorticoids on bone formation (14,15).

Insulin-like growth factor-binding proteins

In addition to IGF I, bone cultures contain one or more IGF-binding proteins which probably modulate the availability of IGF and its effects on bone formation. IGF-binding proteins have been found in serum and extraskeletal tissues, but it is not known whether those found in bone cultures are the same as those already described. Studies from our laboratory indicated the existence of IGF-binding proteins in calvarial cultures (9). When medium conditioned by calvariae was fractionated by gel filtration chromatography in 1 M acetic acid, material eluting with an approximate molecular mass (Mr) higher than 25,000 daltons bound radiolabeled IGF I when incubated at neutral pH. Affinity labeling, cross-linking and polyacrylamide gel analysis of the material fractionated by gel filtration revealed at least two bands migrating with approximate Mr of 20,000 to 35,000 (Figure 2). While IGF-binding proteins from calvarial cultures have not been characterized, work performed in other cellular systems indicates that IGF-binding proteins play a critical role in IGF biological actions. These proteins may prolong the half-life of IGF, neutralize or enhance its biological activity, or participate in the transport of IGF to its target cells (16). We expect that IGF I-binding proteins play a similar role in the skeletal system.

Effects of insulin-like growth factors on bone remodeling

Although information about the in vivo effects of IGF's on bone remodeling is still limited, a considerable amount of work has been performed to elucidate their effects in in vitro systems. This work was carried out primarily in cultures of 21-day fetal rat calvariae and in cultures of calvarial-derived osteoblasts. IGF I and II stimulate DNA and collagen synthesis in calvarial cultures (Figure 3). The effects of both factors are identical, but while IGF I is effective at 10 nM, concentrations of IGF II that are 4 to 7 times higher are necessary to observe the same response (17).

Studies using autoradiography and bone histomorphometric analysis revealed that IGF I stimulates primarily preosteoblast cell replication (18). In addition, IGF I stimulates type I collagen and bone matrix synthesis. This effect is not changed in the presence of DNA synthesis inhibitors, indicating that it is independent of the IGF effect on DNA synthesis and cell replication. It is therefore apparent that IGF's have separate effects on bone cell replication and matrix synthesis.

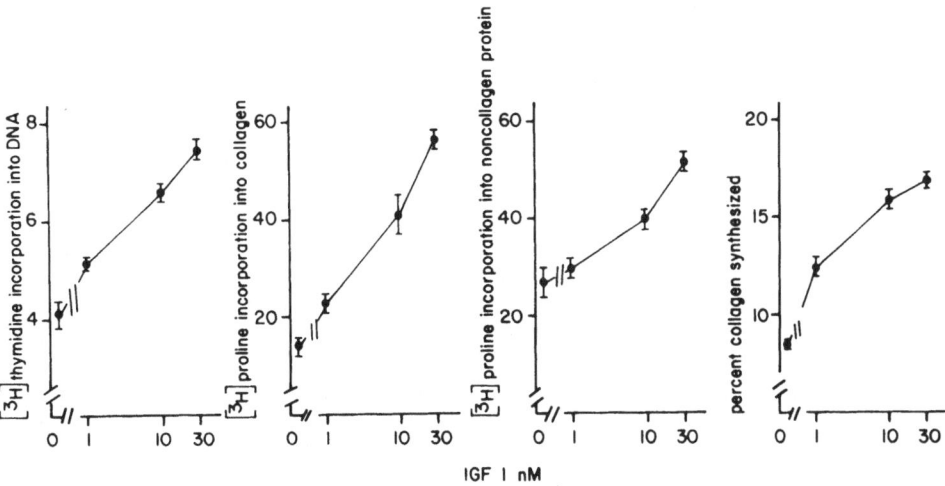

Figure 3. Effect of IGF I on [³H]thymidine incorporation into DNA, and [³H]proline incorporation into collagen and noncollagen protein in intact calvariae. Symbols represent means ± SE for 5 half calvariae cultured for 24 h in the continuous presence of IGF I. Data are expressed as dpm/μg bone weight.

IGF I and II appear to regulate osteoblastic type I collagen synthesis at a transcriptional level. Northern blot analysis of total cellular. RNA extracts using a rat cDNA probe for the α1 chain of type I collagen revealed that 24 h treatment with IGF I or II increased type I collagen-mRNA in Ob-cell cultures (Figure 4). This change was observed in the absence of changes in DNA content, confirming that the effects of IGF I and II on collagen synthesis are independent of those on cell replication. To substantiate that IGF's had a direct action on the differentiated function of the osteoblast, the effects of IGF I on osteocalcin synthesis were examined in calvarial cultures. IGF I was found to increase osteocalcin concentrations and to be synergistic with 1,25-dihydroxyvitamin D3, the major regulator of osteocalcin synthesis (19).

In addition to their effects on cell replication and collagen synthesis, IGF I and II prevent the degradation of collagen in calvarial cultures (17). When calvariae were "pulsed" with radiolabeled proline and "chased" in the presence of excess unlabeled proline in the presence and absence of IGF I or II, both factors decreased the release of labeled hydroxyproline, representing newly synthesized collagen, from the bone compartment to the culture medium. This indicates that both IGF I and II decreased the degradation of newly synthesized collagen and suggests an additional role for these factors in the regulation of bone collagen. The mechanism of this effect is not known, but IGF's may block the production of osteoblast-derived collagenase or stimulate the production of collagenase inhibitors. Since IGF's increase collagen synthesis and decrease collagen degradation, it is likely that these two functions are critical in the maintenance of normal bone mass.

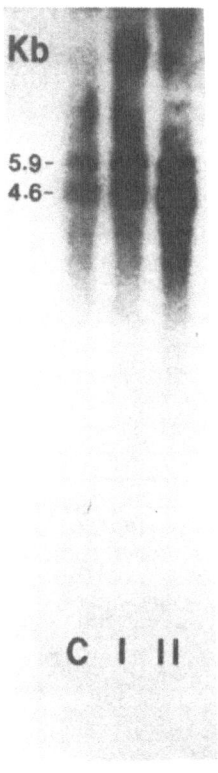

Kb

5.9-
4.6-

C I II

Figure 4. Northern blot analysis of
the effect of IGF I and II on type I
collagen-mRNA levels in osteoblast-
enriched (Ob) cells. Ob-cells were
grown to confluence, and serum deprived prior to a 23 h treatment
period with IGF I or II at 100 nM in
serum-free medium. Total RNA from
control and treated cultures was extracted, separated on a 1.5% agarose,
2.2 M formaldehyde gel, transferred
to a charged nylon membrane, immobilized and hybridized with a [32P]-
labeled α1R1 (rat α1 type I collagen,
kindly provided by Drs. D. Rowe and
B. Kream, Farmington, CT) cDNA probe
and displayed by autoradiography.
The two bands observed consist of
transcripts containing 6.4 and 5.4
kilobases. Lane C represents a control culture, and lanes I and II,
cultures treated with IGF I and II
respectively.

While osteoblasts have type 1 and 2 IGF receptors (20), it is presently unclear which function of IGF on bone remodeling is mediated by either receptor. It is possible that the effects on collagen synthesis and cell replication are mediated by the type 1 receptor as it has been shown in other systems, whereas the effects on collagen degradation may be mediated by the type 2 receptor.

In summary, IGF I and II are present in significant quantitites in the bone matrix and IGF I, and likely, II are synthesized by skeletal cells. The synthesis of skeletal IGF I is regulated by systemic hormones such as GH and PTH, but less is known about the regulation of IGF II synthesis. Both IGF I and II stimulate preosteoblastic cell replication and also have an independent stimulatory effect on bone collagen synthesis. IGF I and II also prevent bone collagen degradation and through these mechanisms they are important in the regulation and maintenance of normal bone matrix.

ACKNOWLEDGMENTS

This work was supported by Grant AR21707 from the National Institute of Arthritis and Musculoskeletal and Skin Diseases. The authors thank Mrs. Sheila Rydziel, Miss Sandra Casinghino, Mrs. Bari Gabbitas and Mrs. Michele Kervick for expert technical assistance and Mrs. Nancy Brignoli for valuable secretarial assistance.

REFERENCES

1. E. Canalis, The hormonal and local regulation of bone formation, Endocr. Rev. 4:62 (1983).
2. E. Canalis, T. McCarthy and M. Centrella, Growth factors and the regulation of bone remodeling, J. Clin. Invest. 81:277 (1988).
3. P. V. Hauschka, A. E. Mavrakos, M. D. Iafrati, S. E. Doleman and M. Klagsbrun, Growth factors in bone matrix, J. Biol. Chem. 261:12665 (1986)
4. S. M. Seyedin, A. Y. Thompson, H. Bentz, D. M. Rosen, J. M. McPherson, A. Conti, N. R., Siegel, G. R. Galluppi, and K. A. Piez, Cartilage inducing factor. Apparent identity to transforming growth factor-β, J. Biol. Chem. 261:5693 (1986).
5. C. A. Frolik, L. F. Ellis and D. C. Williams, Isolation and characterization of insulin-like growth factor-II from human bone, Biochem. Biophys. Res. Commun. 151:1011 (1988).
6. S. Mohan, J. C. Jennings, T. A. Linkhart and D. J. Baylink, Primary structure of human skeletal growth factor: homology with human insulin-like growth factor II, Biochem. Biophys. Acta 966:44 (1988).
7. E. Canalis, T. McCarthy and M. Centrella, Isolation of growth factors from adult bovine bone, Calcif. Tissue. Int. (1988) (In Press).
8. M. R. Urist, R. J. DeLange and G. A. M. Finerman, Bone cell differentiation and growth factors, Science 220:680 (1983).
9. E. Canalis, T. McCarthy and M. Centrella, Isolation and characterization of insulin-like growth factor I (Somatomedin-C) from cultures of fetal rat calvariae, Endocrinology 122:22 (1988).
10. P. Gehron Robey, M. F. Young, K. C. Flanders, N. S. Roche, P. Kondaiah, A. H. Reddi, J. D. Termine, M. B. Sporn and A. B. Roberts, Osteoblasts synthesize and respond to transforming growth factor-type β (TGF-β) in vitro, J. Cell Biol. 105:457 (1987).

11. T. L. McCarthy, M. Centrella, and E. Canalis, Parathyroid hormone enhances the transcript and polypeptide levels of insulin-like growth factor I in osteoblast-enriched cultures from fetal rat bone, Endocrinology (1989) (In Press)

12. M. Ernst and E. R. Froesch, Growth hormone dependent stimulation of osteoblast-like cells in serum-free cultures via local synthesis of insulin-like growth factor I, Biochem. Biophys. Res. Commun. 151:142 (1988).

13. M. Ernst, C. Schmid, J. Zapf and E. R. Froesch, Osteoblasts synthesize specific carrier proteins for insulin-like growth factor in response to growth hormone and estradiol in vitro, J. Bone Min. Res. 3 Suppl 1 S206 (1988).

14. A. Bennett, T. Chen, D. Feldman, R. Hintz and R. G. Rosenfeld, Characterization of insulin-like growth factor I receptors on cultured rat bone cells: Regulator of receptor concentration by glucocorticoids, Endocrinology 115:1577 (1984).

15. E. Canalis, Effect of glucocorticoids on type I collagen synthesis, alkaline phosphatase activity and deoxyribonucleic acid content in cultured rat calvariae, Endocrinology 112:931 (1983).

16. M. Binoux, P. Hossenlopp, S. Hardouin, D. Seurin, C. Lassarre and M. Gourmelen, Somatomedin (insulin-like growth factor)-binding proteins. Molecular forms and regulation, Horm. Res. 24:141 (1986).

17. T. L. McCarthy, M. Centrella, and E. Canalis, Regulatory effects of insulin-like growth factor I and II on bone collagen synthesis in rat calvarial cultures, Endocrinology (1989) (In Press)

18. J. M. Hock, M. Centrella and E. Canalis, Insulin-like growth factor I has independent effects on bone matrix formation and cell replication, Endocrinology 122:254 (1988)

19. E. Canalis, and J. Lian, Effects of bone associated growth factors on DNA, collagen and osteocalcin synthesis in cultured fetal rat calvariae, Bone 9:243 (1988)

20. M. Centrella, T. L. McCarthy and E. Canalis, Discrete high molecular weight IGF-I and IGF-II receptors in osteoblast-enriched cultures from fetal rat bone, J. Cell Biol. (1989) Abstract (In Press)

EFFECTS OF INSULIN-LIKE GROWTH FACTORS ON CHROMAFFIN CELLS

Mary K. Dahmer and Robert L. Perlman

Departments of Pediatrics and Pharmacological and
Physiological Sciences and the Joseph P. Kennedy, Jr.
Mental Retardation Research Center, The University of
Chicago, Chicago, IL 60637

INTRODUCTION

In 1966, Levi-Montalcini and her colleagues reported that insulin
stimulates RNA and lipid synthesis in sensory ganglia (Angeletti et al.,
1966). Because these workers used high concentrations of insulin in their
studies, the significance of their observations was not clear. The
subsequent discovery of insulin-like growth factors (IGFs) and IGF
receptors, and the recognition that high concentrations of insulin can
activate IGF-I receptors, raised the possibility that the actions of
insulin on sensory ganglia may have been due to the stimulation of IGF-I
receptors in these ganglia. Other neural crest-derived cells may also be
targets of IGF action. Wilson et al. (1985) have reported that high
concentrations of insulin increase the content of opioid peptides and
enhance catecholamine secretion in adrenal chromaffin cells; these
actions of insulin may reflect a physiological role of IGF-I in the
adrenal medulla. The finding of IGF mRNA (Han et al., 1987) and IGF-I
like immunoreactivity (Hansson et al., 1988) in the adrenal gland
suggests that IGF-I may function as an autocrine or paracrine regulatory
factor in this tissue.

To understand the actions of insulin and IGFs in the adrenal medulla
and sympathetic nervous system, we have investigated the effects of these
peptides on PC12 pheochromocytoma cells and normal chromaffin cells. PC12
cells have many of the properties of sympathetic neuroblasts and
chromaffin cell precursors. These cells synthesize, store and secrete
catecholamines, and also retain the ability to replicate. In response to
nerve growth factor, PC12 cells differentiate into cells which more
closely resemble sympathetic neurons (Greene and Tischler, 1982). Normal
chromaffin cells have been widely used to study the mechanism and
regulation of catecholamine secretion (Livett et al., 1983).

INSULIN-LIKE GROWTH FACTORS STIMULATE PC12 CELL REPLICATION

PC12 cells are routinely grown in medium containing 10% horse serum
and 5% fetal bovine serum (Greene and Tischler, 1982). Although this
medium supports optimal growth of the cells, the serum factors that
regulate PC12 cell replication have not been identified. We assayed the
mitogenic activity of a number of growth factors on serum-deprived PC12

cells. In these studies, we cultured cells in medium containing 0.2%
horse serum and 0.1% fetal bovine serum, stimulated the cells with
various growth factors, and measured the incorporation of [³H]thymidine
into the cells (Dahmer and Perlman, 1988a). Insulin, IGF-I and IGF-II all
stimulated [³H]thymidine incorporation into serum-deprived PC12 cells
(Fig. 1). IGF-I (EC$_{50}$ $\cong$ 0.3 nM) was the most potent of these peptides. The
relative potency of these peptides is consistent with the interpretation
that their effects are due to the stimulation of IGF-I receptors on the
cells (Rechler and Nissley, 1985), and that the activation of these
receptors leads to PC12 cell replication. Additional studies demonstrated
that insulin treatment of growth-restricted PC12 cells results in an
increase in cell number and a decrease in the number of cells in the
G$_0$/G$_1$ phase of the cell cycle (Dahmer and Perlman, 1988a). These
observations provide additional evidence that insulin, and presumably
also IGFs, stimulate proliferation of PC12 cells. Nielsen and Gammeltoft
(1988) have also reported that insulin and the IGFs can act as mitogens
for PC12 cells and that IGF-I is the most potent of the three peptides.
Moreover, IGF-I has been shown to stimulate the replication of
sympathetic neuroblasts (DiCicco-Bloom and Black, 1988), which are
thought to arise from the same neural crest precursor cells as do
chromaffin cells (LeDouarin, 1982). The effect of IGF-I on PC12 cells may
reflect the physiological role of this peptide as a mitogen for
developing sympathoadrenal cells.

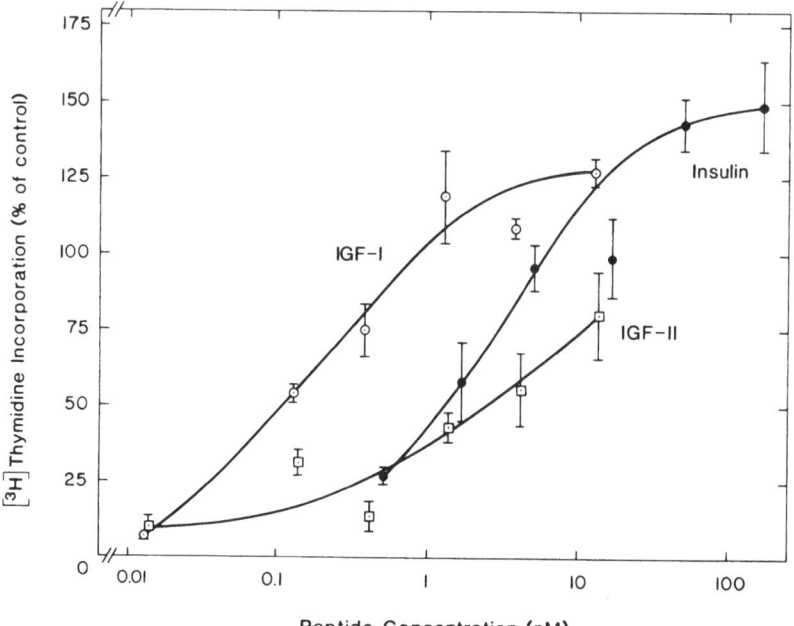

Fig. 1. Concentration-depndence of the effects of insulin and IGFs on the
incorporation of [³H]thymidine into PC12 cells. PC12 cells were
cultured for 2 days in medium containing 0.2% horse serum and 0.1%
fetal bovine serum and were then incubated for 24 h in medium
supplemented with various concentrations of insulin or IGFs, or in
medium containing 10% horse serum and 5% fetal bovine serum. The
incorporation of [³H]thymidine into the cells was measured; the
effect of growth factors on [³H]thymidine incorporation is expressed
as a percentage of the effect produced by incubation in high-serum
medium. Reproduced from Dahmer and Perlman, 1988a.

IDENTIFICATION OF IGF-I RECEPTORS ON PC12 CELLS

The studies described above suggest that PC12 cells contain IGF-I receptors. To identify IGF-I receptors on PC12 cells directly, we incubated the cells with ^{125}I-IGF-I in the presence of various concentrations of unlabeled IGF-I or insulin. Both IGF-I and insulin competed for binding of ^{125}I-IGF-I to PC12 cells. IGF-I ($IC_{50} \cong 8$ nM) was a more potent competitor than was insulin ($IC_{50} \cong 130$ nM). Scatchard analysis of a number of competition curves indicated that there were 11,000 ± 1,500 IGF-I binding sites per cell and that these sites bound IGF-I with a K_D of 7.2 ± 0.6 nM. Crosslinking of ^{125}I-IGF-I to PC12 cells followed by SDS-polyacrylamide gel electrophoresis demonstrated that the IGF-I binding protein had a M_r >200,000 in the absence of β-mercaptoethanol, and a M_r of ~130,000 in the presence of β-mercaptoethanol. The labeling of this protein was competed for by the addition of unlabeled IGF-I (100 nM) but not by unlabeled insulin (100 nM). The M_r of the labeled protein and the specificity of labeling suggest that the ^{125}I-IGF-I was bound to the α subunit of the IGF-I receptor on PC12 cells.

IGF-I receptors on other cells have protein tyrosyl kinase activity (Rechler and Nissley, 1985). To determine if IGF-I receptors on PC12 cells also had protein kinase activity, we incubated wheat germ agglutinin-purified preparations of PC12 cell membrane proteins with $[\gamma$-$^{32}P]$ATP and the synthetic substrate poly(glu:tyr, 4:1). IGF-I produced a concentration-dependent stimulation of the phosphorylation of this synthetic polypeptide, with 1 nM IGF-I causing a significant increase in phosphorylation. Our data suggest that PC12 cells have IGF-I receptors with characteristics very similar to IGF-I receptors on other cells.

IDENTIFICATION OF IGF-I RECEPTORS ON NORMAL CHROMAFFIN CELLS

Since PC12 cells have IGF-I receptors it was of interest to determine whether normal chromaffin cells also had IGF-I receptors. Chromaffin cells were prepared by collagenase digestion of bovine adrenal glands (Greenberg and Zinder, 1982). Chromaffin cells bound ^{125}I-IGF-I; as with PC12 cells, unlabeled IGF-I was a more potent competitor of ^{125}I-IGF-I binding than was insulin. Scatchard analysis of several competition curves yielded estimates of 111,000 ± 40,000 IGF-I receptors per cell and 1.1 ± 0.3 nM for the K_D of the receptor for IGF-I (Dahmer and Perlman, 1988b). Crosslinking experiments with ^{125}I-IGF-I showed labeling of a band that had an M_r >200,000 under non-reducing conditions and ~130,000 under reducing conditions. The labeling of this band was competed for by 30 nM IGF-I but not by 30 nM insulin. Purified receptor preparations from bovine adrenal medullae exhibited IGF-I stimulated protein tyrosyl kinase activity against the synthetic substrate poly(glu:tyr, 4:1); 10 nM IGF-I produced maximal stimulation of this kinase activity. Thus, normal chromaffin cells, like PC12 cells, appear to have characteristic IGF-I receptors.

IGF-I ENHANCES SECRETAGOGUE-STIMULATED CATECHOLAMINE SECRETION FROM NORMAL CHROMAFFIN CELLS

As mentioned above, Wilson et al. (1985) have reported that insulin enhances catecholamine secretion from bovine chromaffin cells. Because the high concentrations of insulin used by these workers might have stimulated IGF-I receptors, we compared the effects of IGF-I and of insulin on catecholamine secretion stimulated by medium containing elevated K^+ (55 mM). Incubation of chromaffin cells for 4 days with IGF-I or insulin had no significant effect on the catecholamine content of the cultures or on the basal release of catecholamine from the cells (not shown; Wilson et al., 1985). When stimulated by high K^+, however, cells that had been cultured in the presence of IGF-I secreted a larger percentage of their stored catecholamine than did cells that had been cultured in the absence of this factor (Fig. 2; Dahmer and Perlman, 1988b). IGF-I (10 nM) caused an almost two-fold increase in high K^+ evoked catecholamine secretion. Insulin was much less potent than IGF-I in enhancing catecholamine secretion; thus, the effect of IGF-I is almost certainly mediated by IGF-I receptors. This is the first demonstration that the activation of IGF-I receptors may regulate neurotransmitter release.

The effect of IGF-I on catecholamine secretion required long-term treatment of the cells with IGF-I. Acute treatment with 10 nM IGF-I did not stimulate catecholamine secretion and did not enhance high K^+-evoked secretion. The effect of IGF-I on high K^+ stimulated catecholamine release become apparent after 1 day of IGF-I treatment and reached maximal levels after 3 days of treatment.

IGF-I not only enhanced high K^+-evoked catecholamine secretion, but also augmented catecholamine secretion elicited by the nicotinic agonist dimethylphenylpiperazinium, the dihydropyridine agonist Bay K 8644, or Ba^{2+}. Since IGF-I enhances dimethylphenylpiperazinium-stimulated catecholamine secretion, it presumably also would enhance secretion evoked by the physiological chromaffin cell secretagogue, acetylcholine.

Calcium is known to be an important regulator of catecholamine secretion. Secretagogues increase the uptake of Ca^{2+} into chromaffin cells and thereby increase the concentration of intracellular Ca^{2+} in the

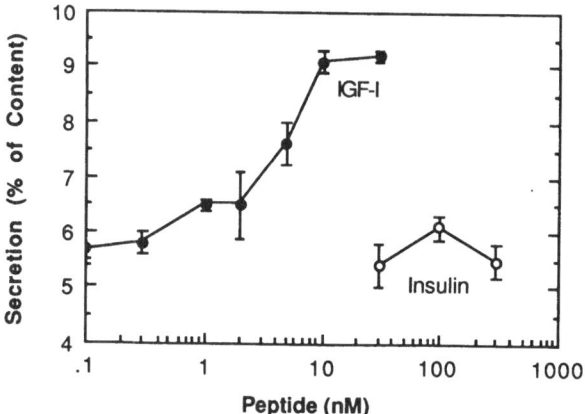

Fig. 2. Effects of IGF-I and insulin on high K^+-evoked catecholamine secretion. Chromaffin cells were cultured for 3 days in serum-free medium supplemented with various concentrations of IGF-I or insulin and were then incubated for 10 min in medium containing 55 mM K^+. Catecholamine secretion was measured and is expressed as the percentage of the catecholamine content of the cultures released into the medium, mean ± S.E.M. of 3 wells.

cells; this increase in intracellular Ca^{2+} then stimulates exocytosis. To study the effect of IGF-I on the Ca^{2+}-sensitivity of catecholamine secretion, we cultured cells for 3 days in serum-free medium in the presence or absence of IGF-I and then incubated the cells in medium containing 55 mM K^+ in the presence of various concentrations of Ca^{2+}. IGF-I enhanced catecholamine secretion at all Ca^{2+} concentrations tested, but did not appear to affect the Ca^{2+}-sensitivity of the secretory process;. the concentration of extracellular Ca^{2+} that resulted in half-maximal secretion was ~0.8 mM in both untreated and IGF-I treated cells.

IGF-I could enhance catecholamine secretion by increasing secretagogue-induced Ca^{2+} influx and thereby increasing the intracellular free Ca^{2+} concentration, or by stimulating some step in the secretory process which occurs subsequent to the Ca^{2+}-dependent event. To distinguish between these possibilities, we studied the effect of IGF-I on the uptake of $^{45}Ca^{2+}$ into chromaffin cells. In these experiments, we cultured cells in the presence or absence of IGF-I and then incubated the cells for 5 min with $^{45}Ca^{2+}$ in the presence of 55 mM K^+ or dimethylphenylpiperazinium (30 μM). IGF-I treatment had little effect on the uptake of Ca^{2+} into unstimulated cells. IGF-I treated cells, however, showed greater secretagogue-induced Ca^{2+} uptake than did untreated cells in response to high K^+ (2.81 ± 0.05 vs. 1.25 ± 0.07 nmol/well) or dimethylphenylpiperazinium (1.28 ± 0.04 vs 0.07 ± 0.06 nmol/well). These observations are consistent with the idea that IGF-I enhances catecholamine secretion by enhancing Ca^{2+} uptake into chromaffin cells and suggest that IGF-I may affect the number or regulation of Ca^{2+} channels on the cells.

If the effect of IGF-I on catecholamine secretion is due solely to an effect on Ca^{2+} channels, then IGF-I should not enhance catecholamine secretion under conditions in which Ca^{2+} entry is not regulated by Ca^{2+} channels. To address this question, we cultured cells in the presence or absence of IGF-I, permeabilized the cells by brief exposure to 0.01% saponin (Brooks and Treml, 1983), and then measured catecholamine release in the presence of various concentrations of free Ca^{2+}. After permeabilization, catecholamine released from untreated and IGF-I treated cells was nearly identical. This finding supports the idea that IGF-I does not regulate catecholamine secretion at a step subsequent to Ca^{2+} entry.

Our data are consistent with the hypothesis that IGF-I enhances catecholamine secretion by enhancing the secretagogue-stimulated uptake of Ca^{2+} into the cells and thereby augmenting the secretagogue-induced rise in intracellular free Ca^{2+} in the cells. Chromaffin cells are known to contain voltage-dependent Ca^{2+} channels (Fenwick et al., 1982). IGF-I may regulate the number or activity of these channels. In future studies, it will be important to study the effects of IGF-I on Ca^{2+} fluxes in chromaffin cells and to evaluate the effects of this agent on the Ca^{2+} channels in the cells.

SUMMARY

We have shown that both PC12 pheochromocytoma cells and normal chromaffin cells contain IGF-I receptors, and that IGF-I affects the functions of both of these cell types. IGF-I stimulates the replication of PC12 cells. Taken together with the demonstration that IGF-I stimulates replication of sympathetic neuroblasts (DiCicco-Bloom and Black, 1988), our observations suggest that IGF-I may act as a mitogen for chromaffin cell precursors and other cells derived from the neural crest. PC12 cells may be useful for studies on the mechanism by which

IGF-I stimulates the replication of neural crest-derived cells. IGF-I enhances catecholamine secretion from normal adrenal chromaffin cells. IGF-I appears to be a novel type of trophic factor for these cells, in that it does not stimulate secretion itself but rather acts to make the cells more responsive to secretory stimuli. Since chromaffin cells are exposed to IGF-I in vivo, this peptide is likely to be a physiologically important regulator of chromaffin cell function. It will be of interest to determine whether IGF-I also affects catecholamine secretion or other differentiated properties of neurons in the central nervous system.

REFERENCES

Angeletti, P.U., Liuzzi, A. and Levi-Montalcini, R.(1966) Effetti dell'insulina sulla sintesi di RNA e di lipidi nei gangli sensitivi embrionali. Ann Ist. Super. Sanità 2, 420-422.

Brooks, J.C. and Treml, S. (1984) Catecholamine secretion by chemically skinned cultured chromaffin cells. J. Neurochem. 40, 468-473.

Dahmer, M.K. and Perlman, R.L. (1988a) Insulin and insulin-like growth factors stimulate DNA synthesis in PC12 pheochromocytoma cells. Endocrinology 122, 2109-2113.

Dahmer, M.K. and Perlman , R.L. (1988b) Bovine chromaffin cells have insulin-like growth factor-I (IGF-I) receptors: IGF-I enhances catecholamine secretion. J. Neurochem. 51, 321-323.

DiCicco-Bloom, E. and Black, I.B. (1988) Insulin growth factors regulate the mitotic cycle in cultured rat sympathetic neuroblasts. Proc. Natl. Acad. Sci. USA 85, 4066-4070.

Fenwick, E.M., Marty, A. and Neher, E. (1982) Sodium and calcium channels in bovine chromaffin cells. J. Physiol. 331, 599-635.

Greenberg, A. and Zinder, O. (1982) α- and β-receptor control of catecholamine secretion from isolated adrenal medulla cells. Cell Tissue Res. 226, 655-665.

Greene, L.A. and Tischler, A.S. (1982) PC12 pheochromocytoma cultures in neurobiological research. In: Advances in Cellular Neurobiology (S. Fedoroff and L Hertz, eds.), New York: Academic Press, vol. 3, pp. 373-414.

Han, V.K.M., D'Ercole, A.J. and Lund, P.K. (1987) Cellular localization of somatomedin (insulin-like growth factor) messenger RNA in the human fetus. Science 236, 193-197.

Hansson, H.A., Nilsson, A., Isgaard, J., Billig, H., Isaksson, O, Skottner, A., Andersson, I.K. and Rozell, B. (1988) Immunohistochemical localization of insulin-like growth factor I in the adult rat. Histochemistry 89, 403-410.

LeDouarin, N.M. (1982) The Neural Crest. Cambridge: Cambridge University Press.

Livett B.G., Boksa, P., Dean, D.M., Mizobe, F. and Lindenbaum, M.H. (1983) Use of isolated chromaffin cells to study basic release mechanisms. J. Autonom. Nerv. Syst. 7, 59-86.

Nielsen, F.C. and Gammeltoft, S. (1988) Insulin-like growth factors are mitogens for rat pheochromocytoma PC12 cells. Biochem. Biophys. Res. Commun. 154, 1018-1023.

Rechler, M.M. and Nissley, S.P. (1985) The nature and regulation of receptors for insulin-like growth factors. Ann. Rev. Physiol. 47, 425-442.

Wilson, S.P., Viveros, O.H. and Kirshner, N. (1985) Relationship between the regulation of enkephalin-containing peptide and dopamine β-hydroxylase levels in cultured adrenal chromaffin cells. J. Neurochem. 45, 1363-1370.

INTERACTIONS OF INSULIN AND IGFS WITH CELLULAR COMPONENTS OF THE ARTERIAL WALL: POTENTIAL IMPACT ON ATHEROSCLEROSIS

Robert S. Bar, William L. Lowe, Jr. and Robert G. Spanheimer

Veterans Administration Hospital
Diabetes and Endocrinology Research Center
Department of Internal Medicine
The University of Iowa
Iowa City, IA

Insulin and the insulin—like growth factors (IGFs) have a diverse and complicated array of functions in endothelial cells, smooth muscle cells and fibroblasts, three major cell components of the arterial wall. In this chapter, we will review the *in vitro* studies assessing insulin/IGF effects on these three cell types and introduce other factors which may modulate the action of insulin/IGF on the arterial wall both in normal physiology and in the disease process atherosclerosis.

1. Receptors for Insulin and the IGFs

Fibroblasts, smooth muscle cells and endothelial cells are the three cell types of all arteries, and each has specific and distinct receptors for insulin, IGF—I and IGF—II.[1-3] These specific binding sites have been characterized by several independent means including competition binding studies, interaction with antireceptor antibodies and crosslinking studies. To illustrate such studies, data will be shown for cultured endothelial cells although similar results have been obtained for both vascular smooth muscle cells and fibroblasts.

a. Insulin Receptors

Using [125]I—insulin as tracer, unlabelled peptides compete for binding in the order of insulin > proinsulin > IGF—I = IGF—II.[1,4] Maximal tracer binding of [125]I—insulin is ~2—4% per 10^6 endothelial cells and there are ~10,000—15,000 receptors per cell. Ambient insulin will downregulate receptors with significant decreases in surface binding at insulin concentrations of ~0.6—1 ng/ml. As in other cell types, the holoreceptor consists of two extracellular alpha subunits of ~130—140,000 kDa which bind insulin and two transmembrane beta subunits of ~90,000 kDa which are phosphorylated in the presence of insulin.

b. IGF Receptors

IGF—I and IGF—II receptors display features characteristic of type I and type II IGF receptors. Using [125]I—IGF—I as tracer, unlabelled peptides bind to endothelial cells in the order IGF–I $\geq$ IGF–II > insulin[1] (Figure 1). With ^{125}I–IGF–II as tracer, unlabelled hormones compete in the order IGF–II $\geq$ IGF–I, while insulin does not compete for the IGF–II binding site (Figure 2). The IGF–I receptor, like the insulin receptor, consists of two binding alpha subunits of ~130,000 kDa and two transmembrane beta subunits of ~90,000 kDa. The IGF–II receptor is a single chain glycoprotein of ~250,000 kDa. Maximal binding of ^{125}I–IGF–I or ^{125}I–IGF–II is 20–40% per 10^6 endothelial cells, i.e. substantially greater than insulin binding to the same cells. Vascular smooth muscle cells also demonstrate markedly greater binding of IGF than insulin, whereas fibroblasts have relatively similar quantitative binding of insulin and the IGFs.

That the insulin receptors present on endothelial cells are truly distinct from the IGF receptors, particularly the IGF–I receptor, is suggested by several types of studies. These include the aforementioned analog binding studies as well as experiments using specific antireceptor antibodies. When endothelial cells are exposed to the monoclonal antibody 47–9, a specific inhibitor of binding to the insulin receptor, insulin binding can be decreased by >80% when IGF–I and IGF–II binding are not affected[5] (Figure 3). Further, when endothelial cells incubated with the anti IGF–I receptor monoclonal antibody αIR$_3$, IGF–I binding can be decreased by >80% while insulin and IGF–II binding are unaffected (Figure 3).

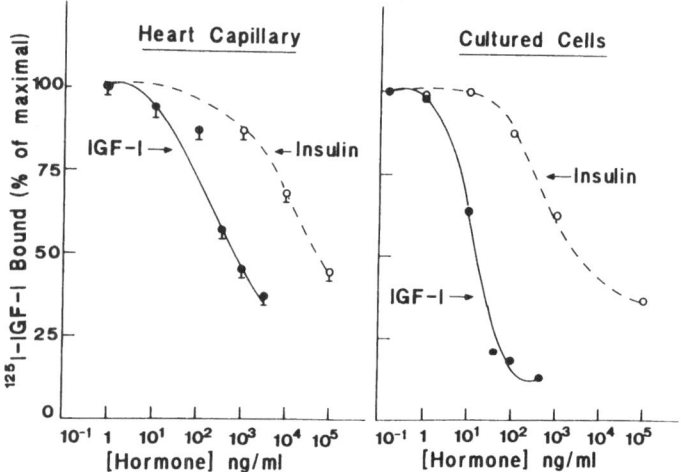

Fig. 1.   The binding of ^{125}I–IGF–I in intact heart capillary endothelium (left panel) and cultured endothelial cells (right panel).

Although the majority of insulin and IGF binding is to the well–characterized "classical" insulin, type I and type II IGF receptors a subpopulation of receptors has been identified in endothelial cells which have equal affinity for insulin and the IGFs.[6] The function of these hybrid or "atypical" receptors is not known, nor is it clear that they are present in vascular smooth muscle cells or fibroblasts.

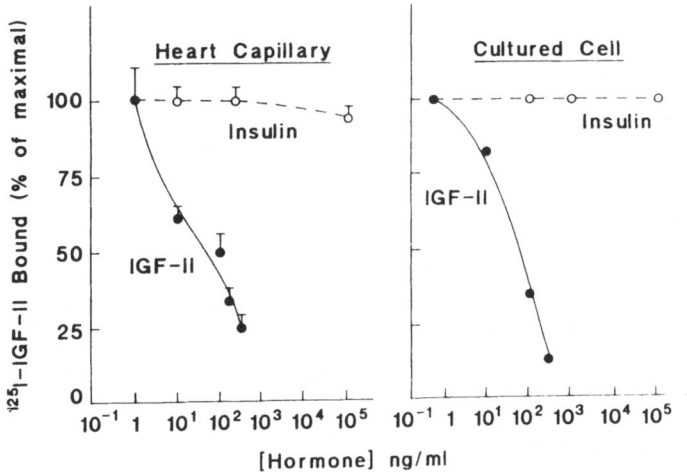

Fig. 2.   IGF−II binding to intact heart capillary endothelium and cultured endothelial cells.

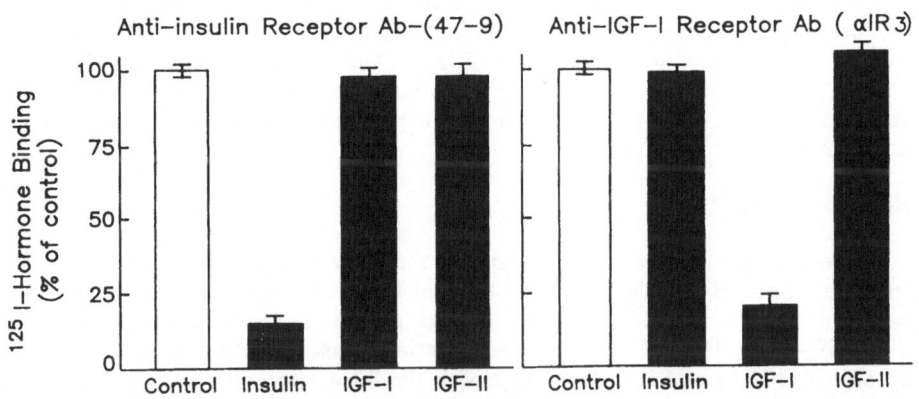

Fig. 3.   The effect of the anti−insulin receptor antibody 47−9 (left panel) and the anti−IGF−I receptor antibody αIR₃ (right panel) on the binding of insulin, IGF−I and IGF−II to cultured endothelial cells.

2. Functions of Insulin/IGFs

Insulin and the IGFs modulate a diverse group of properties in cultured endothelial cells, vascular smooth muscle cells and fibroblasts. For most processes studied, insulin and IGFs have overlapping functions in all three cell types.

a. Endothelial Cells

In endothelial cells, insulin and the IGFs stimulate a range of cell functions ranging from the metabolism of carbohydrates, proteins and lipids to the stimulation of cell proliferation.[5,7] Both insulin and the IGFs stimulate glucose uptake, glucose oxidation, alpha amino isobutyric acid (AIB) uptake, protein synthesis, lipid synthesis and thymidine incorporation into DNA. Importantly, these stimulatory functions occur in endothelial cells cultured from capillaries and microvessels including microvessel cells cultured from retina, fat, cerebral tissue and cardiac muscle.[7,8] Such effects of insulin and the IGFs are not observed in endothelial cells cultured from larger diameter blood vessels such as the aorta, pulmonary artery, pulmonary vein, umbilical vein or umbilical artery.[5,9] There is one process that departs from these general concepts, i.e. the synthesis of proteoglycans. IGF—I and II, but not insulin, stimulate the incorporation of sulfate into proteoglycans in both microvessel and macrovessel endothelial cells. In macrovessel cells, the IGFs uniformly stimulate sulfate incorporation into all proteoglycan classes, whereas in microvessel cells there is preferential stimulation of sulfate incorporation into heparan sulfate and heparin—containing proteoglycans.[10]

b. Vascular Smooth Muscle and Fibroblasts

In general, insulin and the IGFs have biological effects in cultured vascular smooth muscle cells and fibroblasts which are similar to those previously described in the microvessel endothelial cells. Dose response curves for individual processes may differ, particularly those related to mitogenic properties such as the effect on thymidine incorporation into DNA. In addition, a few properties have been described that may be specific for a given cell type such as the induction by IGF—I of tropoelastin mRNA and tropoelastin protein expression in vascular smooth muscle cells.[11]

c. Receptor Specificity[7]

The overlapping effects of insulin and the IGFs that are observed in cultured microvessel endothelial cells have also been observed in other cell types, including smooth muscle cells and fibroblasts, which each possess distinct surface receptors for insulin, IGF—I and IGF—II. There is substantial crossreactivity among these three receptors, i.e. IGF—I and IGF—II interact with the insulin receptor (although at lower affinity), insulin is recognized by the IGF—I receptor (again with lower affinity) and both IGFs crossreact with each others' specific receptor. It is, therefore, not always clear through which specific receptor a given peptide hormone is mediating its biological effect. One indirect way to explore this question is to perform detailed dose response curves for insulin, IGF—I and IGF—II for a given biological response and compare the dose response curves for the bioassay with dose—response curves for the radioreceptor assay. Thus, in cells with all three receptors, such as the endothelial cells, biological responses observed at "low" hormone concentrations (less than 10 ng/ml) are presumed to be mediated through the homologous receptor, since at these concentrations the hormone does not significantly crossreact with the other surface receptors. At higher hormone concentrations, however, it remains unclear whether biological effects are mediated through one receptor or reflect the net effects of the ligand interacting with two, or, for the IGFs, even three distinct receptors. This is especially true for processes such as smooth muscle cell proliferation, in which insulin responses are usually not observed until "higher" hormonal concentrations, i.e. ≥ 50 ng/ml, are used. A second, perhaps more direct approach, involves the use of a true receptor antagonist, i.e. a compound which specifically blocks ligand binding to one of the three receptors and has no intrinsic biological activity relative to the functional property being studied. After cells are exposed to such an antagonist, dose response bioassays are repeated. If the specific receptor is mediating the biological effect over a certain range of hormone concentrations, then that portion of the dose response curve should be totally inhibited or at least right–shifted, even when dealing with biological properties that demonstrate "spare" receptors. We have utilized such an approach to evaluate the contribution of the insulin receptor to the effects of insulin and the IGFs on glucose uptake, neutral amino acid (AIB) uptake and thymidine incorporation into DNA in cultured microvessel endothelium. In collaboration with Professor K. Siddle, we have found that the monoclonal antibody 47–9 specifically inhibited insulin binding to bovine microvessel endothelial cells, did not inhibit IGF–I or IGF–II binding, and had no intrinsic agonist activity relative to glucose uptake, AIB uptake or thymidine incorporation into DNA. When endothelial cells were exposed to antibody 47–9 under the exact conditions of the bioassay, substantial inhibition of insulin binding occurred without significant effect on the closely related IGF–I binding sites. After exposure to antibody 47–9, full dose response curves for insulin and IGF–I were repeated. For glucose uptake, AIB uptake, and thymidine incorporation into DNA the dose response curves after antibody exposure were right–shifted at insulin concentrations less than 100 ng/ml (Figure 4). At higher insulin concentration, inhibition of insulin binding by antibody 47–9 was without effect on biological responses. In contrast, antibody

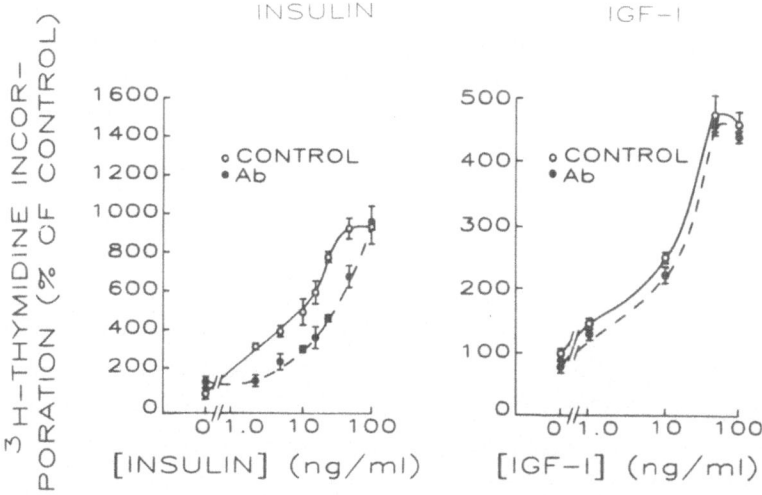

Fig. 4. Dose–response curves for microvessel endothelial cells exposed to insulin (left panel) and IGF–I (right panel) under control conditions (0 —— 0) and after exposure to the anti–insulin–receptor antibody 47–9 (● – – – ●).

exposure had no effect on IGF—I dose—response curves for any of the three biological properties. These findings suggest that the effects of insulin at concentrations less than 100 ng/ml are at least partially mediated via the insulin receptor whereas the IGF—I effects are not mediated through IGF—I binding to the insulin receptor, regardless of the concentration of IGF—I being evaluated. Similar studies, using epitope specific antireceptor antibodies, have been performed by King, et al. in fibroblasts[12] and vascular smooth muscle cells[13] and have yielded somewhat different conclusions. When vascular smooth muscle cells were exposed to antibody specific for the IGF—I receptor, the mitogenic activities (thymidine incorporation into DNA and cell proliferation) were completely inhibited, suggesting that both the insulin and IGF—I responses on cell growth were mediated via IGF—I receptors. In earlier work, King, et al.[12] reached similar conclusions in cultured fibroblasts, utilizing fragments of anti—insulin receptor antibodies to specifically block insulin receptor binding. Thus, in cells expressing insulin and IGF receptors and possessing overlapping functional effects of the hormones, a given process may be mediated through one or several receptors.

3. Insulin/IGFs and Atherosclerosis

A fundamental component in the early lesion of atherosclerosis is disordered smooth muscle cell proliferation. Since insulin and the IGFs can stimulate the proliferation of vascular smooth vessel cells in vitro,[14] a potential role for these growth factors in the pathogenesis of atherosclerosis has been suggested. For insulin, it is thus conceivable that in conditions such as type II diabetes mellitus and insulin—treated diabetics, the elevated circulating insulin would have direct access to vascular smooth muscle through a discontinuous endothelium or a damaged endothelium (Figure 5). The increased levels of insulin in the media of the arterial wall would then stimulate smooth muscle proliferation with insulin acting through its own receptor or through the IGF—I receptor. However, based on cell culture studies, it is perhaps more likely that the IGFs (especially IGF—I), and not insulin, would influence smooth muscle cell proliferation. Specifically, in vitro studies indicate that a) smooth muscle

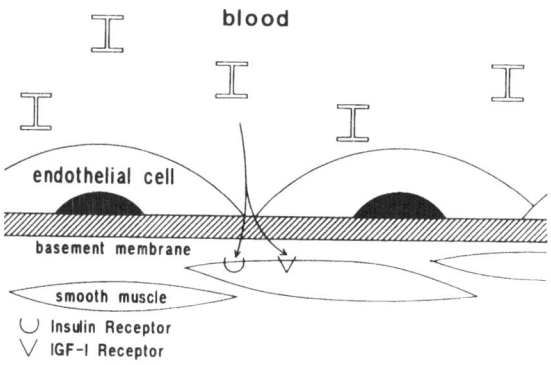

Fig. 5

cells (as well as endothelial cells) have substantially greater high affinity binding sites for IGF—I than insulin and b) IGF—I is a more potent stimulator of smooth muscle cell proliferation than insulin. In evaluating a potential role for IGF—I (and, indirectly, perhaps insulin) in the proliferative response associated with atherosclerosis, two additional components should be considered that may directly have impact on IGF stimulation of vascular smooth muscle proliferation *in vivo*: 1) factors affecting IGF binding and action in vascular smooth muscle, and 2) determinants of IGF concentrations in the vascular media.

a. Factors Affecting IGF—I Binding and Action in Vascular Smooth Muscle Cells

The effect of IGF—I on the proliferation of smooth muscle cells can be modified by many *in vitro* conditions ranging from cell confluency/cell cycle to the presence of other specific growth factors. This discussion will focus on two growth factors that markedly amplify the IGF—I effect and are especially relevant to any consideration of vessel wall physiology, platelet—derived growth factor (PDGF) and basic fibroblast growth factor (bFGF).

1. Platelet—derived Growth Factor (PDGF)

PDGF is a cationic glycoprotein of ~30,000 kDa composed of 2 subunits, an A and B chain (homodimers of A—A and B—B may also be expressed).[15] It is a basic protein with a pI of ~10 and has a strong affinity for the glycosaminoglycans heparin. The B chain of PDGF is structurally similar to the v—sis oncogene product of the simian sarcoma virus. When injected into the circulation, PDGF has a brief serum half—life of < 2 minutes suggesting that potential PDGF effects *in vivo* are likely to involve autocrine/paracrine action rather than endocrine effects. Further supporting autocrine/paracrine function for PDGF are the large number of cell types which are capable of producing PDGF—like molecules. These include endothelial cells (micro— and macrovessel) and vascular smooth muscle cells as well as platelets, monocytes/macrophages, and cytotrophoblasts. Control of PDGF expression in each of these cells has already been shown to be complex. For example, cultured endothelial cells express mRNAs for both the A chain and B chain of PDGF, but conditions have been described which result in differential regulation of the two mRNAs.[16]

Both PDGF and IGF—I can stimulate thymidine incorporation into DNA and cellular proliferation in vascular smooth muscle cells, with PDGF generally having a greater effect. However, when cells are pretreated with PDGF then exposed to IGF—I or coincubated with both PDGF and IGF—I, the effects of both agents are greatly amplified. Cell cycle kinetic studies by Pledger and Clemmons[17] suggest that PDGF acts as a competence factor which stimulates resting cells to enter the GI phase of the cell cycle. IGF—I as well as other agents, then serve as progression factors enabling cells to progress through GI and enter the S phase in which DNA synthesis occurs. Even in vascular smooth muscle cells exposed to only PDGF, the mitogenic effects of PDGF may be partially mediated by IGF—I. Smooth muscle cells secrete IGF—I into the medium and PDGF can stimulate IGF—I production by these cells.[18] When vascular smooth muscle cells are incubated with anti—IGF—I antibodies, which neutralize the effect of any secreted IGF—I, PDGF—induced mitogenic response is decreased by 85%.[19]

The synergism between PDGF and IGF—I in vascular smooth muscle cells may also be mediated on the cell surface at the receptor level.[20] Exposure of vascular smooth muscle cells to PDGF increases subsequent binding of ^{125}I—IGF—I in a dose—dependent manner, with maximal increases of 60—70% of control.[20] Likewise, exposure of the same cells to IGF—I can increase ^{125}I—PDGF binding by ~70%.

b. basic Fibroblast Growth Factor (bFGF)

Basic FGF is a single chain cationic polypeptide (pI=9.6) that stimulates the proliferation of several cell types including vascular smooth muscle cells. Basic FGF is synthesized by many cell types, including endothelial cells in which it is retained by the cell, probably bound to specific endothelial heparin/heparan sulfate proteoglycans.[21] Like PDGF, bFGF is synergistic with IGF—I in stimulating smooth muscle cell proliferation. Furthermore, as with PDGF, exposing cells to bFGF results in a substantial increase of IGF—I binding to vascular smooth muscle cells in culture.[20]

In summary, there is a complex synergism among IGF—I, PDGF and bFGF in stimulating DNA synthesis and proliferation of vascular smooth muscle cells. Since all three proteins are synthesized by cellular components of the arterial wall, an understanding of the *in vivo* environmental regulation of the synthesis and expression of each of these growth factors will be central to an understanding of autocrine/paracrine factors involved in the pathogenesis of atherosclerosis.

2. Determinants of IGF—I Concentrations in the Vascular Media

The effects of IGF—I on the proliferation of smooth muscle cells *in vivo* will obviously be influenced by the concentration of biologically active IGF—I within the media of the arterial wall. Two recently described factors are likely to be important in regulating media IGF—I concentrations: cellular production of IGF—I and IGF—binding proteins.

a. Cellular Production of IGF—I

IGF—I and IGF—II are present in the circulation bound to large molecular weight proteins.[22] The majority of the circulating IGFs, as well as their binding proteins, are thought to originate in the liver. However, in recent years it has become apparent that several cells and tissues, in addition to the liver, are capable of making IGF I/II and/or IGF binding proteins.[23] For IGF—I, at least two of the three cellular components of the arterial wall synthesize this growth factor *in vitro*. Both vascular smooth muscle cells and fibroblasts secrete immunoassayable IGF—I *in vitro*. It is not yet clear whether endothelial cells synthesize IGF—I, although after vascular injury, IGF—I can be localized to endothelial cells by immunocytochemical analysis;[24] however, it remains to be determined whether the endothelial cell produced the IGF—I or internalized IGF—I which had been synthesized at another site.

As noted above, the various growth factors produced by the cellular components of the cell wall appear to be synergistic in stimulating DNA synthesis and in regulating the level of each other's binding. Similarly, these growth factors may also regulate the local synthesis of each other. To date, PDGF has been shown to increase IGF—I production by cultured vascular smooth muscle cells and fibroblasts[18,25] while basic FGF increases IGF—I production in cultured fibroblasts.[25] Interestingly, insulin in the presence of PDGF is also capable of increasing IGF—I production by cultured vascular smooth muscle cells to levels greater than those produced by stimulation with PDGF alone.

b. IGF—binding Proteins

IGF—I and IGF—II are present in the circulation, bound to serum binding proteins which presumably render the IGF biologically inactive.[22] The major circulating complex containing IGF consists of a 53 kDa binding protein, an acid dissociable protein and IGF. The structure of the 53 kDa binding protein and the acid dissociable component have both been recently elucidated.[26] In addition to the 53 kDa protein, a 30—40 kDa serum protein, and perhaps other minor binding proteins, are present in the circulation (Figure 6). In recent years it has become apparent that many cell types, in addition to liver cells, synthesize low molecular weight IGF—binding proteins.[27] Cells and tissues shown to produce IGF—binding proteins in vitro include endothelial cells, vascular smooth muscle cells and fibroblasts, as well as myoblasts, MDCK cells, endometrium, hypothalamic tissue, decidua and several transformed cell lines. Cellular binding proteins produced by these cells and tissues have been classified on the basis of their structure, size, immunologic properties and relative affinities for IGF—I versus IGF—II. Where sufficient information permits detailed comparison, there is striking overlap among the binding proteins from different cell types, suggesting that these proteins will likely comprise a limited group of proteins, despite the diversity of cell types capable of producing the IGF binding proteins.[28] All of the cell binding proteins described thus far have molecular sizes of ~25—42 kDa, bind only IGF—I and IGF—II, have a cysteine—rich region near the N—terminal portion of the molecule and an RGD (arg—gly—asp) sequence near the C—terminal portion of the protein. Despite these overlapping features, there are already important functional differences among the binding proteins. First, although most of the binding proteins inhibit IGF effects, at least two (amniotic fluid binding proteins[29] and endothelial cell binding proteins)[30] do not inhibit IGF effects and, in certain conditions, actually amplify IGF action. Second, at least one of the binding proteins

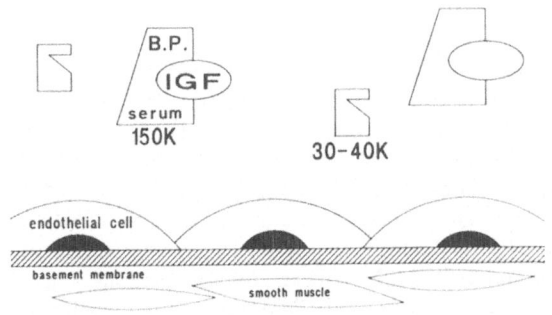

Fig. 6

produced by endothelial cells, but apparently not other binding proteins, has intrinsic biological activity, being capable of stimulating the uptake of glucose and AIB in microvessel endothelial cells.[30] Third, some of the binding proteins may have selective adsorptive properties toward specific cell surfaces.[29] In this regard, it is of interest that a purified amniotic fluid binding protein appears to selectively adhere to the surface of vascular smooth muscle cells.

The finding that cells of the arterial wall can produce both IGF–binding proteins and IGFs themselves has important implications for the potential autocrine/paracrine effects of the IGFs in the vessel wall and should certainly be relevant in determining the concentration of bioactive IGF within the vascular wall. We believe that an understanding of the *in vivo* regulation of arterial wall IGF/IGF–binding proteins by circulating factors, such as insulin, growth hormone, glucocorticoids, IGFs and glucose, as well as by specific regional factors, such as PDGF and bFGF will provide new insights into diseases of the arterial vessel wall, especially those characterized by abnormalities of smooth muscle proliferation.

ACKNOWLEDGEMENTS

This work was supported by research funds from Veterans Administration, and grants DK25421 and DK25295 (Diabetes and Endocrinology Research Center) from the National Institutes of Health.

REFERENCES

1.   R. S. Bar and M. Boes, Distinct receptors for IGF–I, IGF–II, and insulin are present on bovine capillary endothelial cells and large vessel endothelial cells. Biochem. Biophys. Res. Comm. 124:203 (1984).

2.   B. Pfeifle and H. Ditschuneit, Receptors for insulin and insulin–like growth factor in cultured arterial smooth muscle cells depend on their growth state. J. Endocr. 96:251 (1983).

3.   M. M. Rechler and J. M. Podskalny, Insulin receptors in cultured human fibroblasts, Diabetes 25:250 (1976).

4.   G. L. King, A. D. Goodman, S. Buzney, A. Moses, and C. R. Kahn, Receptors and growth–promoting effects of insuiln and insulin–like growth factors on cells from bovine retinal capillaries and aorta. J. Clin. Invest. 75:1028 (1985).

5.   R. S. Bar, K. Siddle, S. Dolash, M. Boes, and B. Dake, Actions of insulin and insulin–like growth factors I and II in cultured microvessel endothelial cells from bovine adipose tissue. Metabolism 37:714 (1988).

6.   A. J. Cox, R. S. Bar, H. A. Jonas, and J. D. Newman, Atypical receptors for insulin–like peptides in endothelial cells. Diabetes 36:54A (1987).

7.   R. S. Bar, M. Boes, B.L. Dake, B. A. Booth, and S. A. Henley, Insulin, insulin–like growth factors and vascular endothelium, Am. J. Med. 85 (Suppl. 5A):59 (1988).

8.   M. E. Gerritsen and T. M. Burke, Insulin binding and effects of insuiln on glucose uptake and metabolism in cultured rabbit coronary microvessel endothelium. Proc. Soc. Exp. Biol. Med. 180:17 (1985).

9.   G. L. King, S. Buzney, C. R. Kahn, N. Hetu, S. Buchwald, S. G. Macdonald, and L. I. Rand, Differential responsiveness to insulin in endothelial and support cells from micro– and macrovessels. J. Clin. Invest. 71:974 (1983).

10. R. S. Bar, B. L. Dake, and S. Stueck, Stimulation of proteoglycans by IGF—I and IGF—II in microvessel and large vessel endothelial cells. Am. J. Physiol. 253:E21 (1987).

11. J. Foster, C. B. Rich, and J. R. Florini, Insulin—like growth factor I, somatomedin C, induces the synthesis of tropoelastin in aortic tissue, Collagen Rel. Res. 7:161 (1987).

12. G. L. King, C. R. Kahn, M. M. Rechler, and S. P. Nissley, Direct demonstration of separate receptors for growth and metabolic activities of insulin and multiplication—stimulating activity (an insulin—like growth factor) using antibodies to the insulin receptor. J. Clin. Invest. 66:130 (1980).

13. G. L. King, N. B. Banskota, and R. Taub, Additivity effects of PDGF and insulin on the stimulation of proto—oncogene c—myc and cellular growth in aortic smooth muscle cells, Abstract presented at the American Diabetes Association's 23rd International Research Symposium "Diabetes, Lipoproteins, and Atherosclerosis" (1989).

14. D. R. Clemmons, Interaction of circulating cell—derived and plasma growth factors in stimulating cultured smooth muscle cell replication, J. Cell. Physiol. 121:425 (1984).

15. R. Ross, E. W. Raines and D. F. Bowen—Pope, The biology of platelet derived growth factor, Cell 46:155 (1986).

16. T. O. Daniel, V. C. Gibbs, D. F. Milfay, and L. T. Williams, Agents that increase cAMP accumulation block endothelial c—sis induction by thrombin and transforming growth factor—β, J. Biol. Chem. 262:11893 (1987).

17. D. R. Clemmons, J. J. Van Wyke, and W. J. Pledger, Sequential addition of platelet factor and plasma to BALB/C3T3 fibroblast cultures stimulates somatomedin—C binding early in cell cycle. Proc. Natl. Acad. Sci. USA 77:6644 (1980).

18. D. R. Clemmons, Variables controlling the secretion of somatomedin—like peptide by cultured porcine smooth muscle cells. Circ. Res. 56:418, (1985).

19. D. R. Clemmons and J. J. van Wyk, Evidence for a functional role of endogenously produced somatomedinlike peptides in the regulation of DNA synthesis in cultured human fibroblasts and procine smooth muscle cells. J. Clin. Invest. 75:1914, (1985).

20. B. Pfeifle, H. Boeder, and H. Ditschuneit, Interaction of receptors for insulin—like growth factor I, platelet—derived growth factor, and fibroblast growth factor in rat aortic cells, Endocrinology 120:2251 (1987).

21. I. Vlodavsky, R. Fridman, R. Sullivan, J. Sasse, and M. Klagsbrun, Aortic endothelial cells synthesize basic fibroblast growth factor which remains cell associated and platelet—derived growth factor—like protein which is secreted, J. Cell. Physiol. 131:402 (1987).

22. J. Zapf and E. R. Froesch, Insulin—like growth factors/somatomedins: Structure, secretion, biological actions and physiological role, Hormone Res. 24:121 (1986).

23. V. K. M. Han, A. J. D'Ercole and P. K. Lund, Cellular localization of somatomedin (insulin—like growth factor) messenger RNA in the human fetus, Science 236:193 (1987).

24. H—A Hansson, E. Jennische, and A. Skottner, Regenerating endothelial cells express insulin—like growth factor—I immunoreactivity after arterial injury, Cell Tissue Res. 250:499 (1987).

25. D. R. Clemmons, L. E. Underwood, and J. J. Van Wyk, Hormonal control of immunoreactive somatomedin production by cultured human fibroblasts. J. Clin. Invest. 67:10 (1981).

26. W. I. Wood, G. Cachianes, W. J. Henzel, G. A. Winslow, S. A. Spencer, R. Hellmiss, J. L. Martin, and R. C. Baxter, Cloning and expression of the growth hormone—dependent insulin—like growth factor—binding protein, Mol. Endo. 2:1176 (1988).

27. R. S. Bar, L. C. Harrison, R. C. Baxter, M. Boes, B. L. Dake, B. Booth, and A. Cox, Production of IGF—binding proteins by vascular endothelial cells. Biochem. Biophys. Res. Comm. 148:734 (1987).

28. A. Brinkman, C. Groffen, D. J. Kortleve, A. G. van Kessel, and S. L. S. Drop, Isolation and characterization of a cDNA encoding the low molecular weight insulin—like growth factor binding protein (IPB—1), EMBO Journal 7:2417 (1988).

29. R. G. Elgin, W. H. Busby, Jr., and D. R. Clemmons, An insulin—like growth factor (IGF) binding protein enhances the biologic response to IGF—I, Proc. Natl. Acad. Sci. USA 84:3254 (1987).

30. R. S. Bar, B. A. Booth, M. Boes, and B. L. Dake, IGF—binding proteins from vascular endothelial cells: Purification, characterization and intrinsic biological activities, Endocrinology, submitted (1989).

INTERACTIONS OF INSULIN-LIKE GROWTH FACTOR-I (IGF-I) WITH MULTIPLE

SIGNAL TRANSDUCTION PATHWAYS IN FRTL5 THYROID FOLLICULAR CELLS

Albert G. Frauman, Donatella Tramontano and Alan C. Moses

Diabetes and Metabolism Unit, Department of Medicine
Beth Israel Hospital and Harvard Medical School
Boston, MA, 02215

INTRODUCTION

Insulin-like growth factor-I (IGF-I) binds to a specific cell surface receptor that displays intrinsic tyrosine kinase activity (1). The type I IGF receptor is activated by autophosphorylation of its beta subunit, however the postreceptor pathways involved in IGF-I stimulated DNA synthesis have not been elucidated completely. The IGFs classically act on mesenchymal tissue to promote cell growth and differentiated function (2), however studies of IGF action on endocrine epithelial cells have been relatively limited. In rat and porcine ovarian granulosa cells, IGF-I synergizes with FSH to enhance progesterone synthesis and LH receptor upregulation (3-6). This synergy occurs through IGF-I's ability to potentiate FSH-stimulated cAMP accumulation (7). IGF-I by itself does not stimulate adenylate cyclase. Ovarian granulosa cells also have been demonstrated to secrete IGFs that appear to act as autocrine growth (or differentiation) factors (8). The rat thyroid follicular cell line, FRTL5, is another example of endocrine tissue of epithelial origin which responds to IGF-I (and to a lesser extent IGF-II and insulin) by increasing DNA synthesis and cell proliferation. In FRTL5 cells, the pituitary glycoprotein thyrotropin (TSH) acts both to stimulate thyroid differentiated function and cellular metabolism and replication (9). Thus, for FRTL5, TSH is a tissue specific mitogen. Similar to the ovarian granulosa cell, FRTL5 cells display synergistic interaction between a pituitary glycoprotein hormone (TSH) and IGF-I (10). Unlike the ovarian granulosa cell, this synergistic interaction affects cell replication but not thyroid differentiated functions. The mechanism(s) of the interaction between IGF-I and TSH remains unclear but offers a useful model for dissecting the pathways of IGF-I signal transduction.

485

FRTL5 CELLS: CHARACTERISTICS

Ambesi-Impiombato et al first developed the FRTL cell line, in 1980 (11). These are normal thyroid follicular cells, derived from the Fischer rat, and maintain thyroid differentiated function (thyroglobulin synthesis and iodide transport) in the presence of low concentrations of serum (0.1-0.5%) (11). Tramontano et al have cloned from FRTL5 a subclone known as FRTL5 (clone 2, DOC) that are characterized by dormancy in 5% calf serum with the capacity to grow and undergo thyroid differentiated function when exposed to TSH (9, 12). This effect of TSH at least in part is due to activation of adenylate cyclase (13). In addition to TSH, other hormones appear necessary for optimal FRTL5 growth; these are insulin, transferrin, somatostatin, glycyl-histidine-lysine and hydrocortisone. Insulin is included at supraphysiologic concentrations and is believed to act through the type I IGF receptor in these cells (14). Although capable of growing continuously in cell culture, the FRTL5 fulfill a number of criteria which distinguish them as 'nontransformed'; they possess the normal rat chromosome number of 46, do not grow in soft agar, do not form tumours when injected into nude mice, and exhibit contact inhibition of growth (9). The FRTL5 cells also have the advantage of a high signal-to-noise ratio in bioassays, and a growth response to a restricted number of mitogens.

GROWTH RESPONSES IN FRTL5

FRTL5 cells respond to TSH exposure by increasing cell number, ^{3}H-thymidine incorporation into DNA, nuclear labeling of thymidine (9, 13), and nutrient transport (^{14}C-amino-isobutyric acid and glucose transport) (15, 16). Moreover, TSH enhances thyroid differentiated function in these cells. The growth responses to TSH are associated with an increase in cellular cAMP generation and are mimicked by Graves' IgG, dibutyryl cAMP and other agents which increase effective cAMP levels, such as forskolin, and isobutylmethylxanthine (10, 13). The maximal growth stimulating effect of TSH is evident at a concentration of 1 nM (1 mU/ml).

IGF-I also potently stimulates ^{3}H-thymidine incorporation, cell replication and nutrient transport (14), but does not stimulate iodide transport in FRTL5. The growth promoting effects of IGF-I are independent of alterations in cAMP levels and are maximal at approximately 50 ng/ml (6.5 nM) (14). IGF-II and insulin produce similar mitogenic effects in FRTL5 cells; IGF-II is about 20% as potent as IGF-I while insulin is only approximately 1% as potent on a molar basis (14). Importantly, IGF-I

486

dramatically enhances the ability of TSH to stimulate DNA synthesis (10) and nutrient transport (unpublished observations) in FRTL5 cells; this effect is more than additive and is not associated with any enhancement of TSH-stimulated differentiated function nor with any potentiation of cAMP generation (13). IGF-I increases the maximal response of FRTL5 to a given concentration of TSH and shifts the TSH-dose response curve 1-2 log concentrations to the left; this effect is best seen with submaximal concentrations of IGF-I (10 ng/ml) and TSH (0.01 nM). The precise pathway(s) subserving this synergistic response are unknown, however. This interaction depends on increased intracellular levels of cAMP. Dibutyryl cAMP, a cAMP analog which enters the cell and bypasses adenylate cyclase activation, forskolin, which indirectly activates adenylate cyclase, and isobutylmethylxanthine all increase effective intracellular cAMP content and mimic the effects of TSH on IGF-I stimulated ^{3}H-thymidine incorporation (13). This synergistic effect also is seen when IGF-II or insulin are coincubated with TSH (17). It is of interest that preincubation of FRTL5 cells with TSH or dibutyryl cAMP for 24 hours followed by their removal and subsequent incubation with IGF-I markedly potentiates the growth response to IGF-I (13). There is, however, no potentiation of thymidine incorporation when a 24 hour incubation with IGF-I precedes incubation with TSH or dibutyryl cAMP (13). Thus, the stimulation of intracellular cAMP in FRTL5 cells potentiates processes mediated by IGF-I, and these processes are maintained for 12-24 hours following the removal of the agents that increase cAMP. In the case of IGF-I, the continued presence of IGF-I is essential for a full synergistic interaction with TSH to occur. The replicative response of FRTL5 cells to IGF-I and to TSH also is associated with the rapid but transient expression of the nuclear proto-oncogenes c-myc and c-fos (18, 19), which are believed to be important in the regulation of growth and differentiation in many cell types (20).

TSH-stimulated DNA synthesis in FRTL5 depends, in part, on the presence of IGF produced by these cells (17). We have demonstrated that bioactive IGF-II is found in conditioned media of FRTL5 cells, and that Sm 1.2, a monoclonal antibody that recognizes both IGF-I and IGF-II, shifts the TSH-dose response curve for mitogenesis to the right (17). The effects of TSH on iodide transport are not affected by IGF-I or by anti-IGF antibody. Thus, IGF-II is an autocrine factor in FRTL5 cells, and is capable of modifying the growth response to TSH. Other evidence favouring this hypothesis is the observation that PP-12, an IGF-binding protein derived from human decidua and also found in normal human serum (21), is capable of inhibiting the mitogenic effects of both IGF-I and TSH in FRTL5 cells (22). The effect of PP-12 upon IGF-I stimulated DNA synthesis occurs with a concomitant reduction in IGF-I binding to FRTL5 cells, while the effect upon TSH-stimulated DNA synthesis occurs without any alteration in TSH binding to the cells (22). Thus, by binding and reducing the mitogenic effects of locally produced IGFs, PP-12 also inhibits the mitogenic effects of exogenously added TSH.

FRTL5 cells express high affinity TSH, IGF-I and IGF-II receptors on their cell surface (14, 15). TSH (or dibutyryl cAMP or forskolin), induces a dose and time dependent down regulation of TSH and IGF-I receptors (23). In contrast, IGF-I alone results in down regulation of IGF-I binding only, without affecting TSH binding (23). Interestingly, the combination of IGF-I and TSH results in the down regulation of both TSH and IGF-I receptors. Thus, receptor down regulation appears to be another end point of the synergistic interaction between IGF-I and TSH. These data also demonstrate that the synergistic interaction between IGF-I and TSH can not be explained by upregulation of their respective receptors, as has been demonstrated in ovarian granulosa cells for IGF-I (4).

The regulation of thyroid cell growth and function is complex as evidenced by the pleiotropic effects of agents involving other pathways of signal transduction. Adenosine is a ubiquitous nucleoside which acts upon many cell types, via at least 2 cell surface receptors, A_1 and A_2 (24). The A_1 receptor classically is associated with inhibition of cAMP generation and the A_2 receptor with an increase in cAMP generation (24). In FRTL5 cells, adenosine causes a dose dependent inhibition of TSH-stimulated DNA synthesis and cAMP generation (25). Using A_1 ligands, this effect is likely to be through activation of the A_1 receptor, although post-cAMP mechanisms are also operative since an A_1 receptor antagonist does not block TSH-stimulated cAMP generation but does block both TSH- and dibutyryl cAMP-stimulated ^{3}H-thymidine incorporation (26). Moreover, removal of endogenous adenosine (by enzamatic conversion to inosine, using adenosine deaminase) enhances TSH-stimulated DNA synthesis in these cells (27).

Importantly, adenosine potentiates IGF-I stimulated ^{3}H-thymidine incorporation, most likely via A_2 receptors (25, 27); however, this effect independent of cAMP generation (unpublished observations). The locus of interaction between IGF-I and adenosine remains an active area of investigation.

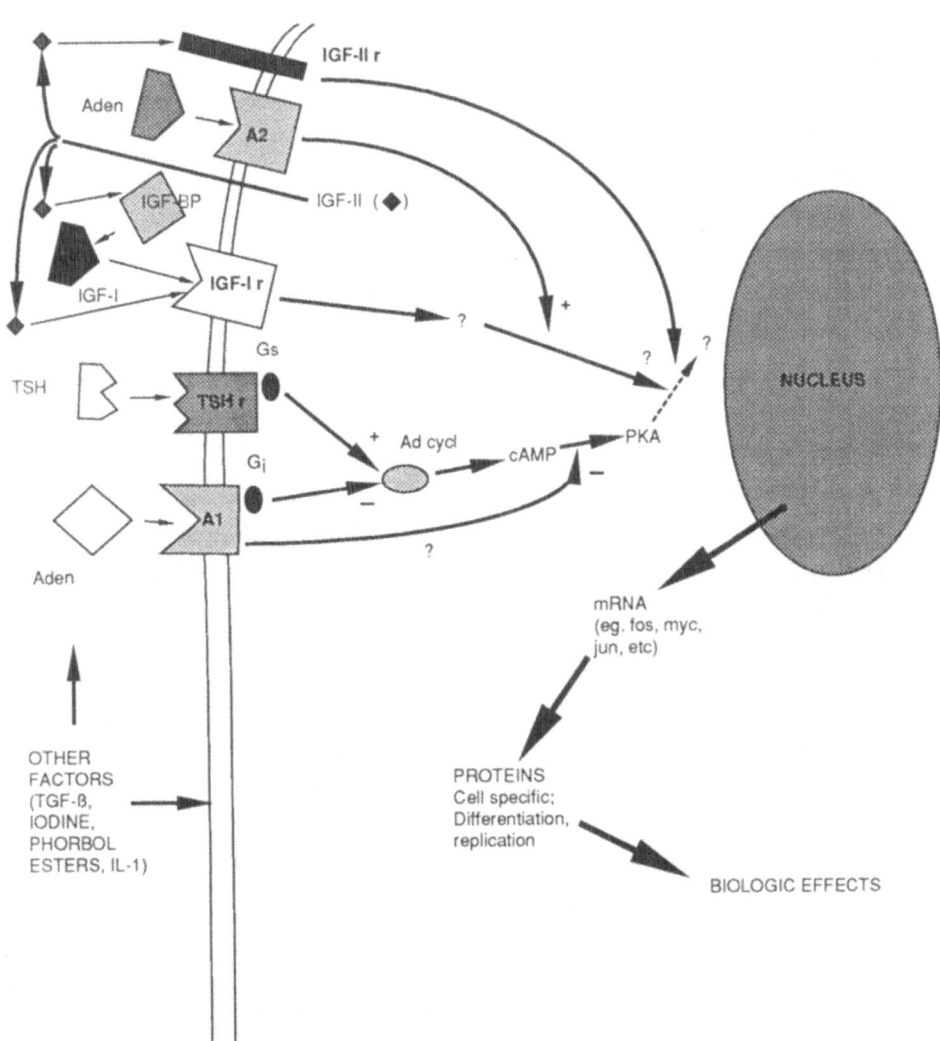

Figure 1. Summary of various agents which interact with IGF-I, in the FRTL5 cell line (IGF-I r/IGF-II r: type I and II IGF receptors. TSH r: TSH receptor. IGF-BP: IGF binding protein. A_1/A_2: Type 1 and 2 adenosine receptors. Aden: Adenosine. Ad cycl: Adenylate cyclase. G_s/G_i: G protein subunits. PKA: Protein kinase A (cAMP-dependent protein kinase).

489

Other factors involved in the regulation of FRTL5 cell growth and function are Transforming Growth Factor-beta (TGF-beta), phorbol esters (activating protein kinase C), organic iodine and the cytokine, interleukin-1 (IL-1). TGF-beta inhibits IGF-I and TSH-stimulated DNA synthesis and TSH-stimulated iodine metabolism, at a post-cAMP level (28). The phorbol ester, tetradecanoyl phorbol acetate (TPA), which activates protein kinase C, inhibits TSH-stimulated DNA synthesis, cAMP generation and iodine uptake in FRTL5 cells (29). Moreover, TPA enhances IGF-I stimulated DNA synthesis in FRTL5 cells (29). Organic iodine exhibits a specific inhibitory effect on TSH-stimulated cell proliferation in FRTL5 cells (30). IL-1 has been shown to enhance serum and IGF-I stimulated DNA synthesis in these cells (31). The physiologic significance of these agents which modulate thyroid cell growth and function remains unanswered.

In summary (figure 1), it is evident that several pathways may be of importance in modulating the biological effects of IGF-I. Pathways involving TSH, adenosine, phorbol esters and other growth factors all interact with IGF-I in the FRTL5 cell, suggesting multiple mechanisms of signal transduction of IGF-I or, alternatively, "cross-talk" with a common IGF-I stimulated pathway. The FRTL5 cell line is a useful model for the elucidation of the mechanisms involved in IGF-I signal transduction, and should yield important insights into growth factor action on cell growth.

REFERENCES

1. Zick Y, Sasaki N, Rees-Jones RW, Greenberger G, Nissley SP, Rechler MM 1984. Insulin-like growth factor-I (IGF-I) stimulates tyrosine kinase activity in purified receptors from a rat liver cell line. Biochem Biophys Res Comm 119:6-13.
2. Froesch ER, Schmid C, Schwander J, Zapf F 1985. Actions of insulin-like growth factors. Ann Rev Physiol 47:425-442.
3. Adashi EY, Resnick CE, Svoboda ME, Van Wyk JJ 1985. Somatomedin-C synergizes with follicle-stimulating hormone in the acquisition of progestin biosynthetic capacity by cultured rat granulosa cells. Endocrinology 116:2135-2142.
4. Adashi EY, Resnick CE, Svoboda ME, Van Wyk JJ 1985. Somatomedin-C enhances induction of luteinizing hormone receptors by follicle-stimulating

hormone in cultured rat granulosa cells. Endocrinology 116:2369-2375.

5. Baranao JL, Hammond JM 1984. Comparative effects of insulin and insulin-like growth factors on DNA synthesis and differentiation of porcine granulosa cells. Biochem Biophys Res Comm 124:484-490.

6. Velduis JD, Furlanetto R 1985. Trophic actions of human somatomedin C/insulin-like growth factor-I on ovarian cells: In vitro studies with swine granulosa cells. Endocrinology 116:1235-1242.

7. Adashi EY, Resnick CE, Hernandez ER, May JV, Knecht M, Svoboda ME, Van Wyk JJ 1988. Insulin-like growth factor-I as an amplifier of follicle-stimulating hormone action: Studies on mechanism(s) and site(s) of action in cultured rat granulosa cells. Endocrinology 122:1583-1591.

8. Hammond JM, Baranao JLS, Skaleris D, Knight AB, Romanus JA, Rechler MM 1985. Production of insulin-like growth factors by ovarian granulosa cells. Endocrinology 117:2553-2555.

9. Ambesi-Impiombato FS, Picone R, Tramontano D 1982. Influence of hormones and serum on growth and differentiation of the thyroid cell strain FRTL. In: Sirbasku DA, Sato AH, Pardee A (eds). Growth of Cells in Hormonally Defined Media. Cold Spring Harbor Conference on Cell Proliferation, Cold Spring Harbor Laboratory, Vol 9:483-492.

10. Tramontano D, Cushing GW, Moses AC, Ingbar SH 1986. Insulin-like growth factor-I stimulates the growth of rat thyroid cells in culture and synergizes the growth-promoting effect of thyrotropin. Endocrinology 119: 940-942.

11. Ambesi-Impiombato FS, Parks LAM, Coon HG 1980. Culture of hormone-dependent functional epithelial cells from rat thyroids. Proc Natl Acad Sci USA 77:3455-3459.

12. Weiss JS, Philp NJ, Grollman EF 1984. Iodide transport in a continuous line of cultured cells from rat thyroid. Endocrinology 114:1090-1098.

13. Tramontano D, Moses AC, Veneziani BM, Ingbar SH 1988. Adenosine 3',5'-monophosphate mediates both the mitogenic effect of thyrotropin and its ability to amplify the response to insulin-like growth factor I in FRTL5 cells. Endocrinology 122:127-132.

14. Tramontano D, Moses AC, Picone R, Ingbar SH 1987. Characterization and regulations of the receptor for insulin-like growth factor-I in the FRTL5 rat thyroid follicular cell line. Endocrinology 120:785-790.

15. Tramontano D, Ingbar SH 1986. Properties and regulation of the thyrotropin receptor in the FRTL5 rat thyroid cell line. Endocrinology 118:1945-1951.

16. Filetti S, Dimante G, Foti D 1987. Thyrotropin stimulates glucose transport in cultured rat thyroid cells. Endocrinology 120:2576-2581.

17. Maciel RMB, Moses AC, Villone G, Tramontano D, Ingbar SH 1988. Demonstration of the production and physiological role of insulin-like growth factor-II in rat thyroid follicular cells in culture. J Clin Invest 82:1546-1553.

18. Tramontano D, Chin WW, Moses AC, Ingbar SH 1986. Thyrotropin and dibutyryl cyclic AMP increase levels of c-myc and c-fos mRNAs in cultured rat thyroid cells. J Biol Chem 261: 3919-3922.

19. Tsuzaki O, Kohn LD 1987. Control of c-fos anc c-myc proto oncogene induction in rat thyroid cells in culture.   Mol Endo 1:839-848.

20. Stiles CD 1983. The molecular biolgoy of platelet-derived growth factor. Cell 33:653-5.

21. Hardouin S, Hossenlopp P, Segovia B, Seurin D, Portolan G, Lassarre C, Binoux M 1987. Heterogeneity of insulin-like growth factor binding proteins and relationships between structure and affinity. Eur J Biochem 170: 121-132.

22. Frauman AG, Tsuzaki S, Moses AC 1989. The binding characteristics and biological effects in FRTL5 cells of placental protein-12, an insulin-like growth factor-binding protein purified from human amniotic fluid. Endocrinology (In press).

23. Tramontano D, Moses AC, Ingbar SH 1988. The role of adenosine 3', 5' - monophosphate in the regulation of receptors for thyrotropin and insulin-like growth factor I in the FRTL5 rat thyroid follicular cell. Endocrinology 122:133-136.

24. Jacobson KA 1988. Chemical Approaches to the Definition of Adenosine Receptors, pp 1-26. Alan R. Liss, Inc, New York.

25, Moses AC, Adelman B, Tramontano D 1987. Adenosine modulates cellular proliferation: Inhibition of thyrotropin (TSH), but potentiation of insulin-like growth factor I (IGF-I), stimulated DNA synthesis. Clinical Research 35:399A.

26. Frauman AG, Moses AC 1989. The A_1 adenosine receptor antagonist 1, 3 dipropyl-8-cyclopentylxanthine (DPCPX) displays adenosine agonist properties in the FRTL5 thyroid cell line. Biochem Biophys Res Comm. 159:355-362.

27. Moses AC, Frauman AG, Adenosine modulates the mitogenic effects of thyrotropin and insulin-like growth factor I in FRTL5 cells through two distinct adenosine receptor subtypes 1988. 70th Ann Meeting of the Endocrine Society, New Orleans, LA. Abstr 896.

28. Morris JC, III, Ranganathan G, Hay ID, Nelson RE, Jiang N-S 1988. The effects of transforming grwoth factor-B on growth and differentiation

of the continuous rat thyroid follicular cell line, FRTL5. Endocrinology
123: 1385-1394.

29. Lombardi A, Veneziani BM, Tramontano D, Ingbar SH 1988. Independent
and interactive effects of tetradecanoyl phorbol acetate on growth and
differentiated functions of FRTL5 cells. Endocrinology 123: 1544-1552.

30. Becks GP, Eggo MC, Burrow GN 1988. Organic iodine inhibits deoxyribonucleic acid synthesis and growth in FRTL-5 thyroid cells.
Endocrinology 123:545-551

31. Mine M, Tramontano D, Chin WW, Ingbar SH 1987. Interleukin-1
stimulates thyroid cell growth and increases the concentration of the
c-myc proto-oncogene mRNA in thyroid follicular cells in culture.
Endocrinology 120: 1212-1214.

CONTRIBUTORS

MARTIN ADAMO
Diabetes Branch
NIDDK, NIH
Bethesda, MD   20892

ELI Y. ADASHI
Div. Reproductive Endocrinology
Dept. of Obstetrics and Gynecology
Univ. of Maryland Sch. of Medicine
Frank C. Bressler Research Bldg.
655 West Baltimore Street
Baltimore, MD   21201

ROBERT S. BAR
Diabetes and Endocrinology
   Research Center
Dept. of Internal Medicine
VA Hospital, Room 3E-21
University of Iowa
Iowa City, IA   52240

DENIS G. BASKIN
Depts. of Biological Structure
   and Medicine
Univ. of Washington
   and the V.A. Med. Center
1660 South Columbian Way
Seattle, WA   98108

MARVIN L. BAYNE
Dept. of Growth Factor Research
Merck Sharp & Dohme Research
   Laboratories
P.O. Box 2000
Rahway, NJ   07065

P. BECHTEL
Washington Univ. Sch. of Medicine
Depts. of Medicine and Genetics
St. Louis, MO   63110

MICHAEL BERELOWITZ
State University of New York
Endocrine Division
HSC T15, 060
Stony Brook, NY   11794-8154

KEITH D. BOEHM
Dept. of Reproductive Biology
Case Western Reserve University
Cleveland, OH   44106

ALEXANDRA L. BROWN
Molecular, Cellular and Nutritional
   Endocrinology Branch
National Institute of Diabetes and
   Digestive and Kidney Diseases
National Institutes of Health
Bethesda, MD   20892

CARMELO B. BRUNI
Centro di Endocrinologica ed
   Oncologia Sperimentale del
   Consiglio Nazionale delle
   Ricerche
Dipartimento di Biologia e Patologia
   Cellulare e Molecolare
Universita degli Studi di Napoli
ITALY   80122

ERNESTO CANALIS
St. Francis Hospital
   and Medical Center
114 Woodland Street
Hartford, CT   06105

MARGARET A. CASCIERI
Dept. of Biochemical Endocrinology
Merck Sharp & Dohme Research
   Laboratories
P.O. Box 2000
Rahway, NJ   07065

MICHAEL CENTRELLA
St. Francis Hospital
   and Medical Center
114 Woodland Street
Hartford, CT   06105

LORENZO CHIARIOTTI
Centro di Endocrinologica ed
  Oncologia Sperimentale del
  Consiglio Nazionale delle
  Ricerche
Dipartimento di Biologia e Patologia
  Cellulare e Molecolare
Universita degli Studi di Napoli
ITALY 80122

DAVID R. CLEMMONS
University of North Carolina
  School of Medicine
Div. Endocrinology, CB #7170
Chapel Hill, NC 27599-7170

KEVIN J. CULLEN
Lombardi Cancer Research Center
Georgetown Univ. Medical Center
3800 Reservoir Road N.W.
Washington, DC 20007

MARY K. DAHMER
Department of Pediatrics
Kennedy Center
The University of Chicago
5841 S. Maryland Ave., Box 413
Chicago, IL 60637

WILLIAM H. DAUGHADAY
Washington Univ. Sch. of Medicine
Department of Internal Medicine
Box 8127
660 South Euclid Avenue
St. Louis, MO 63110

ALBERTA DAVIS
Dept. of Microbiology and
  Immunology
University of South Carolina
Columbia, SC 29208

SHERIN DEVASKAR
The Pediatric Research Institute
Cardinal Glennon Children's
  Hospital
1465 S. Grand Blvd.
St. Louis, MO 63104-1095

D. DeVOL
Washington Univ. Sch. of Medicine
Depts. of Medicine and Genetics
St. Louis, MO 63110

VINCENT DURONIO
Burroughs Wellcome Co.
3030 Cornwallis Road
Research Triangle Park, NC 27709

GARY L. ENGELMANN
Research Institute
Cleveland Clinic Foundation
9500 Euclid Avenue
Cleveland, OH 44106

JAMES A. FAGIN
Dept. of Endocrinology
Cedars Sinai Medical Center
8700 Beverly Blvd., Room B-131
Los Angeles, CA 90048

PAUL FERNYHOUGH
Cell Biophysics Unit
Medical Research Council
Kings College
London WC2B 5RL
ENGLAND

ALBERT G. FRAUMAN
Diabetes and Metabolism Unit
Department of Medicine
Beth Israel Hospital and
  Harvard Medical School
Boston, MA 02215

HENRY G. FRIESEN
Dept. of Physiology
Faculty of Medicine
University of Manitoba
Winnipeg R3E 0W3
CANADA

GORDON W. GLAZNER
Physiology Department
Colorado State University
Fort Collins, CO 80523

S. GOLDSTEIN
Div. Endocrinology and Metabolism
Department of Medicine
Emory Univ. School of Medicine
69 Butler Street S.E.
Atlanta, GA 30303

M. HAMMERMAN
Washington Univ. Sch. of Medicine
Depts. of Medicine and Genetics
St. Louis, MO 63110

JOYCE F. HASKELL
Div. Reproductive Biol. and Endocr.
Dept. of Obstetrics and Gynecology
555 OHB
Univ. of Alabama at Birmingham
Birmingham, AL 35294

KLAUS HAVEMANN
Dept. Internal Medicine
Div. Hematology/Oncology
Philipps-University of Marburg
Baldingerstrasse, 3550 Marburg
WEST GERMANY

LINDA E. HAWS
Dept. of Obstetrics and Gynecology
Univ. of Alabama at Birmingham
Birmingham, AL   35294

ELEUTERIO R. HERNANDEZ
Div. Reproductive Endocrinology
Dept. of Obstetrics and Gynecology
Univ. of Maryland Sch. of Medicine
Frank C. Bressler Research Bldg.
655 West Baltimore Street
Baltimore, MD   21201

ANDREW R. HOFFMAN
Department of Medicine
Stanford University Medical Center
Stanford, CA   94305

PIETERNELLA HOLTHUIZEN
Laboratory for Physiological
   Chemistry
State University of Utrecht
Vondellaan 24a, 3521 GG Utrecht
THE NETHERLANDS

RICHARD HUNT
Dept. of Microbiology and
   Immunology
University of South Carolina
Columbia, SC   29208

JOSEPH ILAN
Dept. of Anatomy
Case Western Reserve University
Cleveland, OH   44106

JUDITH ILAN
Dept. of Reproductive Biology
Case Western Reserve University
Cleveland, OH   44106

DOUGLAS N. ISHII
Physiology Department
Colorado State University
Fort Collins, CO   80523

STEVEN JACOBS
Burroughs Wellcome Co.
3030 Cornwallis Road
Research Triangle Park, NC   27709

GABRIELE JAQUES
Dept. Internal Medicine
Div. Hematology/Oncology
Philipps-University of Marburg
Baldingerstrasse, 3550 Marburg
WEST GERMANY

THOMAS R. JOHNSON
Dept. of Anatomy
Case Western Reserve University
Cleveland, OH   44106

HISANORI KATO
Dept. of Agricultural Chemistry
Faculty of Agriculture
The University of Tokyo
Bunkyo-ku, Tokyo 113
JAPAN

MICHAEL F. KELLEY
Dept. of Reproductive Biology
Case Western Reserve University
Cleveland, OH   44106

WIELAND KIESS
Endocrinology Section
Metabolism Branch
Bldg. 10, Room 4-N-115
National Cancer Institute
National Institutes of Health
Bethesda, MD   20892

MICHAEL G. KING
Dept. of Biological Structure
Univ. of Washington
   and the V.A. Med. Center
1660 South Columbian Way
Seattle, WA   98108

STANISLAUS KINOTA
Department of Neurology
The Children's Hospital and
Harvard Medical School
300 Longwood Ave.
Boston, MA   02115

J. D. KLEIN
Div. Endocrinology and Metabolism
Department of Medicine
Emory Univ. School of Medicine
69 Butler Street S.E.
Atlanta, GA   30303

YONG KO
Dept. of Animal Science and
   Laboratories of Molecular and
   Developmental Biology
Ohio Agricultural Research and
   Development Center
The Ohio State University
Wooster, OH   44691-4096

P. LAJARA
Washington Univ. Sch. of Medicine
Depts. of Medicine and Genetics
St. Louis, MO 63110

EDWARD D. LAMPERTI
Department of Neurology
The Children's Hospital and
Harvard Medical School
300 Longwood Ave.
Boston, MA 02115

DIANNE FIGLEWICZ LATTEMANN
Dept. of Psychology
Univ. of Washington
and the V.A. Med. Center
1660 South Columbian Way
Seattle, WA 98108

DEREK LeROITH
Section of Molecular and Cellular
Physiology
Diabetes Branch, NIDDK
Bldg. 10, Rm. 8S-243
National Institutes of Health
Bethesda, MD 20892

L. ROXANNE LETCHER
Dept. of Animal Science and
Laboratories of Molecular and
Developmental Biology
Ohio Agricultural Research and
Development Center
The Ohio State University
Wooster, OH 44691-4096

MARC E. LIPPMAN
Lombardi Cancer Research Center
Georgetown Univ. Medical Center
3800 Reservoir Road N.W.
Washington, DC 20007

WILLIAM L. LOWE, JR.
Section of Molecular and Cellular
Physiology
Diabetes Branch, NIDDK
Bldg. 10, Rm. 8S-243
National Institutes of Health
Bethesda, MD 20892

P. KAY LUND
Dept. of Physiology
University of North Carolina
School of Medicine
Chapel Hill, NC 27599

BRIAN A. MASTERS
Department of Physiology
Box J-274, JHMHC
University of Florida
Gainesville, FL 32610

THOMAS L. McCARTHY
St. Francis Hospital
and Medical Center
114 Woodland Street
Hartford, CT 06105

JEFFREY F. McKELVY
Abbott Laboratories
1 Abbott Park Road
Abbott Park, IL 60064

SHLOMO MELMED
Dept. of Endocrinology
Cedars Sinai Medical Center
8700 Beverly Blvd., Room B-131
Los Angeles, CA 90048

THOMAS MERIMEE
Division of Endocrinology and
Metabolism
Department of Medicine
Box J-226, JHMHC
University of Florida
Gainesville, FL 32610-0226

YUTAKA MIURA
Dept. of Agricultural Chemistry
Faculty of Agriculture
The University of Tokyo
Bunkyo-ku, Tokyo 113
JAPAN

ALAN C. MOSES
Diabetes and Metabolism Unit
Department of Medicine
Beth Israel Hospital and
Harvard Medical School
Boston, MA 02215

CARY MOXHAM
Burroughs Wellcome Co.
3030 Cornwallis Road
Research Triangle Park, NC 27709

LAURA M. MUDD
Department of Physiology
Box J-274, JHMHC
University of Florida
Gainesville, FL 32610

LIAM J. MURPHY
Dept. of Physiology
Faculty of Medicine
University of Manitoba
Winnipeg R3E 0W3
CANADA

PETER NISSLEY
Endocrinology Section
Metabolism Branch
Bldg. 10, Room 4-N-115
National Cancer Institute
National Institutes of Health
Bethesda, MD 20892

TADASHI NOGUCHI
Dept. of Agricultural Chemistry
Faculty of Agriculture
The University of Tokyo
Bunkyo-ku, Tokyo 113
JAPAN

RICHARD O'BRIEN
Department of Neurology
The Children's Hospital and
Harvard Medical School
300 Longwood Ave.
Boston, MA 02115

IAN OCRANT
Department of Endocrinology
Room S-322
Stanford University Medical Center
Stanford, CA 94305

ASAKO OKOSHI
Dept. of Agricultural Chemistry
Faculty of Agriculture
The University of Tokyo
Bunkyo-ku, Tokyo 113
JAPAN

CRAIG C. ORLOWSKI
Molecular, Cellular and Nutritional
   Endocrinology Branch
National Institute of Diabetes and
   Digestive and Kidney Diseases
National Institutes of Health
Bethesda, MD 20892

SOONMYOUNG PAIK
Lombardi Cancer Research Center
Georgetown Univ. Medical Center
3800 Reservoir Road N.W.
Washington, DC 20007

ROBERT L. PERLMAN
Department of Pediatrics
Kennedy Center
The University of Chicago
5841 S. Maryland Ave., Box 413
Chicago, IL 60637

JEFFREY E. PESSIN
Dept. of Physiology and Biophysics
The University of Iowa
Iowa City, IA 52240

L. S. PHILLIPS
Div. Endocrinology and Metabolism
Department of Medicine
Emory Univ. School of Medicine
69 Butler Street S.E.
Atlanta, GA 30303

CONSTANTIN POLYCHRONAKOS
The Protein and Polypeptide
   Hormone Laboratory
Department of Pediatrics
McGill University
Montreal, Quebec
CANADA

DIANE PRAGER
Dept. of Endocrinology
Cedars Sinai Medical Center
8700 Beverly Blvd., Room B-131
Los Angeles, CA 90048

MOHAN K. RAIZADA
Department of Physiology
Box J-274, JHMHC
University of Florida
Gainesville, FL 32610

MATTHEW M. RECHLER
Molecular, Cellular and Nutritional
   Endocrinology Branch
National Institute of Diabetes and
   Digestive and Kidney Diseases
National Institutes of Health
Bethesda, MD 20892

CAROL E. RESNICK
Div. of Reproductive Endocrinology
Dept. of Obstetrics and Gynecology
Univ. of Maryland Sch. of Medicine
Frank C. Bressler Research Bldg.
655 West Baltimore Street
Baltimore, MD 21201

CHARLES T. ROBERTS, JR.
Section of Molecular and Cellular
   Physiology
Diabetes Branch, NIDDK
Bldg. 10, Rm. 8S-243
National Institutes of Health
Bethesda, MD 20892

MARIA ROJESKI
Diabetes Branch, NIDDK
Bldg. 10, Rm. 8N-244
National Institutes of Health
9000 Rockville Pike
Bethesda, MD 20892

JOYCE A. ROMANUS
Molecular, Cellular and Nutritional
 Endocrinology Branch
National Institute of Diabetes and
 Digestive and Kidney Diseases
National Institutes of Health
Bethesda, MD 20892

KENNETH M. ROSEN
Department of Neurology
The Children's Hospital and
Harvard Medical School
300 Longwood Ave.
Boston, MA 02115

NEAL ROSEN
Lombardi Cancer Research Center
Georgetown Univ. Medical Center
3800 Reservoir Road N.W.
Washington, DC 20007

RON G. ROSENFELD
Department of Pediatrics
Stanford University Medical Center
Stanford, CA 94305

NADIA ROSENTHAL
Department of Biochemistry
Boston University Medical School
Boston, MA 02215

JESSE ROTH
Diabetes Branch, NIDDK
Bldg. 10, Rm. 8N-244
National Institutes of Health
9000 Rockville Pike
Bethesda, MD 20892

P. ROTWEIN
Washington Univ. Sch. of Medicine
Depts. of Medicine and Genetics
St. Louis, MO 63110

H. F. SADIQ
The Pediatric Research Institute
Cardinal Glennon Children's
 Hospital
1465 S. Grand Blvd.
St. Louis, MO 63104-1095

DOUGLAS A. SCHOBER
Dept. of Animal Science and
 Laboratories of Molecular and
 Developmental Biology
Ohio Agricultural Research and
 Development Center
The Ohio State University
Wooster, OH 44691-4096

JOSHUA SHEMER
Section of Molecular and Cellular
 Physiology
Diabetes Branch, NIDDK
National Institutes of Health
Bethesda, MD 20892

FRANK A. SIMMEN
Dept. of Animal Science and
 Laboratories of Molecular and
 Developmental Biology
Ohio Agricultural Research and
 Development Center
The Ohio State University
Wooster, OH 44691-4096

ROSALIA C. M. SIMMEN
Dept. of Animal Science and
 Laboratories of Molecular and
 Developmental Biology
Ohio Agricultural Research and
 Development Center
The Ohio State University
Wooster, OH 44691-4096

MARK SKLAR
Endocrinology Section
Metabolism Branch
Bldg. 10, Room 4-N-115
National Cancer Institute
National Institutes of Health
Bethesda, MD 20892

ROBERT G. SPANHEIMER
Diabetes and Endocrinology
 Research Center
Dept. of Internal Medicine
VA Hospital, Room 3E-21
University of Iowa
Iowa City, IA 52240

MARJORIE E. SVOBODA
Department of Pediatrics
Univ. of North Carolina
 at Chapel Hill
Chapel Hill, NC 27514

PATRICIA SZOT
Dept. of Pharmacology
Univ. of Washington
 and the V.A. Med. Center
1660 South Columbian Way
Seattle, WA 98108

SHIN-ICHIRO TAKAHASHI
Dept. of Agricultural Chemistry
Faculty of Agriculture
The University of Tokyo
Bunkyo-ku, Tokyo 113
JAPAN

DONATELLA TRAMONTANO
Diabetes and Metabolism Unit
Department of Medicine
Beth Israel Hospital and
  Harvard Medical School
Boston, MA   02215

JUDITH L. TREADWAY
Dept. of Physiology and Biophysics
The University of Iowa
Iowa City, IA   52240

TSUTOMU UMEZAWA
Dept. of Agricultural Chemistry
Faculty of Agriculture
The University of Tokyo
Bunkyo-ku, Tokyo 113
JAPAN

KAREN L. VALENTINO
Neurex Corporation
3760 Haven Avenue
Menlo Park, CA   94025-1057

JUDSON J. VAN WYK
Dept. of Pediatrics
Clinical Sciences Bldg., 229-H
University of North Carolina
  School of Medicine
Chapel Hill, NC   27599

LYDIA VILLA-KOMAROFF
Department of Neurology
The Children's Hospital and
Harvard Medical School
300 Longwood Ave.
Boston, MA   02115

RAIMO VOUTILAINEN
Dept. of Pediatrics and Pathology
University of Helsinki
SF-00290 Helsinki
FINLAND

CHIANG WANG
Physiology Department
Colorado State University
Fort Collins, CO   80523

BRUCE M. WENTWORTH
Department of Biochemistry
Boston University Medical School
Boston, MA   02215

TERESA L. WOOD
Dept. of Anatomy and Cell Biology
Columbia University
630 West 168th Street
New York, NY   10032

YVONNE W.-H. YANG
Molecular, Cellular and Nutritional
  Endocrinology Branch
National Institute of Diabetes and
  Digestive and Kidney Diseases
National Institutes of Health
Bethesda, MD   20892

BRUCE YANKNER
Department of Neurology
The Children's Hospital and
Harvard Medical School
300 Longwood Ave.
Boston, MA   02115

DOUGLAS YEE
Lombardi Cancer Research Center
Georgetown Univ. Medical Center
3800 Reservoir Road N.W.
Washington, DC   20007

Myocyte, 352
  growth, 443
  see Cardiomyocyte
Myogenesis, 223, 224
Myoinositol, 431, 432
Myometrium, 155
  and IGF-I, 155

Nephrectomy, 119-121
Neoplasm, *see* Tumor
Nerve, 345, 408-410, 412
  block, *see* Botulinum toxin
  denervation, 408, 410, 411
  growth factor, 220, 403-425
    neuronal, 403-425
  innervation, 408, 411
  re-innervation, 408-410
Nervous system, central, 341-358
Neural-ectodermal tumor, 342
Neurite, 404-407
  growth, mechanism of, 411-417
Neuroblastoma cell, 221, 342, 345,
    347, 412, 467, 468
  line N18, 327-339, 351
    SH-SY5Y, 403-406, 413-417
  phosphorylation in, 327-339
Neurofilament, 413-415
Neuron, 40, 43, 45, 343, 345, 347,
    404, 405
  and IGF, 231-235
  of spinal cord, 403-425
Neuropeptide, 26, 28
Neurotensin, 29
Neurotransmitter, 26, 28-29, 432
  release, 316, 408
Norepinephrine, 29, 343
Northern blot analysis, 15-18,
    65, 66, 98, 100, 101,
    113, 126, 130, 135, 136,
    162, 170, 172, 175, 201,
    211-213, 224, 233, 238,
    239, 241, 249, 414, 461,
    464
Nuclease protection assay, 118
    210-213, 223
Nuclear run-on assay, 131

Olfactory bulb, *see* Bulb
Oligodendrocyte, 316, 342, 343
Oncogene, 445
  see Proto-oncogene
Oophorectomy, 237
  and breast cancer remission,
    237
Osteoblast, 1, 460-465
Osteocalcin, 463
Ovariectomy, 160, 161
Ovary, human
  fetal, 170
  and IGF-I, 141-151

Pancreas
  beta cell, 238
Parathyroid hormone, 461
Phagocytosis, 373
Pheochromocytoma, 222
  cell line PC12, 352, 412-415
    467-469
Phorbol ester, 309, 310, 315-317,
    352, 415, 417, 489, 490
  12, 13-dibutyrate, 352
Phosphodiesterase, 144
Phosphoglycoprotein, 332
Phosphoinositide, 320
  hydrolysis, 316
Phosphoinositol, 26
Phospholipase C, 25, 320
Phosphoprotein, 328-332, 334-337,
    349-352
Phosphorylation, 327-329, 347-351,
    374, 376
  of IGF, 341-358
  of IGF-I, 327-339
    receptor, 309-314
  of insulin, 327-339
Phosphotransferase, 374
Phosphotyrosine, 310
  antibody, 328-336
Phototransduction, retinal, 351
Pig, 196, 202-204, 375
Pit, coated, 374-376
  see Vesicle, coated
Pituitary gland, 25
  adenoma, 60
  anterior, 209, 223, 345
  function, insulin-regulated,
    46-49
  and growth hormone, 57, 58
  and hypothyroid, 66
  and IGF, 39-56
Pituitary hormone, 25
Placenta, human, 179-193, 222, 287,
    291, 445
  and IGF, 179-193
    gene expression, 180-185
    production, 180-188
Plasmid pα 2, 288
Platelet-derived growth factor, 6,
    154, 185-186, 241, 316,
    479, 480
Polylysine, 268, 269
Polyploidation, 445
Potassium, 469-471
Prednisone, 83
Progesterone, 11-13, 147, 159, 173,
    201, 202
Proinsulin, 58, 107, 144, 153, 301
Prolactin, 48, 60, 64, 131, 144,
    145, 187, 201, 202, 374
Proliferin, 374
Proline, 463

Pro-opiomelanocortin, 42
Protamine sulfate, 268
Protein, 10, 62, 351, 381-402, 413
Protein kinase A, 25
Protein kinase C, 26, 309, 316,
    318-320
  and IGF-I regulation, 315-325
  isozyme, 315
  tyrosine-specific, 266-267, 275
Protein kinase, 372, 415-417
Proteinuria, 298
Proteoglycan, 146, 301, 476
Proto-oncogene, 445
Purkinje cell-deficient mouse
        see Mouse strain pcd
Pygmy, African, 73-80
  growth
    hormone, 73-80
      early studies, 73
    pattern, 76
    spurt absent in puberty, 75
  and IGF-I, 73-75
    secretion
      absent in puberty, 75
      defective, 74
  and testosterone, 74

Rabbit
  brain RNA, 233-234
Rat, 32, 83, 88, 91, 118-121,
        125-128, 141, 143, 200
        203, 223, 342, 343, 376
        382, 395-402, 407, 409
        411, 412, 427-442, 445
        460, 461, 485-493
  adrenalectomized, 89, 90
  calvaria, 460, 461
  cell line FRTL-5, 7-10, 485-493
  DES-treated, 144, 145
  diabetic, 82-92, 133, 435-442
  fasted, 83, 88-91
  fibroblast, 220
  granulosa cell, see Granulosa
      cell
  heart, 448-456
  hypophysectomized, 1, 3, 11, 17,
      85, 88, 91, 113, 129-132,
      144, 145, 154-161, 209
  hypothalamus, 212, 213
  IGF-II receptor, 370-371
  kidney, 371
  liver, 211
  muscle, 224
  ovariectomized, 154-157, 160,
      161
  Sertoli cell, 13, 174
  somatotroph cell, 60
  thyroid, 485-493
Receptor, see Insulin, IGF,
      IGF-I, IGF-II
Receptor kinase, 261-284, 318

Relaxin, 107, 153
Renal
  failure, 81-82, 85, 89, 92
  hypertrophy, 117, 119-122
    compensatory, 120, 121
Retina, human
  and IGF receptor, 297-308
Retinopathy, diabetic, 297-298
Retroendocytosis, 309
Rhesus monkey, 170
Ribonuclease (RNase), 19, 61,
    109-113
Ribonuclease protection assay,
    238-241, 249
RNA, 61-66, 120, 130, 131, 134-136,
    181, 183, 186, 187, 211,
    221-223, 233-234, 237, 448,
    460, 464
  messenger (mRNA), 14-18, 58-66,
    98-101, 107, 109, 111, 113,
    114, 117-121, 125-128, 130,
    131, 154-161, 169-177, 195,
    199-202, 209-214, 221-224,
    231-234, 241, 242, 248, 249,
    381, 408-415, 460, 463, 464
see Northern blot analysis
RNA helicase, 109

Saccharomyces cerevisiae
  protease-deficient strain, 287
Scatchard plot analysis, 43-44, 143,
    251-252, 254, 263-265
Serine, 317
Serine/threonine kinase, see
    Protein kinase C
Serotonin, 29
Sertoli cell, 13, 174
Serum binding protein, 287, 290-292
Signal transduction pathway,
    485-493
Signalling transmembrane, 261-284
  ligand-dependent, 261-284
Simian sarcoma virus, 479
Skeletal remodeling, 459
Slot-blot hybridization, 156, 157,
    450
Smooth muscle cell, 473-480
Somatomammotropin (hCS), 57
Somatomedin C, see IGF-I
Somatostatin (SRIF), 27-29, 57, 209
Somatotroph, pituitary, 58, 60
Spermatocyte, pachytene, 13
Sphingomyelinase, 362
Sphingosine, 416, 417
Spinal cord
  cell, 405, 407
  neuron, 403-425
SRIF, see Somatostatin
Stilbestrol, 11
Steroid
  excess and IGF-I, 81-82

Streptozotocin, 125, 126, 133, 436, 437
Substance P, 29
Sulfation factor, 146
Synapse, 407-410
Synaptogenesis, 407-411
Syncytiotrophoblast, 184, 186, 201
System, limbic, 435

Terbutaline sulfate, 146
Testis, 170, 174-176, 299
Testosterone, 74, 174, 214
12-O-Tetradecanoyl-phorbol-*13*-acetate (TPA), 309-313, 315-320, 351, 352, 415
Theca-interstitial cell, 145, 147
Threonine, 317
Thyroglobin, 486
Thyroid follicular cell line FRTL-*5*, 485-493
Thyroid
  hormone, 210
  of rat, 371
Thyrotropin, 485
Thyroxine, 444
TPA, *see* *12*-O-Tetradecanoyl-phorbol-*13*-acetate
Transducin, 351
Transferrin, 298
Transforming growth factor beta, 490
Transmembrane signalling
  and insulin, 270-275
  ligand-dependent, 261-284
  molecular hypothesis, 270-276
Triiodothyronine, 58
Trypsin, 312
Tubulin, 221, 352, 404, 412-415
  gene, 413
  mRNA, 412-415
Tumor, 3
  pituitary, 58
  see separate cancers
Tyrosine, 317
Tyrosine kinase, 299, 301, 309 310, 315-318, 327, 332, 337, 344, 347-351, 376, 430, 469, 485

Ubiquitin, 17
Untranslated region (5'-UTR), 181, 182
Uterine luminal fluid mitogen, 197-199
Uteroferrin, 201, 374
Uterus, 141, 153-159, 195-197
  and IGF-I, 153-159, 162
  mRNA, 155-159, 200

Vanadate, 376
Vesicle, coated, 374

Western blot analysis, 383
Wheat germ agglutinin, 272-274, 328, 332-334, 350, 469
Wilms' tumor, 222

Xenopus sp., 109